Mathematical
Surveys
and
Monographs

Volume 127

Chaotic Billiards

Nikolai Chernov
Roberto Markarian

AMERICAN
MATHEMATICAL
SOCIETY
Providence, Rhode Island

2000 *Mathematics Subject Classification.* Primary 37D50; Secondary 37D25, 37A25, 37N05, 82B99.

For additional information and updates on this book, visit
www.ams.org/bookpages/surv-127

Library of Congress Cataloging-in-Publication Data

Chernov, Nikolai, 1956–.
Chaotic billiards / Nikolai Chernov, Roberto Markarian.
p. cm. — (Mathematical surveys and monographs, ISSN 0076-5376 ; v. 127)
Includes bibliographical references and index.
ISBN 0-8218-4096-7 (alk. paper) | ISBN 1-4704-7442-5 (Softcover)
1. Dynamics. 2. Chaotic behavior in systems. 3. Ergodic theory. 4. Measure theory. 5. Probability theory. 6. Billiards. I. Markarian, Roberto. II. Title. III. Series: Mathematical surveys and monographs ; no. 127.

QA851 .C44 2006
003′.857—dc22 2006042819

Printed in the United States of America.
Reprinted by the American Mathematical Society, 2023.

∞ The paper used in this book is acid-free and falls within the guidelines
established to ensure permanence and durability.
Visit the AMS home page at `http://www.ams.org/`

11 10 9 8 7 6 5 4 3 2 28 27 26 25 24 23

To Yakov Sinai on the occasion of his 70th birthday

Contents

Preface

Billiards are mathematical models for many physical phenomena where one or more particles move in a container and collide with its walls and/or with each other. The dynamical properties of such models are determined by the shape of the walls of the container, and they may vary from completely regular (integrable) to fully chaotic. The most intriguing, though least elementary, are chaotic billiards. They include the classical models of hard balls studied by L. Boltzmann in the nineteenth century, the Lorentz gas introduced to describe electricity in 1905, as well as modern dispersing billiard tables due to Ya. Sinai.

Mathematical theory of chaotic billiards was born in 1970 when Ya. Sinai published his seminal paper [**Sin70**], and now it is only 35 years old. But during these years it has grown and developed at a remarkable speed and has become a well-established and flourishing area within the modern theory of dynamical systems and statistical mechanics.

It is no surprise that many young mathematicians and scientists attempt to learn chaotic billiards in order to investigate some of these and related physical models. But such studies are often prohibitively difficult for many novices and outsiders, not only because the subject itself is intrinsically quite complex, but to a large extent because of the lack of comprehensive introductory texts.

True, there are excellent books covering general mathematical billiards [**Ta95, KT91, KS86, GZ90, CFS82**], but these barely touch upon chaotic models. There are surveys devoted to chaotic billiards as well (see [**DS00, HB00, CM03**]) but those are expository; they only sketch selective arguments and rarely get down to 'nuts and bolts'. For readers who want to look 'under the hood' and become professional (and we speak of graduate students and young researchers here), there is not much choice left: either learn from their advisors or other experts by way of personal communication or read the original publications (most of them very long and technical articles translated from Russian). Then students quickly discover that some essential facts and techniques can be found only in the middle of long dense papers. Worse yet, some of these facts have never even been published – they exist as folklore.

This book attempts to present the fundamentals of the mathematical theory of chaotic billiards in a systematic way. We cover all the basic facts, provide full proofs, intuitive explanations and plenty of illustrations. Our book can be used by students and self-learners. It starts with the most elementary examples and formal definitions and then takes the reader step by step into the depth of Sinai's theory of hyperbolicity and ergodicity of chaotic billiards, as well as more recent achievements related to their statistical properties (decay of correlations and limit theorems).

The reader should be warned that our book is designed for active learning. It contains plenty of exercises of various kinds: some constitute small steps in the proofs of major theorems, others present interesting examples and counterexamples, yet others are given for the reader's practice (some exercises are actually quite challenging). The reader is strongly encouraged to do exercises when reading the book, as this is the best way to grasp the main concepts and eventually master the techniques of billiard theory.

The book is restricted to two-dimensional chaotic billiards, primarily dispersing tables by Sinai and circular-arc tables by Bunimovich (with some other planar chaotic billiards reviewed in the last chapter). We have several compelling reasons for such a confinement. First, Sinai's and Bunimovich's billiards are the oldest and best explored (for instance, statistical properties are established only for them and for no other billiard model). The current knowledge of other chaotic billiards is much less complete; the work on some of them (most notably, hard ball gases) is currently under way and should perhaps be the subject of future textbooks. Second, the two classes presented here constitute the core of the entire theory of chaotic billiards. All its apparatus is built upon the original works by Sinai and Bunimovich, but their fundamental works are hardly accessible to today's students or researchers, as there have been no attempts to update or republish their results since the middle 1970s (after Gallavotti's book [**Ga74**]). Our book makes such an attempt. We do not cover polygonal billiards, even though some of them are mildly chaotic (ergodic). For surveys of polygonal billiards see [**Gut86, Gut96**].

We assume that the reader is familiar with standard graduate courses in mathematics: linear algebra, measure theory, topology, Riemannian geometry, complex analysis, probability theory. We also assume knowledge of ergodic theory. Although the latter is not a standard graduate course, it is absolutely necessary for reading this book. We do not attempt to cover it here, though, as there are many excellent texts around [**Wa82, Man83, KH95, Pet83, CFS82, DS00, BrS02, Dev89, Sin76**] (see also our previous book, [**CM03**]). For the reader's convenience, we provide basic definitions and facts from ergodic theory, probability theory, and measure theory in the appendices.

Acknowledgements. The authors are grateful to many colleagues who have read the manuscript and made numerous useful remarks, in particular P. Balint, D. Dolgopyat, C. Liverani, G. Del Magno, and H.-K. Zhang. It is a pleasure to acknowledge the warm hospitality of IMPA (Rio de Janeiro), where the final version of the book was prepared. We also thank the anonymous referees for helpful comments. Last but not least, the book was written at the suggestion of Sergei Gelfand and thanks to his constant encouragement. The first author was partially supported by NSF grant DMS-0354775 (USA). The second author was partially supported by a Proyecto PDT-Conicyt (Uruguay).

Symbols and notation

$\mathcal{D}$	billiard table	Section	2.1
Γ	boundary of the billiard table		2.1
Γ_+	union of dispersing components of the boundary Γ		2.1
Γ_-	union of focusing components of the boundary Γ		2.1
Γ_0	union of neutral (flat) components of the boundary Γ		2.1
$\tilde{\Gamma}$	regular part of the boundary of billiard table		2.1
Γ_*	Corner points on billiard table		2.1
ℓ	degree of smoothness of the boundary $\Gamma = \partial\mathcal{D}$		2.1
n	normal vector to the boundary of billiard table		2.3
$\mathbf{T}$	tangent vector to the boundary of billiard table		2.6
$\mathcal{K}$	(signed) curvature of the boundary of billiard table		2.1
Φ^t	billiard flow		2.5
Ω	the phase space of the billiard flow		2.5
$\tilde{\Omega}$	part of phase space where dynamics is defined at all times		2.5
π_q, π_v	projections of Ω to the position and velocity subspaces		2.5
ω	angular coordinate in phase space Ω		2.6
η, ξ	Jacobi coordinates in phase space Ω		3.6
μ_Ω	invariant measure for the flow Φ^t		2.6
$\mathcal{F}$	collision map or billiard map		2.9
$\mathcal{M}$	collision space (phase space of the billiard map)		2.9
$\tilde{\mathcal{M}}$	part of $\mathcal{M}$ where all iterations of $\mathcal{F}$ are defined		2.9
$\hat{\mathcal{M}}$	part of $\mathcal{M}$ where all iterations of $\mathcal{F}$ are smooth		2.11
r, φ	coordinates in the collision space $\mathcal{M}$		2.10
μ	invariant measure for the collision map $\mathcal{F}$		2.12
$\mathcal{S}_0$	boundary of the collision space $\mathcal{M}$		2.10
$\mathcal{S}_{\pm 1}$	singularity set for the map $\mathcal{F}^{\pm 1}$		2.10
$\mathcal{S}_{\pm n}$	singularity set for the map $\mathcal{F}^{\pm n}$		2.11
$\mathcal{S}_{\pm\infty}$	same as $\cup_{n\geq 1}\mathcal{S}_{\pm n}$		4.11
$\mathcal{Q}_n(x)$	connected component of $\mathcal{M}\setminus\mathcal{S}_n$ containing x		4.11
$\mathcal{V}$	$(= d\varphi/dr)$ slope of smooth curves in $\mathcal{M}$		3.10
τ	return time (intercollision time)		2.9
$\bar{\tau}$	mean return time (mean free path)		2.12
$\lambda_x^{(i)}$	Lyapunov exponent at the point x		3.1
E_x^s, E_x^u	stable and unstable tangent subspaces at the point x		3.1
$\mathcal{C}_x^s, \mathcal{C}_x^u$	stable and unstable cones at the point x		3.13
Λ	(minimal) factor of expansion of unstable vectors		4.4
$\mathcal{B}$	the curvature of wave fronts		3.7

$\mathcal{R}$	collision parameter	3.6
$\mathbb{H}_k$	homogeneity strips	5.3
$\mathbb{S}_k$	lines separating homogeneity strips	5.3
k_0	minimal nonzero index of homogeneity strips	5.3
$\mathcal{M}_{\mathbb{H}}$	new collision space (union of homogeneity strips)	5.4
$\mathbf{h}$	holonomy map	5.7
$\mathcal{I}$	involution map	2.14
$\mathbf{m}$	Lebesgue measure on lines and curves	5.9
$\|W\|$	length of the curve W	4.5
$W\|_p\ \ \|$	length of the curve W in the p-metric	4.5
$\mathcal{J}_W\mathcal{F}^n(x)$	Jacobian of the restriction of $\mathcal{F}^n$ to the curve W at the point $x \in W$	5.2
$r_W(x)$	distance from $x \in W$ to the nearest endpoint of the curve W	4.12
$r_n(x)$	distance from $\mathcal{F}^n(x)$ to the nearest endpoint of the component of $\mathcal{F}^n(W)$ that contains $\mathcal{F}^n(x)$	5.9
$p_W(x)$	distance from $x \in W$ to the nearest endpoint of W in the p-metric	4.13
$\rho_W(x)$	u-SRB density on unstable manifold W	5.2
$\asymp$	'same order of magnitude'	4.3
L	ceiling function for suspension flows	2.9

CHAPTER 1

Simple examples

We start with a few simple examples of mathematical billiards, which will help us introduce basic features of billiard dynamics. This chapter is for the complete beginner. The reader familiar with some billiards may safely skip it – all the formal definitions will be given in Chapter 2.

1.1. Billiard in a circle

Let $\mathcal{D}$ denote the unit disk $x^2+y^2 \leq 1$. Let a point-like (dimensionless) particle move inside $\mathcal{D}$ with constant speed and bounce off its boundary $\partial\mathcal{D}$ according to the classical rule *the angle of incidence is equal to the angle of reflection*; see below.

Denote by $q_t = (x_t, y_t)$ the coordinates of the moving particle at time t and by $v_t = (u_t, w_t)$ its velocity vector. Then its position and velocity at time $t+s$ can be computed by

$$(1.1)\qquad \begin{aligned} x_{t+s} &= x_t + u_t s \qquad & u_{t+s} &= u_t \\ y_{t+s} &= y_t + w_t s \qquad & w_{t+s} &= w_t \end{aligned}$$

as long as the particle stays inside $\mathcal{D}$ (makes no contact with $\partial\mathcal{D}$).

When the particle collides with the boundary $\partial\mathcal{D} = \{x^2+y^2=1\}$, its velocity vector v gets reflected across the tangent line to $\partial\mathcal{D}$ at the point of collision; see Fig. 1.1.

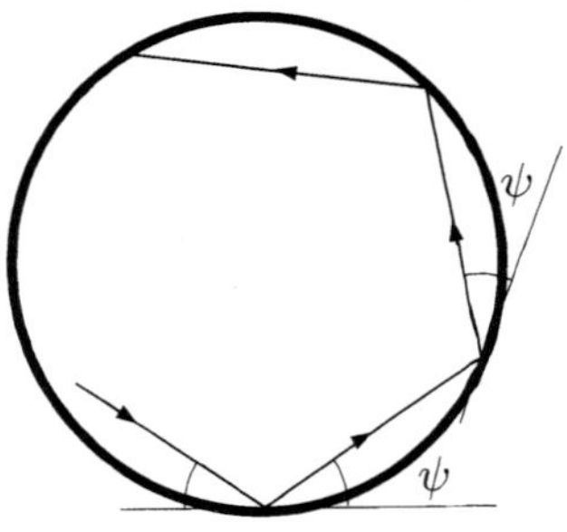

FIGURE 1.1. Billiard motion in a circle.

EXERCISE 1.1. Show that the new (postcollisional) velocity vector is related to the old (precollisional) velocity by the rule

$$(1.2)\qquad v^{\text{new}} = v^{\text{old}} - 2\,\langle v^{\text{old}}, n\rangle\, n,$$

where $n = (x, y)$ is the unit normal vector to the circle $x^2+y^2=1$ and $\langle v, n\rangle = ux + wy$ denotes the scalar product.

After the reflection, the particle resumes its free motion (1.1) inside the disk $\mathcal{D}$, until the next collision with the boundary $\partial\mathcal{D}$. Then it bounces off again, and so on. The motion can be continued indefinitely, both in the future and the past.

For example, if the particle runs along a diameter of the disk, its velocity vector will get reversed at every collision, and the particle will keep running back and forth along the same diameter forever. Other examples of periodic motion are shown in Fig. 1.2, where the particle traverses the sides of some regular polygons.

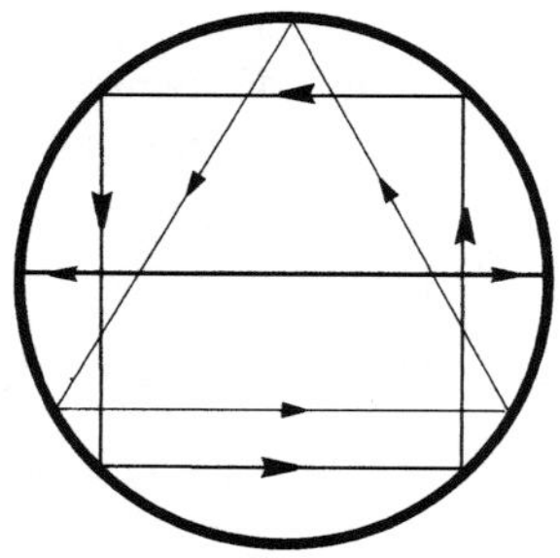

FIGURE 1.2. Periodic motion in a circle.

In the studies of dynamical systems, the primary goal is to describe the evolution of the system over long time periods and its asymptotic behavior in the limit $t \to \infty$. We will focus on such a description.

Let us parameterize the unit circle $x^2 + y^2 = 1$ by the polar (counterclockwise) angle $\theta \in [0, 2\pi]$ (since θ is a cyclic coordinate, its values 0 and 2π are identified). Also, denote by $\psi \in [0, \pi]$ the angle of reflection as shown in Fig. 1.1.

REMARK 1.2. We note that θ is actually an arc length parameter on the circle $\partial\mathcal{D}$; when studying more general billiard tables $\mathcal{D}$, we will always parameterize the boundary $\partial\mathcal{D}$ by its arc length. Instead of ψ, a reflection can also be described by the angle $\varphi = \pi/2 - \psi \in [-\pi/2, \pi/2]$ that the postcollisional velocity vector makes with the inward normal to $\partial\mathcal{D}$. In fact, all principal formulas in this book will be given in terms of φ rather than ψ, but for the moment we proceed with ψ.

For every $n \in \mathbb{Z}$, let θ_n denote the nth collision point and ψ_n the corresponding angle of reflection.

EXERCISE 1.3. Show that

$$\begin{aligned} \theta_{n+1} &= \theta_n + 2\psi_n \qquad (\text{mod } 2\pi) \\ \psi_{n+1} &= \psi_n \end{aligned} \tag{1.3}$$

for all $n \in \mathbb{Z}$.

We make two important observations now:

- All the distances between reflection points are equal.
- The angle of reflection remains unchanged.

COROLLARY 1.4. *Let (θ_0, ψ_0) denote the parameters of the initial collision. Then*

$$\theta_n = \theta_0 + 2n\psi_0 \qquad (\text{mod } 2\pi)$$
$$\psi_n = \psi_0.$$

Every collision is characterized by two numbers: θ (the point) and ψ (the angle). All the collisions make the *collision space* with coordinates θ and ψ on it. It is a cylinder because θ is a cyclic coordinate; see Fig. 1.3. We denote the collision space by $\mathcal{M}$. The motion of the particle, from collision to collision, corresponds to a map $\mathcal{F}\colon \mathcal{M} \to \mathcal{M}$, which we call the *collision map*. For a circular billiard it is given by equations (1.3).

Observe that $\mathcal{F}$ leaves every horizontal level $\mathcal{C}_\psi = \{\psi = \text{const}\}$ of the cylinder $\mathcal{M}$ invariant. Furthermore, the restriction of $\mathcal{F}$ to $\mathcal{C}_\psi$ is a rotation of the circle $\mathcal{C}_\psi$ through the angle 2ψ. The angle of rotation continuously changes from circle to circle, growing from 0 at the bottom $\{\psi = 0\}$ to 2π at the top $\{\psi = \pi\}$ (thus the top and bottom circles are actually kept fixed by $\mathcal{F}$). The cylinder $\mathcal{M}$ is "twisted upward" ("unscrewed") by the map $\mathcal{F}$; see Fig. 1.3.

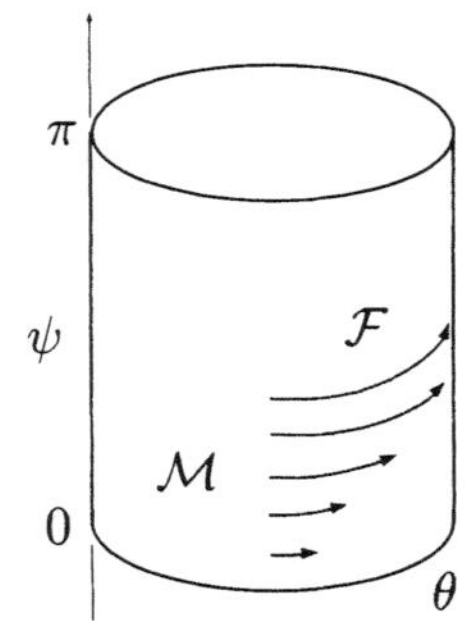

FIGURE 1.3. Action of the collision map $\mathcal{F}$ on $\mathcal{M}$.

Rigid rotation of a circle is a basic example in ergodic theory; cf. Appendix C. It preserves the Lebesgue measure on the circle. Rotations through rational angles are periodic, while those through irrational angles are ergodic.

EXERCISE 1.5. Show that if $\psi < \pi$ is a rational multiple of π, i.e. $\psi/\pi = m/n$ (irreducible fraction), then the rotation of the circle $\mathcal{C}_\psi$ is periodic with (minimal) period n, that is every point on that circle is periodic with period n, i.e. $\mathcal{F}^n(\theta, \psi) = (\theta, \psi)$ for every $0 \le \theta \le 2\pi$.

If ψ/π is irrational, then the rotation of $\mathcal{C}_\psi$ is ergodic with respect to the Lebesgue measure. Furthermore, it is *uniquely ergodic*, which means that the invariant measure is unique. As a consequence, for *every point* $(\psi, \theta) \in \mathcal{C}_\psi$ its images $\{\theta + 2n\psi, \quad n \in \mathbb{Z}\}$ are dense and uniformly distributed[1] on $\mathcal{C}_\psi$; this last fact is sometimes referred to as Weyl's theorem [**Pet83**, pp. 49–50].

[1]A sequence of points $x_n \in C$ on a circle C is said to be uniformly distributed if for any interval $I \subset C$ we have $\lim_{N\to\infty} \#\{n\colon 0 < n < N,\ a_n \in I\}/N = \text{length}(I)/\text{length}(C)$.

EXERCISE 1.6. Show that every segment of the particle's trajectory between consecutive collisions is tangent to the smaller circle $S_\psi = \{x^2 + y^2 = \cos^2 \psi\}$ concentric to the disk $\mathcal{D}$. Show that if ψ/π is irrational, the trajectory densely fills the ring between $\partial\mathcal{D}$ and the smaller circle S_ψ (see Fig. 1.4).

Remark: One can clearly see in Fig. 1.4 that the particle's trajectory looks denser near the inner boundary of the ring (it "focuses" on the inner circle). If the particle's trajectory were the path of a laser ray and the border of the unit disk were a perfect mirror, then it would feel "very hot" there on the inner circle. For this reason, the inner circle is called a *caustic* (which means "burning" in Greek).

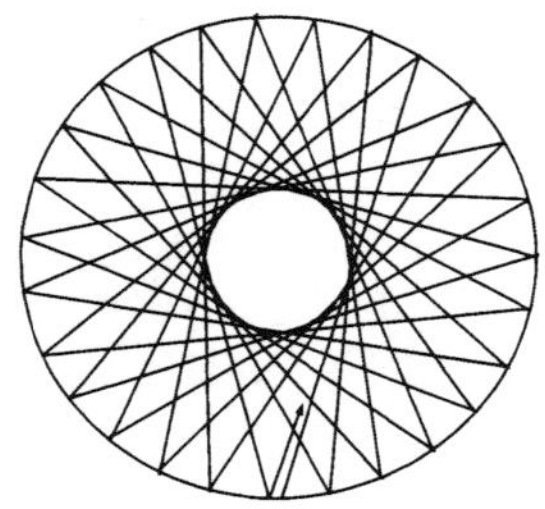

FIGURE 1.4. A nonperiodic trajectory.

EXERCISE 1.7. Can the trajectory of the moving particle be dense in the entire disk $\mathcal{D}$? (Answer: No.)

EXERCISE 1.8. Does the map $\mathcal{F}: \mathcal{M} \to \mathcal{M}$ preserve any absolutely continuous invariant measure $d\mu = f(\theta, \psi)\, d\theta\, d\psi$ on $\mathcal{M}$? Answer: Any measure whose density $f(\theta, \psi) = f(\psi)$ is independent of θ is $\mathcal{F}$-invariant.

Next, we can fix the speed of the moving particle due to the following facts.

EXERCISE 1.9. Show that $\|v_t\| = \text{const}$, so that the speed of the particle remains constant at all times.

EXERCISE 1.10. Show that if we change the speed of the particle, say we set $\|v\|_{\text{new}} = c\,\|v\|_{\text{old}}$ with some $c > 0$, then its trajectory will remain unchanged, up to a simple rescaling of time: $q_t^{\text{new}} = q_{ct}^{\text{old}}$ and $v_t^{\text{new}} = v_{ct}^{\text{old}}$ for all $t \in \mathbb{R}$.

Thus, the speed of the particle remains constant and its value is not important. It is customary to set the speed to one: $\|v\| = 1$. Then the velocity vector at time t can be described by an angular coordinate ω_t so that $v_t = (\cos\omega_t, \sin\omega_t)$ and $\omega_t \in [0, 2\pi]$ with the endpoints 0 and 2π being identified.

Now, the collision map $\mathcal{F}: \mathcal{M} \to \mathcal{M}$ represents collisions only. To describe the motion of the particle inside $\mathcal{D}$, let us consider all possible *states* (q, v), where $q \in \mathcal{D}$ is the position and $v \in S^1$ is the velocity vector of the particle. The space of all states (called the *phase space*) is then a three-dimensional manifold $\Omega\colon = \mathcal{D} \times S^1$, which is, of course, a solid torus (doughnut).

The motion of the billiard particle induces a continuous group of transformations of the torus Ω into itself. Precisely, for every $(q, v) \in \Omega$ and every $t \in \mathbb{R}$ the billiard particle starting at (q, v) will come to some point $(q_t, v_t) \in \Omega$ at time t.

Thus we get a map $(q, v) \mapsto (q_t, v_t)$ on Ω, which we denote by Φ^t. The family of maps $\{\Phi^t\}$ is a *group*; i.e. $\Phi^t \circ \Phi^s = \Phi^{t+s}$ for all $t, s \in \mathbb{R}$. This family is called the *billiard flow* on the phase space.

Let us consider a modification of the circular billiard. Denote by $\mathcal{D}_+$ the upper half disk $x^2 + y^2 \leq 1$, $y \geq 0$, and let a point particle move inside $\mathcal{D}_+$ and bounce off $\partial\mathcal{D}_+$. (A delicate question arises here: what happens if the particle hits $\partial\mathcal{D}_+$ at $(1, 0)$ or $(-1, 0)$, since there is no tangent line to $\partial\mathcal{D}_+$ at those points? We address this question in the next section.)

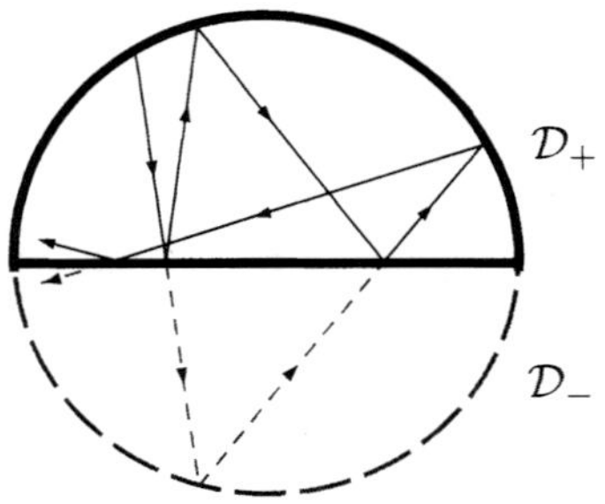

FIGURE 1.5. Billiard in the upper half circle.

A simple trick allows us to reduce this model to a billiard in the full unit disk $\mathcal{D}$. Denote by $\mathcal{D}_-$ the closure of $\mathcal{D} \setminus \mathcal{D}_+$, i.e. the mirror image of $\mathcal{D}_+$ across the x axis $L = \{y = 0\}$. When the particle hits L, its trajectory gets reflected across L, but we will also draw its continuation (mirror image) below L. The latter will evolve in $\mathcal{D}_-$ symmetrically to the real trajectory in $\mathcal{D}_+$ until the latter hits L again. Then these two trajectories will merge and move together in $\mathcal{D}_+$ for a while until the next collision with L, at which time they split again (one goes into $\mathcal{D}_-$ and the other into $\mathcal{D}_+$), etc.

It is important that the second (imaginary) trajectory never actually gets reflected off the line L; it just crosses L every time. Thus it evolves as a billiard trajectory in the full disk $\mathcal{D}$ as described above. The properties of billiard trajectories in $\mathcal{D}_+$ can be easily derived from those discussed above for the full disk $\mathcal{D}$. This type of reduction is quite common in the study of billiards.

EXERCISE 1.11. Prove that periodic trajectories in the half-disk $\mathcal{D}_+$ correspond to periodic trajectories in the full disk $\mathcal{D}$. Note, however, that the period (the number of reflections) may differ.

EXERCISE 1.12. Investigate the billiard motion in a quarter of the unit disk $x^2 + y^2 \leq 1$, $x \geq 0$, $y \geq 0$.

1.2. Billiard in a square

Here we describe another simple example – a billiard in the unit square $\mathcal{D} = \{(x, y) \colon 0 \leq x, y \leq 1\}$; see Fig. 1.6. The laws of motion are the same as before, but this system presents new features.

First of all, when the moving particle hits a vertex of the square $\mathcal{D}$, the reflection rule (1.2) does not apply (there is no normal vector n at a vertex). The particle

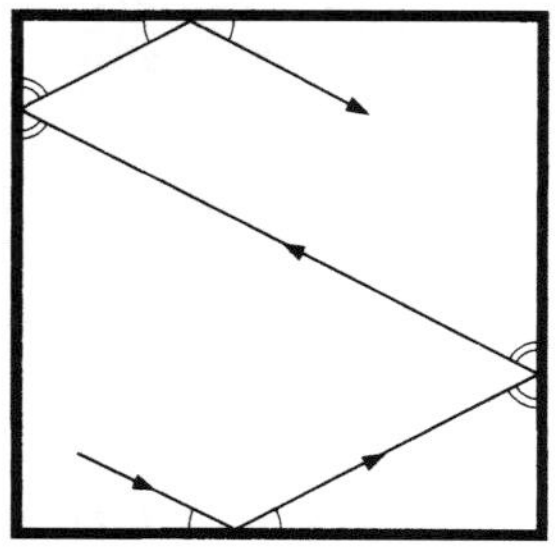

FIGURE 1.6. Billiard in a square.

then stops and its trajectory terminates. We will discuss this exceptional situation later. First we consider regular trajectories that never hit the vertices.

Let $v_t = (u_t, w_t)$ denote the velocity vector of the moving particle at time t (in the x, y coordinates). If it hits a vertical side of $\mathcal{D}$ at time t, then u_t changes sign ($u_{t+0} = -u_{t-0}$) and w_t remains unchanged. If the particle hits a horizontal side of $\mathcal{D}$, then w_t changes sign ($w_{t+0} = -w_{t-0}$) and u_t remains unchanged. Thus,

$$u_t = (-1)^m u_0 \quad \text{and} \quad w_t = (-1)^n w_0, \tag{1.4}$$

where m and n denote the number of collisions with vertical and, respectively, horizontal sides of $\mathcal{D}$ during the time interval $(0, t)$.

EXERCISE 1.13. Show that if $u_0 \neq 0$ and $w_0 \neq 0$ (and assuming the particle never hits a vertex), then all four combinations $(\pm u_0, \pm w_0)$ appear along the particle's trajectory infinitely many times.

Next we make use of the trick shown in Fig. 1.5. Instead of reflecting the trajectory of the billiard particle in a side of $\partial\mathcal{D}$, we reflect the square $\mathcal{D}$ across that side and let the particle move straight into the mirror image of $\mathcal{D}$. If we keep doing this at every collision, our particle will move along a straight line through the multiple copies of $\mathcal{D}$ obtained by successive reflections (the particle "pierces" a chain of squares; see Fig. 1.7). This construction is called the *unfolding* of the billiard trajectory. To recover the original trajectory in $\mathcal{D}$, one *folds* the resulting string of adjacent copies of $\mathcal{D}$ back onto $\mathcal{D}$.

We denote the copies of $\mathcal{D}$ by

$$\mathcal{D}_{m,n} = \{(x, y)\colon\ m \leq x \leq m+1,\ n \leq y \leq n+1\}. \tag{1.5}$$

EXERCISE 1.14. Show that if m and n are even, then the folding procedure transforms $\mathcal{D}_{m,n}$ back onto $\mathcal{D} = \mathcal{D}_{0,0}$ by translations $x \mapsto x - m$ and $y \mapsto y - n$, thus preserving orientation of both x and y. If m is odd, then the orientation of x is reversed (precisely, $x \mapsto m + 1 - x$). If n is odd, then the orientation of y is reversed (precisely, $y \mapsto n + 1 - y$). Observe that these rules do not depend on the particular trajectory that was originally unfolded.

The squares $\mathcal{D}_{m,n}$ with $m, n \in \mathbb{Z}$ tile, like blocks, the entire plane $\mathbb{R}^2$. Any regular billiard trajectory unfolds into a directed straight line on the plane, and any directed line (which avoids the sites of the integer lattice) folds back into a billiard trajectory. A trajectory hits a vertex of $\mathcal{D}$ iff the corresponding line runs into a site of the integer lattice.

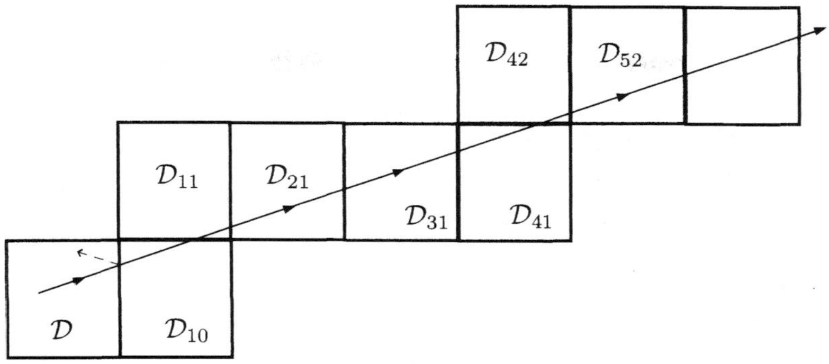

FIGURE 1.7. Unfolding a billiard trajectory.

The structure of blocks $\mathcal{D}_{m,n}$ with the respective folding rules is clearly periodic, in which the 2×2 square

$$\mathbb{K}_2 = \{(x, y) \colon 0 \le x, y \le 2\}$$

plays the role of a fundamental domain – the entire plane is covered by parallel translations of $\mathbb{K}_2$. Thus the standard projection of $\mathbb{R}^2$ onto $\mathbb{K}_2$ transforms unfolded trajectories into directed straight lines on the 2×2 torus Tor^2 (the latter is obtained by identifying the opposite sides of the square $\mathbb{K}_2$). The billiard in the unit square $\mathcal{D}$ thus reduces to the simple linear flow on a flat 2×2 torus Tor^2, in which points move with constant (unit) velocity vectors.

The linear flow on a flat torus is one of the standard examples in ergodic theory; cf. Appendix C and [**KH95, Pet83, Sin76**]. Its main properties are these:

- if a trajectory has rational slope $dy/dx \in \mathbb{Q}$, then it is periodic (it runs along a closed geodesic);
- if a trajectory has irrational slope $dy/dx \notin \mathbb{Q}$, then it is dense (its closure is the whole torus).

This translates into the following alternative for regular billiard trajectories in the unit square $\mathcal{D}$:

COROLLARY 1.15. *If $w_0/u_0 \in \mathbb{Q}$, then the corresponding regular billiard trajectory in the unit square $\mathcal{D}$ is periodic. If $w_0/u_0 \notin \mathbb{Q}$, then the corresponding regular billiard trajectory is dense.*

EXERCISE 1.16. Extend this result to the billiard in a rectangle $\mathcal{R}$ with sides a and b. Answer: a regular billiard trajectory in $\mathcal{R}$ is periodic iff $(aw_0)/(bu_0) \in \mathbb{Q}$; otherwise it is dense. Hint: Transform the rectangle into the unit square by scaling the coordinates: $(x, y) \mapsto (x/a, y/b)$. Argue that the billiard trajectories in $\mathcal{R}$ will be thus transformed into those in $\mathcal{D}$.

EXERCISE 1.17. Extend the above result to billiards in the following polygons: an equilateral triangle, a right isosceles triangle, a right triangle with the acute angle $\pi/6$, and a regular hexagon. What is common about these polygons? (Note that the billiard in a hexagon does not reduce to a geodesic flow on a torus. Does it reduce to a geodesic flow on another manifold?)

The phase space of the billiard system in the unit square $\mathcal{D}$ is the three-dimensional manifold $\Omega = \mathcal{D} \times S^1$; cf. the previous section. The billiard flow Φ^t is defined for all times $-\infty < t < \infty$ on regular trajectories. On exceptional trajectories (which hit a vertex of $\mathcal{D}$ at some time), the flow is defined only until the trajectory terminates in a vertex.

EXERCISE 1.18. Show that the set of exceptional trajectories is a countable union of 2D surfaces in Ω.

We see that the set of exceptional trajectories is negligible in the topological and measure-theoretic sense (it has zero Lebesgue measure and is an F_σ set, i.e. a countable union of nowhere dense closed subsets), but still its presence is bothersome. For the billiard in a square, though, one can get rid of them altogether by extending the billiard flow by continuity.

EXERCISE 1.19. Show that the flow Φ^t can be uniquely extended by continuity to all exceptional trajectories. In that case every trajectory hitting a vertex of $\mathcal{D}$ will simply reverse its course and run straight back; see Fig. 1.8.

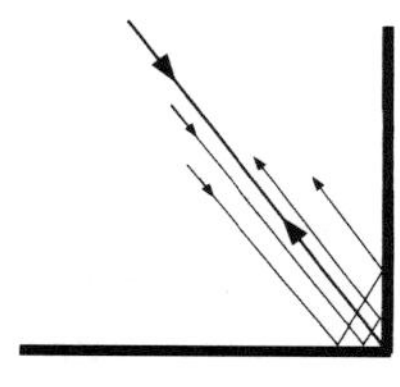

FIGURE 1.8. Extension of the flow near a vertex.

The above extension defines the billiard flow Φ^t on the entire phase space Ω and makes it continuous everywhere. We will assume this extension in what follows. We remark, however, that in generic billiards such nice extensions are rarely possible; see Section 2.8.

Now, the action of the flow Φ^t on the phase space Ω can be fully described as follows. For every unit vector $v_0 = (u_0, w_0) \in S^1$, consider the set

$$\mathcal{L}_{v_0} = \{(q, v) \in \Omega\colon\ q \in \mathcal{D},\ v = (\pm u_0, \pm w_0)\}$$

(the two signs are, of course, independent). Due to (1.4), each set $\mathcal{L}_{v_0}$ remains invariant under the flow Φ^t.

Suppose first that $u_0 \neq 0$ and $v_0 \neq 0$; then $\mathcal{L}_{v_0}$ is the union of four squares obtained by "slicing" Ω at the four "levels" corresponding to the vectors $(\pm u_0, \pm w_0)$; see Fig. 1.9.

EXERCISE 1.20. Check that the four squares constituting the set $\mathcal{L}_{v_0}$ can be glued together along their boundaries, and obtain a smooth closed surface without boundary (a 2×2 torus) $\mathbb{T}^2_{v_0}$ on which the billiard flow will coincide with the linear flow along the vector v_0 (i.e. the flow on $\mathbb{T}^2_{v_0}$ will be defined by differential equations $\dot{x} = u_0$, $\dot{y} = w_0$). Hint: The assembly of the torus $\mathbb{T}^2_{v_0}$ from the squares of $\mathcal{L}_{v_0}$ is very similar to the reduction of the billiard dynamics in $\mathcal{D}$ to the geodesic flow on the 2×2 torus described above (in fact, these two procedures are equivalent).

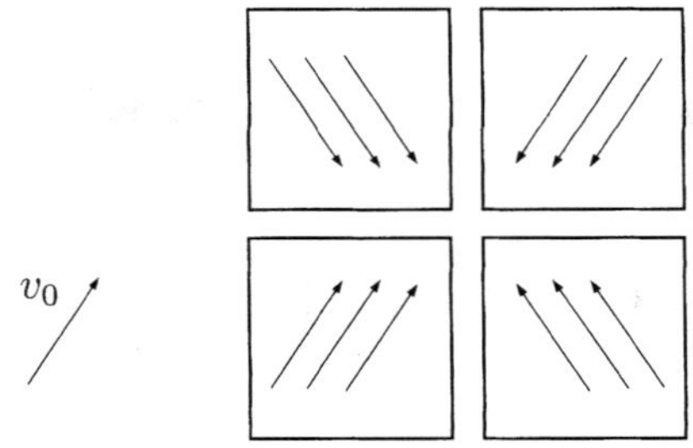

FIGURE 1.9. Four squares constituting $\mathcal{L}_{v_0}$.

Now, it is a standard fact in ergodic theory (cf. Appendix C) that the linear flow on a 2D torus defined by $\dot{x} = u_0$, $\dot{y} = w_0$ is periodic if $w_0/u_0 \in \mathbb{Q}$ and ergodic (furthermore, uniquely ergodic) if $w_0/u_0 \notin \mathbb{Q}$. In the latter case every trajectory is dense and uniformly distributed[2] on the torus.

In the two remaining cases (first $u_0 = 0$, and second $w_0 = 0$) the set $\mathcal{L}_{v_0}$ consists of just two squares. We leave their analysis to the reader as an easy exercise.

This fully describes the action of the flow $\Phi^t\colon \Omega \to \Omega$ for the billiard in the unit square.

1.3. A simple mechanical model

As a motivation for the study of billiards, one usually describes a simple model of two moving particles in a one-dimensional container. It reduces to a billiard in a right triangle, which is similar to a billiard in a square. We describe this model here; see also [**CFS82, CM03**].

Consider a system of two point particles of masses m_1 and m_2 on a unit interval $0 \leq x \leq 1$. The particles move freely and collide elastically with each other and with the 'walls' at $x = 0$ and $x = 1$. Let x_1 and x_2 denote the positions of the particles and u_1 and u_2 their velocities. Since the particles collide upon contact, their positions remain ordered; we assume that $x_1 \leq x_2$ at all times.

FIGURE 1.10. Two particles in a unit interval.

Next we describe collisions. When a particle hits a wall, it simply reverses its velocity. When the two particles collide with each other, we denote by u_i^- the precollisional velocity and by u_i^+ the postcollisional velocity of the ith particle, $i = 1, 2$. The law of elastic collisions requires the conservation of the total momentum, i.e.

$$m_1 u_1^+ + m_2 u_2^+ = m_1 u_1^- + m_2 u_2^-$$

[2]A line x_t on a 2D torus Tor^2 is said to be uniformly distributed if for any rectangle $R \subset \mathrm{Tor}^2$ we have $\lim_{T\to\infty} \mathbf{m}(\{t\colon 0 < t < T,\ x_t \in R\})/T = \mathrm{area}(R)/\mathrm{area}(\mathrm{Tor}^2)$. Here $\mathbf{m}$ is the Lebesgue measure on $\mathbb{R}$.

and the total kinetic energy, i.e.

$$m_1[u_1^+]^2 + m_2[u_2^+]^2 = m_1[u_1^-]^2 + m_2[u_2^-]^2. \tag{1.6}$$

Solving these equations gives

$$u_1^+ = u_1^- + \frac{2m_2}{m_1+m_2}(u_2^- - u_1^-)$$

and

$$u_2^+ = u_2^- + \frac{2m_1}{m_1+m_2}(u_1^- - u_2^-)$$

(we recommend the reader derive these formulas for an exercise). Note that if $m_1 = m_2$, then the particles simply exchange their velocities: $u_1^+ = u_2^-$ and $u_2^+ = u_1^-$.

The variables x_i and u_i are actually inconvenient, so we will work with new variables defined by

$$q_i = x_i\sqrt{m_i} \qquad \text{and} \qquad v_i = dq_i/dt = u_i\sqrt{m_i} \tag{1.7}$$

for $i = 1, 2$. Now the positions of the particles are described by a point $\mathbf{q} = (q_1, q_2) \in \mathbb{R}^2$ (it is called a *configuration point*). The set of all configuration points (called the *configuration space*) is the right triangle

$$\mathcal{D} = \{\mathbf{q} = (q_1, q_2) : 0 \le q_1/\sqrt{m_1} \le q_2/\sqrt{m_2} \le 1\}.$$

The velocities of the particles are described by the vector $\mathbf{v} = (v_1, v_2)$. Note that the energy conservation law (1.6) implies that $\|\mathbf{v}\| = \text{const}$; thus we can set $\|\mathbf{v}\| = 1$.

The state of the system is described by a pair $(\mathbf{q}, \mathbf{v})$. The configuration point $\mathbf{q}$ moves in $\mathcal{D}$ with velocity vector $\mathbf{v}$. When the first particle collides with the wall ($x_1 = 0$), the configuration point hits the left side $q_1 = 0$ of the triangle $\mathcal{D}$. When the second particle collides with the wall ($x_2 = 1$), the point $\mathbf{q}$ hits the upper side $q_2/\sqrt{m_2} = 1$ of $\mathcal{D}$. When the particles collide with each other, the point $\mathbf{q}$ hits the hypotenuse $q_1/\sqrt{m_1} = q_2/\sqrt{m_2}$ of $\mathcal{D}$.

EXERCISE 1.21. Prove that the velocity vector $\mathbf{v}$ changes at collisions so that it gets reflected at $\partial\mathcal{D}$ according to the law 'the angle of incidence is equal to the angle of reflection'.

Thus, the motion of the configuration point $\mathbf{q}$ is governed by the billiard rules. Hence the evolution of the mechanical model of two particles in a unit interval reduces to billiard dynamics in a right triangle.

If $m_1 = m_2$, we obtain a billiard in a right isosceles triangle, which readily reduces to a billiard in a square; see Exercise 1.17. For generic mass ratio m_1/m_2, we obtain a billiard in a generic right triangle, which may be rather complicated (such billiards are not covered in our book).

One complication arises when a billiard trajectory hits a corner point of $\mathcal{D}$. Hitting the vertex of the right angle corresponds to an event when both particles simultaneously collide with opposite walls. Then their further motion is clearly well defined; thus the billiard trajectory can easily be continued (cf. Exercise 1.19).

However, hitting the vertex of an acute angle of $\mathcal{D}$ corresponds to an event when both particles simultaneously collide with the *same* wall (either $x = 0$ or $x = 1$). In this case, for generic m_1 and m_2, the billiard flow cannot be extended by continuity, as nearby trajectories hitting the two adjacent sides in different order will come back to $\mathcal{D}$ along different lines; see Fig. 1.11.

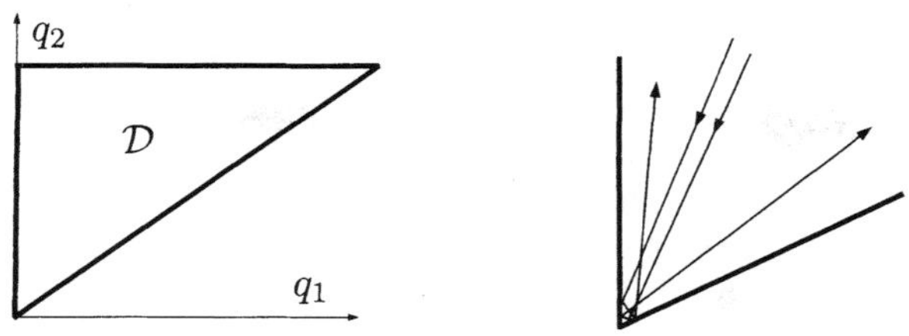

FIGURE 1.11. The right triangle $\mathcal{D}$: hitting the vertex of an acute angle.

In mechanical terms, hitting the vertex of an acute angle of $\mathcal{D}$ corresponds to a *multiple collision*. Such exceptional events usually cannot be resolved by the laws of classical mechanics.

1.4. Billiard in an ellipse

We proceed to yet another simple example that admits a completely elementary analysis – the billiard in an ellipse

$$\frac{x^2}{a^2} + \frac{y^2}{b^2} = 1$$

with some $a > b > 0$. In fact, it was this example that Birkhoff described in the very first book on mathematical billiards in 1927 [**Bi27**, Chapter VIII].

We denote by $\mathcal{D}$ the domain bounded by the ellipse (it will be our billiard table). Let F_1 and F_2 denote the foci of the ellipse, and observe that they lie on the x axis. The ellipse is the locus of points $A \in \mathbb{R}^2$ such that

$$\operatorname{dist}(A, F_1) + \operatorname{dist}(A, F_2) = \text{const}.$$

EXERCISE 1.22. Let $A \in \partial\mathcal{D}$ and L denote the tangent line to the ellipse at A. Prove that the segments AF_1 and AF_2 make equal angles with L. (This fact is known in projective geometry as the Poncelet theorem.) Hint: Reflect the point F_2 across the tangent line L and show that its image will lie on the line AF_1.

Thus, if a billiard trajectory passes through one focus, then it reflects at a point $A \in \partial\mathcal{D}$ on the ellipse and runs straight into the other focus. Such a trajectory will then pass through a focus after every reflection; see Fig. 1.12.

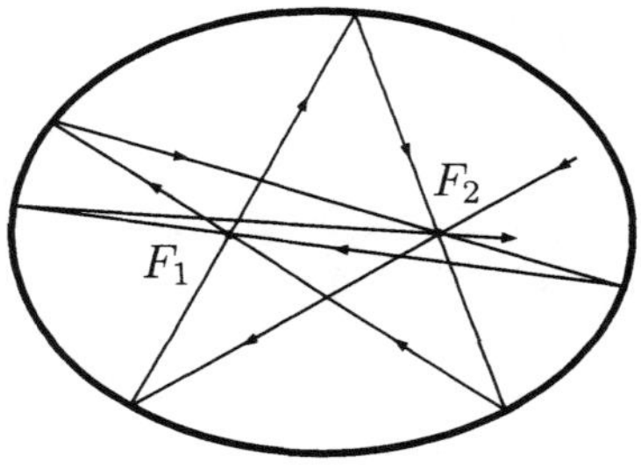

FIGURE 1.12. A trajectory passing through the foci.

EXERCISE 1.23. Show that every trajectory passing through the foci F_1 and F_2 converges to the major axis of the ellipse (the x axis).

By the way, the major and the minor axes of the ellipse are clearly two periodic trajectories – they run back and forth between their endpoints.

In Section 1.1 we used the coordinates ψ and θ to describe collisions in a circular billiard, and the cyclic coordinate θ was actually the arc length parameter on the circle (Remark 1.2). Here we use two coordinates ψ and r, where ψ is the same angle of reflection as in Section 1.1 and r is an arclength parameter on the ellipse. We choose the reference point $r = 0$ as the rightmost point $(a, 0)$ on the ellipse and orient r counterclockwise. Note that $0 \le r \le |\partial\mathcal{D}|$ and $0 \le \psi \le \pi$.

The collision space $\mathcal{M}$ is again a cylinder whose base is the ellipse and whose height is π. It is shown in Fig. 1.13 as a rectangle $[0, |\partial\mathcal{D}|] \times [0, \pi]$, but we keep in mind that the left and right sides of this rectangle must be identified. The motion of the billiard particle, from collision to collision, induces the collision map $\mathcal{F} : \mathcal{M} \to \mathcal{M}$.

EXERCISE 1.24. Verify that the trajectories passing through the foci lie on a closed curve on the surface $\mathcal{M}$. Determine its shape. Answer: It is the ∞-shaped curve in Fig. 1.13 that separates the white and grey areas.

Thus, the trajectories passing through the foci make a special (one-dimensional) family in $\mathcal{M}$.

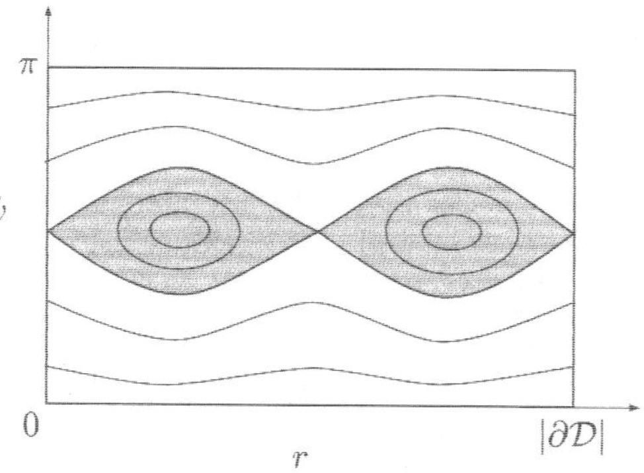

FIGURE 1.13. The collision space of an elliptic billiard.

EXERCISE 1.25. Show that if the trajectory of the billiard particle crosses the segment F_1F_2 joining the foci, then it reflects at $\partial\mathcal{D}$ and crosses this segment again. Similarly, if the trajectory crosses the major axis beyond the segment F_1F_2, say to the left of it, then after one or more reflections at $\partial\mathcal{D}$ it will cross the major axis to the right of this segment, etc.

The previous exercise shows that there are trajectories of two types: those crossing the inner segment F_1F_2 of the major axis after every reflection (we call them *inner* trajectories) and those going around this segment (we call them *outer* trajectories).

EXERCISE 1.26. Verify that the inner trajectories fill the white area in Fig. 1.13, and outer trajectories fill the grey area.

The following is the most important property of elliptic billiards:

THEOREM 1.27. *For every outer trajectory there is an ellipse with foci F_1 and F_2 that is tangent to each link of that trajectory. For every inner trajectory there is a hyperbola with foci F_1 and F_2 that is tangent to each link (or its linear extension) of that trajectory.*

PROOF. We prove only the first statement (about the outer trajectories); the proof of the second is similar. The argument is pretty elementary and illustrated in Fig. 1.14. Here A_1A and A_2A are two successive links of an outer trajectory. The points B_1 and B_2 are obtained by reflecting the foci F_1 and F_2 across the lines A_1A and A_2A, respectively. The four angles $\angle B_1AA_1$, $\angle A_1AF_1$, $\angle F_2AA_2$ and $\angle A_2AB_2$ are equal. Hence the triangles AB_1F_2 and AB_2F_1 are congruent, in particular $|B_1F_2| = |B_2F_1|$. Therefore

$$|F_1C_1| + |F_2C_1| = |F_1C_2| + |F_2C_2|,$$

where C_1 and C_2 are the points of intersection of A_1A with B_1F_2 and A_2A with B_2F_1, respectively. Thus, the points C_1 and C_2 belong to the same ellipse with foci F_1 and F_2, and the lines A_1A and A_2A are tangent to that ellipse. □

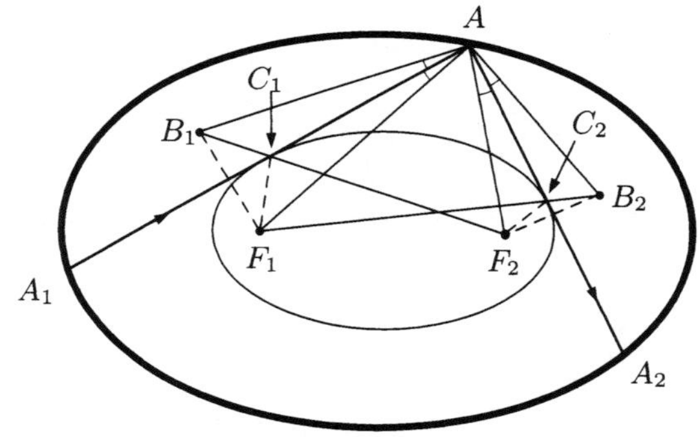

FIGURE 1.14. Proof of Theorem 1.27.

If every link of a billiard trajectory is tangent to a certain given curve, then that curve is called a *caustic*. Fig. 1.15 shows an elliptic caustic for an outer trajectory and a hyperbolic caustic for an inner trajectory. The term 'caustic' is borrowed from optics, where it means a curve on which light rays focus after being reflected off a mirror (we have seen caustics in circular billiards in Section 1.1). Fig. 1.15 demonstrates the concentration of rays on caustics (compare it to Fig. 1.4).

All the trajectories tangent to one elliptic caustic lie on a closed curve in the collision space $\mathcal{M}$. Such curves are shown as 'horizontal waves' in the white area in Fig. 1.13 (remember that the left and right sides of the rectangle need to be identified). Every such curve is obviously invariant under the map $\mathcal{F}$.

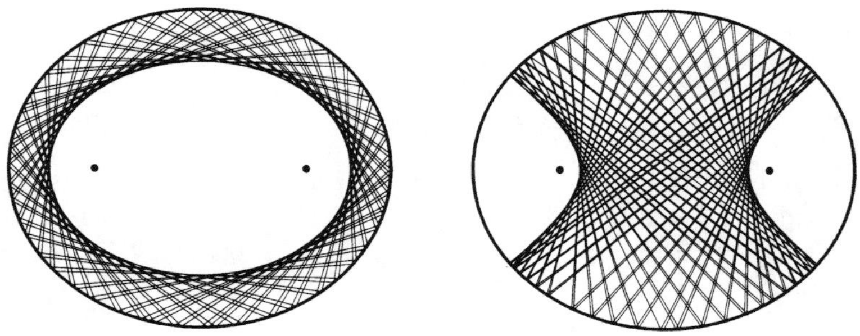

FIGURE 1.15. Elliptic and hyperbolic caustics in the elliptic billiard.

EXERCISE 1.28. On each invariant curve the map $\mathcal{F}$ is conjugate to a rigid circle rotation through some angle (that angle is called the *rotation number*). Show that the rotation number changes continuously and monotonically with the invariant curve. Hint: Consider two outer trajectories starting at the same point $A_0 \in \partial\mathcal{D}$ but with distinct elliptical caustics. Denote by A'_n the reflection points of the trajectory whose elliptical caustic is smaller and by A''_n those of the other trajectory. Observe that the sequence $\{A'_n\}$ will move along the ellipse faster than $\{A''_n\}$ does; see Fig. 1.16.

The action of the map $\mathcal{F}$ on each invariant curve can be analyzed explicitly and the rotation number can be computed analytically (see [**Be01**, Sections 2.5 and 3.2]) but we will not go that far.

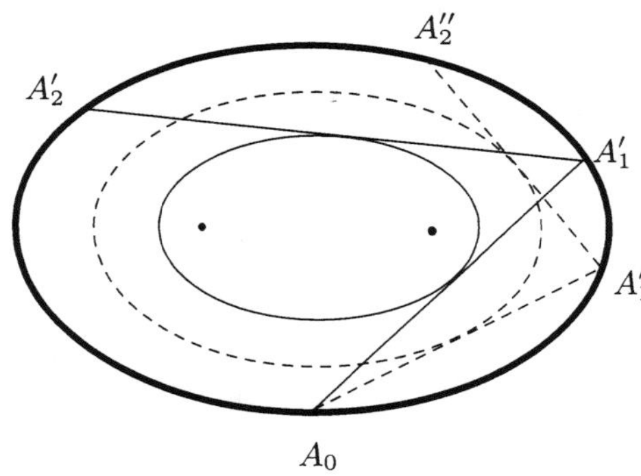

FIGURE 1.16. Exercise 1.28.

Next, all the trajectories tangent to one hyperbolic caustic lie on two closed curves in $\mathcal{M}$, one inside each half of the ∞-shaped grey domain. Such curves appear as ovals in Fig. 1.13. The map $\mathcal{F}$ transforms each oval into an identical oval within the other half of the ∞-shaped grey domain. Thus the union of the two identical (symmetric) ovals will be invariant under $\mathcal{F}$, and each oval separately will be invariant under $\mathcal{F}^2$.

Therefore, the collision space $\mathcal{M}$ of an elliptical billiard is completely foliated by invariant curves. In this sense, the elliptical billiard is similar to those in a circle and in a square. In physics, such models belong to a special class: if the phase space of a system is foliated by one-dimensional invariant submanifolds, the system is said to be *integrable*; the dynamics in such a system is completely regular. Thus, billiards in circles, squares and ellipses are completely regular.

1.5. A chaotic billiard: pinball machine

The simple examples were given in the previous sections for the sake of introducing some basic features of billiards to the novice reader. But they should not be taken as typical; in fact their dynamical characteristics are quite special and in a sense opposite to those of chaotic billiards which will be covered in the rest of the book. Here we will take a glimpse at something that happens in chaotic billiards.

Imagine you are playing a pinball machine. A small ball shoots from a cannon in the right bottom corner of a rectangular table; then it bounces off the edges until it either hits the target (then you win) or falls through an opening in the bottom (goes down the drain; then you lose). The target might be a special figure on the table that registers the hit when the ball touches it. To prevent direct hits, assume the target is screened from the cannon and hitting the screen is forbidden by the rules. Then the ball has to bounce off the edges before reaching the target; see Fig. 1.17. It is quite an unusual pinball machine, but for us it is a good starting example.

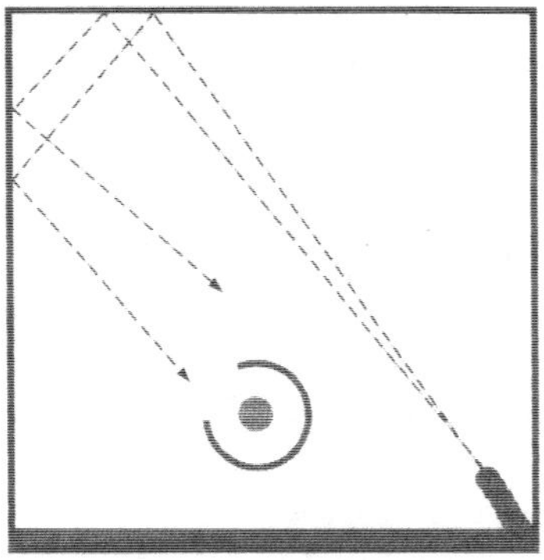

FIGURE 1.17. A pinball machine. The target is the grey disk screened from the cannon by a dark grey arc.

Suppose you can rotate the cannon to change the angle at which the ball shoots out. After you miss once, you can adjust the shooting angle and send the ball more accurately into the target. This is a relatively easy task (illustrated in Fig. 1.17), as the trajectory of the ball (in a rectangular billiard) is very simple and predictable. Also, you do not need to aim with absolute precision.

EXERCISE 1.29. Suppose the target is a disk of radius r and the moving ball is a point particle. Let L denote the distance covered by the ball from the cannon to the target. Show that if the shooting angle is off by less than r/L (radians), then the ball still hits the target.

Now let us make the task more realistic and challenging by installing some bumpers (round pillars) all over the table (see Fig. 1.18) so that our moving ball will bounce between the bumpers on its way to the target (or down the drain). Anyone who has played real pinball machines can easily imagine such a process.

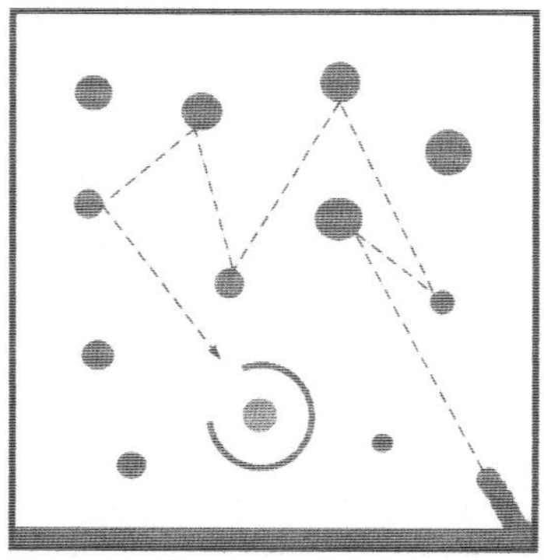

FIGURE 1.18. A pinball machine with bumpers (dark grey disks).

Would it be easy to adjust the cannon on this new table? Obviously not. The route of the ball is complicated and almost unpredictable, as it may bounce off different bumpers. Even finding the right sequence of bumpers that the ball needs to hit before it reaches the target (and avoids the screen) is not a simple task. Professional billiard players solve a similar problem when trying the send a ball to a pocket, so that it hits one or more other balls.

Furthermore, the cannon must be aimed with almost ultimate precision, as a tiny error in the shooting angle may send the ball rolling down along a completely wrong path. This is illustrated in Fig. 1.19, where just two successive bounces are shown. It is quite clear that the instability of the ball's motion increases with every subsequent reflection off a bumper. Again, professional billiard players know that if their ball needs to hit more than one other ball before sending one of them into a pocket, their task is very difficult. Furthermore, if the ball must hit three or more other balls, the task is almost impossible.

A rectangular table with round bumpers installed on it is a classical example of a chaotic billiard. The motion of the billiard particle on such a table is complicated and unpredictable. To the naked eye, it may look like a wild dance between the walls of the table, without any pattern or logic (that is why pinball machines are so attractive!). The lack of predictability is characteristic of chaotic billiards.

Furthermore, slight changes in the initial position and/or velocity of the particle quickly lead to large deviations (such as that in Fig. 1.19), so after just a few collisions with bumpers two trajectories, initially very close together, will separate and move far apart from each other, as if they were unrelated. This instability (also known as *sensitivity to initial conditions*) is another characteristic feature of chaotic billiards (and chaotic dynamics in general).

In practical terms, the best thing the player can do in our game is to shoot randomly and watch the ball running all over the table, bouncing around between bumpers – there will always be a chance that it hits the target 'by accident'. This is essentially a game of chance, just like flipping a coin, rolling a die, or playing cards. We will see in Chapters 6 and 7 that the motion in a chaotic billiard is indeed essentially random and is best described in terms of probability theory.

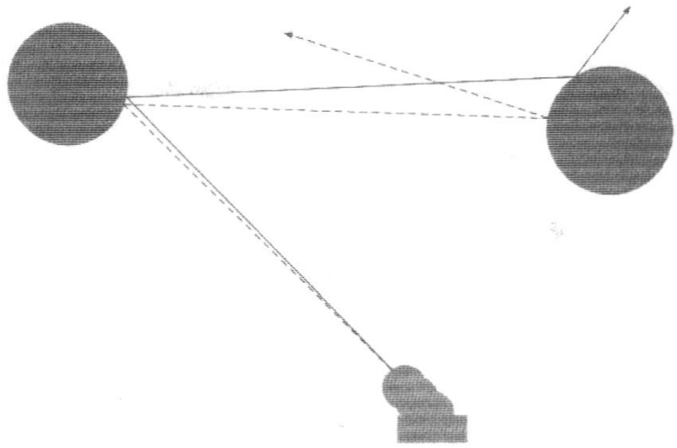

FIGURE 1.19. A ball bouncing off two bumpers: a slight error in the initial shooting angle results in a dramatic deflection in the end.

Our toy example actually has a lot in common with a classical model of statistical physics, called the Lorentz gas. In that model, a small ball (electron) bounces between large fixed disks (molecules) that make a regular periodic (crystalline) structure. We will present it in Chapter 5.6.

We will not attempt to go beyond this very informal introduction to the realm of chaotic billiards, leaving all formalities till future chapters. Interested readers may find a more extensive description of chaotic billiards, including computer illustrations, in [**Be01**, Section 1.1].

CHAPTER 2

Basic constructions

Billiard dynamics in planar regions is commonly defined as follows:

DEFINITION 2.1. Let $\mathcal{D} \subset \mathbb{R}^2$ be a domain with smooth or piece-wise smooth boundary. A billiard system corresponds to a free motion of a point particle inside $\mathcal{D}$ with specular reflections off the boundary $\partial\mathcal{D}$.

This definition is clear and easy-to-use if one has a particular domain $\mathcal{D}$ at hand. In general studies, however, one needs to specify the class of domains to be considered. One has to be especially careful about the boundary $\partial\mathcal{D}$ – whether or not to allow $\partial\mathcal{D}$ to have infinite length, or unbounded curvature, or infinitely many inflection points, etc.

In most research articles on billiards, it is assumed (explicitly or implicitly) that these and other pathologies do not occur. A notable exception is Strelcyn's paper [**St86**], which covers very general billiards, but it clearly demonstrates that irregularities of the boundary may lead to serious complications in the analysis of the dynamics and render its properties barely tractable.

Here we exclude many unpleasant pathologies from our studies by making appropriate assumptions on the domain $\mathcal{D}$ (see Assumptions A1–A3 in Section 2.1 and Assumption A4 in Section 2.4). This allows us to proceed smoothly and avoid many troubles so characteristic of more general studies [**St86**]. Still, our assumptions A1–A4 are general enough to cover practically all presently known chaotic billiards.

2.1. Billiard tables

Let $\mathcal{D}_0 \subset \mathbb{R}^2$ be a bounded open connected domain and $\mathcal{D} = \overline{\mathcal{D}}_0$ denote its closure.

ASSUMPTION A1. The boundary $\partial\mathcal{D}$ is a finite union of smooth (C^ℓ, $\ell \geq 3$) compact curves:

$$\partial\mathcal{D} = \Gamma = \Gamma_1 \cup \cdots \cup \Gamma_r. \tag{2.1}$$

Precisely, each curve Γ_i is defined by a C^ℓ map $f_i\colon [a_i, b_i] \to \mathbb{R}^2$, which is one-to-one on $[a_i, b_i)$ and has one-sided derivatives, up to order ℓ, at the points a_i and b_i. The value of ℓ is the class of smoothness of the billiard table.

We call $\mathcal{D}$ a *billiard table* and $\Gamma_1, \dots, \Gamma_r$ *walls* or *components* of $\partial\mathcal{D}$.

REMARK 2.2. If $f_i(a_i) \neq f_i(b_i)$, then we call Γ_i an *arc* and denote $\partial\Gamma_i = \{f_i(a_i), f_i(b_i)\}$. If $f_i(a_i) = f_i(b_i)$, then Γ_i is a *closed curve*, which may or may not be entirely C^ℓ smooth. If it is, then f_i can be defined on a circle S^1 rather than $[a_i, b_i]$, and we set $\partial\Gamma_i = \emptyset$. It may happen that a closed curve Γ_i lacks smoothness at the point $f_i(a_i) = f_i(b_i)$; then we denote that point by $\partial\Gamma_i$.

ASSUMPTION A2. The boundary components Γ_i can intersect each other only at their endpoints; i.e.

$$\Gamma_i \cap \Gamma_j \subset \partial\Gamma_i \cup \partial\Gamma_j \qquad \text{for } i \neq j. \tag{2.2}$$

We put

$$\Gamma_* = \partial\Gamma_1 \cup \cdots \cup \partial\Gamma_r \quad \text{and} \quad \tilde{\Gamma} = \Gamma \setminus \Gamma_*. \tag{2.3}$$

We call $x \in \Gamma_*$ *corner points* of $\mathcal{D}$ and $x \in \tilde{\Gamma}$ *regular boundary points.*

EXERCISE 2.3. Show that every regular boundary point $x \in \tilde{\Gamma}$ has an open neighborhood $U(x)$ that intersects only one wall Γ_i and is divided by Γ_i into two open connected parts – one lies in the interior of $\mathcal{D}$ and the other in $\mathbb{R}^2 \setminus \mathcal{D}$.

EXERCISE 2.4. Show that every corner point $x \in \Gamma_*$ has an open neighborhood $U(x)$ such that $U(x) \cap \Gamma$ is a disjoint union of $2m$ curves for which x is a common endpoint (here $m = m_x \geq 1$). They divide $U(x)$ into $2m$ open connected domains so that m of them lie in the interior of $\mathcal{D}$ and the other m in $\mathbb{R}^2 \setminus \mathcal{D}$, and these two types of domains alternate as one traverses $U(x)$ around x; see Fig. 2.1.

We call x a *simple corner point* if $m_x = 1$. In our book we mostly deal with simple corner points.

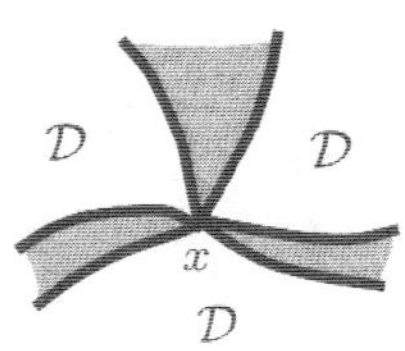

FIGURE 2.1. A corner point x with $m_x = 3$.

In the above exercise, the connected components of $U(x)$ that lie inside $\mathcal{D}$ are called *corners* of the domain $\mathcal{D}$. Each corner is bounded by two curves $\Gamma_i, \Gamma_j \subset \Gamma$ converging to the point x (we call these curves the *sides* of the corner) and characterized by the interior angle γ made by the corresponding tangent lines at x (the angle seen from inside $\mathcal{D}$). If $\gamma = 0$, we call the corner a *cusp.*

We fix an orientation of each Γ_i so that $\mathcal{D}$ lies *to the left* of Γ_i. Then we parameterize every Γ_i by its arclength; thus tangent vectors become unit vectors: $\|f_i'\| = 1$.

ASSUMPTION A3. On every Γ_i, the second derivative f_i'' either never vanishes or is identically zero (thus, every wall Γ_i is either a curve without inflection points or a line segment).

Note that $f_i'' \perp f_i'$, because $\|f_i'\| = \text{const}$. Hence, if $f_i'' \neq 0$, then the pair of vectors f_i', f_i'' is either left or right along the entire curve Γ_i. Accordingly, we distinguish three types of walls:

Flat walls: such that $f'' \equiv 0$;

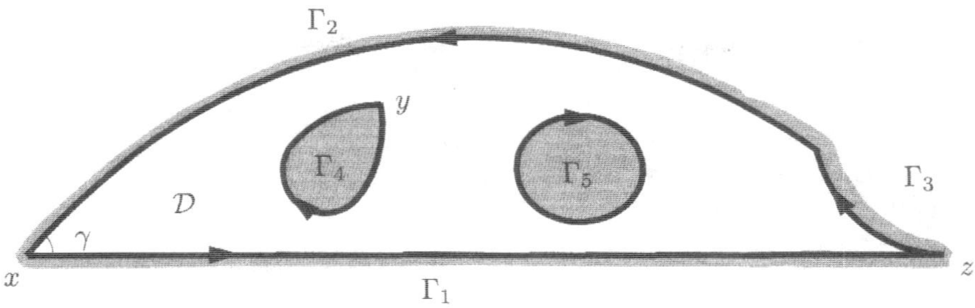

FIGURE 2.2. A billiard table. Here Γ_1 is flat; Γ_2 is focusing; Γ_3 is dispersing; Γ_4 is a closed wall with one corner point, y; Γ_5 is a smooth closed wall; the corner at x has a positive interior angle $\gamma > 0$; and the corner at z is a cusp. The adopted orientation of Γ is shown by arrows.

Focusing walls: such that $f'' \neq 0$ is pointing inside $\mathcal{D}$;
Dispersing walls: such that $f'' \neq 0$ is pointing outside $\mathcal{D}$.

The terms "focusing" and "dispersing" are standard in billiard literature. They refer to the effect the corresponding mirror walls have on reflecting light rays – type (b) walls tend to focus the rays, while type (c) walls tend to disperse them; see Figure 2.3. Focusing walls are also referred to as *convex* and dispersing as *concave*.

We now define the (*signed*) *curvature* on each Γ_i as follows:

$$\mathcal{K} = \begin{cases} 0 & \text{if } \Gamma_i \text{ is flat;} \\ -\|f''\| & \text{if } \Gamma_i \text{ is focusing;} \\ \|f''\| & \text{if } \Gamma_i \text{ is dispersing.} \end{cases} \tag{2.4}$$

Our choice of signs is dictated by a tradition in billiard literature first adopted by Ya. Sinai and his school. However, the opposite choice is sometimes made, too.

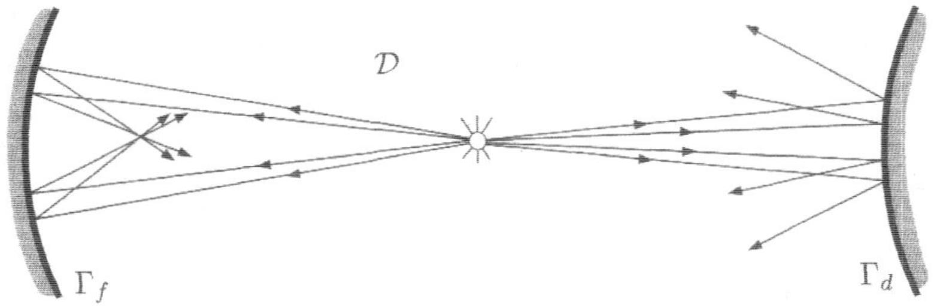

FIGURE 2.3. Reflecting light rays tend to focus near a *focusing* wall Γ_f and disperse near a *dispersing* wall Γ_d.

Accordingly, we define

$$\Gamma_0 = \bigcup_{\mathcal{K}=0} \Gamma_i, \quad \Gamma_- = \bigcup_{\mathcal{K}<0} \Gamma_i, \quad \Gamma_+ = \bigcup_{\mathcal{K}>0} \Gamma_i. \tag{2.5}$$

We note that the curvature of each focusing or dispersing wall Γ_i is bounded away from zero and infinity, due to Assumption A3 and the closedness of Γ_i.

Clearly, each wall Γ_i has finite length. We denote it by $|\Gamma_i|$ and set $|\Gamma| = \sum_i |\Gamma_i|$ to be the total perimeter of $\mathcal{D}$.

2.2. Unbounded billiard tables

We have assumed so far that $\mathcal{D}$ was bounded. Sometimes one deals with unbounded billiard tables $\mathcal{D} \subset \mathbb{R}^2$, and we want to include them in our studies, too. In that case it is reasonable to require that boundary $\partial\mathcal{D}$ be *locally piece-wise smooth*; that is for any large square $\mathbb{K}_B = \{(x_1, x_2) \in \mathbb{R}^2 \colon |x_1| \le B,\ |x_2| \le B\}$ the intersection $\mathcal{D} \cap \mathbb{K}_B$ must have a finitely piece-wise smooth boundary satisfying Assumptions A1–A3.

In many studies of unbounded tables $\mathcal{D}$ one assumes that they have periodic structure. This happens, for example, when there are two orthogonal unit vectors $u, v \in \mathbb{R}^2$ such that

$$q \in \mathcal{D} \iff q + u \in \mathcal{D} \iff q + v \in \mathcal{D}. \tag{2.6}$$

In this case we can choose the coordinate axes parallel to the vectors u and v, so that $\mathbb{R}^2$ can be regarded as the universal cover of the corresponding unit torus $\mathrm{Tor}^2 = \mathbb{R}^2/\mathbb{Z}^2$. Then an unbounded periodic domain $\mathcal{D}$ satisfying (2.6) can be represented by its projection onto Tor^2.

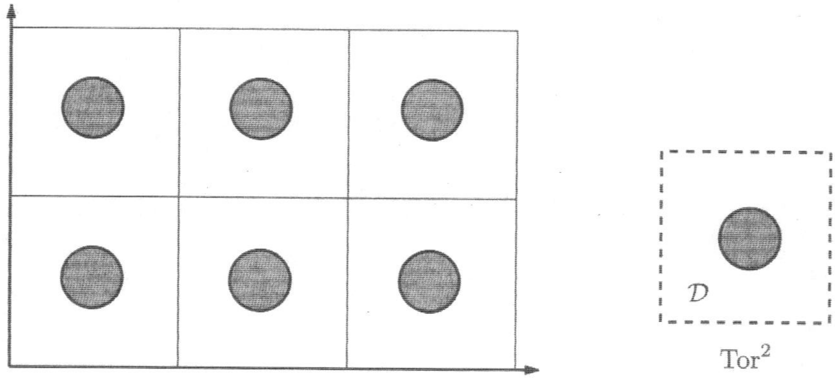

FIGURE 2.4. An unbounded periodic table and its projection on Tor^2.

Slightly abusing our notation, we also denote the above projection of $\mathcal{D}$ onto Tor^2 by $\mathcal{D}$. Obviously, such a domain $\mathcal{D} \subset \mathrm{Tor}^2$ will satisfy Assumptions A1–A3. In addition, we will always assume that $\mathcal{D} \neq \mathrm{Tor}^2$ to avoid trivial "billiards" without walls (or collisions).

We summarize our constructions as follows:

DEFINITION 2.5. A *billiard table* $\mathcal{D}$ is the closure of a bounded open connected domain $\mathcal{D} \subset \mathbb{R}^2$ or $\mathcal{D} \subsetneq \mathrm{Tor}^2$ such that $\partial\mathcal{D}$ satisfies Assumptions A1–A3.

2.3. Billiard flow

Here we construct dynamics on a billiard table. This is not going to be a simple task, since there will be several cases where the construction fails and the trajectory of a billiard particle cannot be defined.

Let $q \in \mathcal{D}$ denote the position of the moving particle and $v \in \mathbb{R}^2$ its velocity vector. Of course, $q = q(t)$ and $v = v(t)$ are functions of time $t \in \mathbb{R}$. When the particle moves inside the table, so that $q \in \operatorname{int} \mathcal{D}$, it maintains a constant velocity:

$$\dot{q} = v \qquad \text{and} \qquad \dot{v} = 0 \tag{2.7}$$

(here the dot denotes the time derivative).

When the particle collides with the regular part of the boundary, i.e. $q \in \tilde{\Gamma}$, its velocity vector instantaneously gets reflected across the tangent line to Γ at the point q. This is specified by the classical rule "the angle of incidence is equal to the angle of reflection" (see Fig. 2.5), and it can be expressed by

$$v^+ = v^- - 2\,\langle v, n\rangle\, n, \tag{2.8}$$

where v^+ and v^- refer to the postcollisional and precollisional velocities, respectively, and n denotes the unit normal vector to $\tilde{\Gamma}$ at the point q.

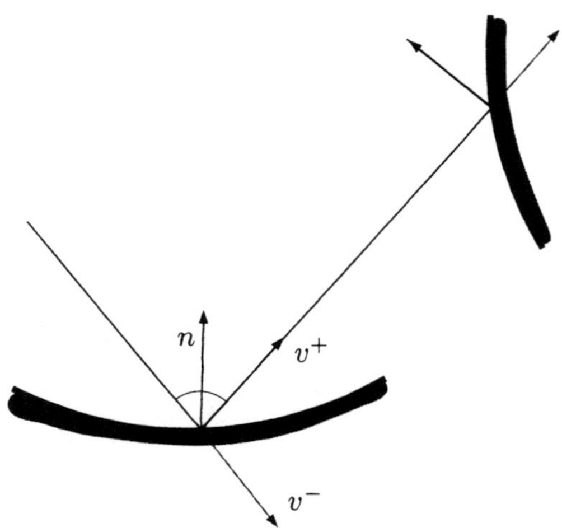

FIGURE 2.5. Velocity vector changes at collisions.

If the moving particle hits a corner point, i.e. $q \in \Gamma_*$, it stops, and its motion will no longer be defined beyond that point. This is one of the complications that make the analysis of billiard dynamics difficult (we will see more later).

The equations of motion (2.7)–(2.8) preserve the norm $\|v\|$, and it is customary to set it to one: $\|v\| = 1$.

DEFINITION 2.6. A collision is said to be *regular* if $q \in \tilde{\Gamma}$ and the vector v^- is not tangent to Γ. In this case $v^+ \neq v^-$ (obviously, v^- must be pointing outside $\mathcal{D}$, and v^+ inside $\mathcal{D}$). If v^- is tangent to Γ at the point of collision, then $v^+ = v^-$, and such a collision is said to be *grazing* or *tangential.*

EXERCISE 2.7. Show that grazing collisions are possible only on dispersing walls (where $\mathcal{K} > 0$). Hint: If the wall is flat, then before a grazing collision the particle should have passed an endpoint of the wall, which is prohibited by our rules.

EXERCISE 2.8. Let the moving particle collide with the regular part of the boundary at time t (that is $q(t) \in \tilde{\Gamma}$). Show that it will move inside $\mathcal{D}$ without collisions during some time interval $(t, t+\varepsilon)$.

We now summarize our observations:

PROPOSITION 2.9. *The trajectory of the particle $(q(t), v(t))$ starting at $q(0) \in \operatorname{int}\mathcal{D}$ is defined at all times $-\infty < t < \infty$ unless one of the two exceptions occurs:*

(a) *The particle hits a corner point of $\mathcal{D}$; i.e. $q(t) \in \Gamma_*$ for some $t \in \mathbb{R}$.*
(b) *Collision times $\{t_n\}$ have an accumulation point in $\mathbb{R}$.*

2.4. Accumulation of collision times

The second exception mentioned in Proposition 2.9 may occur only under very special circumstances. In fact, it never happens in any chaotic billiards discussed in this book. Here we describe the conditions under which (b) is possible.

Suppose the event (b) occurs; i.e. the collision times $\{t_n\}$ have an accumulation point $t_n \to t_\infty \in \mathbb{R}$. Then, obviously, $q(t_n) \to q_\infty \in \Gamma$, and we distinguish two cases:

(b1) $q_\infty \in \Gamma_*$ (a corner point);
(b2) $q_\infty \in \tilde{\Gamma}$ (a regular point).

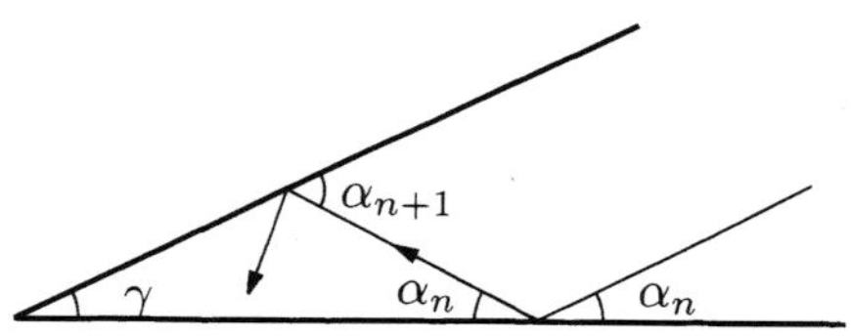

FIGURE 2.6. Proof of Lemma 2.10.

LEMMA 2.10. *Suppose the billiard particle enters the neighborhood of a corner point of the billiard table with a positive interior angle, $\gamma > 0$, and collides with both sides of the corner. Then it must leave that neighborhood after at most $[\pi/\gamma] + 1$ collisions, so that (b1) cannot occur.*

The case where the particle collides with only one side of the corner is equivalent to (b2) and will be discussed later.

PROOF. If the sides of the corner are flat, the angle made by the links of the billiard trajectory with the sides grows as $\alpha_{n+1} = \alpha_n + \gamma$; see Fig. 2.6. Thus, after $[\pi/\gamma]$ collisions it becomes $> \pi$, which proves the lemma. If the sides are not flat, then the result follows by a simple approximation argument, which we leave to the reader. □

Lemma 2.10 shows that (b1) may occur only in a cusp. Now, it is not hard to see that any cusp has at least one dispersing side (check this!).

LEMMA 2.11. *If the billiard particle enters a cusp with two dispersing sides or one dispersing and one flat side, then it must exit the cusp after a finite number of collisions, so that (b1) cannot occur.*

PROOF. Let d denote the distance from the corner point of the cusp to the line along which the particle moves. After every collision with a dispersing side that distance increases (see Fig. 2.7), while at collisions with a flat side it does not change. The rest of the proof requires some simple reasoning, which is left as an exercise. □

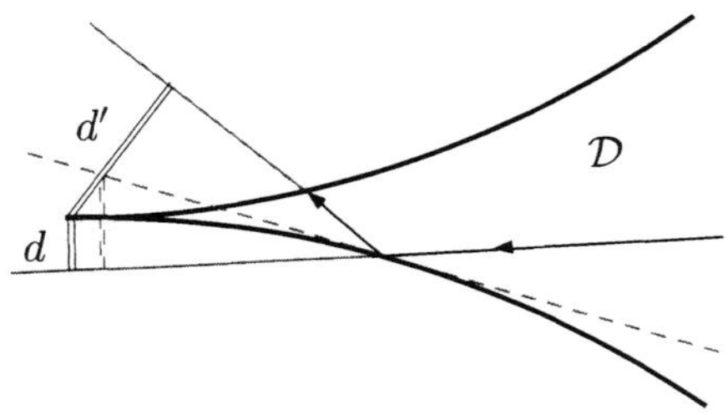

FIGURE 2.7. Proof of Lemma 2.11: $d' > d$.

EXERCISE 2.12. Show that the number of collisions in a cusp is not uniformly bounded; i.e. for any $N \geq 1$ there is a billiard trajectory than experiences more that N collisions in a vicinity of the cusp before leaving it (in contrast to the uniform bound for corners with positive angles provided by Lemma 2.10).

Thus, (b1) may be possible only in a cusp with one dispersing and one focusing side. But is it really possible? To our best knowledge, there are no published examples of this sort. We leave this question to the reader as a (possibly, very difficult) exercise. In any case, billiard tables with such cusps are very special, and we do not want to consider them in this book. To exclude them, we are making our last assumption:

ASSUMPTION A4. Any billiard table $\mathcal{D}$ contains no cusps made by a focusing wall and a dispersing wall.

We now turn to the type (b2) accumulations, i.e. $q_\infty \in \tilde{\Gamma}$. Clearly, (b2) is impossible if q_∞ lies on a flat or dispersing boundary component (check this!).

THEOREM 2.13 (B. Halpern [**Hal77**]). *The type (b2) accumulation of collisions is impossible on any focusing wall with a bounded third derivative and nowhere vanishing curvature.*

Halpern's argument is elementary but rather lengthy, so we omit it and refer the reader to his original paper [**Hal77**]. We note that Halpern found an example of a C^2 (not C^3) focusing billiard wall where the event (b2) does occur.

Due to Halpern's theorem, (b2) never occurs for the billiard tables defined in Section 2.1. Together with Assumption A4 made above, accumulations of collision points are now completely ruled out.

2.5. Phase space for the flow

The *state* of the moving particle at any time is specified by its position $q \in \mathcal{D}$ and the unit velocity vector $v \in S^1$. Thus, the *phase space* of the system is

$$\Omega = \{(q, v)\} = \mathcal{D} \times S^1.$$

This is a three-dimensional manifold with boundary $\partial\Omega = \Gamma \times S^1$. One can picture Ω as a "doughnut" whose cross-section is $\mathcal{D}$.

Furthermore, at each regular boundary point $q \in \tilde{\Gamma}$ it is convenient to identify the pairs (q, v^-) and (q, v^+) related by the collision rule (2.8), which amounts to "gluing" Ω along its boundary. This changes the topology of Ω, of course, but its topological properties will not be essential.

We denote by π_q and π_v the natural projections of Ω onto $\mathcal{D}$ and S^1, respectively, so that $\pi_q(q, v) = q$ and $\pi_v(q, v) = v$.

Let $\tilde{\Omega} \subset \Omega$ denote the set of states (q, v) on which the dynamics of the moving particle is defined at all times $-\infty < t < \infty$ (this set will be accurately described later). Thus we obtain a one-parameter group of transformations (flow)

$$\Phi^t \colon \tilde{\Omega} \to \tilde{\Omega}$$

with continuous time $t \in \mathbb{R}$. That is, $\Phi^0 = \mathrm{Id}$ and $\Phi^{t+s} = \Phi^t \circ \Phi^s$ for all $t, s \in \mathbb{R}$.

Every trajectory of the flow $\{\Phi^t x\}$, $x \in \tilde{\Omega}$, is a continuous curve in Ω (the continuity at collisions is ensured by our identification of v^+ with v^-; see above). It is also customary to call its projection $\pi_q(\Phi^t x)$ onto the table $\mathcal{D}$ a *billiard trajectory*. The latter is a directed polygonal line whose vertices are points of collision. Its segments between successive collisions are called *links*.

If the billiard table is a bounded domain $\mathcal{D} \subset \mathbb{R}^2$, then the free path between collisions cannot exceed the diameter of $\mathcal{D}$; hence every trajectory of the flow experiences infinitely many collisions. This may not be so if $\mathcal{D} \subsetneq \mathrm{Tor}^2$, as in the example shown in Fig. 2.4.

LEMMA 2.14. *If $\mathcal{D}$ is a billiard table on* Tor^2, *then every trajectory of the flow experiences either infinitely many collisions or none at all.*

PROOF. It is a well-known fact that every free trajectory (i.e., a straight line) on Tor^2 is either periodic or dense (the latter means that its closure is Tor^2); see Section 1.2. The rest of the proof is left as an exercise. □

Let $\tilde{\Omega} = \tilde{\Omega}_{\mathrm{c}} \cup \tilde{\Omega}_{\mathrm{f}}$, where $\tilde{\Omega}_{\mathrm{c}}$ contains all trajectories *with* collisions and $\tilde{\Omega}_{\mathrm{f}}$ is the union of all collision-free trajectories (which may exist only if $\mathcal{D} \subsetneq \mathrm{Tor}^2$). Note that both $\tilde{\Omega}_{\mathrm{c}}$ and $\tilde{\Omega}_{\mathrm{f}}$ are invariant under Φ^t. The rest of this section is devoted to billiard tables $\mathcal{D} \subsetneq \mathrm{Tor}^2$, for which $\tilde{\Omega}_{\mathrm{f}}$ may exist.

EXERCISE 2.15. Suppose $\tilde{\Omega}_{\mathrm{f}} \neq \emptyset$. Show that there are only finitely many velocity vectors $v \in S^1$ that correspond to points $x \in \tilde{\Omega}_{\mathrm{f}}$.

DEFINITION 2.16. We say that a billiard table $\mathcal{D} \subsetneq \mathrm{Tor}^2$ has *finite horizon* (or *bounded horizon*) if $\tilde{\Omega}_{\mathrm{f}} = \emptyset$. Otherwise the horizon is said to be *infinite* (or *unbounded*).

EXERCISE 2.17. Suppose $\tilde{\Omega}_{\mathrm{f}} = \emptyset$; i.e. the horizon is finite. Show that the free path between collisions is bounded above by some constant $\tau_{\max}(\mathcal{D}) < \infty$. Hints: Suppose the free path between collisions is unbounded; then there is a sequence of line segments $L_n \subset \mathcal{D}$ of increasing length, $|L_n| \to \infty$. By the compactness of Tor^2, a subsequence L_{n_k} will converge to a line $L \subset \mathcal{D}$, which is obviously a periodic line on Tor^2. Since the horizon is finite, the line L must intersect the boundary $\partial\mathcal{D}$. Now, if at all those intersections the boundary $\partial\mathcal{D}$ lies on one side of L, then a nearby line parallel to L will have no intersections with $\partial\mathcal{D}$ at all, which contradicts the finite horizon assumption. Thus $\partial\mathcal{D}$ lies on both sides of L at their intersections. Now argue that the lengths of L_{n_k} must be bounded.

EXERCISE 2.18. Suppose $\mathcal{D}$ has no cusps; i.e. the interior angles at corner points are all positive. Show that the number of collisions per unit time is bounded by some $M_{\max} < \infty$. Hint: Use Lemma 2.10.

2.6. Coordinate representation of the flow

Here we describe the flow Φ^t in coordinates (x, y, ω) on Ω, where $q = (x, y) \in \mathcal{D}$. Here x and y are the usual Cartesian coordinates on the plane and $\omega \in [0, 2\pi)$ denotes the counterclockwise angle between the positive x axis and the velocity vector v. For every $t \in \mathbb{R}$ the map Φ^t acts on Ω, and we consider an arbitrary point and its image:

$$(x^-, y^-, \omega^-) \xrightarrow{\Phi^t} (x^+, y^+, \omega^+).$$

Our aim is to compute the derivative of the map Φ^t.

If no collisions occur, then by (2.7)

$$x^+ = x^- + t\cos\omega, \quad y^+ = y^- + t\sin\omega, \quad \omega^+ = \omega^-. \tag{2.9}$$

Suppose now there is exactly one (regular) collision at some Γ_i during the interval $(0, t)$. Then a direct relation between (x^-, y^-, ω^-) and (x^+, y^+, ω^+) will be rather complicated. To simplify matters, we will introduce some intermediate variables.

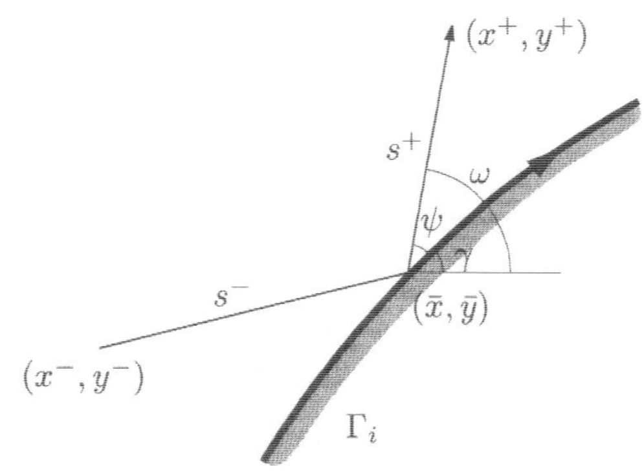

FIGURE 2.8. Action of the flow in coordinates.

Let $(\bar{x}, \bar{y}) \in \Gamma_i$ denote the collision point, $\mathbf{T}$ the tangent vector to Γ_i at that point, and γ the angle between $\mathbf{T}$ and the positive x axis. Let s^- denote the time of collision, $s^+ = t - s^-$, and ψ denote the angle between v^+ and $\mathbf{T}$; see Fig. 2.8.

Then

$$
\begin{aligned}
x^- &= \bar{x} - s^- \cos\omega^-, & x^+ &= \bar{x} + s^+ \cos\omega^+, \\
y^- &= \bar{y} - s^- \sin\omega^-, & y^+ &= \bar{y} + s^+ \sin\omega^+, \\
\omega^- &= \gamma - \psi, & \omega^+ &= \gamma + \psi.
\end{aligned} \tag{2.10}
$$

We are going to differentiate these equations. Let r denote the arc length parameter on Γ_i, then

$$
\begin{aligned}
d\bar{x} &= \cos\gamma\, dr, \\
d\bar{y} &= \sin\gamma\, dr, \\
d\gamma &= -\mathcal{K}\, dr
\end{aligned}
$$

(the sign of $d\gamma$ is completely determined by our orientation of Γ_i, as defined in Section 2.1).

Now differentiating (2.10) gives

$$
\begin{aligned}
dx^+ &= \cos\gamma\, dr + \cos\omega^+\, ds^+ - s^+ \sin\omega^+\, d\omega^+ \\
dy^+ &= \sin\gamma\, dr + \sin\omega^+\, ds^+ + s^+ \cos\omega^+\, d\omega^+ \\
d\omega^+ &= -\mathcal{K}\, dr + d\psi
\end{aligned} \tag{2.11}
$$

and

$$
\begin{aligned}
dx^- &= \cos\gamma\, dr - \cos\omega^-\, ds^- + s^- \sin\omega^-\, d\omega^- \\
dy^- &= \sin\gamma\, dr - \sin\omega^-\, ds^- - s^- \cos\omega^-\, d\omega^- \\
d\omega^- &= -\mathcal{K}\, dr - d\psi.
\end{aligned} \tag{2.12}
$$

EXERCISE 2.19. Prove that

$$
dx^+ \wedge dy^+ \wedge d\omega^+ = \sin\psi\; dr \wedge ds^+ \wedge d\psi \tag{2.13}
$$

and

$$
dx^- \wedge dy^- \wedge d\omega^- = -\sin\psi\; dr \wedge ds^- \wedge d\psi,
$$

where $\wedge$ denotes the exterior multiplication of differential forms [**Ar89**, Chapter 7]. (It is convenient to use the skew-commutativity $du \wedge dv = -dv \wedge du$, and in particular $du \wedge du = 0$.) The reader unfamiliar with exterior forms may, instead, directly compute the Jacobians of the maps $(r, s, \psi) \mapsto (x^\pm, y^\pm, \omega^\pm)$ and show that they are equal to $\pm \sin\psi$, respectively.

Note that $s^- + s^+ = t =$const; hence $ds^+ + ds^- = 0$, and we obtain

$$
dx^+ \wedge dy^+ \wedge d\omega^+ = dx^- \wedge dy^- \wedge d\omega^-. \tag{2.14}
$$

EXERCISE 2.20. Use (2.9) to show that if there are no collisions between the points (x^-, y^-, ω^-) and (x^+, y^+, ω^+), then (2.14) holds as well.

Now an induction on the number of collisions implies the following:

THEOREM 2.21. *The flow Φ^t preserves the volume form $dx \wedge dy \wedge d\omega$; thus it preserves the Lebesgue measure $dx\, dy\, d\omega$ on Ω.*

REMARK 2.22. The Lebesgue measure on Ω is actually the Liouville measure corresponding to the Hamiltonian system (2.7)–(2.8). This fact is often given as a "proof" of the invariance of the Lebesgue measure; however, we avoided this argument because it is hard to check that the collision rule (2.8) preserves the Hamiltonian character of motion.

DEFINITION 2.23. The normalized Lebesgue measure on Ω

$$d\mu_\Omega = (2\pi|\mathcal{D}|)^{-1}\, dx\, dy\, d\omega \tag{2.15}$$

(here $|\mathcal{D}|$ denotes the area of the domain $\mathcal{D}$) is the canonical probability measure preserved by the billiard flow Φ^t.

2.7. Smoothness of the flow

LEMMA 2.24. *The flow Φ^t is $C^{\ell-1}$ smooth at points that experience only regular collisions.*

PROOF. Equations (2.11)–(2.12) show that the derivative of the flow Φ^t is expressed through the curvature $\mathcal{K}$ of the wall Γ_i, which in turn corresponds to the second derivative of the C^ℓ function $f_i(r)$. □

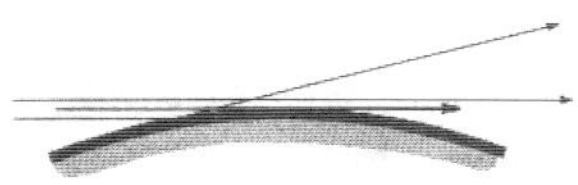

FIGURE 2.9. A grazing collision; here the flow is continuous (but not differentiable).

Now suppose there is a grazing collision of a billiard trajectory between the points (x^-, y^-, ω^-) and (x^+, y^+, ω^+) at a dispersing wall. Then it is not hard to see that Φ^t is not differentiable at (x^-, y^-, ω^-) (even though it is continuous; see Fig. 2.9). We leave this as an exercise.

The set $\Omega \setminus \tilde{\Omega}$ (see Section 2.5 for the definition of $\tilde{\Omega}$) corresponds to states (q, v) on which the dynamics is defined only during a limited time interval until the trajectory hits a corner point $q' \in \Gamma_*$ (in the past or in the future). The states $x \in \Omega$ whose very first contact with Γ (in the past or future) occurs at a corner point make a two-dimensional surface in Ω; see Fig. 2.10.

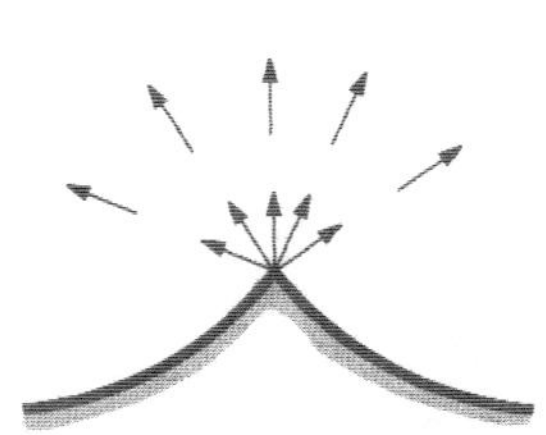

FIGURE 2.10. A "fan" emanating from a corner point of $\mathcal{D}$. These are phase points whose first contact with $\partial\mathcal{D}$ in the past occurs at the corner point.

Since the flow is $C^{\ell-1}$ at regular collisions, the states whose trajectories hit corner points in the distant past or distant future make two-dimensional $C^{\ell-1}$ submanifolds in Ω. Thus we obtain

THEOREM 2.25. *$\Omega \setminus \tilde{\Omega}$ is a countable union of two-dimensional submanifolds in Ω.*

COROLLARY 2.26. *$\tilde{\Omega}$ is a dense G_δ set of full Lebesgue measure in Ω.*

Thus, in terms of the invariant measure μ_Ω, the flow Φ^t is defined almost everywhere.

2.8. Continuous extension of the flow

Here we extend the flow Φ^t to the entire space Ω by continuity.

EXERCISE 2.27. Show that the flow Φ^t is continuous on $\tilde{\Omega}$ (assuming the identification of (q, v^-) and (q, v^+) as above). Warning: Grazing collisions at dispersing walls must be treated carefully; see Fig. 2.9.

Since $\tilde{\Omega}$ is a dense subset of Ω, we can extend the flow Φ^t to the entire space Ω by continuity, but this extension may be multivalued at some points; see an example in Fig. 2.11. Different extensions of Φ^t at the same point $x \in \Omega$ are called *branches* of the (extended) flow. For each $t \in \mathbb{R}$ there are finitely many branches of the extended flow Φ^t at any point $x \in \Omega$. The number of branches may grow indefinitely as $t \to \pm\infty$.

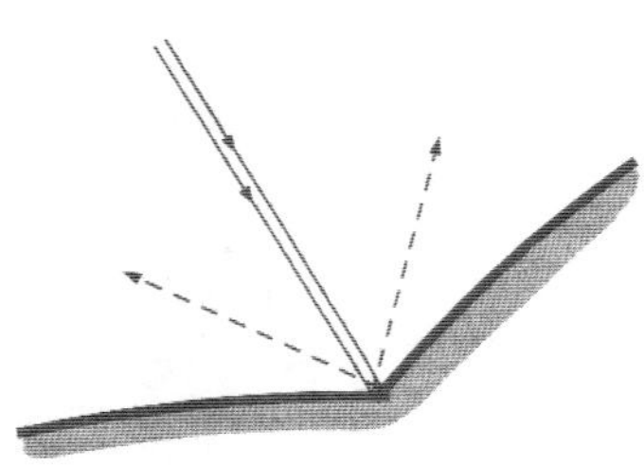

FIGURE 2.11. A multivalued continuous extension of the flow.

EXERCISE 2.28. Show that a trajectory hitting a corner point can be extended, by continuity, to at most two different branches. Under what conditions do these extensions coincide (i.e., when is the flow actually continuous) at a trajectory hitting a corner point? Answer: The branches coincide if and only if the interior angle at the corner point either equals zero or divides π (that is, $\gamma = 0$ or $\gamma = \pi/n$ for some $n \in \mathbb{N}$). Hints: For $\gamma > 0$, see the proof of Lemma 2.10. For $\gamma = 0$, see the proof of Lemma 2.11. In both cases a careful geometric analysis is required to complete the proof.

EXERCISE 2.29. Determine under what conditions the billiard flow Φ^t has a globally continuous extension to the entire space Ω. Answer: This occurs if and only if every interior angle either equals zero or divides π (as an example, recall the billiard in a square in Section 1.2).

2.9. Collision map

It is common in the study of dynamical systems to reduce a flow to a map by constructing a cross-section. Given a flow $\Phi^t \colon \Omega \to \Omega$ on a manifold Ω, one finds a hypersurface $M \subset \Omega$ transversal to the flow so that each trajectory crosses M infinitely many times. Then the flow induces a return map $F \colon M \to M$ and a return time function $L(x) = \min\{s > 0 \colon \Phi^s(x) \in M\}$ on M, so that $F(x) = \Phi^{L(x)}(x)$.

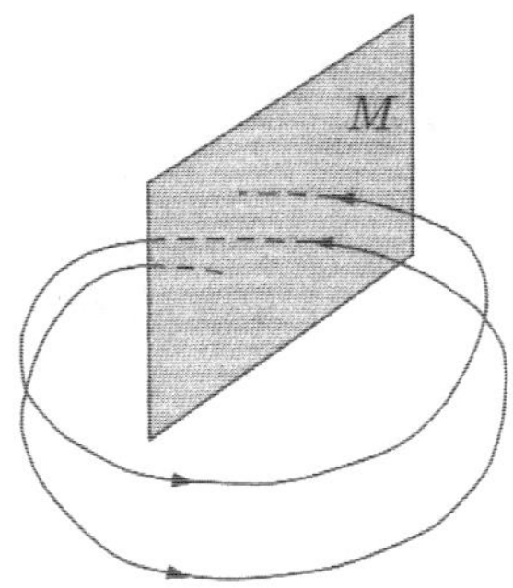

FIGURE 2.12. Cross-section of a flow.

Conversely, given a measurable space M, a measurable map $F \colon M \to M$, and a positive function $L \colon M \to \mathbb{R}_+$, one can construct a space

$$\Omega = \{(x, s) \colon x \in M,\ 0 \le s \le L(x)\}$$

and a flow $\Phi^t \colon \Omega \to \Omega$ defined by $\Phi^t(x, s) = (x, s + t)$ with the identification of points $(x, L(x))$ and $(F(x), 0)$. The flow Φ^t is measurable on Ω. If the map F preserves a probability measure μ on M and

$$\bar{L} = \int_M L(x)\, d\mu(x) < \infty,$$

then the flow preserves the probability measure μ_1 on Ω defined by

$$d\mu_1 = \bar{L}^{-1}\, d\mu \times ds \tag{2.16}$$

(note that this is, locally, a product measure whose invariance under Φ^t is a consequence of the Fubini theorem). The map F is often called *base transformation*, $L(x)$ *ceiling function*, and Φ^t *suspension flow* or *Kakutani flow*. We refer the reader to [**BrS02**, pp. 21–22], [**DS00**, p. 26–28], [**CFS82**, pp. 292–295], and [**Pet83**, p. 11] for more comprehensive discussions of these constructions.

For a billiard system, a hypersurface in Ω is usually constructed on the boundary of the billiard table, i.e. on the set $\Gamma \times S^1$. There is a delicate part, however, in this construction, since we have identified the precollisional and postcollisional velocity vectors v^- and v^+ related by (2.8), thus effectively transforming S^1 into a half-circle. To take this identification into account, it is customary to describe the cross-section as the set of all postcollisional velocity vectors:

$$\mathcal{M} = \cup_i \mathcal{M}_i, \quad \mathcal{M}_i = \{x = (q, v) \in \Omega \colon\ q \in \Gamma_i,\ \langle v, n\rangle \ge 0\}, \tag{2.17}$$

where n denotes the unit normal vector to Γ_i pointing inside $\mathcal{D}$. The set $\mathcal{M}$ is a two-dimensional submanifold in Ω called the *collision space*.

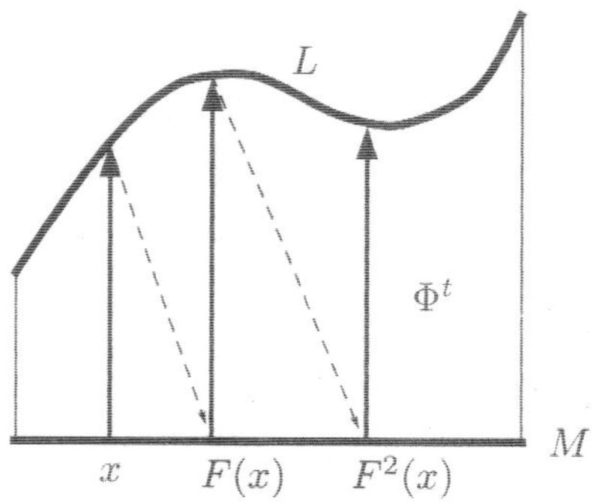

FIGURE 2.13. A suspension flow.

EXERCISE 2.30. Let $x = (q, v) \in \mathcal{M}_i$. Under what conditions is the trajectory $\Phi^t x$ defined during some small interval $0 < t < \varepsilon$? Answer: q must be a regular point ($q \in \tilde{\Gamma}$), and we must have either $\langle v, n\rangle > 0$ (a regular collision) or $\langle v, n\rangle = 0$, but in the latter case Γ_i must be a dispersing wall.

If the trajectory $\Phi^t x$ for $x \in \mathcal{M}$ is defined during some interval of time $(0, \varepsilon)$, then it must intersect the surface $\Gamma \times S^1$ at a future time $\tau(x) > 0$ according to Lemma 2.14, and we call $\tau(x)$ the *return time.* Since the speed is set to one, $\tau(x)$ will also be equal to the distance the billiard trajectory originating at x covers before the next collision.

Any trajectory of the flow $\Phi^t \colon \tilde{\Omega}_{\rm c} \to \tilde{\Omega}_{\rm c}$ (see Section 2.5 for the definition of $\tilde{\Omega}_{\rm c}$) crosses the surface $\mathcal{M}$ infinitely many times. Let $\tilde{\mathcal{M}} = \mathcal{M} \cap \tilde{\Omega}$. This defines the *return map*

$$\mathcal{F} \colon \tilde{\mathcal{M}} \to \tilde{\mathcal{M}} \qquad \text{by} \qquad \mathcal{F}(x) = \Phi^{\tau(x)+0} x. \tag{2.18}$$

We will extend $\mathcal{F}$ to $\mathcal{M} \setminus \tilde{\mathcal{M}}$ later. The map $\mathcal{F}$ is often called the *billiard map* or *collision map.* (Accordingly, $\mathcal{M}$ is sometimes called the *phase space of the billiard map* $\mathcal{F}$.)

2.10. Coordinates for the map and its singularities

We fix an arclength parameter r on each Γ_i, so that r takes values in an interval $[a_i, b_i]$. Of course, $b_i - a_i = |\Gamma_i|$. We assume that the intervals (a_i, b_i) are disjoint in $\mathbb{R}$. On smooth closed curves Γ_i, we identify $a_i = b_i$, thus making r a cyclic parameter. For each point $x \in \mathcal{M}$, let $\varphi \in [-\pi/2, \pi/2]$ denote the angle between v and n oriented as shown in Fig. 2.14.

Then r and φ make coordinates on $\mathcal{M}$. For each smooth closed curve Γ_i, the manifold $\mathcal{M}_i = \Gamma_i \times [-\pi/2, \pi/2]$ is a cylinder (since r is a cyclic coordinate), while for every other boundary component Γ_i, the manifold $\mathcal{M}_i = [a_i, b_i] \times [-\pi/2, \pi/2]$ is a rectangle; see Fig. 2.15.

We denote

$$\mathcal{S}_0 \colon = \partial\mathcal{M} = \{|\varphi| = \pi/2\} \cup \Big(\cup_i\big(\{r = a_i\} \cup \{r = b_i\}\big)\Big), \tag{2.19}$$

where the set $\{r = a_i\} \cup \{r = b_i\}$ is included only for Γ_i's which are *not* smooth closed curves (i.e., where $\{a_i\} \cup \{b_i\}$ is a *true* boundary of the interval $[a_i, b_i]$).

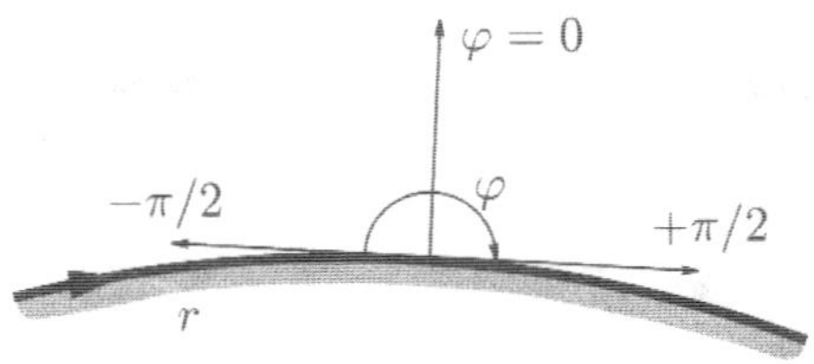

FIGURE 2.14. Orientation of r and φ.

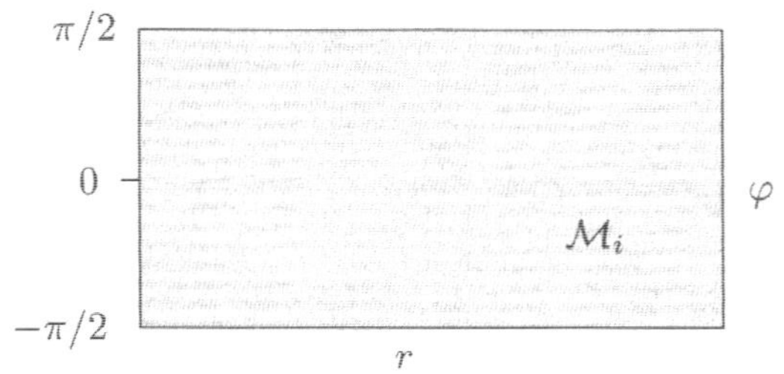

FIGURE 2.15. A component of the collision space $\mathcal{M}$.

For every point $x \in \operatorname{int} \mathcal{M}$ its trajectory $\Phi^t x$ is defined, at least, for $0 < t < \tau(x)$, i.e. until the next intersection with $\mathcal{M}$, at which we have three possible cases:

(a) a regular collision, i.e. $\mathcal{F}(x) \in \operatorname{int} \mathcal{M}$;
(b) a grazing collision at a dispersing wall, i.e. $\mathcal{F}(x) \in \mathcal{S}_0$;
(c) the trajectory hits a corner point and dies.

In the last case $\mathcal{F}(x)$ is not defined. Note that $\mathcal{S}_1 \setminus \mathcal{S}_0$ is the set of points where the event (b) or (c) occurs.

EXERCISE 2.31. Show that the set $\mathcal{M} \setminus \mathcal{S}_1$ is open and at every point of this set the map $\mathcal{F}$ is continuous and in fact is a local homeomorphism. Verify that at every point $x \in \mathcal{S}_1 \setminus \mathcal{S}_0$ the map $\mathcal{F}$ is discontinuous.

EXERCISE 2.32. Check that the above considerations can be applied to the inverse map $\mathcal{F}^{-1}$ and the set

$$\mathcal{S}_{-1} = \mathcal{S}_0 \cup \{x \in \operatorname{int} \mathcal{M} : \mathcal{F}^{-1}(x) \notin \operatorname{int} \mathcal{M}\}. \tag{2.20}$$

Conclude that $\mathcal{F}: \mathcal{M} \setminus \mathcal{S}_1 \to \mathcal{M} \setminus \mathcal{S}_{-1}$ is a homeomorphism.

2.11. Derivative of the map

Here we differentiate the collision map $\mathcal{F}$ at a point $x = (r, \varphi) \in \operatorname{int} \mathcal{M}$ such that $\mathcal{F}(x) = (r_1, \varphi_1) \in \operatorname{int} \mathcal{M}$. Denote by $(\bar{x}, \bar{y}) \in \partial \mathcal{D}$ and $(\bar{x}_1, \bar{y}_1) \in \partial \mathcal{D}$ the coordinates of the boundary points corresponding to r and r_1, respectively, and by ω the angle made by the billiards trajectory between these points and the positive x axis in $\mathbb{R}^2$; see Fig. 2.16. Then

$$\bar{x}_1 - \bar{x} = \tau \cos \omega \qquad \text{and} \qquad \bar{y}_1 - \bar{y} = \tau \sin \omega \tag{2.21}$$

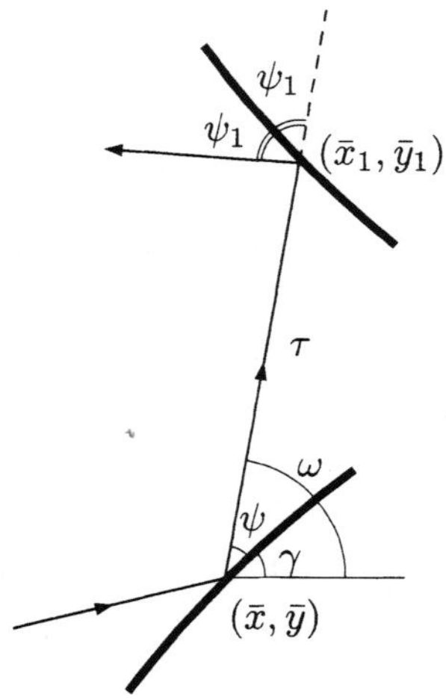

FIGURE 2.16. Action of the map in coordinates.

where $\tau = \tau(x)$. We use the symbols γ and ψ as they were introduced in Section 2.6; see also Fig. 2.16. Note that

$$\psi = \pi/2 - \varphi$$

and recall that

$$\begin{aligned} d\bar{x} &= \cos\gamma\, dr, \\ d\bar{y} &= \sin\gamma\, dr, \\ d\gamma &= -\mathcal{K}\, dr. \end{aligned}$$

Similar notations γ_1 and ψ_1 are used at the point r_1. Note that

$$\omega = \gamma + \psi = \gamma_1 - \psi_1. \tag{2.22}$$

Differentiating the last equation gives

$$d\omega = -\mathcal{K}\, dr + d\psi = -\mathcal{K}_1\, dr_1 - d\psi_1. \tag{2.23}$$

Differentiating (2.21) yields

$$\begin{aligned} \cos\gamma_1\, dr_1 - \cos\gamma\, dr &= \cos\omega\, d\tau - \tau\sin\omega\, d\omega \\ \sin\gamma_1\, dr_1 - \sin\gamma\, dr &= \sin\omega\, d\tau + \tau\cos\omega\, d\omega. \end{aligned}$$

Eliminating $d\tau$ and using (2.22) give

$$\sin\psi_1\, dr_1 + \sin\psi\, dr = \tau\, d\omega. \tag{2.24}$$

Solving the equations (2.23) and (2.24) for dr_1 and $d\psi_1$ and replacing ψ and ψ_1 by $\pi/2 - \varphi$ and $\pi/2 - \varphi_1$, respectively, give

$$\begin{aligned} -\cos\varphi_1\, dr_1 &= (\tau\mathcal{K} + \cos\varphi)\, dr + \tau\, d\varphi \\ -\cos\varphi_1\, d\varphi_1 &= (\tau\mathcal{K}\mathcal{K}_1 + \mathcal{K}\cos\varphi_1 + \mathcal{K}_1\cos\varphi)\, dr \\ &\quad + (\tau\mathcal{K}_1 + \cos\varphi_1)\, d\varphi. \end{aligned} \tag{2.25}$$

Thus we obtain the derivative $D\mathcal{F}$ at the point $x = (r, \varphi)$ as a 2×2 matrix

$$D_x\mathcal{F} = \frac{-1}{\cos\varphi_1}\begin{bmatrix} \tau\mathcal{K} + \cos\varphi & \tau \\ \tau\mathcal{K}\mathcal{K}_1 + \mathcal{K}\cos\varphi_1 + \mathcal{K}_1\cos\varphi & \tau\mathcal{K}_1 + \cos\varphi_1 \end{bmatrix}. \tag{2.26}$$

Recall that the map $\mathcal{F}\colon \mathcal{M} \setminus \mathcal{S}_1 \to \mathcal{M} \setminus \mathcal{S}_{-1}$ is a homeomorphism. Now we claim more:

THEOREM 2.33. *The map $\mathcal{F}\colon \mathcal{M} \setminus \mathcal{S}_1 \to \mathcal{M} \setminus \mathcal{S}_{-1}$ is a $C^{\ell-1}$ diffeomorphism.*

PROOF. The derivative $D\mathcal{F}$ is expressed through the curvature $\mathcal{K}$ and $\mathcal{K}_1$ of the boundary $\partial\mathcal{D}$, which corresponds to the second derivative of the C^ℓ functions $f_i\colon [a_i, b_i] \to \mathbb{R}^2$. □

The expression (2.26) shows that the derivatives of $\mathcal{F}$ are unbounded: they blow up as $\cos\varphi_1 \to 0$, i.e. when x_1 is near $\mathcal{S}_0$ and x is near $\mathcal{S}_1$.

We regard $\mathcal{S}_1$ as the *singularity set* for the map $\mathcal{F}$. Similarly, $\mathcal{S}_{-1}$ is the *singularity set* for the map $\mathcal{F}^{-1}$. These sets are unions of compact smooth curves (their exact degree of smoothness will be determined later); cf. Theorem 2.25.

EXERCISE 2.34. Suppose the horizon is bounded. Show that $\mathcal{S}_1$ and $\mathcal{S}_{-1}$ are *finite* unions of compact smooth curves. If the horizon is unbounded, then $\mathcal{S}_1$ and $\mathcal{S}_{-1}$ are *finite* or *countable* unions of compact smooth curves.

We define, inductively,

$$\mathcal{S}_{n+1} = \mathcal{S}_n \cup \mathcal{F}^{-1}(\mathcal{S}_n) \quad\text{and}\quad \mathcal{S}_{-(n+1)} = \mathcal{S}_{-n} \cup \mathcal{F}(\mathcal{S}_{-n}). \tag{2.27}$$

It is easy to see that $\mathcal{S}_{n+1}$ and $\mathcal{S}_{-(n+1)}$ are the singularity sets for the maps $\mathcal{F}^{n+1}$ and $\mathcal{F}^{-(n+1)}$, respectively. These sets are unions of compact smooth curves. Thus, on the set

$$\hat{\mathcal{M}}\colon = \mathcal{M} \setminus \cup_{n=-\infty}^{\infty}\mathcal{S}_n \tag{2.28}$$

all the iterations of $\mathcal{F}$ are defined and are $C^{\ell-1}$ differentiable.

As a result, the map $\mathcal{F}$ is well defined by (2.18) on a dense G_δ subset $\hat{\mathcal{M}} \subset \mathcal{M}$ of full Lebesgue measure. It can be extended by continuity to the entire $\mathcal{M}$, but this extension may be multivalued at some points.

2.12. Invariant measure of the map

By using the explicit formula for $D_x\mathcal{F}$ (see (2.26)), one easily obtains

$$\det D_x\mathcal{F} = \cos\varphi/\cos\varphi_1. \tag{2.29}$$

LEMMA 2.35. *The map $\mathcal{F}$ preserves the measure $\cos\varphi\, dr\, d\varphi$ on $\mathcal{M}$.*

PROOF. Using (2.29) and changing variables give

$$\iint_{\mathcal{F}(A)} \cos\varphi_1\, dr_1\, d\varphi_1 = \iint_A \cos\varphi\, dr\, d\varphi$$

for any Borel set $A \subset \mathcal{M}$. □

Note that

$$\iint_{\mathcal{M}} \cos\varphi\, dr\, d\varphi = \int_{-\pi/2}^{\pi/2} \cos\varphi\, d\varphi \int_\Gamma dr = 2\,|\Gamma|.$$

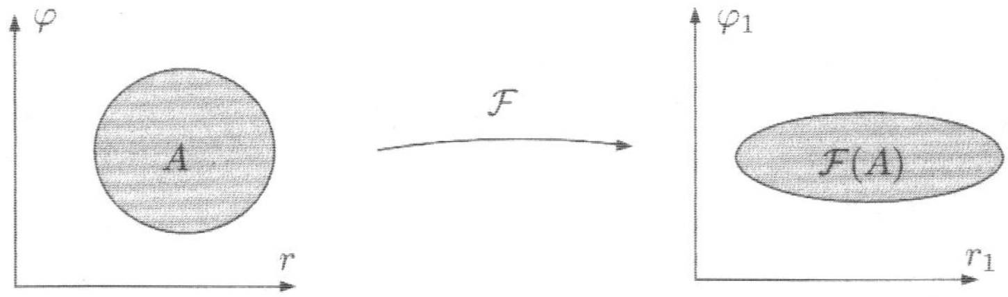

FIGURE 2.17. Proof of Lemma 2.35.

DEFINITION 2.36. The normalized measure on $\mathcal{M}$,

$$d\mu = (2\,|\Gamma|)^{-1}\cos\varphi\,dr\,d\varphi,$$

is the canonical probability measure preserved by the billiard map $\mathcal{F}$.

According to the general fact (2.16), the flow Φ^t must preserve the probability measure

$$\begin{aligned} d\mu_1 &= \bar{\tau}^{-1}\,d\mu \times ds \\ &= (2\,|\Gamma|)^{-1}\,\bar{\tau}^{-1}\cos\varphi\,dr\,d\varphi\,ds, \end{aligned} \tag{2.30}$$

where

$$\bar{\tau} = \int_{\mathcal{M}} \tau(x)\,d\mu(x)$$

is the *mean return time*. The latter is more often called the *mean free path* on the billiard table $\mathcal{D}$ (note that time=path, since the velocity equals one). Due to (2.13) we can rewrite (2.30) as

$$d\mu_1 = (2\,|\Gamma|)^{-1}\,\bar{\tau}^{-1}dx\,dy\,d\omega$$

(here we used the fact $\sin\psi = \cos\varphi$). Comparing this formula with (2.15) we conclude that $\mu_1 = \mu_\Omega$ and, therefore, the normalizing coefficients for these measures coincide:

$$2\pi|\mathcal{D}| = 2\,|\Gamma|\,\bar{\tau} \tag{2.31}$$

(by the way, this also proves that $\bar{\tau} < \infty$).

EXERCISE 2.37. Show that for μ-almost every point $x \in \mathcal{M}$ there exist limits

$$\bar{\tau}_x = \lim_{n\to\infty}\frac{1}{n}\sum_{i=0}^{n}\tau(\mathcal{F}^i x) = \lim_{n\to\infty}\frac{1}{n}\sum_{i=0}^{n}\tau(\mathcal{F}^{-i} x),$$

where $\bar{\tau}_x$ is an $\mathcal{F}$-invariant function on $\mathcal{M}$ satisfying

$$\int_{\mathcal{M}} \bar{\tau}_x\,d\mu = \bar{\tau}.$$

Hint: Use the Birkhoff ergodic theorem.

EXERCISE 2.38. Show that for μ_Ω-almost every point $X \in \Omega$ there exist limits

$$\bar{\tau}_X = \lim_{T\to\infty}\frac{T}{n_X(0,T)} = \lim_{T\to\infty}\frac{T}{n_X(-T,0)}$$

where $n_X(a,b)$ denotes the number of collisions of the trajectory $\Phi^t(X)$ during the time interval $a < t < b$. Moreover, show that $\bar{\tau}_X = \bar{\tau}_x$ if the points $X \in \Omega$ and $x \in \mathcal{M}$ belong to the same trajectory.

2.13. Mean free path

Equation (2.31) gives us a simple and elegant formula for the mean free path on a billiard table $\mathcal{D}$:

$$\bar{\tau} = \frac{\pi|\mathcal{D}|}{|\Gamma|}. \tag{2.32}$$

Thus, the mean free path depends only on the area and perimeter of the domain $\mathcal{D}$, but not on its shape. This equation is well known in integral geometry and geometric probability, where it is proved by different methods [**Mat75, Sa76**].

EXERCISE 2.39. Derive (2.32) by a direct integration of the identity (2.13). Hint: Note that $\int ds^+ = \tau$.

EXERCISE 2.40. Compute the mean free path on a billiard table $\mathcal{D} = \mathrm{Tor}^2 \setminus B_{\mathbf{r}}$ where $B_{\mathbf{r}}$ is a ball (disk) of radius $\mathbf{r}$; see Fig. 2.4. Answer: $\bar{\tau} = (1 - \pi\mathbf{r}^2)/(2\mathbf{r})$. Note that if $\mathbf{r}$ is small, then $\bar{\tau} \approx 1/(2\mathbf{r})$. This can be understood in probabilistic terms: the billiard particle moving freely on the unit torus may randomly hit the fixed disk $B_{\mathbf{r}}$ with probability equal to its diameter $2\mathbf{r}$; thus the average distance between collisions is $\approx 1/(2\mathbf{r})$.

If the horizon is bounded, then $\tau(x) \le \tau_{\max} < \infty$; hence all the moments of the function $\tau(x)$ are finite. This may not be so when the horizon is unbounded, i.e. when $\sup \tau(x) = \infty$.

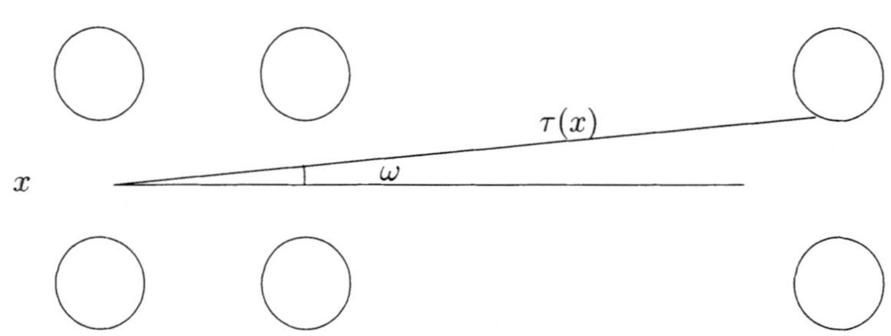

FIGURE 2.18. Hint to Exercise 2.41.

EXERCISE 2.41. Consider the billiard table of Exercise 2.40. Show that

$$\int_{\mathcal{M}} [\tau(x)]^2 \, d\mu(x) = \infty.$$

Hints: First, use (2.13) to replace the above integral with

$$(2\,|\Gamma|)^{-1} \int_{\mathcal{M}} [\tau(r,\varphi)]^2 \cos\varphi \, dr \, d\varphi = (2\,|\Gamma|)^{-1} \int_{\Omega} \tau(x,y,\omega) \, dx \, dy \, d\omega,$$

where $\tau(x,y,\omega)$ denotes the length of the link of the billiard trajectory passing through the point $(x,y,\omega) \in \Omega$ (the link between the first future collision and the last past collision). Now consider "almost horizontal trajectories" with $\omega \approx 0$

passing through points $(x, y) \in \mathcal{D}$ located above or below the disk $B_{\mathbf{r}}$. Observe that $\tau(x, y, \omega) \sim 1/\omega$; thus the above integral diverges.

The function $\tau(x)$ may have a positive lower bound, as in the last two exercises. However, if the billiard table has a focusing boundary component or a corner point with interior angle $\gamma < \pi$, then $\min_x \tau(x) = 0$. In that case the distribution of small values of $\tau(x)$ is important, and the reader is invited to verify the following facts.

EXERCISE 2.42. Show that if the billiard table has a focusing boundary component but no corner points, then $\mu\{x \colon \tau(x) < \varepsilon\} \leq \text{const} \cdot \varepsilon^2$. If the billiard table has a corner point with interior angle $\gamma < \pi$ but no cusps, then $\mu\{x \colon \tau(x) < \varepsilon\} \leq \text{const} \cdot \varepsilon$. If the table has a cusp, then $\mu\{x \colon \tau(x) < \varepsilon\} \leq \text{const} \cdot \varepsilon^{1/2}$. The last case is quite difficult; the reader should not expect an easy solution.

EXERCISE 2.43. Determine the set of positive numbers $a > 0$ for which the function $F(x) = [\tau(x)]^{-a}$ is integrable over $\mathcal{M}$.

Answer: For the three cases described in the previous exercise, these intervals are $(0, 2)$, $(0, 1)$, and $(0, 1/2)$, respectively.

2.14. Involution

Billiard dynamics has an important *involution* property: for any $x = (q, v) \in \Omega$ the point $\mathcal{I}_\Omega(x) = (q, -v)$ satisfies

$$\Phi^{-t}(\mathcal{I}_\Omega(x)) = \mathcal{I}_\Omega(\Phi^t x)$$

whenever the flow is defined. Hence, the involution $\mathcal{I}_\Omega \colon \Omega \to \Omega$ anticommutes with the flow Φ^t, and this fact can be written as

$$\Phi^{-t} \circ \mathcal{I}_\Omega = \mathcal{I}_\Omega \circ \Phi^t.$$

This means, plainly, that if we reverse the particle's velocity, it will retrace its past trajectory 'backwards' (this fact is known as *time reversibility* of the billiard dynamics). Note that the map $\mathcal{I}_\Omega$ also preserves the measure μ_Ω.

The collision map $\mathcal{F}$ also admits an involution, $\mathcal{I}$, defined by $(r, \varphi) \mapsto (r, -\varphi)$. It anticommutes with $\mathcal{F}$; i.e.

$$\mathcal{F}^{-k} \circ \mathcal{I} = \mathcal{I} \circ \mathcal{F}^k, \qquad k \in \mathbb{Z},$$

whenever the map $\mathcal{F}^k$ is defined. We also note that the map $\mathcal{I} \colon \mathcal{M} \to \mathcal{M}$ preserves the measure μ.

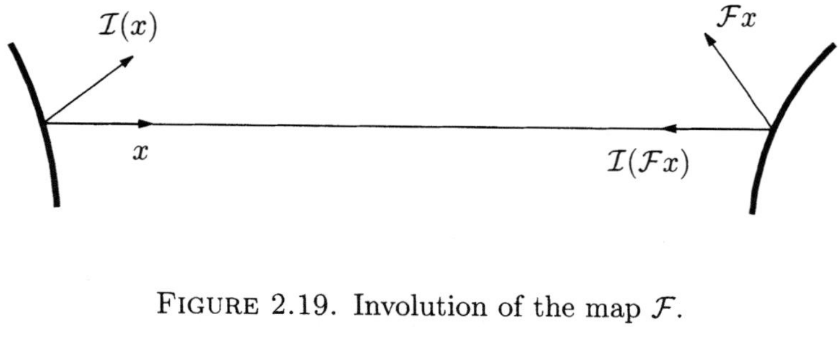

FIGURE 2.19. Involution of the map $\mathcal{F}$.

EXERCISE 2.44. Consider the vector function $\mathbf{L}\colon \mathcal{M} \to \mathbb{R}^2$ defined by

$$\mathbf{L}(r,\varphi) = (\bar{x}_1 - \bar{x}, \bar{y}_1 - \bar{y}) = (\tau\cos\omega, \tau\sin\omega)$$

as illustrated by Fig. 2.16. This is the first directed link of the billiard trajectory originating from (r,φ). Show that

$$\int_{\mathcal{M}} \mathbf{L}\, d\mu = 0.$$

Hint: Note that $\mathbf{L}(x) = -\mathbf{L}(\mathcal{I} \circ \mathcal{F}(x))$ and then use the invariance of the measure μ under both $\mathcal{F}$ and $\mathcal{I}$.

CHAPTER 3

Lyapunov exponents and hyperbolicity

In the first section of this chapter, we recall basic definitions and facts about Lyapunov exponents. We give them in the form suitable for billiards, but they actually apply to much more general classes of maps with singularities.

3.1. Lyapunov exponents: general facts

One of the main tools in the studies of differentiable conservative dynamical systems is characteristic Lyapunov exponents; they measure the stability (or instability) of trajectories under small perturbations.

Let M be a compact Riemannian manifold (perhaps, with boundary and corners), $N \subset M$ an open and dense subset and $F\colon N \to M$ a C^r (with $r \geq 2$) diffeomorphism of N onto $F(N)$. Note that all the iterations of F are defined on the set

$$\tilde{N} = \cap_{n=-\infty}^{\infty} F^n(N).$$

Assume that F preserves a probability measure μ on M and $\mu(\tilde{N}) = 1$.

THEOREM 3.1 (Oseledets,[1] **[Os68, Ru79, LS82, BP01]**). *Suppose*

$$\int_M \log^+ \|D_x F\| \, d\mu(x) < \infty \qquad \text{and} \qquad \int_M \log^+ \|D_x F^{-1}\| \, d\mu(x) < \infty,$$

where $\log^+ s = \max\{\log s, 0\}$. *Then there exists an F-invariant set $H \subset \tilde{N}$, $\mu(H) = 1$, such that for all $x \in H$ there is a DF-invariant decomposition of the tangent space*

$$\mathcal{T}_x M = E_x^{(1)} \oplus \cdots \oplus E_x^{(m)} \tag{3.1}$$

with some $m = m(x)$, such that for all nonzero vectors $v \in E_x^{(i)}$

$$\lim_{n \to \pm\infty} \frac{1}{n} \log \|D_x F^n \, v\| = \lambda_x^{(i)} \tag{3.2}$$

where $\lambda_x^{(1)} > \cdots > \lambda_x^{(m)}$.

The values of $\lambda_x^{(i)}$ are called *Lyapunov exponents* of the map F at the point x and $k_i = \dim E_x^{(i)}$ their *multiplicities.* Of course, Lyapunov exponents may be defined at any point $x \in \tilde{N}$ where a DF-invariant decomposition (3.1) and the limits (3.2) exist (regardless of invariant measures).

It immediately follows from (3.2) that the Lyapunov exponents and their multiplicities are invariant under the map F. If the map F is ergodic, then the Lyapunov exponents $\lambda^{(1)} > \cdots > \lambda^{(m)}$ (as well as their multiplicities) are almost everywhere constant.

[1]This name is often spelled (or rather misspelled) *Oseledec.*

Equation (3.2) can be loosely restated as $\|D_xF^n v\| \sim e^{n\lambda_x^{(i)}}$ (see also Exercise 3.5 for a more precise formula). If $\lambda_x^{(i)} > 0$, then nonzero tangent vectors $v \in E_x^{(i)}$ grow exponentially (at a rate of $\lambda_x^{(i)}$) in the future and contract exponentially (at the same rate) in the past. If $\lambda_x^{(i)} < 0$, it is vice versa. In the language of differential equations, vectors $v \in E_x^{(i)}$ with $\lambda_x^{(i)} > 0$ correspond to *unstable* perturbations (of the initial condition x) and vectors $v \in E_x^{(i)}$ with $\lambda_x^{(i)} < 0$ correspond to *stable* perturbations.

If $\lambda_x^{(i)} = 0$, then the corresponding tangent vectors do not expand or contract at any exponential rate (but they may grow or shrink more slowly, say, linearly in n, see Exercise 3.8 below).

DEFINITION 3.2. A point $x \in M$ is said to be *hyperbolic* if Lyapunov exponents exist at x and none of them equals zero.

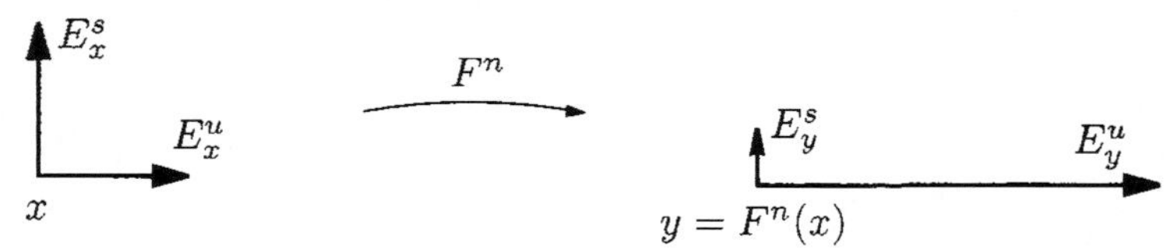

FIGURE 3.1. Tangent space decomposition $\mathcal{T}_xM = E_x^u \oplus E_x^s$.

For a hyperbolic point $x \in M$, we have $\mathcal{T}_xM = E_x^u \oplus E_x^s$, where

$$E_x^u = \oplus_{\lambda_x^{(i)}>0} E_x^{(i)} \qquad \text{and} \qquad E_x^s = \oplus_{\lambda_x^{(i)}<0} E_x^{(i)}$$

(the superscripts u and s stand for *unstable* and *stable*, respectively).

Obviously, the subspaces E_x^u and E_x^s are transversal to each other, thus the angle $\gamma_0(x)$ between them must be positive. Certain complications may arise when this angle is small, though. The following addition to the Oseledets theorem provides some control over this angle.

Pick $a \in \mathbb{R}$ such that Lyapunov exponents are different from a at μ-almost every point $x \in M$. Denote

$$E_x^{a+} = \oplus_{\lambda_x^{(i)}>a} E_x^{(i)}, \qquad E_x^{a-} = \oplus_{\lambda_x^{(i)}<a} E_x^{(i)}$$

(this is a generalization of E_x^u and E_x^s defined above). Let $\gamma_a(x)$ denote the angle between the spaces E_x^{a+} and E_x^{a-}.

PROPOSITION 3.3 ([**KS86**]). *For μ-almost every point $x \in M$ we have*

$$\lim_{n\to\pm\infty} \frac{1}{n} \log \gamma_a(F^n(x)) = 0. \tag{3.3}$$

In other words, the angle $\gamma_a(F^n(x))$ may approach zero, as $n \to \infty$, but it does so more slowly than any exponential function. This is a useful addition to the Oseledets theorem; see Lemma 3.9 below.

DEFINITION 3.4. A map F is said to be *hyperbolic* if μ-almost every point $x \in M$ is hyperbolic.

Lastly, we make an important distinction between *uniform* and *nonuniform* hyperbolicity. The following exercise will help us to approach this issue. Suppose $x \in M$ is a hyperbolic point and

$$\lambda_x = \min_i |\lambda_x^{(i)}| > 0.$$

EXERCISE 3.5. Show that for any small $\varepsilon > 0$ there is a $C(x, \varepsilon) > 0$ such that for all $n \geq 1$

$$\|D_x F^{-n}(v)\| \leq C(x, \varepsilon)\, e^{-n(\lambda_x - \varepsilon)} \|v\| \qquad \forall v \in E_x^u \tag{3.4}$$

and

$$\|D_x F^{n}(v)\| \leq C(x, \varepsilon)\, e^{-n(\lambda_x - \varepsilon)} \|v\| \qquad \forall v \in E_x^s. \tag{3.5}$$

Hint: First show that for any given nonzero vector $v \in E_x^u$ (resp., $v \in E_x^s$), there is a $C(x, \varepsilon)$ satisfying (3.4)–(3.5), but it might depend on v. Then pick an orthonormal basis $e_1, \dots, e_k$ in E_x^u (resp., E_x^s), ensure that (3.4)–(3.5) hold with the same constant $C(x, \varepsilon)$ for all the basis vectors $e_1, \dots, e_k$, and then use the triangle inequality to derive (3.4)–(3.5) for arbitrary vectors v.

If F is a hyperbolic map, then $\lambda_x > 0$ and $C(x, \varepsilon) > 0$ are measurable functions on M. We say that the hyperbolicity is *uniform* if these functions can be made constant. Equivalently we say that a hyperbolic map F is *uniformly hyperbolic* if there are $\lambda > 0$ and $C > 0$ such that for every $x \in M$ and all $n \geq 1$

$$\|D_x F^{-n}(v)\| \leq C\, e^{-n\lambda} \|v\| \qquad \forall v \in E_x^u \tag{3.6}$$

and

$$\|D_x F^{n}(v)\| \leq C\, e^{-n\lambda} \|v\| \qquad \forall v \in E_x^s \tag{3.7}$$

(wherever E_x^u and E_x^s exist). We will encounter uniformly hyperbolic billiards in Chapter 4 and nonuniformly hyperbolic billiards in Chapter 8.

3.2. Lyapunov exponents for the map

Here we apply the general theory of Lyapunov exponents to the billiard map $\mathcal{F}\colon \mathcal{M} \to \mathcal{M}$ preserving the measure μ.

THEOREM 3.6. *The Oseledets theorem applies to the billiard map $\mathcal{F}$, so Lyapunov exponents exist at μ-almost every point $x \in \mathcal{M}$.*

PROOF. We need to verify that the functions $\log^+ \|D_x \mathcal{F}\|$ and $\log^+ \|D_x \mathcal{F}^{-1}\|$ are integrable over $\mathcal{M}$. By the involution property, it is enough to do this only for $D_x \mathcal{F}$. It follows from (2.26) that

$$\|D_x \mathcal{F}\| \leq \frac{C}{\cos \varphi_1},$$

where $C > 0$ is a constant (note that the curvature $\mathcal{K}$ of the wall Γ is bounded). Then

$$\begin{aligned}
\int_{\mathcal{M}} \log^+ \|D_x \mathcal{F}\|\, d\mu &\leq \int_{\mathcal{M}} |\log C + \log \cos \varphi_1|\, d\mu \\
&\leq |\log C| + (2\,|\Gamma|)^{-1} \int_{\mathcal{M}} |\log \cos \varphi_1| \cos \varphi \; d\varphi\, dr \\
&= |\log C| + (2\,|\Gamma|)^{-1} \int_{\mathcal{M}} |\log \cos \varphi| \cos \varphi \; d\varphi\, dr,
\end{aligned}$$

where at the last step we have used the invariance of the measure μ. Finally,

$$\int_{\mathcal{M}} |\log \cos\varphi|\, \cos\varphi\; d\varphi\, dr = |\Gamma| \int_{-\pi/2}^{\pi/2} |\log\cos\varphi|\, \cos\varphi\; d\varphi$$

$$= |\Gamma|\,(2-\log 4) < \infty. \tag{3.8}$$

The theorem is proved. □

REMARK 3.7. It is essential in the above proof that the curvature of Γ is bounded. Strelcyn [**St86**, Sections 5.2–5.4] gives an example of a billiard table whose boundary is C^∞ everywhere, except for one point, at which the curvature turns infinite, for which $\int_{\mathcal{M}} \log^+ \|D_x\mathcal{F}\|\, d\mu = \infty$.

The proof of the above theorem demonstrates the role of singularities in billiard dynamics. On the one hand, their union has zero measure and as such may be neglected. On the other hand, the blow-up of the derivatives in the vicinity of the singularities has a crucial effect on the billiard dynamics, and we will see that this blow-up will dominate our analysis of chaotic billiards.

EXERCISE 3.8. Show that the Lyapunov exponents for the billiard map in a circle (see Section 1.1) are zero at every point. Hint: Work in the (ω,ψ) coordinates introduced in Section 1.1, let $(d\omega, d\psi)$ denote a tangent vector and $(d\omega_n, d\psi_n) = \mathcal{F}^n(d\omega, d\psi)$ its image. Differentiate two equations in Corollary 1.4 and conclude that $D\mathcal{F}^n = \begin{bmatrix} 1 & 2n \\ 0 & 1 \end{bmatrix}$. Thus, the images of tangent vectors can grow at most linearly.

Since $\dim \mathcal{M} = 2$, we have, at every point $x \in \mathcal{M}$, either one Lyapunov exponent λ of multiplicity two or two Lyapunov exponents $\lambda_x^{(1)} > \lambda_x^{(2)}$ of multiplicity one each.

LEMMA 3.9. $\lambda_x^{(1)} + \lambda_x^{(2)} = 0$ *(up to a set of zero measure).*

PROOF. Let Π be a parallelogram in $\mathcal{T}_x\mathcal{M}$ with sides a and b parallel to $E_x^{(1)}$ and $E_x^{(2)}$, respectively. Its area is $|\Pi| = ab\sin\gamma$, where γ denotes the angle between $E_x^{(1)}$ and $E_x^{(2)}$. Now $D_x\mathcal{F}^n(\Pi)$ is a similar parallelogram in $\mathcal{T}_{\mathcal{F}^n x}\mathcal{M}$ whose sides we denote by a_n and b_n, and the angle between them by γ_n. Therefore,

$$|\det D_x\mathcal{F}^n| = \frac{\sin\gamma_n}{\sin\gamma}\,\frac{a_n}{a}\,\frac{b_n}{b}. \tag{3.9}$$

Due to the definition of Lyapunov exponents, we have

$$n^{-1}\log a_n/a \to \lambda_x^{(1)}, \qquad n^{-1}\log b_n/b \to \lambda_x^{(2)},$$

as $n \to \infty$. On the other hand, by using (2.29) one obtains

$$|\det D_x\mathcal{F}^n| = \cos\varphi/\cos\varphi_n, \tag{3.10}$$

where φ and φ_n are the φ-coordinates of the points x and $\mathcal{F}^n x$, respectively. Observe that $\frac{1}{n}\log\cos\varphi_n \to 0$ almost everywhere (it is a simple consequence of the Birkhoff ergodic theorem that for any integrable function $g\colon \mathcal{M} \to \mathbb{R}$ we have $\frac{1}{n}\, g(\mathcal{F}^n x) \to 0$ almost everywhere). Now taking the logarithms in (3.9) and (3.10), dividing by n, taking a limit as $n \to \infty$, and using Proposition 3.3, we prove the lemma. □

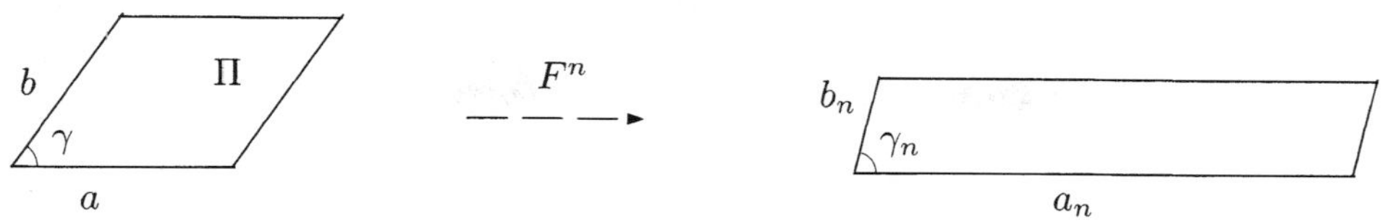

FIGURE 3.2. Proof of Lemma 3.9.

EXERCISE 3.10. Prove that if a point $x \in \mathcal{M}$ has one Lyapunov exponent λ_x of multiplicity two, then $\lambda_x = 0$ (up to a set of zero measure).

We see that at almost every point $x \in \mathcal{M}$ the set of Lyapunov exponents (which is called the *Lyapunov spectrum*) is symmetric about zero: either zero is the only Lyapunov exponent (of multiplicity two), or the two distinct Lyapunov exponents have equal absolute values but opposite signs. In the latter case x is hyperbolic.

3.3. Lyapunov exponents for the flow

Lyapunov exponents for a flow $\Phi^t \colon \Omega \to \Omega$ on a manifold Ω are defined in a way similar to (3.1)–(3.2): if there exists a Φ^t-invariant set $H \subset \Omega$ and for every point $X \in H$ there is a $D\Phi^t$-invariant decomposition of the tangent space

$$\mathcal{T}_X\Omega = E_X^{(0)} \oplus \cdots \oplus E_X^{(m)}$$

with some $m = m(X)$ such that for all nonzero vectors $v \in E_X^{(i)}$

$$\lim_{T\to\pm\infty} \frac{1}{T} \log \|D_X\Phi^T v\| = \lambda_X^{(i)}, \tag{3.11}$$

then $\lambda_X^{(0)}, \dots, \lambda_X^{(m)}$ are called the Lyapunov exponents of the flow at the point $X \in \Omega$ and $m_i = \dim E_X^{(i)}$ are their multiplicities. The values of $\lambda_X^{(i)}$ and their multiplicities are invariant under the flow. If the flow is ergodic, they are constant almost everywhere.

Here we derive formulas for the Lyapunov exponents of the billiard flow Φ^t. We continue using the coordinates introduced in Section 2.6. Consider a point $X = (x, y, \omega) \in \Omega$ and a tangent vector $dX = (dx, dy, d\omega) \in \mathcal{T}_X\Omega$. We denote by $X_t = (x_t, y_t, \omega_t) = \Phi^t(X)$ and $dX_t = (dx_t, dy_t, d\omega_t) = D_X\Phi^t(DX)$ their respective images at time $t \in \mathbb{R}$ (wherever Φ^t is defined and smooth at X).

Suppose first that for some $c \neq 0$

$$dx = c\,\cos\omega, \quad dy = c\,\sin\omega, \quad d\omega = 0; \tag{3.12}$$

i.e. the tangent vector dX is parallel to the direction of the flow. If this happens before a collision, then it easily follows from (2.12) that

$$dr = d\psi = 0 \qquad \text{and} \qquad ds^- = -c. \tag{3.13}$$

Recall that $ds^+ = -ds^-$; hence $ds^+ = c$. Now (2.11) imply

$$dx^+ = c\,\cos\omega, \quad dy^+ = c\,\sin\omega, \quad d\omega^+ = 0;$$

hence the vector $D\Phi^t(dX)$ will be parallel to the flow direction after the collision as well, and its length will not change. By an obvious inductive argument,

$$dx_t = c\,\cos\omega_t, \quad dy_t = c\,\sin\omega_t, \quad d\omega_t = 0,$$

for all $t \in \mathbb{R}$. In other words, if we perturb the point $X = (q, v)$ so that q is displaced along the vector v, and v remains unchanged, then the new point $X' = (q + cv, v)$ will satisfy $\Phi^t(X') = \Phi^{t+c}(X)$ for all $t \in \mathbb{R}$.

EXERCISE 3.11. Verify that the one-dimensional subspace $\mathcal{T}_X^0\,\Omega \subset \mathcal{T}_X\Omega$ defined by

$$\mathcal{T}_X^0\,\Omega = \{dX\colon\ \sin\omega\,dx = \cos\omega\,dy \ \ \&\ \ d\omega = 0\}$$

is $D\Phi^t$-invariant and the Lyapunov exponent corresponding to its vectors equals zero. (In fact, these vectors do not expand or contract under the flow; their images have a constant length. We also note that this property is common for all flows where a particle moves with constant speed.)

Another important consequence of (2.11) is that

$$\cos\omega^+\,dx^+ + \sin\omega^+\,dy^+ = \cos\psi\,dr + ds^+,$$

and, similarly, it is a consequence of (2.12) that

$$\cos\omega^-\,dx^- + \sin\omega^-\,dy^- = \cos\psi\,dr - ds^-.$$

Since $ds^- = -ds^+$, it follows that the derivative of the flow $D\Phi^t$ preserves the quantity $\cos\omega\,dx + \sin\omega\,dy$, which is the projection of the vector (dx, dy) onto the velocity vector $v = (\cos\omega, \sin\omega)$. That is,

$$\cos\omega_t\,dx_t + \sin\omega_t\,dy_t \equiv \text{const}$$

for all $t \in \mathbb{R}$.

COROLLARY 3.12. *If (dx, dy) is orthogonal to v at time $t = 0$, then the orthogonality will be preserved at all times.*

Consider the 2-dimensional subspace $\mathcal{T}_X^\perp\,\Omega \subset \mathcal{T}_X\Omega$ defined by

$$\mathcal{T}_X^\perp\,\Omega = \{dX\colon\ \cos\omega\,dx + \sin\omega\,dy = 0\}. \tag{3.14}$$

It follows from our previous analysis that the orthogonal decomposition of the tangent space

$$\mathcal{T}_X\Omega = \mathcal{T}_X^0\,\Omega \oplus \mathcal{T}_X^\perp\,\Omega$$

is $D\Phi^t$-invariant.

EXERCISE 3.13. Prove that if $\lambda_X^{(i)} \neq 0$ is a nonzero Lyapunov exponent with the corresponding subspace $E_X^{(i)}$, then $E_X^{(i)} \subset \mathcal{T}_X^\perp\,\Omega$. Hint: If the orthogonal projection of a vector $v \in E_X^{(i)}$ onto $\mathcal{T}_X^0\,\Omega$ were not zero, then its length would be constant in time, thus $\|D_X\Phi^t(v)\|$ would fail to converge to zero as $t \to \pm\infty$.

Our further analysis is restricted to the 2D tangent space $\mathcal{T}_X^\perp\,\Omega$. For every phase point $X = (q, v) \in \Omega$ let

$$t^+(X) = \min\{t > 0\colon\ \Phi^t(X) \in \mathcal{M}\} \tag{3.15}$$

and

$$t^-(X) = \max\{t < 0\colon\ \Phi^t(X) \in \mathcal{M}\} \tag{3.16}$$

denote the first positive and negative times of collisions, respectively. Then we have two projections of Ω onto $\mathcal{M}$:

$$\mathbf{P}^\pm(X) = \Phi^{t^\pm(X)}X. \tag{3.17}$$

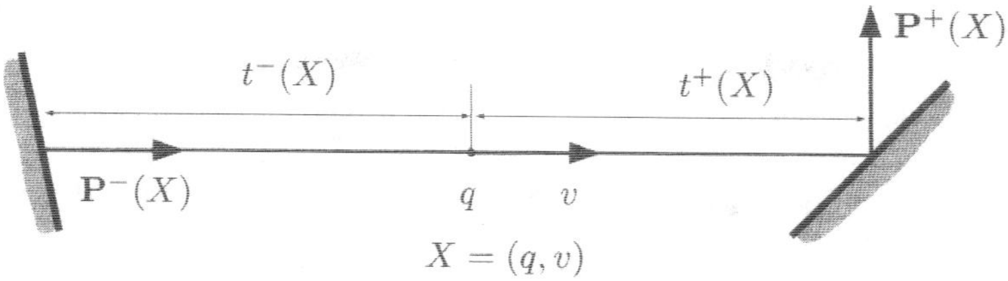

FIGURE 3.3. Two projections $\mathbf{P}^{\pm}: \Omega \to \mathcal{M}$.

EXERCISE 3.14. Verify that for $x \in \mathcal{M}$

$$t^+(x) = \tau(x), \qquad t^-(x) = \tau(\mathcal{F}^{-1}x), \qquad \mathcal{F}^{\pm 1}(x) = \mathbf{P}^{\pm}(x).$$

The derivative $D\mathbf{P}^+(X)$ maps the 3-dimensional tangent space $\mathcal{T}_X\Omega$ onto the 2-dimensional space $\mathcal{T}_z\mathcal{M}$, where $z = \mathbf{P}^+(X)$. Due to (3.13), the kernel of $D\mathbf{P}^+(X)$ is $\mathcal{T}_X^0\,\Omega$; hence the linear map

$$D\mathbf{P}^+(X)\colon \mathcal{T}_X^{\perp}\,\Omega \to \mathcal{T}_z\mathcal{M} \tag{3.18}$$

is one-to-one.

This allows us to compute the Lyapunov exponents for the flow via those of the map. Let $X \in \Omega$ and $z = \mathbf{P}^+(X)$. Suppose λ is a Lyapunov exponent of the map $\mathcal{F}$ at the point z corresponding to a tangent vector $dz = (dr, d\varphi) \in \mathcal{T}_z\mathcal{M}$. Denote by $dX = (dx, dy, d\omega) \in \mathcal{T}_X^{\perp}\,\Omega$ the unique preimage of $dz = (dr, d\varphi)$ under the projection (3.18).

First we need to compare the lengths $\|dX\| = \sqrt{(dx)^2 + (dy)^2 + (d\omega)^2}$ and $\|dz\| = \sqrt{(dr)^2 + (d\varphi)^2}$. Since dX, by our construction, belongs to the space (3.14), we have $\|dX\| = \sqrt{(d\xi)^2 + (d\omega)^2}$, where

$$d\xi = -\sin\omega\, dx + \cos\omega\, dy.$$

It follows from (2.12) that

$$d\xi = \cos\varphi\, dr - s\, d\omega, \qquad d\omega = -\mathcal{K}\, dr + d\varphi, \tag{3.19}$$

where $s = t^+(X)$; thus

$$dr = \frac{d\xi + s\, d\omega}{\cos\varphi}, \qquad d\varphi = \frac{\mathcal{K}\, d\xi + (s\,\mathcal{K} + \cos\varphi)\, d\omega}{\cos\varphi}.$$

EXERCISE 3.15. Denote $\hat{t}_X = \max\{t^+(X), 1\}$. Show that

$$\|dz\| \le \text{const}\,\frac{\hat{t}_X\,\|dX\|}{|\cos\varphi|}$$

and

$$\|dX\| \le \text{const}\,\hat{t}_X\,\|dz\|,$$

where constant factors are determined by the domain $\mathcal{D}$. As a corollary, we obtain

$$\log\|dz\| = \log\|dX\| + \mathcal{O}(\log\cos\varphi) + \mathcal{O}(\log\hat{t}_X). \tag{3.20}$$

Now consider the trajectory $\Phi^t(X)$ during a time interval $(0, T)$ and denote by $n = n_X(0, T)$ the number of collisions it experiences. According to (3.20),

$$\log \|D_X\Phi^T(dX)\| = \log \|D_z\mathcal{F}^n(dz)\| \\ + \mathcal{O}(\log\cos\varphi_n) + \mathcal{O}(\log\hat{t}_{\Phi^T X}), \tag{3.21}$$

where φ_n is the φ-coordinate of the point $\mathcal{F}^n z$. We now divide (3.21) by T and take the limit $T \to \pm\infty$. By the result of Exercise 2.38,

$$\lim_{T\to\pm\infty} \frac{1}{T} \log \|D_z\mathcal{F}^n(dz)\| = \bar{\tau}_X^{-1}\lambda.$$

Since the function $|\log\cos\varphi|$ is integrable on $\mathcal{M}$ (see (3.8)) we have $\frac{1}{T}\log\cos\varphi_n \to 0$ (see the explanation in the proof of Lemma 3.9). For the same reason $\frac{1}{T}\log\hat{t}_{\Phi^T X} \to 0$, because the function $\log\hat{t}_X$ is also integrable, since even $\hat{t}_X$ is integrable due to (2.32). Thus we arrive at

$$\lim_{T\to\pm\infty} \frac{1}{T} \log \|D_X\Phi^T(dX)\| = \bar{\tau}_X^{-1}\lambda \tag{3.22}$$

(up to a set of measure zero). This gives us a Lyapunov exponent for the flow, as defined by (3.11).

We summarize our results:

THEOREM 3.16. *The flow Φ^t has a zero Lyapunov exponent in the subspace $\mathcal{T}_X^0\Omega$ and two more Lyapunov exponents proportional to those of the map $\mathcal{F}$:*

$$\lambda_X^{(0)} = 0, \qquad \lambda_X^{(i)} = \bar{\tau}_X^{-1}\lambda_z^{(i)}, \qquad i = 1, 2,$$

where $z = \mathbf{P}^+(X)$ and $\bar{\tau}_X$ was introduced in Exercise 2.38. If $\lambda_X^{(i)} \neq 0$ for $i = 1, 2$, then the corresponding subspaces $E_X^{(i)}$ satisfy

$$E_X^{(i)} \subset \mathcal{T}_X^{\perp}\Omega \quad \textit{and} \quad D\mathbf{P}^+\big(E_X^{(i)}\big) = E_z^{(i)}.$$

EXERCISE 3.17. Verify that $\lambda_X^{(0)} + \lambda_X^{(1)} + \lambda_X^{(2)} = 0$, up to a set of zero measure. Hint: Use Lemma 3.9.

DEFINITION 3.18. A point $X \in \Omega$ is said to be *hyperbolic* if its Lyapunov exponents exist and $\lambda_X^{(i)} \neq 0$ for $i = 1, 2$.

According to Theorem 3.16 a point $X \in \Omega$ is hyperbolic if and only if its projection $\mathbf{P}^+(X) \in \mathcal{M}$ is hyperbolic; hence, the hyperbolicity of the billiard flow Φ^t is equivalent to that of the collision map $\mathcal{F}$.

3.4. Hyperbolicity as the origin of chaos

The billiard dynamics is *deterministic*. This means that given the position $q_0 \in \mathcal{D}$ and velocity $v_0 \in S^1$ of the billiard particle at time $t = 0$, one can determine its position q_t and velocity v_t at any time $t \in \mathbb{R}$ by using equations (2.7) and (2.8). In other words, the present state of the system completely determines its future and past.

Let us look at the determinism from a practical standpoint: does it allow us to *evaluate analytically* or *compute numerically* the future and the past from the present state? For very simple domains, such as circles and squares (Chapter 1), a complete analytic solution is possible. But for more general domains we run into serious obstacles. Analytic evaluation of q_t and v_t may be relatively easy for small

t, as long as the particle experiences very few collisions with the walls. For large t, the corresponding expressions for q_t and v_t become increasingly complicated and ultimately intractable.

Numerical calculations can be easily done on a computer, due to the recursive character of the equations (2.7) and (2.8), but the results will necessarily involve round-off errors, which we need to assess. At small times, the errors are usually of order of the available machine precision (say, $\sim 10^{-15}$ for the double-precision arithmetic on modern computers), but at large times they may grow, as we describe next.

Let $X_s \in \Omega$ denote the true state of the moving particle at time $s > 0$ and $\tilde{X}_s = X_s + \delta X_s$ be its computed state. For small s, we may assume that $\|\delta X_s\| \sim 10^{-15}$. At time $s + t$, we may expect that

$$\delta X_{s+t} \approx D_{X_s}\Phi^t(\delta X_s) + \delta X_t' \tag{3.23}$$

where $\delta X_t'$ is the additional round-off error made during the calculations of $\Phi^u(X_0)$, $s < u < s + t$. For simplicity, we will assume again that $\|\delta X_t'\| \sim 10^{-15}$, but the first term in (3.23) may grow fast, depending on the derivative $D_{X_s}\Phi^t$.

If the initial point $X_0 \in \Omega$ has only zero Lyapunov exponents, the tangent vectors would grow slowly (if at all), and the first term in (3.23) may remain small for large t. But suppose that X_0 has a positive Lyapunov exponent $\lambda > 0$; then

$$\|D_{X_s}\Phi^t(\delta X_s)\| \sim e^{\lambda t}\|\delta X_s\|.$$

The exponential function $e^{\lambda t}$ may increase rapidly with t. For example, if $\lambda = 1$, then at time $t = \ln 10^{15} \approx 35$ the errors will be of order one! Then the errors would quickly exceed the size of the phase space and render further computations meaningless.

Therefore, if the initial state $X_0 \in \Omega$ has a positive Lyapunov exponent, its distant future is practically inaccessible, even approximately! Similarly, the existence of a negative Lyapunov exponent at X_0 prevents approximative analysis of its remote past. If the set of hyperbolic points has full (or at least positive) measure, the failure to predict their long-term behavior will be a typical (or at least probable) event.

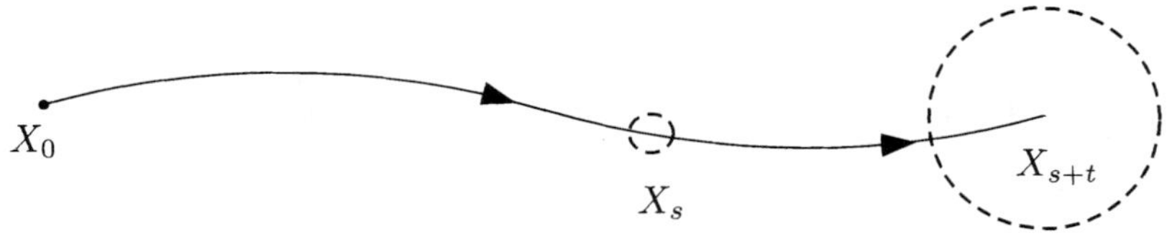

FIGURE 3.4. Growing uncertainty of approximations.

The implications of these facts are far reaching and can be characterized by a unifying term, *chaos*. The most important feature of a chaotic system is that its distant future and remote past become essentially *independent* of its present state and can be described only "on average", in probabilistic terms. On a large time scale, the evolution of the system resembles a purely random process; in all practical aspects, it behaves as stochastically and should be studied by means of probability theory, involving probability measures and averaging.

This approach is best known to physicists dealing with gases and fluids, where the molecular dynamics, albeit theoretically deterministic, is practically chaotic. Many basic laws of statistical mechanics have been derived by assuming that the positions and velocities of the moving particles (molecules and atoms) are chosen randomly, according to some probability distribution, and then averaging the results. The same approach is successfully used in the studies of chaotic billiards, and it will dominate the later chapters of our book.

3.5. Hyperbolicity and numerical experiments

Our analysis in the previous section may suggest (wrongly) that numerical calculations on systems with hyperbolic points in phase space are meaningless. This is not so; in fact many extensive computer experiments have been performed on various chaotic models, and they accurately described their long-term behavior. The reason is that in numerical experiments one normally wants to generate a *typical* trajectory (with the initial point X_0 selected randomly) rather than the trajectory of any *particular* given point X_0. And here the following *shadowing lemma* comes to into play. We state it in the context of maps rather than flows.

DEFINITION 3.19. Let $F\colon M \to M$ be an invertible map on a metric space. For a small $\delta > 0$ a finite or infinite sequence of points $\{x_n\}$, $n_1 < n < n_2$ ($n_1 = -\infty$ or $n_2 = \infty$ or both are permitted), is called a δ-pseudo-trajectory if

$$\operatorname{dist}\big(F(x_n), x_{n+1}\big) < \delta$$

for all $n_1 < n < n_2 - 1$.

If $\delta = 0$, then $x_{n+1} = F(x_n)$, so that $\{x_n\}$ is the true trajectory of the point x_0. For small $\delta > 0$ one gets a "nearly right" point $x_{n+1} \approx F(x_n)$ at every iteration of F. This is exactly what one observes in a numerical experiment where round-off errors are involved. The following fact is known in the study of hyperbolic maps:

Fact (Shadowing Lemma): For any $\varepsilon > 0$ there is a $\delta > 0$ such that for every δ-pseudo-trajectory $\{x_n\}$, $n_1 < n < n_2$, there exists $y_0 \in M$ such that

$$\operatorname{dist}\big(x_n, F^n(y_0)\big) < \varepsilon$$

for all $n_1 < n < n_2$. One says that the trajectory of y_0 ε-*shadows* the pseudo-trajectory $\{x_n\}$.

Note that there are no restrictions on n_1 and n_2 here; shadowing takes place for arbitrary long, as well as infinite, time intervals. The shadowing lemma, as stated, is proved in [**Bow75, KH95**] for Anosov and Axiom A diffeomorphisms. It holds for pseudo-trajectories in chaotic billiards, too, under certain additional conditions (which are beyond the scope of our book). Curiously, the shadowing lemma holds for hyperbolic systems but fails for many simpler systems with zero Lyapunov exponents (as an easy exercise, check that it fails for the identity map on any manifold!).

Now, the shadowing lemma asserts that the computed δ-pseudo-trajectory $\{x_n\}$ of the point x_0 remains ε-close to the real trajectory of another phase point y_0 (in fact, y_0 is itself ε-close to x_0). In experiments where the initial point x_0 is picked randomly anyway, one can argue that the new point y_0 is just "as good" as the randomly chosen x_0; hence the overall results of the experiment should be perfectly valid. And the broad success of computer experiments well supports this logic.

The fact that the true trajectory of x_0 may be very far from the computed pseudo-trajectory $\{x_n\}$ is of no concern here. What matters is that the sequence $\{x_n\}$ is, approximately, a true trajectory of the system.

Back to the determinism, though, the issue there was to predict the future of a *given* phase point x_0 (representing the true present state of the system). For that, the shadowing lemma is of no help, because we derive no comfort from predicting the trajectory of *some* nearby point $y_0 \neq x_0$ (at the same time knowing that the true trajectory of x_0 can be virtually anywhere in the phase space).

3.6. Jacobi coordinates

Here we return to our discussion of the billiard flow $\Phi^t\colon \Omega \to \Omega$.

Our analysis in Section 3.3 suggests a more convenient coordinate system $(d\eta, d\xi, d\omega)$ in the 3D tangent space $\mathcal{T}_X\Omega$ with

$$d\eta = \cos\omega\, dx + \sin\omega\, dy, \qquad d\xi = -\sin\omega\, dx + \cos\omega\, dy. \tag{3.24}$$

We call these *Jacobi coordinates* (they are induced by the so-called transversal Jacobi fields [**Wo94**], but the latter are not used in our book). Note that $d\eta$ is the component of the vector (dx, dy) along the velocity vector v and $d\xi$ is its orthogonal component.

For any tangent vector $dX = (d\eta, d\xi, d\omega)$ we denote by $dX_t = (d\eta_t, d\xi_t, d\omega_t) = D_X\Phi^t(dX)$ its image at time $t \in \mathbb{R}$. The linear map $(d\eta, d\xi, d\omega) \mapsto (d\eta_t, d\xi_t, d\omega_t)$ is given by the 3×3 matrix

$$D_X\Phi^t = \begin{bmatrix} 1 & 0 & 0 \\ 0 & \times & \times \\ 0 & \times & \times \end{bmatrix}$$

according to the results of Section 3.3. The bottom right 2×2 block (yet to be determined; see below) corresponds to the transformation $(d\xi, d\omega) \mapsto (d\xi_t, d\omega_t)$. From now on we restrict the map $D_X\Phi^t$ to the subspace $\mathcal{T}_X^\perp\Omega$ with Jacobi coordinates $d\xi$ and $d\omega$. We denote this restriction by $D_X^\perp\Phi^t\colon \mathcal{T}_X^\perp\Omega \to \mathcal{T}_{\Phi^tX}^\perp\Omega$.

EXERCISE 3.20. Show that if there are no collisions between X and Φ^tX, then

$$d\xi_t = d\xi + t\, d\omega, \qquad d\omega_t = d\omega; \tag{3.25}$$

hence the map $D_X\Phi^t$ on $\mathcal{T}_X^\perp\Omega$ is given by the 2×2 upper triangular matrix

$$U_t = \begin{bmatrix} 1 & t \\ 0 & 1 \end{bmatrix}. \tag{3.26}$$

Next we consider the action of $D_X\Phi^t$ at a collision using the notation of Section 2.6. Equations (2.11)–(2.12) imply

$$\begin{aligned} d\xi^- &= +\cos\varphi\, dr - s^-\, d\omega^- & \qquad d\omega^- &= -\mathcal{K}\, dr + d\varphi \\ d\xi^+ &= -\cos\varphi\, dr + s^+\, d\omega^+ & \qquad d\omega^+ &= -\mathcal{K}\, dr - d\varphi \end{aligned}$$

(recall that $\psi + \varphi = \pi/2$; hence $d\psi = -d\varphi$). We set $s^- = s^+ = 0$ to express the action of the flow "at the time of collision". Now eliminating dr and $d\varphi$ from the above equations gives

$$\begin{aligned} d\xi^+ &= -d\xi^- \\ d\omega^+ &= -\mathcal{R}\, d\xi^- - d\omega^-, \qquad \mathcal{R} = \tfrac{2\mathcal{K}}{\cos\varphi}; \end{aligned} \tag{3.27}$$

hence the action of the flow at the collision is given by a lower triangular matrix

$$L_{\mathcal{R}} = -\begin{bmatrix} 1 & 0 \\ \mathcal{R} & 1 \end{bmatrix}. \tag{3.28}$$

(we will call $\mathcal{R}$ the *collision parameter*). Therefore, the map $D_X^\perp \Phi^t$ (recall that this is the restriction of $D_X\Phi^t$ to $\mathcal{T}_X^\perp \Omega$) is, generally, given by a product of alternating upper and lower triangular matrices:

$$D_X^\perp \Phi^t = (-1)^n U_{t-t_n} L_{\mathcal{R}_n} U_{t_n - t_{n-1}} \cdots U_{t_2 - t_1} L_{\mathcal{R}_1} U_{t_1}, \tag{3.29}$$

where t_i is the time of the ith collision during the time interval $(0, t)$ and $\mathcal{R}_i = 2\mathcal{K}_i / \cos\varphi_i$ is the parameter of the ith collision. The alternation of the factors U and L corresponds to the natural alternation of intervals of free motion and collisions in the billiard dynamics.

EXERCISE 3.21. Let $\mathcal{D}$ be a polygon. Show that, in this special case, $D_X^\perp \Phi^t = \pm \begin{bmatrix} 1 & t \\ 0 & 1 \end{bmatrix}$ for *all* $t \in \mathbb{R}$ and *all* points $X \in \Omega$ where Φ^t is defined. Then prove that the Lyapunov exponents for the flow and the map are zero everywhere.

3.7. Tangent lines and wave fronts

For hyperbolic points $X \in \Omega$, the subspaces $E_X^{(i)} \subset \mathcal{T}_X\Omega$, $i = 1, 2$, corresponding to nonzero Lyapunov exponents are one-dimensional and belong in $\mathcal{T}_X^\perp\Omega$; cf. Section 3.3. Here we describe the action of the derivative $D\Phi^t$ on tangent lines $L \subset \mathcal{T}_X^\perp\Omega$. Any such line is completely specified by its slope

$$\mathcal{B} = d\omega / d\xi \tag{3.30}$$

in the Jacobi $d\xi, d\omega$ coordinates on the plane $\mathcal{T}_X^\perp\Omega$. If $d\xi = 0$, we set $\mathcal{B} = \infty$.

FIGURE 3.5. Geometric meaning of the sign of $\mathcal{B}$.

EXERCISE 3.22. The sign of $\mathcal{B}$ has the following geometric meaning. Recall that the vector $(0, d\xi)$ corresponds to a displacement $dq = (dx, dy)$ of the point $q = (x, v)$ in the direction orthogonal to v (see (3.24)) and the value $d\xi$ corresponds to a displacement dv of a vector v. Now show that $\mathcal{B} > 0$ if and only if dq and dv point in the same direction; see Fig. 3.5.

In differential geometry, it is standard to associate a line L in the tangent space to a manifold Ω at a point $X \in \Omega$ with smooth (C^1) curves on Ω passing through X that are tangent to the line L at X. This will be helpful in our case, too.

Let $\gamma' \subset \Omega$ be a curve passing through a point $X = (q, v)$ and tangent to a line $L \subset \mathcal{T}_X^\perp\Omega$. Assume first that $\mathcal{B} \neq \infty$. Then the projection $\sigma' = \pi_q(\gamma')$ of the curve γ' onto the table $\mathcal{D}$ is a curve, which is orthogonal to v at the point q.

Every point $X' = (q', v') \in \gamma'$ moves along a billiard trajectory, $(q' + sv', v')$, which traces a segment of the line $q' + sv'$ on the billiard table (here s must be small, so that $q' + sv'$ will stay inside $\mathcal{D}$). This gives us a C^1 smooth family of directed lines $\{q' + sv' \colon (q', v') \in \gamma'\}$ in the domain $\mathcal{D}$. This family is two-dimensional, since $v' = v'(q')$ is a function of $q' \in \sigma'$.

Denote by σ the orthogonal cross-section of this family passing through q (this is a C^1 smooth curve that crosses every line in our family orthogonally; see Fig. 3.6). If we equip every point $q'' \in \sigma$ with a unit normal vector v'' to σ pointing in the direction of motion, we obtain a curve $\gamma \subset \Omega$.

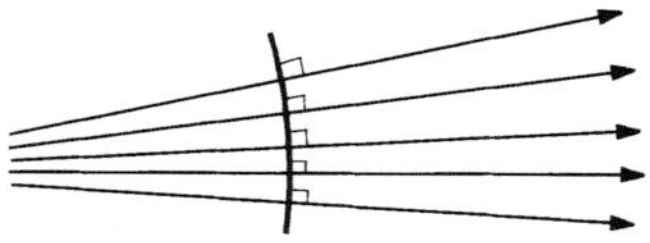

FIGURE 3.6. An orthogonal cross-section of a family of directed lines.

EXERCISE 3.23. Verify that the curve γ contains the point $X = (q, v)$ and is tangent to γ', hence to L, at the point X.

DEFINITION 3.24. A smooth curve $\sigma \subset \mathcal{D}$ equipped with a continuous family of unit normal vectors is called a *wave front*.

Since every point $q \in \sigma$ has a velocity vector v_q attached to it, we get a phase state $(q, v_q) \in \Omega$; thus, a wave front becomes a curve in the phase space

$$\gamma = \{(q, v_q) \in \Omega \colon \; q \in \sigma\}.$$

Note that the tangent line $\mathcal{T}_Y\gamma$ to γ lies in $\mathcal{T}_Y^{\perp}\Omega$ at every point $Y \in \gamma$, which makes wave fronts a convenient tool in the study of the billiard flow Φ^t.

EXERCISE 3.25. Show that if $\gamma \subset \Omega$ is a wave front, then for small enough t, the curve $\Phi^t(\gamma)$ is a wave front, too. Hint: Recall that $D_X\Phi^t(\mathcal{T}_X^{\perp}\Omega) = \mathcal{T}_{\Phi^t X}^{\perp}\Omega$.

For example, if $\sigma \subset \mathcal{D}$ is a circle equipped with the outer unit normal vectors, then the Φ^t-image of this wave front will be a concentric circle whose radius grows with t (this demonstrates that wave fronts move like ripples on the surface of water); see Fig. 3.7.

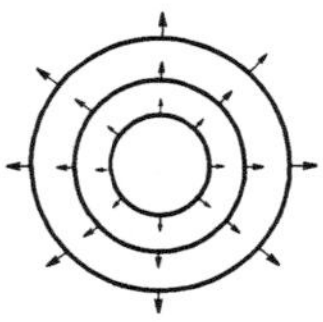

FIGURE 3.7. A moving wave front.

The Φ^t-image of a wave front γ for large t's may develop singularities. First, the smoothness of $\Phi^t(\gamma)$ will be broken at the points of reflection at $\partial\mathcal{D}$, but it will be recovered immediately after the reflection; see below. Second, the front may collapse (*focus*) when the trajectories of its points intersect each other within the table $\mathcal{D}$. For example, if $\sigma \subset \mathcal{D}$ is a circle of radius $r > 0$ equipped with the *inner* unit normal vectors, then at time $t = r$ the entire front will converge to the center of this circle (we say that it focuses). But it will immediately recover (*defocus*) after the collapse and then spread out as shown in Fig. 3.7.

We have left out the degenerate case $\mathcal{B} = \infty$ from our analysis. In that case there are no smooth wave fronts tangent to L at X, but L can be represented by a "collapsed" front as follows:

EXERCISE 3.26. Let $L \subset \mathcal{T}_X^{\perp}\Omega$ be a tangent line with $\mathcal{B} = \infty$. Let $\gamma \subset \Omega$ be a smooth curve such that $X = (q, v) \in \gamma$ and $\pi_q(\gamma) = \{q\}$; i.e. the curve γ entirely focuses at q. Show that γ is tangent to L.

Let $X \in \Omega$ and $L \subset \mathcal{T}_X^{\perp}\Omega$ be a tangent line with a finite slope $\mathcal{B} \neq \infty$. Suppose $\gamma \subset \Omega$ is a wave front tangent to L at the point $X = (q, v)$ and $\sigma = \pi_q(\gamma)$.

EXERCISE 3.27. Check that $\mathcal{B}$ is the curvature of σ at the point q with respect to the normal vector v. (In fact, $\mathcal{B}$ is traditionally introduced in the literature as the curvature of wave fronts. We chose to introduce $\mathcal{B}$ by (3.30) because it suits our general ideas better.)

According to Exercise 3.22 and Fig. 3.5, the sign of $\mathcal{B}$ allows us to distinguish between the following types of wave fronts:

(a) If $\mathcal{B} > 0$, the front is divergent (*dispersing*).
(b) If $\mathcal{B} < 0$, the front is convergent (*focusing*).
(c) If $\mathcal{B} = 0$, the front is *flat* (its velocity vectors are parallel).

The case $\mathcal{B} = \infty$ would correspond to a degenerate (collapsed or focused) wave front; in that case we call X a *focusing point* of the front.

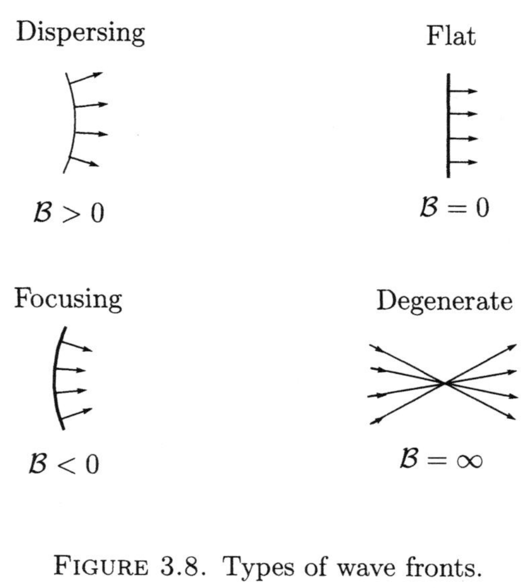

FIGURE 3.8. Types of wave fronts.

3.8. Billiard-related continued fractions

Here we introduce a special tool for the study of billiards, first used by Ya. Sinai [**Sin70, BS73**] – continued fractions.

Let $L \subset \mathcal{T}_X^{\perp}\Omega$ be a tangent line with slope $\mathcal{B}$. Let $L_t = D_X\Phi^t(L) \subset \mathcal{T}_{\Phi^t X}^{\perp}\Omega$ denote its image at time t and $\mathcal{B}_t$ its slope.

EXERCISE 3.28. Show that if there are no collisions between X and $\Phi^t X$, then

$$\mathcal{B}_t = \frac{\mathcal{B}}{1 + t\mathcal{B}} = \frac{1}{t + \dfrac{1}{\mathcal{B}}}. \tag{3.31}$$

Hint: Use (3.26).

Note that $\mathcal{B}_t = \infty$ at time $t = t^* = -1/\mathcal{B}$ according to (3.31). Thus, any wave front tangent to the line L at the point X will collapse (focus) at time t^* (if $t^* < 0$, then the front has collapsed at time t^*) unless it hits $\partial\mathcal{D}$ before that, of course. If the wave front collapses at t^*, we call the point $\Phi^{t^*}X$ the *focusing point* in the evolution of the wave front. Of course, the equation (3.31) remains valid beyond the focusing time t^* as well, as the wave front will defocus and move on smoothly.

EXERCISE 3.29. Show that if there are no collisions or focusing points between X and $\Phi^t X$, then (in the notation of Exercise 3.28)

$$\int_0^t \mathcal{B}_t \, dt = \log(1 + t\mathcal{B}). \tag{3.32}$$

Also check that if there *is* a focusing point, i.e. $\mathcal{B}_s = \infty$ for some $0 \le s \le t$, then the above integral diverges.

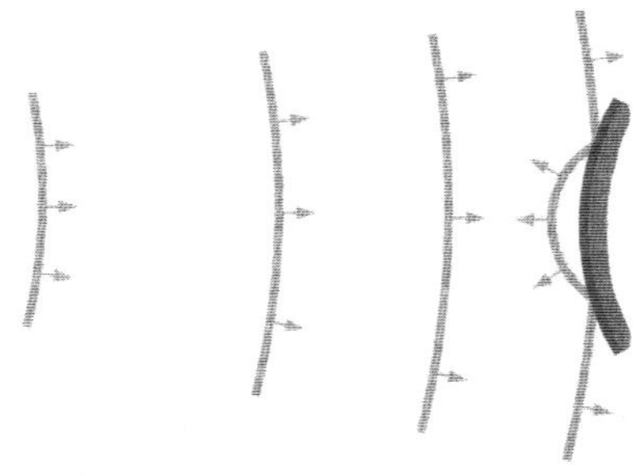

FIGURE 3.9. A wave front hitting a wall.

LEMMA 3.30. *Let t be a time of collision, i.e. $\Phi^t X \in \mathcal{M}$. Then the precollisional and postcollisional slopes $\mathcal{B}^- = \mathcal{B}_{t-0}$ and $\mathcal{B}^+ = \mathcal{B}_{t+0}$ are related by*

$$\mathcal{B}^+ = \mathcal{R} + \mathcal{B}^-, \qquad \mathcal{R} = \tfrac{2\mathcal{K}}{\cos\varphi} \tag{3.33}$$

($\mathcal{R}$ is the collision parameter).

PROOF. This easily follows from (3.27). We leave it as an exercise. □

The formula (3.33) is known in geometrical optics as *mirror equation* (cf. [**Wo86**, p. 396]). It describes the transformation of a wave front at a collision with a curved wall; see Fig. 3.9 (note that if the wall is flat, then $\mathcal{R} = 0$ and so $\mathcal{B}^+ = \mathcal{B}^-$).

Combining equations (3.31) and (3.33) gives a continued fraction formula for the slope $\mathcal{B}_t$:

$$\mathcal{B}_t = \cfrac{1}{t - t_n + \cfrac{1}{\mathcal{R}_n + \cfrac{1}{t_n - t_{n-1} + \cfrac{1}{\mathcal{R}_{n-1} + \cfrac{1}{\ddots + \cfrac{1}{t_1 + \cfrac{1}{\mathcal{B}}}}}}}} \tag{3.34}$$

in the notation of (3.29).

EXERCISE 3.31. Let $\mathcal{D}$ be a polygon. Show that if $\mathcal{B} = 0$, then $\mathcal{B}_t = 0$ for all $t \in \mathbb{R}$; i.e. flat wave fronts remain flat at all times. Furthermore, show that if $\mathcal{B} \neq 0$, then $\mathcal{B}_t \to 0$ as $t \to \pm\infty$; i.e. all wave fronts flatten out eventually. Lastly, check that the convergence here is of order $1/|t|$; that is $|\mathcal{B}_t| = \mathcal{O}(1/|t|)$ for large $|t|$.

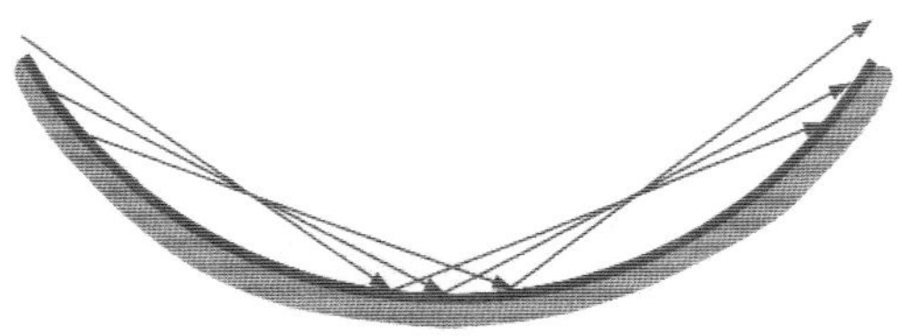

FIGURE 3.10. A stable wave front in a disk focuses half-way between collisions.

EXERCISE 3.32. Let $\mathcal{D}$ be a unit disk; see Section 1.1. For any $X \in \Omega$ consider its first collision point $\mathbf{P}^+(X) = (r, \varphi) \in \mathcal{M}$ and denote by t_n the collision times along the trajectory of X (observe that $\tau = t_n - t_{n-1} = 2\cos\varphi$ for all n). Denote $\mathcal{B}_* = -1/\cos\varphi = -2/\tau$.

(a) Show that if $\mathcal{B}_{t_0+0} = \mathcal{B}_*$, then $\mathcal{B}_{t_n+0} = \mathcal{B}_*$ for all $n \in \mathbb{Z}$ (i.e., if a wave front focuses half-way between two collisions, its image will keep focusing half-way between collisions; see Fig. 3.10).

(b) Verify that either $\mathcal{B}_{t_0+0} \leq -1/\tau$ or else $\mathcal{B}_{t_1} < -3/\tau$. Conclude that there is at least one focusing point (where $\mathcal{B}_s = \infty$) within the time interval $t_0 < s < t_1 + \tau/3$.

(c) Show that if $\mathcal{B}_{t_0+0} \neq \mathcal{B}_*$, then $\mathcal{B}_{t_n+0} \to \mathcal{B}_*$ as $n \to \pm\infty$; i.e. all wave fronts approach the only 'stable regime' where they focus half-way between collisions.

(d) Check that the convergence in (c) is of order $1/|n|$; that is $|\mathcal{B}_{t_n+0} - \mathcal{B}_*| = \mathcal{O}(1/|n|)$ for large $|n|$.

Hint: First show that $\mathcal{B}_{t_{n+1}+0} = 4/\tau + 1/(\tau + 1/\mathcal{B}_{t_n+0})$. Then denote $z_n = \tau\mathcal{B}_{t_n+0}$ and rewrite the previous formula as $z_{n+1} = -3 - 1/(1+z_n)$. This defined a rational transformation $\mathbb{R} \to \mathbb{R}$ with the only fixed point $z = -2$, which is attractive.

3.9. Jacobian for tangent lines

The expression (3.34) gives the slope of the line $L_t = D_X\Phi^t(L)$ in terms of the slope of the initial line $L \subset \mathcal{T}_X^\perp\Omega$. However, the map $D_X\Phi^t \colon L \to L_t$ is not an isometry; it may expand or contract vectors $dX \in L$. Here we compute its Jacobian.

Let $(d\xi, d\omega) \in L$ and $(d\xi_t, d\omega_t) = D_X\Phi^t(d\xi, d\omega) \in L_t$. If there are no collisions between X and $\Phi^t X$, then

$$d\xi_t = (1 + t\mathcal{B})\, d\xi, \qquad d\omega_t = d\omega$$

according to (3.25). At the time of a collision, $|d\xi^+| = |d\xi^-|$ due to (3.27). Combining these equations gives an important formula

$$\frac{|d\xi_t|}{|d\xi|} = \prod_{i=0}^{n} |1 + \tau_i\mathcal{B}_i^+| \tag{3.35}$$

where $\tau_0 = t_1$, $\tau_i = t_{i+1} - t_i$ for $1 \le i \le n-1$, and $\tau_n = t - t_n$, all in the notation of (3.34), and $\mathcal{B}_0^+ = \mathcal{B}$ and $\mathcal{B}_i^+ = \mathcal{B}_{t_i+0}$ for $1 \le i \le n$.

If we define a norm (metric) on the line $L \subset \mathcal{T}_X^\perp\,\Omega$ by $|d\xi|$, then (3.35) gives the Jacobian of the linear transformation $L \to L_t$. The Jacobian in the Euclidean metric $\|dX\| = \sqrt{(d\xi)^2 + (d\eta)^2}$ will then be given by

$$\frac{|dX_t|}{|dX|} = \sqrt{\frac{1 + \mathcal{B}_t^2}{1 + \mathcal{B}^2}}\,\prod_{i=0}^{n} |1 + \tau_i\mathcal{B}_i|. \tag{3.36}$$

EXERCISE 3.33. Show that if the ith reflection occurs at a flat wall (where $\mathcal{K} = 0$), then in the expression (3.35)

$$|1 + \tau_{i-1}\mathcal{B}_{i-1}^+|\,|1 + \tau_i\mathcal{B}_i^+| = |1 + (\tau_{i-1} + \tau_i)\mathcal{B}_{i-1}^+|;$$

i.e. the ith reflection may be effectively ignored in the computation of the Jacobian.

EXERCISE 3.34. Show that if $\mathcal{B}_s \neq \infty$ for all $0 \le s \le t$ (i.e. there is no focusing points on the trajectory of X), then

$$\log\frac{|d\xi_t|}{|d\xi|} = \int_0^t \mathcal{B}_s\, ds. \tag{3.37}$$

On the other hand, if there is at least one focusing point, then the integral diverges. Hint: Recall (3.32).

Lastly we note that the metric $|d\xi|$ is defined on all nondegenerate lines (where $\mathcal{B} \neq \infty$). We will not need to compute the Jacobian for degenerate lines, so we prefer the metric $|d\xi|$ for the sake of simplicity of the corresponding Jacobian formulas; see (3.35) and (3.37).

3.10. Tangent lines in the collision space

Here we describe the action of the derivative $D_x\mathcal{F}$ of the collision map on tangent lines $L \subset \mathcal{T}_x\mathcal{M}$ to the collision space. Any such line is completely specified by its slope

$$\mathcal{V} = d\varphi/dr. \tag{3.38}$$

For vertical lines, where $dr = 0$, we set $\mathcal{V} = \infty$.

We will relate the lines $L \subset \mathcal{T}_x\mathcal{M}$ to tangent lines in $\mathcal{T}_X^\perp\,\Omega$ treated in Section 3.7 by using the projections $\mathbf{P}^\pm$ introduced by (3.17). Given a point $x = (r,\varphi) \in \mathcal{M}$ we consider two corresponding phase states

$$X^+ = \lim_{t\downarrow 0}\Phi^t(x), \qquad X^- = \lim_{t\uparrow 0}\Phi^t(x)$$

immediately *after* and *before* the collision at x, respectively. (Even though we identified such pairs in Section 2.5, here we treat them as distinct states without confusion). We note that $t^+(X^-) = t^-(X^+) = 0$ and $\mathbf{P}^+(X^-) = \mathbf{P}^-(X^+) = x$, in the notation of (3.15)–(3.17).

Now the linear map $D\mathbf{P}^+(X^-)\colon \mathcal{T}_{X^-}^\perp\,\Omega \to \mathcal{T}_x\mathcal{M}$ is one-to-one (see (3.18)), and we denote by $L^- \subset \mathcal{T}_{X^-}^\perp\,\Omega$ the preimage of the line L under this map. Similarly, the map $D\mathbf{P}^-(X^+)\colon \mathcal{T}_{X^+}^\perp\,\Omega \to \mathcal{T}_x\mathcal{M}$ is one-to-one, and we denote by $L^+ \subset \mathcal{T}_{X^+}^\perp\,\Omega$ the preimage of the line L under that map.

Thus, every tangent line $L \subset \mathcal{T}_x\mathcal{M}$ in the collision space $\mathcal{M}$ corresponds to two special tangent lines in the phase space Ω: the "precollisional" line $L^- \subset \mathcal{T}_{X^-}^\perp\,\Omega$ and the "postcollisional" line $L^+ \subset \mathcal{T}_{X^+}^\perp\,\Omega$, as constructed above. Let $\mathcal{B}^-$ and $\mathcal{B}^+$ denote the slopes of the lines L^- and L^+, respectively, as defined by (3.30) and related by (3.33).

EXERCISE 3.35. Show that

$$\mathcal{V} = \mathcal{B}^-\cos\varphi + \mathcal{K} = \mathcal{B}^+\cos\varphi - \mathcal{K}, \tag{3.39}$$

where $\mathcal{K}$ denotes the curvature of the boundary $\partial\mathcal{D}$ at the point x. Hint: The first identity follows from (3.19), where one should notice that $s = 0$.

The norm $|d\xi|$ on the lines L^+ and L^- induces (under the maps $\mathbf{P}^-(X^+)$ and $\mathbf{P}^+(X^-)$, respectively) the same norm on the line L, which is due to (3.19),

$$\|(dr,d\varphi)\|_p = \cos\varphi\,|dr|.$$

This is often called the *p-norm* (or *p-metric*) on tangent lines in the collision space $\mathcal{M}$. The p-norm is defined on any line L unless it is vertical, i.e. unless $\mathcal{V} = \infty$. Due to (3.39), the condition $\mathcal{V} = \infty$ is equivalent to $\mathcal{B}^- = \mathcal{B}^+ = \infty$ at all regular collisions (where $\varphi \neq \pm\pi/2$). Thus the p-norm is defined on all nondegenerate lines.

Lastly we find the Jacobian of the map $D_x\mathcal{F}^n$ in the p-norm. Let $L \subset \mathcal{T}_x\mathcal{M}$ and $L_n = D_x\mathcal{F}^n(L)$ denote the image of L. We put $\tau_n = \tau(\mathcal{F}^n(x))$ and let $\mathcal{B}_n^+$ denote the slope of the postcollisional line L_n^+ corresponding to L_n. Then for any tangent vector $0 \neq dx \in L$

$$\frac{\|D_x\mathcal{F}^n(dx)\|_p}{\|dx\|_p} = \prod_{i=0}^{n-1}\left|1 + \tau_i\mathcal{B}_i^+\right|, \tag{3.40}$$

as immediately follows from (3.35).

The Euclidean norm $\|dx\| = \sqrt{(dr)^2 + (d\varphi)^2}$ is related to the p-norm by

$$\begin{aligned} \|dx\| &= |dr|\sqrt{1+\mathcal{V}^2} \\ &= \frac{\|dx\|_p}{\cos\varphi}\sqrt{1+(\mathcal{B}^+\cos\varphi-\mathcal{K})^2}. \end{aligned} \tag{3.41}$$

The p-metric is easier to use. We will often derive certain relations in the p-metric and then convert them into the Euclidean metric by (3.41).

3.11. Stable and unstable lines

Let $X \in \Omega$ be a hyperbolic point and E^u_X and E^s_X denote the one-dimensional subspaces corresponding to the nonzero Lyapunov exponents; see Section 3.3. We denote their slopes by $\mathcal{B}^u(X)$ and $\mathcal{B}^s(X)$, respectively. We also set $\mathcal{B}^u(X) = \mathcal{B}^s(X) = 0$ at all nonhyperbolic points. Then $\mathcal{B}^u$ and $\mathcal{B}^s$ become measurable functions on Ω.

For $\alpha = u, s$, we set

$$\Omega^\alpha_\infty = \{X \in \Omega\colon \mathcal{B}^\alpha(\Phi^t(X)) = \infty \text{ for some } t \in \mathbb{R}\}.$$

Obviously, Ω^u_∞ and Ω^s_∞ are Φ^t-invariant sets.

EXERCISE 3.36. Show that if $\mu_\Omega(\Omega^\alpha_\infty) > 0$, then the function $\mathcal{B}^\alpha(X)$ is not integrable. Hint: Recall Exercise 3.29.

EXERCISE 3.37. Let the function $\mathcal{B}^u(X)$ be integrable. Show that at almost every hyperbolic point X the positive Lyapunov exponent is

$$\lambda^+_X = \lim_{t\to\pm\infty}\frac{1}{t}\log\frac{|d\xi_t|}{|d\xi|} \tag{3.42}$$

where $0 \neq (d\xi, d\eta) \in E^u_X$. Hint: Use (3.35)–(3.36).

THEOREM 3.38. *Suppose that the function $\mathcal{B}^u(X)$ is integrable. Then the positive Lyapunov exponent λ^+_X is the time average*

$$\lambda^+_X = \lim_{T\to\pm\infty}\frac{1}{T}\int_0^T \mathcal{B}^u(\Phi^t(X))\,dt$$

(up to a set of μ_Ω-measure zero). Furthermore,

$$\int_\Omega \lambda^+_X\,d\mu_\Omega(X) = \int_\Omega \mathcal{B}^u(X)\,d\mu_\Omega(X) \tag{3.43}$$

(here we set $\lambda^+_X = 0$ at all nonhyperbolic points $X \in \Omega$).

PROOF. The first identity follows from (3.42) and (3.37). The second one is a consequence of the Birkhoff ergodic theorem. □

REMARK 3.39. Similarly, the negative Lyapunov exponent λ^-_X can be computed in terms of the function $\mathcal{B}^s(X)$ when the latter is integrable.

We should remark that the function $\mathcal{B}^u(X)$ is not integrable for certain large classes of chaotic billiards (see below), so its integrability is a somewhat restrictive condition.

Next we consider the collision space $\mathcal{M}$. Let $x \in \mathcal{M}$ be a hyperbolic point and E^u_x and E^s_x denote the subspaces corresponding to the nonzero Lyapunov exponents; see Section 3.1. We denote their slopes by $\mathcal{V}^u(x)$ and $\mathcal{V}^s(x)$, as defined by (3.38), and the slopes of the corresponding precollisional and postcollisional lines in the phase space by $\mathcal{B}^{u\pm}(x)$ and $\mathcal{B}^{s\pm}(x)$ (cf. Section 3.10). We also set $\mathcal{V}^\alpha(x) = 0$ and $\mathcal{B}^{\alpha\pm}(x) = 0$ ($\alpha = u, s$) at all nonhyperbolic points. Then $\mathcal{V}^\alpha$ and $\mathcal{B}^{\alpha\pm}$ become measurable functions on $\mathcal{M}$.

EXERCISE 3.40. Let the function $\log^+[\mathcal{B}^{u+}(x)]$ be integrable; i.e.

$$\int_{\mathcal{M}} \log^+[\mathcal{B}^{u+}(x)]\, d\mu < \infty \tag{3.44}$$

(here $\log^+ t = \max\{\log t, 0\}$, as in the Oseledets theorem). Show that at almost every hyperbolic point x the positive Lyapunov exponent is

$$\lambda_x^+ = \lim_{n\to\pm\infty} \frac{1}{n} \log \frac{\|D_x\mathcal{F}^n(dx)\|_p}{\|dx\|_p} \tag{3.45}$$

where $0 \neq dx \in E_x^u$. Hint: Use (3.41) and the integrability of the function $|\log\cos\varphi|$.

THEOREM 3.41. *If (3.44) holds, then the positive Lyapunov exponent λ_x^+ at $x \in \mathcal{M}$ is the time average*

$$\lambda_x^+ = \lim_{n\to\infty} \frac{1}{n} \sum_{i=0}^{n-1} \log\bigl|1 + \tau(\mathcal{F}^i x)\,\mathcal{B}^{u+}(\mathcal{F}^i x)\bigr|$$

(up to a set of μ-measure zero). Furthermore,

$$\int_{\mathcal{M}} \lambda_x^+\, d\mu(x) = \int_{\mathcal{M}} \log\bigl|1 + \tau(x)\,\mathcal{B}^{u+}(x)\bigr|\, d\mu(x) \tag{3.46}$$

(here we set $\lambda_x^+ = 0$ at all nonhyperbolic points $x \in \mathcal{M}$).

PROOF. The first identity follows from (3.45) and (3.40). The second one is a consequence of the Birkhoff ergodic theorem. □

The last theorem has a wider range of applicability than Theorem 3.38, because the integrability of $\mathcal{B}^u(X)$ is not required.

3.12. Entropy

Entropy is a measure of the complexity of a dynamical system. Here we deal with the so-called *measure-theoretic entropy* (also referred to as *Kolmogorov-Sinai entropy*, because it was introduced by A. Kolmogorov and Ya. Sinai in the late 1950s). We refer the reader to Appendix C for basic definitions and to standard textbooks [**Pet83, Wa82, BrS02, ME81**] for more detailed coverage.

The entropy is closely related to Lyapunov exponents via the following formula, which we state here in the context of billiards:

THEOREM 3.42. *The measure-theoretic entropy of the flow Φ^t with respect to the measure μ_Ω is*

$$h(\Phi^t) = \int_\Omega \lambda_X^+\, d\mu_\Omega(X) \tag{3.47}$$

where λ_X^+ is the positive Lyapunov exponent of the flow Φ^t at the point $X \in \Omega$ (if there is none, we set $\lambda_X^+ = 0$).

The measure-theoretic entropy of the collision map $\mathcal{F}$ with respect to the measure μ is

$$h(\mathcal{F}) = \int_{\mathcal{M}} \lambda_x^+ \, d\mu(x) \tag{3.48}$$

where λ_x^+ is the positive Lyapunov exponent of the map $\mathcal{F}$ at the point $x \in \mathcal{M}$ (if there is none, we set $\lambda_x^+ = 0$).

REMARK 3.43. Our formulas (3.47) and (3.48) hold for 2D billiards, where every phase point has at most one positive Lyapunov exponent. In the multidimensional case, one replaces it with the sum of all positive Lyapunov exponents.

The identity (3.48) is known as Pesin's formula. It was proved by Pesin for smooth maps with absolutely continuous invariant measures and extended by Katok and Strelcyn [**KS86**] to more general maps with singularities, including billiards. The entropy of the flow Φ^t is defined to be the entropy of its time-one map Φ^1 (obtained by setting $t = 1$); hence (3.47) is a particular case of a general Pesin formula for maps with singularities derived in [**KS86**].

REMARK 3.44. As we explained in Section 3.4, a billiard system becomes chaotic whenever it has at least one positive Lyapunov exponent (on a set of positive measure). Now we see that chaos in billiards is equivalent to the entropy being positive.

The entropy of the map and that of the flow are related by a classical Abramov formula

$$h(\mathcal{F}) = \bar{\tau}\, h(\Phi^t). \tag{3.49}$$

This is valid, more generally, for any suspension flow and its base transformation (as defined in Section 2.9); see [**Ab59**]. Due to (2.32) we obtain another useful relation,

$$h(\mathcal{F}) = \pi h(\Phi^t)\, |\mathcal{D}|/|\Gamma|.$$

EXERCISE 3.45. Derive Abramov's formula (3.49) for billiards assuming that the map $\mathcal{F}$ and hence the flow Φ^t are ergodic. Hint: Note that λ_x^+ is constant almost everywhere on $\mathcal{M}$, λ_X^+ is constant a.e. on Ω, and the function $\bar{\tau}_X$ (see Exercise 2.38) is constant a.e. on Ω; then use Theorem 3.16.

EXERCISE 3.46. Derive Abramov's formula (3.49) assuming that the map $\mathcal{F}$ is not ergodic but has finitely many ergodic components, i.e. $\mathcal{M} = \cup_{i=1}^k \mathcal{M}_i$, so that for each $i = 1, \dots, k$ we have $\mathcal{F}(\mathcal{M}_i) = \mathcal{M}_i$ (mod 0), $\mu(\mathcal{M}_i) > 0$, and the restriction of $\mathcal{F}$ to $\mathcal{M}_i$ is ergodic. Hint: Use Theorem 3.16 and the relation $d\mu_\Omega(X) = \bar{\tau}^{-1}\, d\mu \times ds$; see (2.30).

The verification of Abramov's formula (3.49) in a more general case (where the ergodic components of $\mathcal{F}$ may have zero measure) is based on ergodic decomposition; see Appendix C.

REMARK 3.47. Combining (3.48) and (3.46) gives us another integral formula for the entropy of the map,

$$h(\mathcal{F}) = \int_{\mathcal{M}} \log\bigl|1 + \tau(x)\, \mathcal{B}^{u+}(x)\bigr|\, d\mu(x).$$

(provided (3.44) holds). Combining (3.47) with (3.43) and (3.49) gives us two equivalent integral formulas for the entropy of the flow

$$\begin{aligned} h(\Phi^t) &= \int_\Omega \mathcal{B}^u(X)\, d\mu_\Omega(X) \\ &= \bar{\tau}^{-1} \int_\mathcal{M} \log\bigl|1 + \tau(x)\, \mathcal{B}^{u+}(x)\bigr|\, d\mu(x). \end{aligned} \tag{3.50}$$

Note, however, that the second formula is valid under more general conditions: it requires only the mild assumption (3.44), whereas the first formula needs the integrability of the function $\mathcal{B}^u(X)$ on Ω, which is a rather restrictive requirement. We will see later that $\mathcal{B}^u$ is integrable for dispersing chaotic billiards (Chapter 4) but not for focusing chaotic billiards (Chapter 8). In fact for the latter $\mu_\Omega(\Omega^u_\infty) = \mu_\Omega(\Omega^s_\infty) = 1$; thus $\mathcal{B}^u$ and $\mathcal{B}^s$ are never integrable. See Exercise 3.36.

EXERCISE 3.48. Suppose a billiard table $\mathcal{D}$ has at least one focusing boundary component $\Gamma_i \subset \Gamma_-$ (see (2.5)) and almost every point $X \in \Omega$ is hyperbolic. Show that

$$\mu_\Omega(\Omega^u_\infty) > 0;$$

thus the function $\mathcal{B}^u(X)$ cannot be integrable. Hint: Assume first that Γ_i is a circular arc; i.e. its curvature is constant. Then Exercise 3.32 (b) implies that every trajectory experiencing at least three successive collisions at Γ_i contains a focusing point, hence belongs in Ω^u_∞. For an arbitrary focusing wall, use an approximating argument (remember that the curvature of Γ_i is C^1 smooth).

3.13. Proving hyperbolicity: cone techniques

We conclude this chapter by discussing a general method for establishing hyperbolicity for phase points $x \in \mathcal{M}$ and constructing their stable and unstable subspaces E^s_x and E^u_x. We use the billiard notation $\mathcal{F}\colon \mathcal{M} \to \mathcal{M}$, but the method applies without change to any type of maps introduced in Section 3.1 as long as $\dim M = 2$ (it can also be extended to any dimension [**Wo85**], but we do not pursue this goal). We put $x_n = \mathcal{F}^n(x)$ for all $n \in \mathbb{Z}$.

The basic idea of constructing the unstable space E^u_x is to choose some nonzero vectors $v_{-n} \in \mathcal{T}_{x_{-n}}\mathcal{M}$ for $n \geq 0$ and then take the limit

$$E^u_x = \lim_{n\to\infty} \operatorname{span}\bigl\{D_{x_{-n}}\mathcal{F}^n(v_{-n})\bigr\}. \tag{3.51}$$

Of course, (3.51) need not hold for *all* choices of $\{v_{-n}\}$, but it should hold for typical choices, as the following (semi-heuristic) argument suggests. Suppose the point x, and hence x_n for all n, is indeed hyperbolic. Then $v_{-n} = c_{-n,u}v_{-n,u} + c_{-n,s}v_{-n,s}$, where $v_{-n,u} \in E^u_{x_{-n}}$ and $v_{-n,s} \in E^s_{x_{-n}}$ are some *unit* vectors. Due to (3.2) one expects that

$$D_{x_{-n}}\mathcal{F}^n(v_{-n}) \sim e^{\lambda_x n} c_{-n,u} w_{-n,u} + e^{-\lambda_x n} c_{-n,s} w_{-n,s},$$

where $\lambda_x > 0$ is the positive Lyapunov exponent and $w_{-n,u} \in E^u_x$ and $w_{-n,s} \in E^s_x$ are some *unit* vectors. Now, if the coefficients $c_{-n,u}$ and $c_{-n,s}$ are bounded away from zero and infinity, i.e.

$$0 < c_{\min} \leq |c_{-n,u}|, |c_{-n,s}| \leq c_{\max} < \infty,$$

then (3.51) obviously holds. In this case one can estimate the Lyapunov exponent $\lambda_x > 0$ by

$$\lambda_x = \lim_{n\to\infty} \frac{1}{n} \log \|D_{x_{-n}}\mathcal{F}^n(v_{-n})\|/\|v_{-n}\|. \tag{3.52}$$

Furthermore, both (3.51) and (3.52) hold unless the ratio $|c_{-n,s}/c_{-n,u}|$ grows, as $n \to \infty$, as fast as $e^{2\lambda_x n}$ or faster. Thus we need only to choose the initial vectors v_{-n} not too close to the stable spaces $E^s_{x_{-n}}$ (a very mild requirement, but it still needs to be met!).

To choose v_{-n} properly, Alekseev [**Al69**] proposed the method of cones that later became almost universal.

DEFINITION 3.49. Let $L \subset \mathcal{T}_x\mathcal{M}$ be a line and $\alpha \in (0, \pi/2)$. A *cone* $\mathcal{C}$ with axis L and opening α is the set of all tangent vectors $v \in \mathcal{T}_x\mathcal{M}$ that make angle $\leq \alpha$ with the line L.

The boundary $\partial\mathcal{C}$ consists of vectors making angle α with L (this includes the zero vector).

Suppose one finds cones $\mathcal{C}^u_x \subset \mathcal{T}_x\mathcal{M}$ at all (or almost all) points $x \in \mathcal{M}$ (with axes L_x and openings α_x being measurable functions of x) such that

$$D_x\mathcal{F}(\mathcal{C}^u_x) \subset \mathcal{C}^u_{\mathcal{F}x} \qquad \text{(invariance)} \tag{3.53}$$

and for all $v \in \mathcal{C}^u_x$

$$\|D_x\mathcal{F}(v)\| \geq \Lambda\|v\| \qquad \text{(expansion)}, \tag{3.54}$$

where $\Lambda > 1$ is a constant (independent of x and v).

EXERCISE 3.50. Assume that almost every point $x \in \mathcal{M}$ is hyperbolic and the cones $\mathcal{C}^u_x$ satisfy (3.53). Show that $E^u_x \subset \mathcal{C}^u_x$ and $E^s_x \setminus \{0\} \subset \mathcal{T}_x\mathcal{M} \setminus \mathcal{C}^u_x$ at almost every $x \in \mathcal{M}$. Then show that

$$E^u_x = \cap_{n\geq 0} D_{x_{-n}}\mathcal{F}^n(\mathcal{C}^u_{x_{-n}}). \tag{3.55}$$

This result justifies (3.51) with any choice of $v_{-n} \in \mathcal{C}^u_{x_{-n}}$.

EXERCISE 3.51. Assume that the cones $\mathcal{C}^u_x$ satisfy (3.53) and (3.54). Show that almost every point $x \in \mathcal{M}$ has a positive Lyapunov exponent $\lambda_x \geq \ln\Lambda$.

In order to approximate stable subspaces E^s_x and estimate the negative Lyapunov exponents, one can, similarly, find another system (family) of cones $\mathcal{C}^s_x \subset \mathcal{T}_x\mathcal{M}$ such that

$$D_x\mathcal{F}^{-1}(\mathcal{C}^s_x) \subset \mathcal{C}^s_{\mathcal{F}^{-1}x} \qquad \text{(invariance)} \tag{3.56}$$

and for all $v \in \mathcal{C}^s_x$

$$\|D_x\mathcal{F}^{-1}(v)\| \geq \Lambda\|v\| \qquad \text{(contraction)}. \tag{3.57}$$

The results of Exercises 3.50 and 3.51 apply (with obvious modifications); in particular

$$E^s_x = \cap_{n\geq 0} D_{x_n}\mathcal{F}^{-n}(\mathcal{C}^s_{x_n}). \tag{3.58}$$

One calls $\mathcal{C}^u_x$ *unstable* cones and $\mathcal{C}^s_x$ *stable* cones.

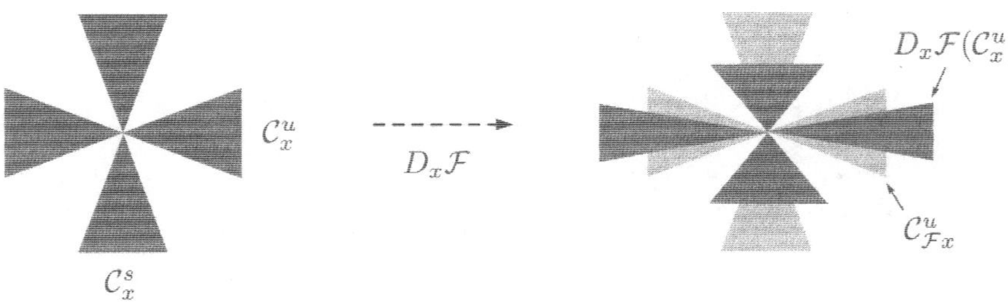

FIGURE 3.11. Transformation of unstable and stable cones.

REMARK 3.52. We note that the invariance properties (3.53) and (3.56) alone do not imply hyperbolicity. For example, invariant cones $\mathcal{C}^u_x$ and $\mathcal{C}^s_x$ can be constructed for billiards in circles and squares (Chapter 1), which are not hyperbolic. We leave this construction as an exercise; see also Section 8.4.

REMARK 3.53. M. Wojtkowski [**Wo85, Wo86**] proved that if the cones are *strictly invariant*, i.e.

$$D_x\mathcal{F}(\mathcal{C}^u_x) \subset \operatorname{int} \mathcal{C}^u_{\mathcal{F}x} \cup \{0\} \tag{3.59}$$

and

$$D_x\mathcal{F}^{-1}(\mathcal{C}^s_x) \subset \operatorname{int} \mathcal{C}^s_{\mathcal{F}^{-1}x} \cup \{0\}, \tag{3.60}$$

then hyperbolicity follows without the expansion and contraction assumptions (3.54) and (3.57). However, this is true only for symplectic maps (in the 2D case, the symplectic property is equivalent to the preservation of an absolutely continuous measure).

REMARK 3.54. Suppose there is a nondegenerate quadratic form Q_x defined on $\mathcal{T}_x\mathcal{M}$, for every $x \in \mathcal{M}$, such that

$$Q_{\mathcal{F}(x)}\big(D_x\mathcal{F}(u)\big) > Q_x(u) \qquad \forall u \neq 0$$

(i.e., Q monotonically increases under $\mathcal{F}$). Then the cones

$$\mathcal{C}^u_x = \{Q_x(u) \geq 0\} \qquad \text{and} \qquad \mathcal{C}^s_x = \{Q_x(u) \leq 0\}$$

are strictly invariant and the hyperbolicity of $\mathcal{F}$ follows by Wojtkowski's results; see above. We note that in some cases it is more convenient to find a monotonically increasing quadratic form rather than construct invariant cones; see for example [**Mar88**].

Lastly, it is clear that the conditions (3.53)–(3.54) and (3.56)–(3.57) can hold only for uniformly hyperbolic maps (see Section 3.1), which is quite restrictive. For nonuniformly hyperbolic maps, the cone method can be adapted as follows. Suppose there is a measurable function $n_x \colon \mathcal{M} \to \mathbb{N}$ such that for all $n \geq n_x$

$$D_x\mathcal{F}^n(\mathcal{C}^u_x) \subset \mathcal{C}^u_{\mathcal{F}^n x} \qquad \text{(eventual invariance)} \tag{3.61}$$

and for all $v \in \mathcal{C}^u_x$

$$\|D_x\mathcal{F}^n(v)\| \geq \Lambda\|v\| \qquad \text{(eventual expansion)}, \tag{3.62}$$

and similar modifications are made to (3.56) and (3.57). Then the map $\mathcal{F}\colon \mathcal{M} \to \mathcal{M}$ is hyperbolic and the formulas (3.55) and (3.58) hold.

We note that if the function n_x on $\mathcal{M}$ is bounded (that is, equivalently, constant), the map $\mathcal{F}$ is actually uniformly hyperbolic.

CHAPTER 4

Dispersing billiards

4.1. Classification and examples

Here we unveil the main class of chaotic billiards. It was introduced and first studied extensively by Ya. Sinai [**Sin70**].

DEFINITION 4.1. A billiard table $\mathcal{D} \subset \mathbb{R}^2$ (or $\mathcal{D} \subsetneq \mathrm{Tor}^2$) is said to be *dispersing* if all the walls $\Gamma_i \subset \Gamma = \partial\mathcal{D}$ are dispersing, i.e. $\Gamma = \Gamma_+$ (cf. Section 2.1).

One can further classify dispersing billiards according to the following attributes:

(a) whether or not $\mathcal{D}$ has corners; if yes, whether it has cusps;
(b) if $\mathcal{D} \subsetneq \mathrm{Tor}^2$, whether or not the horizon is bounded.

Accordingly, we define 6 subclasses (categories) of dispersing billiards:

	no corners	some corners, no cusps	some cusps
bounded horizon	A	C	E
unbounded horizon	B	D	F

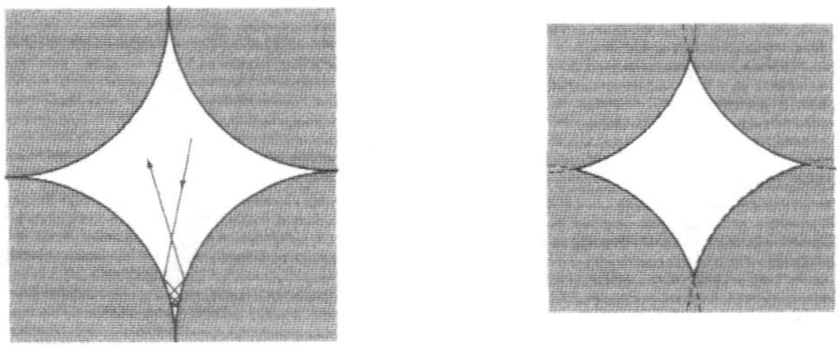

FIGURE 4.1. Billiards of categories E (left) and C (right).

EXERCISE 4.2. Let $\mathcal{D}_{\mathbf{r}} = \mathrm{Tor}^2 \setminus B_{\mathbf{r}}$, where $B_{\mathbf{r}}$ is a disk of radius $\mathbf{r} > 0$. What category does $\mathcal{D}_{\mathbf{r}}$ belong to? Answer: category B for $\mathbf{r} < 1/2$ (see Fig. 2.4), E for $\mathbf{r} = 1/2$ and C for $\mathbf{r} > 1/2$ (Fig. 4.1). The diamond-shaped billiard table in Fig. 4.1 (right) is quite popular in numerical experiments [**CCG, GG94**].

EXERCISE 4.3. Show that if $\mathcal{D}$ belongs to category A or B, then its boundary $\partial\mathcal{D}$ is a finite union of smooth closed curves, each of which encloses a convex region. Then argue that $\mathcal{D} \subsetneq \mathrm{Tor}^2$; i.e. $\mathcal{D}$ cannot be a bounded domain in $\mathbb{R}^2$.

Each convex region enclosed by a smooth convex curve $\Gamma_i \subset \Gamma$ is called a *scatterer*. In billiards of categories A and B, the particle moves between scatterers and bounces off them (like in a pinball machine).

A billiard table of type A is shown on Figure 4.2 (left); it has two scatterers (a larger disk and a smaller disk) that are positioned to block all possible directions in which the particle could move freely without collisions. Fig. 4.2 (right) shows the universal cover of Tor^2.

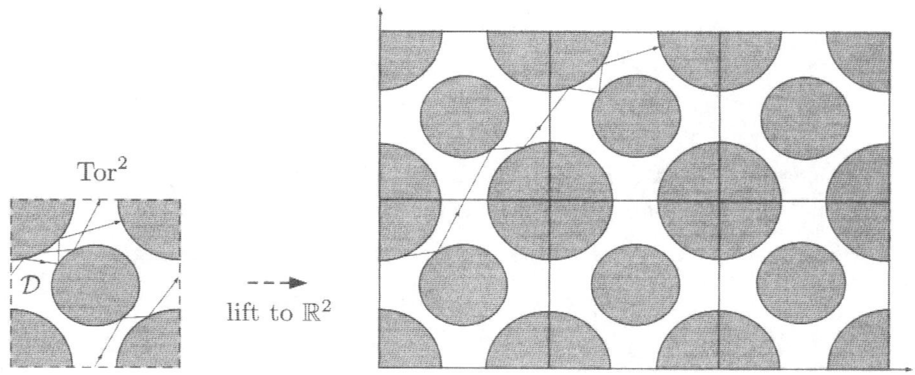

FIGURE 4.2. A billiard of category A.

EXERCISE 4.4. Can a billiard table of category A can be constructed with a single scatterer?

Every billiard trajectory on a table $\mathcal{D} \subset \mathrm{Tor}^2$ of type A or B can be lifted from the torus to its universal cover $\mathbb{R}^2$; see an example in Fig. 4.2. Then we get a trajectory moving on an unbounded table with infinitely many periodically located scatterers.

EXERCISE 4.5. Find a periodic trajectory on the table $\mathcal{D} \subset \mathrm{Tor}^2$ shown in Fig. 4.2 (left) such that its "lifted sister" in Fig. 4.2 (right) is not periodic.

In physics, a model consisting of a point particle moving between infinitely many disjoint convex bodies in space is known as *Lorentz gas*. It was introduced by H. Lorentz in 1905 [**Lo05**] to describe the motion of electrons in metals (the bodies represented heavy molecules). Here we have a 2D version of this model, in which the bodies (scatterers) are positioned periodically. This version is called a *planar periodic Lorentz gas*.

EXERCISE 4.6. Show that for the billiards of categories A and B the function $\tau(x)$ defined in Section 2.9 has a positive lower bound

$$\tau(x) \geq \tau_{\min} > 0 \qquad \forall x \in \mathcal{M} \tag{4.1}$$

($\tau_{\min}$ is the minimal distance between the scatterers). On the other hand, for the categories A, C, and E this function has an upper bound

$$\tau(x) \leq \tau_{\max} < \infty \qquad \forall x \in \mathcal{M}. \tag{4.2}$$

Even though *all* types of dispersing billiards are technically quite involved, some of them are more complex than others. In our opinion, their level of difficulty is roughly represented by the following diagram:

"easy"	A
medium	B, C
hard	D, E, F

Historically, all the major properties of dispersing billiards (hyperbolicity, ergodicity, Bernoulli mixing, bounds of correlations, central limit theorem, etc.) have first been proved for category A. Most of them have been later extended to categories B and C as well, but that always took extra effort and required special approaches. Categories D, E, and F (as the most complicated and, arguably, least interesting) remain virtually unexplored.

In our book we mostly discuss the main category, A, but some remarks and exercises are devoted to other categories. To avoid confusion, we adopt a simple convention: all our major results will be related to category A unless stated otherwise.

4.2. Another mechanical model

Here we describe a mechanical system of two hard disks that reduces to a dispersing billiard. This mechanical model is actually a prototype of all dispersing billiard tables and is the model which motivated Sinai to introduce dispersing billiards after he had studied the motion of hard disks and hard balls in the 1960s.

Consider two identical disks of unit mass $m = 1$ and a small radius r moving on the unit torus Tor^2 without obstacles or walls of any kind. The disks move freely with constant velocities and collide with each other elastically, as described below.

Let the disks collide and denote their velocity vectors before the collision by u_- and v_-. Let L denote the common tangent line to the disks at the moment of collision. For every vector v denote by $v^\perp$ and $v^\parallel$ its components perpendicular and parallel to the line L, respectively. We call $v^\parallel$ the *tangential* component of v and $v^\perp$ the *normal* component of v.

According to the law of elastic collision, the postcollisional velocity vectors of the disks are

$$u_+ = u_-^\parallel + v_-^\perp \qquad \text{and} \qquad v_+ = v_-^\parallel + u_-^\perp. \tag{4.3}$$

In other words, *the velocity vectors keep their tangential components but exchange their normal components*; see Fig. 4.3. Observe that this rule implies the preservation of the total kinetic energy

$$\|u_+\|^2 + \|v_+\|^2 = \|u_-\|^2 + \|v_-\|^2$$

and the total momentum

$$u_+ + v_+ = u_- + v_-$$

(recall that the disks have unit mass).

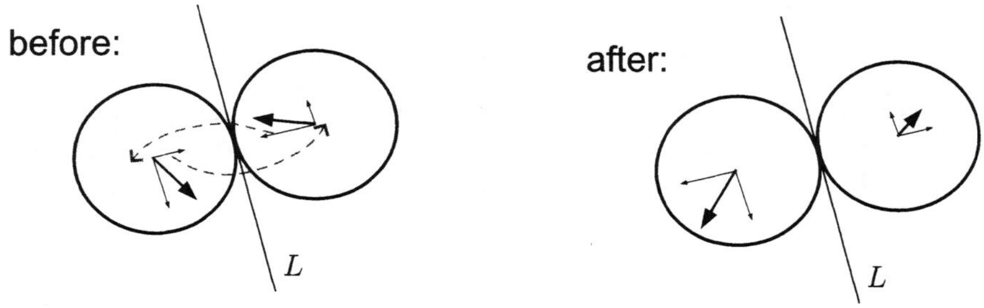

FIGURE 4.3. Elastic collision of two disks.

The conservation of the total momentum implies that the center of mass (x_c, y_c) moves on the torus with a constant velocity vector. This is a simple periodic or quasiperiodic motion (see Appendix C), and we eliminate it from the picture by assuming that the total momentum is zero; i.e. $u + v = 0$ at all times.

Now the center of mass is at rest. More precisely, let $0 \le x, y \le 1$ denote rectangular coordinates on Tor^2 and (x_1, y_1) and (x_2, y_2) the centers of our disks. Choose the coordinate system so that, initially, the center of mass

$$x_c = \tfrac{1}{2}(x_1 + x_2) \qquad \text{and} \qquad y_c = \tfrac{1}{2}(y_1 + y_2)$$

coincides with the center of the torus $(\frac{1}{2}, \frac{1}{2})$; equivalently,

$$x_1 + x_2 = 1 \qquad \text{and} \qquad y_1 + y_2 = 1. \tag{4.4}$$

EXERCISE 4.7. Verify that the equations (4.4) will hold at all times; hence the center of mass remains constant: $x_c = \frac{1}{2}$ and $y_c = \frac{1}{2}$. Hint: Remember that x and y are cyclic coordinates, so they can occasionally increase or decrease by one. Argue that when x_1 increases by one, for example, then x_2 must decrease by one, etc.

Next, the disks collide with each other whenever the distance between their centers is $2r$, i.e. whenever

$$(x_1 - x_2 + m)^2 + (y_1 - y_2 + n)^2 = (2r)^2$$

for some $m = 0, \pm 1$ and $n = 0, \pm 1$. Due to (4.4), this is equivalent to

$$\left(x_1 - \tfrac{1-m}{2}\right)^2 + \left(y_1 - \tfrac{1-n}{2}\right)^2 = r^2. \tag{4.5}$$

EXERCISE 4.8. Show that equation (4.5) specifies four circles of radius r with centers at $(0,0)$, $(0, \frac{1}{2})$, $(\frac{1}{2}, 0)$, and $(\frac{1}{2}, \frac{1}{2})$.

Let $\mathbb{O}_i$, $1 \le i \le 4$, denote the circles (4.5). We now examine the transformation of the velocity vectors u and v at a collision. Since $u + v = 0$, we obviously have $u_-^{\|} = -v_-^{\|}$ and $u_-^{\perp} = -v_-^{\perp}$ in the notation of (4.3); therefore

$$u_+ = u_-^{\|} - u_-^{\perp} \qquad \text{and} \qquad v_+ = v_-^{\|} - v_-^{\perp}.$$

In other words, each velocity vector simply gets reflected off the common tangent line L to the colliding disks. Observe that L is parallel to the tangent line to the circle $\mathbb{O}_i$ on which the point (x_1, y_1) lies at the moment of collision. Also note that $\|u\| = \|v\| = \sqrt{2}/2$ remain constant in time.

Therefore, the center (x_1, y_1) of the first disk moves on Tor^2 with a constant velocity vector until it hits one of the circles $\mathbb{O}_i$. Then its velocity gets reflected across the tangent line to that circle at the point of hit, and so on. This is exactly a billiard trajectory[1] on the table $\mathrm{Tor}^2 \setminus \cup \mathbb{O}_i$.

EXERCISE 4.9. Use the symmetry to reduce this model to a billiard on a 'quarter-torus' $[0, 1/2] \times [0, 1/2]$ with a single circular scatterer of radius r.

EXERCISE 4.10. Determine how the type of the resulting dispersing billiard changes with r and what it means in mechanical terms. Hint: Use Exercise 4.2. Answer: It is of type B for $r < 1/4$ (the disks are small enough to move around freely), type E for $r = 1/4$ and type C for $r > 1/4$ (the disks are too big and block each other's way on the torus).

4.3. Dispersing wave fronts

In dispersing billiards, the curvature of the boundary $\partial \mathcal{D}$ is positive $\mathcal{K} > 0$, and due to our Assumption A3 we actually have upper and lower bounds on it:

$$0 < \mathcal{K}_{\min} \le \mathcal{K} \le \mathcal{K}_{\max} < \infty.$$

Recall that tangent lines $L \subset \mathcal{T}_X^{\perp}\Omega$ at phase points $X \in \Omega$ can be characterized by slope $\mathcal{B} = d\omega/d\xi$ and represented by wave fronts (whose curvature equals $\mathcal{B}$); cf. Sect. 3.7. Dispersing wave fronts have positive curvature $\mathcal{B} > 0$.

LEMMA 4.11. *If a tangent line $L \subset \mathcal{T}_X^{\perp}\Omega$ has positive slope $\mathcal{B} > 0$, then its image $L_t = D_X\Phi^t(L)$ will have positive slope $\mathcal{B}_t > 0$ for all $t > 0$. In other words, dispersing wave fronts remain dispersing in the future.*

PROOF. A simple inductive argument based on (3.31) and (3.33) suffices, because $\mathcal{R} = 2\mathcal{K}/\cos\varphi > 0$ at every collision. □

Dispersing wave fronts play an important role in the study of dispersing billiards. In the first few sections of this chapter, we develop necessary technical tools for the analysis of wave fronts. In many respects, dispersing wave fronts resemble unstable horocycles in geodesic flows on manifolds of negative curvature.

The collision parameter $\mathcal{R} = 2\mathcal{K}/\cos\varphi$ is bounded away from zero:

$$\mathcal{R} \ge \mathcal{R}_{\min} := 2\mathcal{K}_{\min} > 0. \tag{4.6}$$

Furthermore, for nearly grazing collisions, where $\cos\varphi \approx 0$, there is a better estimate,

$$C_1/\cos\varphi \le \mathcal{R} \le C_2/\cos\varphi, \tag{4.7}$$

where $0 < C_1 < C_2$ are constants. For example, if a nearly grazing collision occurs at time t, then $\mathcal{B}_{t+0} > \mathcal{R} \ge C_1/\cos\varphi$ may be arbitrarily large.

[1]To be precise, the point (x_1, y_1) moves at a constant, but not unit, speed. One can rescale the time (Exercise 1.10) to get a conventional unit speed.

REMARK 4.12. We see that certain quantities, such as $\mathcal{R}$ and $\mathcal{B}_{t+0}$ in the above example, grow to infinity (blow-up) as $\cos\varphi \to 0$. This is typical for dispersing billiards. Nearly grazing collisions (where $\cos\varphi \approx 0$) often cause trouble and require special treatment.

Convention. We will write inequalities like (4.7) in a shorter way, as $\mathcal{R} \asymp 1/\cos\varphi$. More generally,

$$F \asymp G \qquad \Longleftrightarrow \qquad C_1 < F/G < C_2$$

for some constants $0 < C_1 < C_2$ depending only on the table $\mathcal{D}$.

Next, note that a flat front ($\mathcal{B} = 0$) will remain flat until the first collision with $\partial\mathcal{D}$, upon which it will become dispersing and stay dispersing forever; i.e. $\mathcal{B}_t > 0$ for all $t > t^+(X)$; see (3.15).

We will consider long-term evolution of dispersing wave fronts, so we always assume that our fronts have been dispersing for some time in the past.

The graph of $\mathcal{B}_t$ is a saw-toothed curve: at every collision it jumps by the corresponding $\mathcal{R}$. Then between collisions it goes down along a hyperbola (whose horizontal asymptote is $\mathcal{B} = 0$), etc.; see Fig. 4.4.

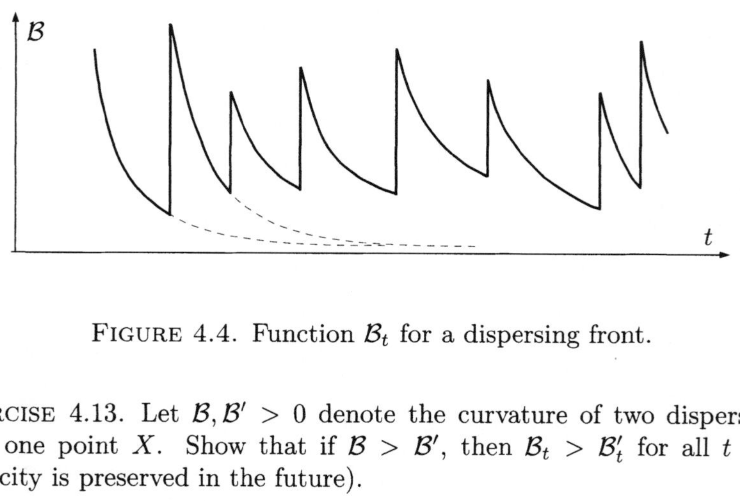

FIGURE 4.4. Function $\mathcal{B}_t$ for a dispersing front.

EXERCISE 4.13. Let $\mathcal{B}, \mathcal{B}' > 0$ denote the curvature of two dispersing wave fronts at one point X. Show that if $\mathcal{B} > \mathcal{B}'$, then $\mathcal{B}_t > \mathcal{B}'_t$ for all $t > 0$ (the monotonicity is preserved in the future).

For dispersing wave fronts in billiards with finite horizon (categories A, C, and E) the function $\mathcal{B}_t$ stays bounded away from zero:

$$\mathcal{B}_t \geq \mathcal{B}_{\min} := \frac{1}{\tau_{\max} + 1/\mathcal{R}_{\min}} > 0. \tag{4.8}$$

LEMMA 4.14. *If a tangent vector $(d\xi, d\omega) \in \mathcal{T}_X^\perp\Omega$ has positive slope $d\omega/d\xi > 0$, then its image $(d\xi_t, d\omega_t) = D_X\Phi^t(d\xi, d\omega)$ will monotonically grow; i.e. for any $0 < t' < t''$ we have*

$$|d\xi_{t'}| < |d\xi_{t''}| \qquad \text{and} \qquad |d\omega_{t'}| \leq |d\omega_{t''}|$$

(in other words, dispersing wave fronts monotonically expand in the future).

PROOF. Due to the previous lemma, $\mathcal{B}_t = d\omega_t/d\xi_t > 0$ for all $t > 0$, and by (3.37) we have $\frac{d}{dt}\log|d\xi_t| = \mathcal{B}_t > 0$. The (nonstrict) monotonicity of $|d\omega_t|$ now follows from (3.25) and (3.27). $\square$

REMARK 4.15. For categories A, C, and E we have $\frac{d}{dt}\log|d\xi_t| \geq \mathcal{B}_{\min}$; hence

$$\frac{|d\xi_t|}{|d\xi|} \geq e^{t\mathcal{B}_{\min}}. \tag{4.9}$$

Thus dispersing wave fronts grow exponentially at all times.

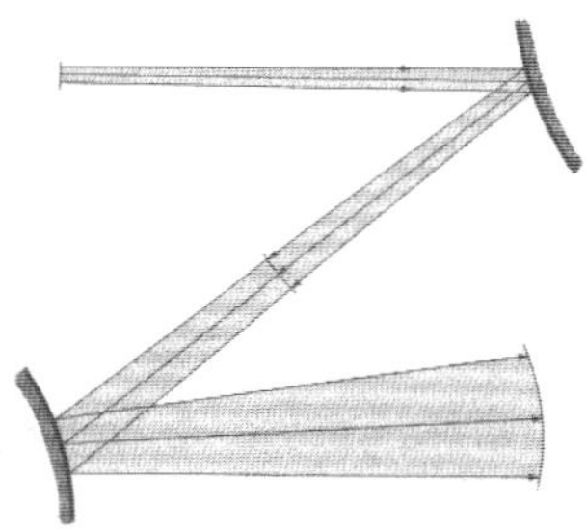

FIGURE 4.5. Expansion of a dispersing wave front.

It is most essential to estimate $\mathcal{B}_t$ for wave fronts at the times of collisions. We will use the following notation: given a point $x = (r, \varphi) \in \mathcal{M}$, we denote by $x_n = (r_n, \varphi_n) = \mathcal{F}^n(x)$ its images and by $\mathcal{K}_n$, $\mathcal{R}_n$, etc., the corresponding parameters at x_n. Given a wave front moving along the trajectory of x, its image at the nth collision induces a tangent line $L_n \subset \mathcal{T}_{x_n}\mathcal{M}$, and we denote by $\mathcal{V}_n$, $\mathcal{B}_n^{\pm}$, etc., the corresponding quantities defined in the previous chapter. We denote by t_n the time of the nth collision and by $\tau_n = \tau(x_n) = t_{n+1} - t_n$ the intervals between collisions.

The following estimates are immediate: for any dispersing wave front we have

$$\mathcal{B}_n^+ \geq \mathcal{R}_{\min}/\cos\varphi_n \tag{4.10}$$

and, moreover, for the billiards of categories A and B,

$$\mathcal{B}_n^- \leq \mathcal{B}_{\max}^- : = 1/\tau_{\min} \tag{4.11}$$

and

$$\mathcal{B}_n^+ \asymp 1/\cos\varphi_n. \tag{4.12}$$

REMARK 4.16. The involution introduced in Section 2.14 establishes a simple and explicit conjugacy between Φ^t and Φ^{-t}, as well as between $\mathcal{F}$ and $\mathcal{F}^{-1}$. This allows us to 'reverse the time' and relate the properties of Φ^{-t} and $\mathcal{F}^{-1}$ to those of Φ^t and $\mathcal{F}$, respectively. For example, Lemma 4.11 has its 'time reversal' counterpart: $\mathcal{B} < 0$ implies $\mathcal{B}_t < 0$ for all $t < 0$; i.e. focusing wave fronts will remain focusing in the past (note that the involution transforms dispersing wave fronts into focusing ones and vice versa). Nearly every statement about $\mathcal{F}$ or Φ^t has its 'time reversal', which we will not always state explicitly.

4.4. Hyperbolicity

Lemma 4.11 allows us to construct unstable cones $\mathcal{C}_x^u$ for the map $\mathcal{F}$: all wave fronts that are dispersing or flat upon arriving at the point x (i.e. such that $\mathcal{B}^- \geq 0$) make a cone

$$\mathcal{C}_x^u = \{(dr, d\varphi) \in \mathcal{T}_x\mathcal{M}\colon\ \mathcal{K} \leq d\varphi/dr \leq \infty\},$$

in accordance with (3.39). Lemma 4.14 readily implies the invariance of $\mathcal{C}_x^u$ in the sense of (3.53), but we need more.

EXERCISE 4.17. For every point $x_0 \in \mathcal{M}$, use (3.31) and (3.33) to compute the upper edge $\mathcal{V}_{\rm up}$ and the lower edge $\mathcal{V}_{\rm lo}$ of the cone $D_{x_0}\mathcal{F}(\mathcal{C}_{x_0}^u) \subset \mathcal{T}_{x_1}\mathcal{M}$ at the point $x_1 = \mathcal{F}(x_0)$. Answer:

$$\mathcal{V}_{\rm up} = \mathcal{K}_1 + \frac{\cos\varphi_1}{\tau_0}, \qquad \mathcal{V}_{\rm lo} = \mathcal{K}_1 + \frac{\cos\varphi_1}{\tau_0 + \cos\varphi_0/2\mathcal{K}_0};$$

see the notation in the previous section.

LEMMA 4.18. *The cones $\mathcal{C}_x^u$ are strictly invariant:*

$$D_{x_0}\mathcal{F}(\mathcal{C}_{x_0}^u) \subset \operatorname{int}\mathcal{C}_{x_1}^u \cap \{0\}.$$

PROOF. By the previous exercise, $\mathcal{K}_1 < \mathcal{V}_{\rm lo} < \mathcal{V}_{\rm up} < \infty$. □

EXERCISE 4.19. Show that the more narrow unstable cones defined by

$$\hat{\mathcal{C}}_x^u = \{(dr, d\varphi) \in \mathcal{T}_x\mathcal{M}\colon\ \mathcal{K} \le d\varphi/dr \le \mathcal{K} + \cos\varphi/\tau_{-1}\} \tag{4.13}$$

are also strictly invariant. Also check that $D_{x_0}\mathcal{F}(\mathcal{C}_{x_0}^u) \subset \hat{\mathcal{C}}_{x_1}^u$.

For billiards of categories A and B, the narrow cones $\hat{\mathcal{C}}_x^u$ give uniform upper and lower bounds on the slopes of unstable vectors:

$$0 < \mathcal{V}_{\min} < d\varphi/dr < \mathcal{V}_{\max} < \infty \tag{4.14}$$

with $\mathcal{V}_{\min} = \mathcal{K}_{\min}$ and $\mathcal{V}_{\max} = \mathcal{K}_{\max} + 1/\tau_{\min}$.

Wojtkowski's results (see Remark 3.53) now imply

COROLLARY 4.20. *The map $\mathcal{F}$ is hyperbolic: almost every point $x \in \mathcal{M}$ has one positive and one negative Lyapunov exponent. Therefore, the flow Φ^t is hyperbolic, too; see Section* 3.3.

Next we establish uniform hyperbolicity for main categories of dispersing billiards. Let $dx = (dr, d\varphi) \in \mathcal{C}_x^u$ be an unstable tangent vector and $dx_n = (dr_n, d\varphi_n) = D_x\mathcal{F}^n(dx)$ its image. The expansion factor of dx, in the p-metric, is given by the formula (3.40). Observe that

$$\frac{\|dx_{n+1}\|_p}{\|dx_n\|_p} = 1 + \tau_n\mathcal{B}_n^+ \ge 1 + \tau_n\mathcal{R}_{\min}. \tag{4.15}$$

Furthermore, at nearly grazing collisions a better estimate holds:

$$\frac{\|dx_{n+1}\|_p}{\|dx_n\|_p} \asymp \frac{\tau_n}{\cos\varphi_n}. \tag{4.16}$$

Now for billiards of categories A and B, we have

$$\Lambda = 1 + \tau_{\min}\mathcal{R}_{\min} > 1; \tag{4.17}$$

hence the hyperbolicity is uniform:

$$\frac{\|dx_n\|_p}{\|dx_0\|_p} \ge \Lambda^n \qquad \forall n \ge 1.$$

EXERCISE 4.21. Show that the hyperbolicity is also uniform for billiards of categories C and D. Hint: Use the result of Exercise 2.18.

REMARK 4.22. For billiards of categories E and F, the hyperbolicity is *not* uniform. Cusps act as 'traps' where a billiard trajectory may experience arbitrary many collisions during an arbitrarily short time interval; see Fig. 4.1 (left). Our recent studies show **[CM06]** that if a trajectory has n collisions inside a cusp, the factor of expansion of unstable fronts is only $\sim n^a$, where $a > 0$ is a constant.

We have used the p-metric to prove uniform hyperbolicity for billiards of type A and B. We now make respective calculations in the Euclidean metric $\|dx\| = \sqrt{(dr)^2 + (d\varphi)^2}$. Due to (3.41)

$$\frac{\|dx_n\|}{\|dx_0\|} = \frac{\|dx_n\|_p}{\|dx_0\|_p} \frac{\cos\varphi_0}{\cos\varphi_n} \frac{\sqrt{1+\mathcal{V}_n^2}}{\sqrt{1+\mathcal{V}_0^2}}. \tag{4.18}$$

The last fraction is uniformly bounded away from zero and infinity due to (4.14). To deal with the middle fraction, we recall that

$$\frac{\|dx_1\|_p}{\|dx_0\|_p} \geq \frac{\text{const}}{\cos\varphi_0}$$

by (4.16). Therefore,

$$\frac{\|dx_n\|}{\|dx_0\|} \geq \text{const} \times \frac{\|dx_n\|_p}{\|dx_1\|_p} \geq \hat{c}\,\Lambda^n, \tag{4.19}$$

where $\hat{c} = \hat{c}(\mathcal{D}) > 0$ is a constant.

Thus, the hyperbolicity is uniform in both metrics. Note, however, that the expansion of unstable vectors is always monotone in the p-metric, but it may not be so in the Euclidean metric:

EXERCISE 4.23. Show that there exists a dispersing billiard table $\mathcal{D}$ and an unstable vector $dx \in \hat{C}_x^u$ such that $\|\mathcal{F}(dx)\| < \|dx\|$. (Hint: You need that $\cos\varphi_0 \ll 1$, $\cos\varphi_1 \approx 1$, $\mathcal{K}_0 \approx 0$, $\mathcal{K}_1 \approx 0$, and $\mathcal{B}^- \approx 0$.)

Lastly, (4.16) and (3.41) imply a useful relation:

$$\frac{\|dx_{n+1}\|}{\|dx_n\|} \asymp \frac{\tau_n}{\cos\varphi_{n+1}}. \tag{4.20}$$

4.5. Stable and unstable curves

Here we introduce stable and unstable curves that play an instrumental role in the studies of ergodic and statistical properties of dispersing billiards.

When we use wave fronts to represent tangent lines $L \subset \mathcal{T}_X^{\perp}\Omega$, they need only be infinitesimally small. In this section we consider dispersing wave fronts of positive size. Then $\mathcal{B}(X)$ becomes a function on the front.

If $\gamma \subset \Omega$ is a dispersing wave front, then its image $\gamma_t = \Phi^t(\gamma)$ must grow (Lemma 4.14). But the space Ω has finite size; hence γ_t has to break down and fold up as t increases. The way this happens is illustrated in Fig. 4.6.

EXERCISE 4.24. Let γ be a dispersing wave front. Show that in billiards of categories A, B, C, and D, its image $\gamma_t = \Phi^t(\gamma)$ is a finite union of dispersing wave fronts for every $t > 0$. In billiards of categories A and B (where the flow Φ^t is continuous, according to Exercise 2.29), γ_t is a continuous curve for every $t > 0$. It looks like a zig-zag line, a fragment of which is shown in Fig. 4.6.

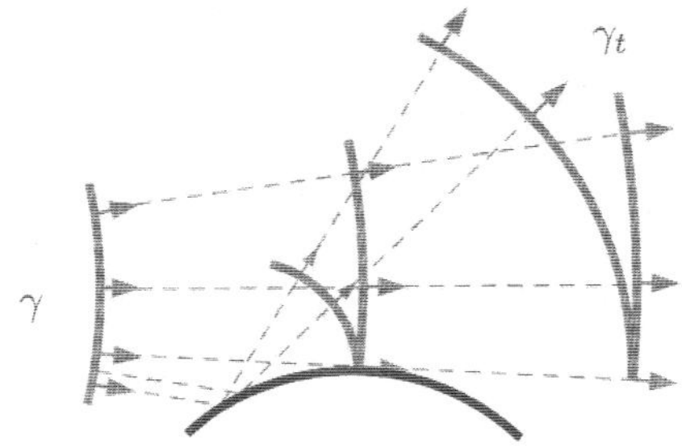

FIGURE 4.6. A dispersing front folding up.

If a dispersing wave front γ is C^m smooth, $m \geq 1$, then γ_t is a union of $C^{m'}$ smooth wave fronts, where $m' = \min\{m, \ell - 1\}$, because the flow Φ^t is $C^{\ell-1}$ smooth (Section 2.7). For this reason, we will consider only $C^{\ell-1}$ smooth wave fronts.

Every dispersing wave front γ can be projected onto the collision space $\mathcal{M}$ by the map $\mathbf{P}^+$ defined by (3.17). We call $\mathbf{P}^+(\gamma)$ the *trace* the wave front γ leaves on $\mathcal{M}$ during its evolution.

EXERCISE 4.25. Let γ be a dispersing wave front. Show that in billiards of categories A and C its trace $\mathbf{P}^+(\gamma)$ is a *finite* union of smooth curves. On the other hand, find a dispersing wave front γ in the billiard of category B shown in Fig. 2.4 whose trace is an *infinite* union of smooth curves.

Suppose a curve $W \subset \mathcal{M}$ is defined by a function $\varphi = \varphi_W(r)$ for some $r'_W \leq r \leq r''_W$. We say that W is C^m smooth ($m \geq 1$) if the function $\varphi_W(r)$ is C^m smooth, up to the endpoints r'_W and r''_W.

We will say that a C^m smooth curve $W \subset \mathcal{M}$ is *unstable* (or *increasing*) if at every point $x \in W$ the tangent line $\mathcal{T}_xW$ belongs in the unstable cone $\hat{\mathcal{C}}^u_x$; see (4.13).

Similarly, a C^m smooth curve $W \subset \mathcal{M}$ is said to be *stable* (or *decreasing*) if at every point $x \in W$ the tangent line $\mathcal{T}_xW$ belongs in the stable cone

$$\hat{\mathcal{C}}^s_x = \{(dr, d\varphi) \in \mathcal{T}_x\mathcal{M}\colon\ -\mathcal{K} - \cos\varphi/\tau \leq d\varphi/dr \leq -\mathcal{K}\}, \tag{4.21}$$

which is the 'time-reversal' counterpart of $\hat{\mathcal{C}}^u_x$.

Note that stable and unstable curves are monotonic in the coordinates r and φ. Their slope is bounded away from zero and for billiards of categories A and B, from infinity.

For any smooth curve $W \subset \mathcal{M}$ we denote by $|W|$ its length in the Euclidean metric and by $|W|_p$ its length in the p-metric

$$|W|_p = \int_{r'_W}^{r''_W} \cos\varphi_W(r)\, dr.$$

EXERCISE 4.26. Show that in billiards of categories A and B, for any stable or unstable curve W we have

$$|W|_p < |W| < C\,|W|_p^{1/2} \tag{4.22}$$

where $C = C(\mathcal{D})$ is a constant. Thus, these metrics are equivalent on stable and unstable curves. (The bounds (4.22) also hold for billiards of categories C and D, but their proofs are quite involved; see [**BSC91**].)

The following simple facts deserve a note:

REMARK 4.27. Stable and unstable curves cannot self-intersect. Their lengths (in both metrics, $|\cdot|$ and $|\cdot|_p$) are uniformly bounded. If W^u is an unstable curve and W^s a stable curve, then their intersection $W^u \cap W^s$ consists of at most one point.

Given a smooth unstable curve $W \subset \mathcal{M}$, the trajectories of its points $x \in W$ make a dispersing wave front whose trace on $\mathcal{M}$ will be the image $\mathcal{F}(W)$. Thus, $\mathcal{F}(W)$ is a union of unstable curves.

EXERCISE 4.28. Show that $\mathcal{F}(W)$ consists of a finite number of unstable curves for billiards of categories A and C, but that number may be countable for categories B and D. By induction, show that the same is true for further images $\mathcal{F}^n(W)$, $n \geq 1$.

If W is C^m smooth, then $\mathcal{F}(W)$ consists of $C^{m'}$ smooth curves, where $m' = \min\{m, \ell-1\}$, because the map $\mathcal{F}$ is $C^{\ell-1}$ smooth. For this reason, we will consider only $C^{\ell-1}$ smooth stable and unstable curves.

We have not imposed (yet) any restrictions on the derivatives

$$\varphi_W^{(\nu)}(r) = d^\nu \varphi_W(r)/dr^\nu, \qquad 1 \leq \nu \leq \ell - 1 \tag{4.23}$$

for stable and unstable curves. However, it is intuitively clear that if W is unstable, then its images $\mathcal{F}^n(W)$, $n \geq 1$, stretch out exponentially fast (in n), and so all irregularities along these curves should smooth out. Eventually, all the derivatives (4.23) corresponding to the curves $W' \subset \mathcal{F}^n(W)$ should be bounded by an absolute constant, which leads us to

PROPOSITION 4.29. *Let $\mathcal{D}$ be a billiard of category A or B. Then for each $\nu = 1, \ldots, \ell - 1$ there exists a constant $C_\nu = C_\nu(\mathcal{D}) > 0$ such that for every $C^{\ell-1}$ smooth unstable curve $W \subset \mathcal{M}$ there is an $n_W \geq 1$ such that for all $n > n_W$ every smooth curve $W' \subset \mathcal{F}^n(W)$ has its νth derivative bounded by C_ν:*

$$|\varphi_{W'}^{(\nu)}(r)| \leq C_\nu. \tag{4.24}$$

The proof of (4.24) for $\nu > 2$ has never been published before (the case $\nu = 2$ is treated in [**Ch01**], and the case $\nu = 1$ is trivial). Since this is a new result, we provide a complete proof in the next section. However, that proof is rather technical, and the reader may skip it without any harm to the further reading of the book.

4.6. Proof of Proposition 4.29

The first derivative $d\varphi/dr$ is bounded by (4.14). Differentiating (3.39) gives

$$d^2\varphi/dr^2 = d\mathcal{K}/dr - \mathcal{B}^- \sin\varphi \; d\varphi/dr + \cos\varphi \; d\mathcal{B}^-/dr;$$

hence the second derivative would be bounded if we had a uniform bound on $d\mathcal{B}^-/dr$. Further differentiation (we leave it as an exercise) allows us to reduce Proposition 4.29 to the following:

LEMMA 4.30. *For each $\nu = 1, \dots, \ell-2$ there exists a constant $C'_\nu = C'_\nu(\mathcal{D}) > 0$ such that for every $C^{\ell-1}$ smooth unstable curve $W \subset \mathcal{M}$ there is $n_W \geq 1$ such that for all $n > n_W$ every smooth curve $W' \subset \mathcal{F}^n(W)$ satisfies*

$$|d^\nu \mathcal{B}^-_{W'}/dr^\nu| \leq C'_\nu$$

where $\mathcal{B}^-_{W'}(r) = \bigl(d\varphi_{W'}(r)/dr - \mathcal{K}(r)\bigr)/\cos\varphi_{W'}(r)$, according to (3.39).

Now the value $\mathcal{B}^-_{W'}$ is the curvature of the dispersing wave front corresponding to the unstable curve W' immediately before the collision. Since $\mathcal{B} = d\omega/d\xi$ by (3.30), it will be more convenient to differentiate $\mathcal{B}$ with respect to ξ rather than r. This change of variables requires certain analysis, which we do next.

Let γ be a dispersing wave front, in the notation of Section 3.7, and s a smooth parameter on the curve $\pi_q(\gamma) \subset \mathcal{D}$. For every point $X_s = (x_s, y_s, \omega_s) \in \gamma$ denote by $X_{st} = (x_{st}, y_{st}, \omega_{st}) = \Phi^t(X_s)$ its image at time $t > 0$ (here we use the coordinates (x, y, ω) introduced in Section 2.6). Now the points $\{X_{st}\}$ fill a 2D surface $\Sigma \subset \Omega$ parameterized by s and t; see Fig. 4.7.

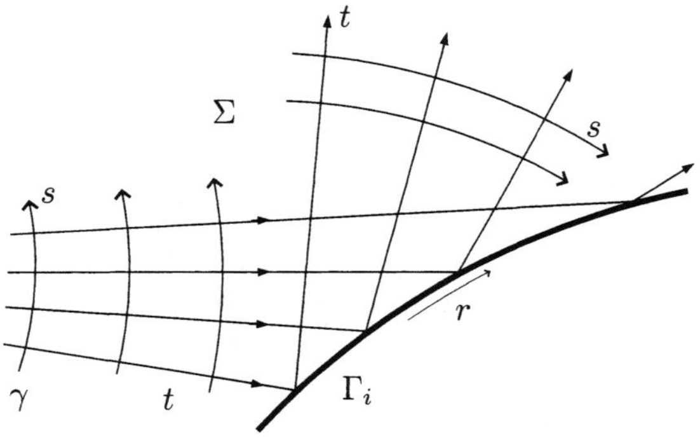

FIGURE 4.7. Parametrization of Σ by s and t.

In our calculations below, we denote differentiation with respect to s by primes and that with respect to t by dots. For example,

$$(\dot{x}, \dot{y}, \dot{\omega}) = (\cos\omega, \sin\omega, 0).$$

The vector (x', y') remains orthogonal to the velocity vector $(\cos\omega, \sin\omega)$ at all times (see Section 3.7); hence $x'\cos\omega + y'\sin\omega = 0$. Put

$$u = -x'\sin\omega + y'\cos\omega.$$

Note that $u^2 = (x')^2 + (y')^2$; hence $|u|$ equals the distance on the table $\mathcal{D}$ corresponding to the unit increment of the parameter s.

EXERCISE 4.31. Check that for any smooth function $F\colon \Sigma \to \mathbb{R}$

$$\frac{dF}{d\xi} = \frac{1}{u}\frac{dF}{ds}. \tag{4.25}$$

In particular,

$$\mathcal{B} = \frac{d\omega}{d\xi} = \frac{\omega'}{u}. \tag{4.26}$$

Check also that

$$\dot{u} = u\mathcal{B}. \tag{4.27}$$

Combining (4.25) with (4.27) gives

$$\frac{d}{dt}\frac{d}{d\xi}F = -\mathcal{B}\frac{d}{d\xi}F + \frac{d}{d\xi}\frac{d}{dt}F. \tag{4.28}$$

In particular, letting $F = \omega$ and using (4.26) give

$$\frac{d\mathcal{B}}{dt} = -\mathcal{B}^2; \qquad \text{hence} \qquad \frac{d}{dt}\left(\frac{1}{\mathcal{B}}\right) = 1, \tag{4.29}$$

which agrees with (3.31). Denote

$$\mathcal{E}_\nu = \frac{d^\nu \mathcal{B}}{d\xi^\nu}, \qquad \nu = 1, \dots, \ell - 2.$$

Then (4.28) implies

$$\dot{\mathcal{E}}_\nu = -\mathcal{E}_\nu\mathcal{B} + \frac{d}{d\xi}\dot{\mathcal{E}}_{\nu-1}, \qquad \nu = 1, \dots, \ell - 2$$

(assuming $\mathcal{E}_0 = \mathcal{B}$, of course). In particular, one easily checks that

$$\dot{\mathcal{E}}_1 = -3\mathcal{E}_1\mathcal{B}, \qquad \dot{\mathcal{E}}_2 = -4\mathcal{E}_2\mathcal{B} - 3\mathcal{E}_1^2, \qquad \text{etc.}$$

Observe that $\dot{\mathcal{E}}_\nu$ is a homogeneous quadratic polynomial of $\mathcal{B}, \mathcal{E}_1, \dots, \mathcal{E}_\nu$ for every $\nu \geq 1$.

Our previous analysis dealt with the free motion between collisions. Next we consider what happens at a collision with a wall Γ_i; see Fig. 4.7. For every collision point $\big(\bar{x}(r), \bar{y}(r)\big) \in \Gamma_i$ (again we use the notation of Section 2.6) there is a unique pair (s, t) such that

$$x_{st} = \bar{x}(r) \qquad \text{and} \qquad y_{st} = \bar{y}(r); \tag{4.30}$$

hence s and t, constrained to the collision, become functions of r. Differentiating (4.30) and using some elementary trigonometry (we leave this as an exercise) yield

$$\frac{dt}{dr} = \sin\varphi \qquad \text{and} \qquad \frac{ds}{dr} = \frac{\cos\varphi}{u^-} = -\frac{\cos\varphi}{u^+} \tag{4.31}$$

(note that $u^+ = -u^-$ because the (s, t) coordinate system changes orientation after the collision; see Fig. 4.7).

Now for any smooth function $F\colon \Sigma \to \mathbb{R}$ we denote by F^- and F^+ its restrictions to the parts of Σ before and after the collision, respectively. Then, due to (4.31),

$$\begin{aligned} \frac{dF^-(s(r), t(r))}{dr} &= \frac{(F')^- \cos\varphi}{u^-} + (\dot{F})^- \sin\varphi \\ &= \left(\frac{dF}{d\xi}\right)^- \cos\varphi + (\dot{F})^- \sin\varphi \end{aligned} \tag{4.32}$$

and, similarly,

$$\frac{dF^+(s(r), t(r))}{dr} = -\left(\frac{dF}{d\xi}\right)^+ \cos\varphi + (\dot{F})^+ \sin\varphi. \tag{4.33}$$

We now return to the proof of Lemma 4.30. Using (4.32) and (4.29) yields

$$d\mathcal{B}^-/dr = \mathcal{E}_1^- \cos\varphi - (\mathcal{B}^-)^2 \sin\varphi, \tag{4.34}$$

so $d\mathcal{B}^-/dr$ would be uniformly bounded if we had a uniform bound on $\mathcal{E}_1^-$. Differentiating further gives

$$d^2\mathcal{B}^-/dr^2 = \mathcal{E}_2^- \cos^2\varphi - 5\,\mathcal{E}_1^-\mathcal{B}^- \cos\varphi \sin\varphi - \mathcal{E}_1^-\mathcal{V}\sin\varphi \\ - (\mathcal{B}^-)^2\mathcal{V}\cos\varphi + 2(\mathcal{B}^-)^3\sin^2\varphi,$$

where $\mathcal{V} = d\varphi/dr = \mathcal{B}^- \cos\varphi + \mathcal{K}$; see (3.39). Hence $d^2\mathcal{B}^-/dr^2$ would be uniformly bounded if we had a uniform bound on $\mathcal{E}_2^-$. Subsequent differentiation (we leave it as an exercise) reduces Lemma 4.30 to the following:

LEMMA 4.32. *For each $\nu = 1, \dots, \ell-2$ there exists a constant $C''_\nu = C''_\nu(\mathcal{D}) > 0$ such that for every $C^{\ell-1}$ smooth dispersing wave front $\gamma \subset \Omega$ there is $n_\gamma \geq 1$ such that for all $n \geq n_\gamma$ the νth derivative $\mathcal{E}_\nu = d^\nu\mathcal{B}/d\xi^\nu$ of the curvature $\mathcal{B}$ of its image $\Phi^t(\gamma)$ before the nth collision satisfies*

$$|\mathcal{E}_\nu^-| \leq C''_\nu.$$

The reader should notice that dispersing wave fronts play the same role for the flow Φ^t as unstable curves play for the map $\mathcal{F}$. In this sense, Lemma 4.32 is a "flow" version of the "map" statement 4.29.

PROOF. First, we let $\nu = 1$, and then use induction on ν. Suppose that F is a smooth function on the surface Σ introduced above such that $dF/dt = 0$; i.e. the function F is independent of t between collisions. Then (4.29) and (4.28) imply

$$\frac{d}{dt}\left(\frac{1}{\mathcal{B}}\frac{d}{d\xi}F\right) = \frac{\mathcal{B}^2}{\mathcal{B}^2}\frac{d}{d\xi}F - \frac{\mathcal{B}}{\mathcal{B}}\frac{d}{d\xi}F = 0; \tag{4.35}$$

hence the function $G = \mathcal{B}^{-1}\,dF/d\xi$ is also independent of t between collisions.

Due to (4.29), the function $F_0 = t - 1/\mathcal{B}$ satisfies $\dot{F}_0 = 0$. Therefore, the function

$$F_1 = \frac{1}{\mathcal{B}}\frac{dF_0}{d\xi} = \frac{\mathcal{E}_1}{\mathcal{B}^3} \tag{4.36}$$

is independent of t between collisions; i.e. $\dot{F}_1 = 0$. To compute its increment at collisions, we recall that

$$\mathcal{B}^+ = \mathcal{B}^- + \mathcal{R},$$

where $\mathcal{R} = 2\mathcal{K}(r)/\cos\varphi$. Differentiating this equation with respect to r and using (4.32)–(4.33) give

$$-\mathcal{E}_1^+ = \mathcal{E}_1^- + \mathcal{R}_1$$

where

$$\mathcal{R}_1 = (\mathcal{B}^+)^2\tan\varphi - (\mathcal{B}^-)^2\tan\varphi + \frac{1}{\cos\varphi}\frac{d\mathcal{R}}{dr} \\ = \frac{6\mathcal{K}^2\sin\varphi + 6\mathcal{K}\mathcal{B}^-\sin\varphi\cos\varphi + 2\mathcal{K}'\cos\varphi}{\cos^3\varphi},$$

where $\mathcal{K}' = d\mathcal{K}/dr$. Therefore,

$$-F_1^+ = \left(\frac{\mathcal{B}^-}{\mathcal{B}^+}\right)^3 F_1^- + H_1, \tag{4.37}$$

where $H_1 = \mathcal{R}_1/(\mathcal{B}^+)^3$, i.e.

$$H_1 = \frac{6\mathcal{K}^2\sin\varphi + 6\mathcal{K}\mathcal{B}^-\sin\varphi\cos\varphi + 2\mathcal{K}'\cos\varphi}{(2\mathcal{K} + \mathcal{B}^-\cos\varphi)^3}.$$

Observe that H_1 is represented by a fraction whose terms are all uniformly bounded and whose denominator is larger than a positive constant $8\mathcal{K}^3_{\min}$. Thus $|H_1| \le \tilde{C}_1 = \tilde{C}_1(\mathcal{D})$, an absolute constant. Also note that

$$\frac{\mathcal{B}^-}{\mathcal{B}^+} = \frac{\mathcal{B}^-}{\mathcal{B}^- + \mathcal{R}} \le \theta, \qquad \theta : = \frac{\mathcal{B}^-_{\max}}{\mathcal{B}^-_{\max} + \mathcal{R}_{\min}} < 1.$$

Since F_1 is constant between collisions, let us denote by $F_1(n)$ its value between the nth and $(n+1)$st collision. Then (4.37) implies

$$|F_1(n)| \le \theta^3 |F_1(n-1)| + \tilde{C}_1$$

and by a simple induction on n we obtain

$$\begin{aligned} |F_1(n)| &\le \theta^{3n}|F_1(0)| + \tilde{C}_1\big(1 + \theta^3 + \ldots + \theta^{3(n-1)}\big) \\ &\le \theta^{3n}|F_1(0)| + \tilde{C}_1/(1-\theta^3). \end{aligned} \tag{4.38}$$

Therefore, for all

$$n > n_{\gamma,1} : = \ln(|\tilde{C}_1|/|F_1(0)|)/\ln\theta^3$$

we have $|F_1(n)| < \tilde{C}_1' : = 2\tilde{C}_1/(1-\theta^3)$. Lastly, (4.36) implies $|\mathcal{E}_1| < \tilde{C}_1'\mathcal{B}^3$; hence

$$|\mathcal{E}_1^-| < \tilde{C}_1'(\mathcal{B}^-)^3 \le \tilde{C}_1'(\mathcal{B}^-_{\max})^3 =: C_1''$$

for all collisions past the $n_{\gamma,1}$th collision. This proves Lemma 4.32 for $\nu = 1$.

We now turn to the inductive step. First we make inductive assumptions. Suppose for some $1 \le \nu \le \ell - 3$ we have a function F_ν on Σ with the following properties (the reader is advised to check that they have been proved for $\nu = 1$):

(a) it is given by

$$F_\nu = \frac{Q_\nu(\mathcal{E}_\nu, \mathcal{E}_{\nu-1}, \ldots, \mathcal{E}_1, \mathcal{B})}{\mathcal{B}^{2\nu+1}}$$

where Q_ν is a homogeneous polynomial of degree ν of the variables $\mathcal{E}_\nu, \ldots, \mathcal{E}_1$ and $\mathcal{B}$, in which $\mathcal{E}_\nu$ appears in one term, namely in $\mathcal{E}_\nu\mathcal{B}^{\nu-1}$. Moreover, if we replace $\mathcal{E}_i$ by x^i and $\mathcal{B}$ by $x^0 = 1$, then every term of $Q_\nu(x)$ will be proportional to x^ν.

(b) $|F_\nu| < \tilde{C}_\nu'$ for some constant $\tilde{C}_\nu'(\mathcal{D}) > 0$ for all times past the $n_{\gamma,\nu}$-th collision.

(c) $\dot{F}_\nu = 0$; i.e. F_ν is independent of t between collisions.

(d) The increments of F_ν at collisions satisfy the equation

$$(-1)^\nu F_\nu^+ = \left(\frac{\mathcal{B}^-}{\mathcal{B}^+}\right)^{\nu+2} F_\nu^- + H_\nu \tag{4.39}$$

where

$$H_\nu = \frac{G_\nu}{(2\mathcal{K} + \mathcal{B}^- \cos\varphi)^{2\nu+1}}$$

and G_ν is an algebraic expression (a polynomial) whose terms ("variables") are $\cos\varphi$, $\sin\varphi$, $\mathcal{B}^-$, $\mathcal{E}_1^-, \ldots, \mathcal{E}_{\nu-1}^-$ and $\mathcal{K}$, $d\mathcal{K}/dr, \ldots, d^\nu\mathcal{K}/dr^\nu$.

We also assume that it has been proven that $|\mathcal{E}_i| \le \tilde{C}_\nu''\mathcal{B}^{i+2}$ for some constants $\tilde{C}_\nu''(\mathcal{D}) > 0$ and all $1 \le i \le \nu$ at all times past the $n_{\gamma,\nu}$th collision.

We now proceed to the inductive step per se. Due to (4.35), the function

$$F_{\nu+1} = \frac{1}{\mathcal{B}}\frac{dF_\nu}{d\xi}$$

is independent of t between collisions; i.e. $\dot{F}_{\nu+1} = 0$. Property (a) of F_ν implies

$$F_{\nu+1} = \frac{\mathcal{B}\, dQ_\nu/d\xi - (2\nu+1)Q_\nu \mathcal{E}_1}{\mathcal{B}^{2\nu+3}} = \frac{Q_{\nu+1}(\mathcal{E}_{\nu+1},\dots,\mathcal{E}_1,\mathcal{B})}{\mathcal{B}^{2\nu+3}} \tag{4.40}$$

where $Q_{\nu+1}$ is a homogeneous polynomial of degree $\nu+1$ in which $\mathcal{E}_{\nu+1} = d\mathcal{E}_\nu/d\xi$ appears in one term, namely in $\mathcal{E}_{\nu+1}\mathcal{B}^\nu$.

EXERCISE 4.33. Check that

$$Q_2 = \mathcal{E}_2\mathcal{B} - 3\mathcal{E}_1^2, \qquad Q_3 = \mathcal{E}_3\mathcal{B}^2 - 10\mathcal{E}_2\mathcal{E}_1\mathcal{B} + 15\mathcal{E}_1^3.$$

Show by induction on ν that if we replace $\mathcal{E}_i$ by x^i and $\mathcal{B}$ by $x^0 = 1$, then every term of $Q_\nu(x)$ will be proportional to x^ν. Moreover, if we replace $\mathcal{E}_i$ by x^{i+2} and $\mathcal{B}$ by x, then every term of $Q_\nu(x)$ will have degree $\geq 2\nu+1$.

To estimate the change of $F_{\nu+1}$ at collisions, we differentiate (4.39) with respect to r and use (4.32)–(4.33), which gives

$$(-1)^{\nu+1}F_{\nu+1}^+ = \left(\frac{\mathcal{B}^-}{\mathcal{B}^+}\right)^{\nu+3} F_{\nu+1}^- + H_{\nu+1} \tag{4.41}$$

where

$$H_{\nu+1} = \frac{1}{\mathcal{B}^+\cos\varphi}\frac{dH_\nu}{dr} + (\nu+2)F_\nu^-(\mathcal{B}^-)^{\nu+1}\frac{\mathcal{R}\, d\mathcal{B}^-/dr - \mathcal{B}^-\, d\mathcal{R}/dr}{(\mathcal{B}^+)^{\nu+4}\cos\varphi}.$$

EXERCISE 4.34. Show that

$$H_{\nu+1} = \frac{G_{\nu+1}}{(2\mathcal{K} + \mathcal{B}^-\cos\varphi)^{2\nu+3}}$$

where $G_{\nu+1}$ is an algebraic expression (a polynomial) whose terms ("variables") are $\cos\varphi$, $\sin\varphi$, $\mathcal{B}^-$, $\mathcal{E}_1^-,\dots,\mathcal{E}_\nu^-$ and $\mathcal{K}$, $d\mathcal{K}/dr,\dots,d^{\nu+1}\mathcal{K}/dr^{\nu+1}$. Conclude that $|H_{\nu+1}| \leq \tilde{C}_{\nu+1}$, where $\tilde{C}_{\nu+1}(\mathcal{D}) > 0$ is a constant.

We note that at the last step, when $\nu = \ell - 3$, the function $H_{\nu+1} = H_{\ell-2}$ will involve $d^{\ell-2}\mathcal{K}/dr^{\ell-2}$, which is a continuous but not necessarily differentiable function. Hence $H_{\ell-2}$ may not be differentiable anymore, but it is still uniformly bounded.

The rest of the proof is similar to the case $\nu = 1$. Let $F_{\nu+1}(n)$ denote the value of $F_{\nu+1}$ between the nth and $(n+1)$st collision. Then by (4.41) we have

$$|F_{\nu+1}(n)| \leq \theta^{\nu+3}|F_{\nu+1}(n-1)| + \tilde{C}_{\nu+1}$$

and by induction on n

$$|F_{\nu+1}(n)| \leq \theta^{(\nu+3)n}|F_{\nu+1}(0)| + \tilde{C}_{\nu+1}/(1-\theta^{\nu+3});$$

hence for all sufficiently large n we have $|F_{\nu+1}(n)| < \tilde{C}'_{\nu+1} := 2\tilde{C}_{\nu+1}/(1-\theta^{\nu+3})$.

Lastly, we estimate $\mathcal{E}_{\nu+1}$. Due to (4.40) and the subsequent comments

$$|\mathcal{E}_{\nu+1}| \leq \tilde{C}'_{\nu+1}|\mathcal{B}|^{\nu+3} + \hat{Q}_{\nu+1}/\mathcal{B}^\nu$$

where $\hat{Q}_{\nu+1}$ is a homogeneous polynomial of degree $\nu+1$ of variables $\mathcal{E}_\nu,\dots,\mathcal{E}_1$ and $\mathcal{B}$. Since $|\mathcal{E}_i| \leq \tilde{C}''_i\mathcal{B}^{i+2}$ for all $1 \leq i \leq \nu$, the last part of Exercise 4.33 implies that $|\hat{Q}_{\nu+1}/\mathcal{B}^\nu| \leq \text{const}\cdot\mathcal{B}^{\nu+3}$; thus $|\mathcal{E}_{\nu+1}| \leq \tilde{C}''_{\nu+1}\mathcal{B}^{\nu+3}$ for some constant $\tilde{C}''_{\nu+1}(\mathcal{D}) > 0$, and

$$|\mathcal{E}_{\nu+1}^-| \leq \tilde{C}''_{\nu+1}(\mathcal{B}^-)^{\nu+3} \leq \tilde{C}''_{\nu+1}(\mathcal{B}^-_{\max})^{\nu+3} =: C''_{\nu+1}.$$

This proves Lemma 4.32 by induction on ν. Proposition 4.29 is proved. □

4.7. More continued fractions

Let $X \in \Omega$ be a hyperbolic point. Here we compute the slope $\mathcal{B}^u(X)$ of the unstable subspace E^u_X by using Alekseev's cone techniques described in Section 3.13. We derive a remarkable continued fraction formula first discovered by Sinai [**Sin70**].

Let $x = \mathbf{P}^+(X)$, as defined by (3.17), and $x_n = \mathcal{F}^n(x)$ for $n \in \mathbb{Z}$. We also use notation $\tau_n = \tau_n(x)$ and $\mathcal{R}_n = \mathcal{R}_n(x)$ introduced in Section 4.3.

PROPOSITION 4.35. *Let $X \in \Omega$ be a hyperbolic point. The slope $\mathcal{B}^u(X)$ of the unstable subspace E^u_X is given by an infinite continued fraction:*

$$\mathcal{B}^u(X) = \cfrac{1}{|t^-(X)| + \cfrac{1}{\mathcal{R}_{-1} + \cfrac{1}{\tau_{-2} + \cfrac{1}{\mathcal{R}_{-2} + \cfrac{1}{\tau_{-3} + \cfrac{1}{\ddots}}}}}}. \tag{4.42}$$

where $|t^-(X)|$ is defined by (3.16).

PROOF. Due to (3.31), it is enough to prove this for $X \in \mathcal{M}$, in which case $x = X$ and $|t^-(X)| = \tau_{-1}$, and then (4.42) will represent the precollisional curvature $\mathcal{B}^{u-}(X)$.

The value of an infinite continued fraction is, by definition, the limit of its convergents. The even-numbered convergents

$$\mathcal{B}^{\mathrm{e}}_{-n} = \cfrac{1}{\tau_{-1} + \cfrac{1}{\mathcal{R}_{-1} + \cfrac{1}{\ddots + \cfrac{1}{\mathcal{R}_{-n}}}}},$$

correspond to wave fronts whose preimages before the nth collision in the past are flat (i.e., $\mathcal{B}^-_{-n} = 0$), recall (3.34). The odd-numbered convergents

$$\mathcal{B}^{\mathrm{o}}_{-n} = \cfrac{1}{\tau_{-1} + \cfrac{1}{\mathcal{R}_{-1} + \cfrac{1}{\ddots + \cfrac{1}{\tau_{-n}}}}}$$

correspond to wave fronts whose preimages before the nth collision in the past focus (i.e. $\mathcal{B}^{\pm}_{-n} = \infty$). Since both $\mathcal{B}^{\mathrm{e}}_{-n}$ and $\mathcal{B}^{\mathrm{o}}_{-n}$ correspond to the images of dispersing wave fronts constructed in the remote past, they converge to $\mathcal{B}^u(X)$, as $n \to \infty$, due to (3.51). □

Interestingly, the convergence of $\mathcal{B}^{\mathrm{e}}_{-n}$ and $\mathcal{B}^{\mathrm{o}}_{-n}$ to a common limit (given by (4.42)) can also be verified directly and regardless of the hyperbolicity of the point X.

LEMMA 4.36. *The sequences* $\{\mathcal{B}^{\mathrm{e}}_{-n}\}$ *and* $\{\mathcal{B}^{\mathrm{o}}_{-n}\}$ *converge to a common limit (4.42) whenever* $t_{-n} \to -\infty$ *as* $n \to \infty$, *i.e. whenever the past trajectory of the point* X *is defined at all times. Moreover,*

$$|\mathcal{B}^u(X) - \mathcal{B}^{\mathrm{e}}_{-n}| \le 1/|t_{-n}|, \qquad |\mathcal{B}^u(X) - \mathcal{B}^{\mathrm{o}}_{-n}| \le 1/|t_{-n}|. \tag{4.43}$$

PROOF. Since $\mathcal{R}_n > 0$ and $\tau_n > 0$ for all n, one can easily check that

$$\mathcal{B}^{\mathrm{e}}_{-1} < \mathcal{B}^{\mathrm{e}}_{-2} < \cdots < \mathcal{B}^{\mathrm{e}}_{-n} < \cdots < \mathcal{B}^{\mathrm{o}}_{-n} < \cdots < \mathcal{B}^{\mathrm{o}}_{-2} < \mathcal{B}^{\mathrm{o}}_{-1}. \tag{4.44}$$

Next we estimate the difference $\mathcal{B}^{\mathrm{o}}_{-n} - \mathcal{B}^{\mathrm{e}}_{-n}$. These are two finite continued fractions that differ only in the last term. We use the identity

$$\frac{1}{a+b_1} - \frac{1}{a+b_2} = \frac{1}{a+b_1}\,\frac{1}{a+b_2}\,(b_2 - b_1),$$

then transform $b_2 - b_1$ similarly (since b_2 and b_1 are continued fractions, too), etc. In the end we arrive at

$$\mathcal{B}^{\mathrm{o}}_{-n} - \mathcal{B}^{\mathrm{e}}_{-n} = W^{\mathrm{o}}_1 W^{\mathrm{e}}_1 W^{\mathrm{o}}_2 W^{\mathrm{e}}_2 \cdots W^{\mathrm{o}}_{n-1} W^{\mathrm{e}}_{n-1} \left[\frac{1}{\tau_{-n}} - \frac{1}{\tau_{-n} + 1/\mathcal{R}_{-n}}\right]$$

where

$$W^{\mathrm{o}}_i = \cfrac{1}{1 + \tau_{-i}\left(\mathcal{R}_{-i} + \cfrac{1}{\tau_{-i-1} + \cfrac{1}{\ddots + \cfrac{1}{\tau_{-n}}}}\right)}$$

and W^{e}_i is defined similarly, but ending with $\mathcal{R}_{-n}$ instead of τ_{-n}. Since all terms of our fractions are positive, it is easy to check that $W^{\mathrm{e}}_i \le 1$ and

$$W^{\mathrm{o}}_i \le \frac{\tau_{-i-1} + \cdots + \tau_{-n}}{\tau_{-i} + \cdots + \tau_{-n}}.$$

Combining all our estimates gives

$$\begin{aligned}\mathcal{B}^{\mathrm{o}}_{-n} - \mathcal{B}^{\mathrm{e}}_{-n} &\le \frac{1}{\tau_{-n}} \times \prod_{i=1}^{n-1} \frac{\tau_{-i-1} + \cdots + \tau_{-n}}{\tau_{-i} + \cdots + \tau_{-n}} \\ &= \frac{1}{\tau_{-1} + \cdots + \tau_{-n}} = \frac{1}{|t_{-n}|},\end{aligned}$$

which proves the lemma. □

COROLLARY 4.37. *Let* $L \subset \mathcal{T}^{\perp}_X\Omega$ *be a tangent line with slope* $\mathcal{B} > 0$. *Suppose that* $\mathcal{B}_{t_{-n}-0} \ge 0$, *where* t_{-n} *is the time of the nth collision in the past. Then* $|\mathcal{B} - \mathcal{B}^u(X)| \le 1/|t_{-n}|$.

PROOF. The tangent line $D_X\Phi^{t_{-n}-0}(L)$ corresponds to a wave front that has curvature between zero (which was used to construct $\mathcal{B}^{\mathrm{e}}_{-n}$; see the proof of Proposition 4.35) and infinity (which was used to construct $\mathcal{B}^{\mathrm{o}}_{-n}$). Therefore, $\mathcal{B}^{\mathrm{e}}_{-n} \le \mathcal{B} \le \mathcal{B}^{\mathrm{o}}_{-n}$, due to the result of Exercise 4.13. □

Lemma 4.36 shows that $\mathcal{B}^u(X)$ is defined by (4.42) on the set $\tilde{\Omega}^- = \{X\colon \Phi^t(X)$ exists $\forall t < 0\}$ of phase points whose past trajectories exist (which is definitely larger than the set of hyperbolic points).

EXERCISE 4.38. Let $\hat{\Omega}^- \subset \tilde{\Omega}^-$ denote the set of points whose past trajectories experience regular collisions only (no grazing collisions). Show that the function $\mathcal{B}^u(X)$ is continuous at every point $X \in \hat{\Omega}^-$ but discontinuous at every $X \in \tilde{\Omega}^- \setminus \hat{\Omega}^-$. Hint: Observe that the continued fraction $\mathcal{B}^u(X)$ depends continuously on its terms and use Corollary 4.37.

Due to the time reversibility (Remark 4.16), the slope $\mathcal{B}^s(X)$ of the stable subspace E^s_X, $X \in \Omega$, is expressed by the continued fraction

$$\mathcal{B}^s(X) = \cfrac{1}{t^+(X) + \cfrac{1}{\mathcal{R}_0 + \cfrac{1}{\tau_0 + \cfrac{1}{\mathcal{R}_1 + \cfrac{1}{\tau_1 + \cfrac{1}{\ddots}}}}}} \tag{4.45}$$

where $t^+(X)$ was defined by (3.15) and the symbols τ_i and $\mathcal{R}_i$ again refer to the point $x = \mathbf{P}^+(X)$.

The function $\mathcal{B}^s(X)$ is well defined on the set $\tilde{\Omega}^+ = \{X\colon \Phi^t(X) \text{ exists } \forall t > 0\}$ and continuous on the subset $\hat{\Omega}^+ \subset \tilde{\Omega}^+$ of points whose forward trajectories experience regular collisions only.

The unstable subspace $E^u_x \subset \mathcal{T}_x\mathcal{M}$ has slope

$$\mathcal{V}^u(x) = \mathcal{B}^{u-}(x)\cos\varphi + \mathcal{K} = \mathcal{B}^{u+}(x)\cos\varphi - \mathcal{K}; \tag{4.46}$$

see (3.38) and Section 3.12 for the definition of $\mathcal{B}^{u\pm}(x)$. The above analysis shows that the function $\mathcal{V}^u(x)$ is well defined and continuous on the set

$$\hat{\mathcal{M}}^-\colon = \mathcal{M} \setminus \cup_{n=-\infty}^{0} \mathcal{S}_n \tag{4.47}$$

of points at which all the past iterations of $\mathcal{F}$ are smooth; cf. (2.28). Similarly, the stable subspace $E^s_x \subset \mathcal{T}_x\mathcal{M}$ has slope

$$\mathcal{V}^s(x) = \mathcal{B}^{s-}(x)\cos\varphi + \mathcal{K} = \mathcal{B}^{s+}(x)\cos\varphi - \mathcal{K}, \tag{4.48}$$

which is well defined and continuous on the set

$$\hat{\mathcal{M}}^+\colon = \mathcal{M} \setminus \cup_{n=0}^{\infty} \mathcal{S}_n \tag{4.49}$$

of points where all the forward iterations of $\mathcal{F}$ are smooth.

We note that the slopes $\mathcal{V}^u$ and $\mathcal{V}^s$ are continuous on their domains, but they are not Hölder continuous (unlike their counterparts in Anosov and Axiom A systems).

We conclude this section by answering one question raised in Section 3.12:

PROPOSITION 4.39. *The function $\mathcal{B}^u(X)$ is integrable on Ω.*

PROOF. Using the notation of Section 2.12 and Eq. (3.31) we have

$$\begin{aligned}\int_\Omega \mathcal{B}^u(X)\,d\mu_\Omega &= C\int_{\mathcal{M}}\int_0^{\tau(x)} \frac{1}{t + 1/\mathcal{B}^{u+}(x)}\,dt\,d\mu \\ &= C\int_{\mathcal{M}} \ln\bigl[1 + \tau(x)\mathcal{B}^{u+}(x)\bigr]\,d\mu \\ &\le C\int_{\mathcal{M}} \ln\bigl[1 + \tau(x)\bigr]\,d\mu + C\int_{\mathcal{M}} \ln\bigl[1 + \mathcal{B}^{u+}(x)\bigr]\,d\mu\end{aligned}$$

where $C > 0$ is a constant. Now the first integral is finite because $\tau(x)$ is integrable (Section 2.13). The second one equals

$$\int_{\mathcal{M}} \ln\bigl[1 + \mathcal{R}(x) + \mathcal{B}^{u-}(x)\bigr]\, d\mu$$
$$\leq \int_{\mathcal{M}} \ln\bigl[1 + \mathcal{R}(x)\bigr]\, d\mu + \int_{\mathcal{M}} \ln\bigl[1 + \mathcal{B}^{u-}(x)\bigr]\, d\mu.$$

The first integral here is finite since $\int |\ln\cos\varphi|\, d\mu < \infty$ (see (3.8)), and for the second integral we note that $\mathcal{B}^{u-}(x) \leq 1/\tau_{-1}(x)$. Thus

$$\int_{\mathcal{M}} \ln\bigl[1 + \mathcal{B}^{u-}(x)\bigr]\, d\mu \leq \int_{\mathcal{M}} \ln\bigl[1 + \tau_{-1}(x)\bigr]\, d\mu + \int_{\mathcal{M}} |\ln \tau_{-1}(x)|\, d\mu,$$

where the first integral on the right is finite (see above), and the second one is finite due to Exercise 2.43. □

4.8. Singularities (local analysis)

Here we investigate the singularity set $\mathcal{S}_n$ of the map $\mathcal{F}^n$, $n \in \mathbb{Z}$; recall the relevant definitions (2.19), (2.20) and (2.27).

EXERCISE 4.40. Use the involution (Section 2.14) to show that if $x = (r, \varphi) \in \mathcal{S}_n$, then $(r, -\varphi) \in \mathcal{S}_{-n}$.

This symmetry allows us to restrict ourselves to the case $n < 0$; i.e. we consider the singularities of the past iterations of the map $\mathcal{F}$.

PROPOSITION 4.41. *For each $n < 0$, the set $\mathcal{S}_n \setminus \mathcal{S}_0$ consists of C^1 smooth unstable curves $S \subset \mathcal{M}$ (i.e., the tangent line to S at every point $x \in S$ belongs in the unstable cone $\hat{C}^u_x$).*

PROOF. It is enough to check this for $n = -1$ and then use the invariance of the cones (Section 4.4). Every curve $S \subset \mathcal{S}_{-1} \setminus \mathcal{S}_0$ is made by trajectories either (i) coming from grazing collisions or (ii) emanating from corners; see Fig. 4.8. Both types of trajectories obviously make dispersing wave fronts when they collide with $\partial\mathcal{D}$ and form the curve S. The preimages of these dispersing wave fronts focus at the previous collision, which either is a grazing collision or occurs at a corner.

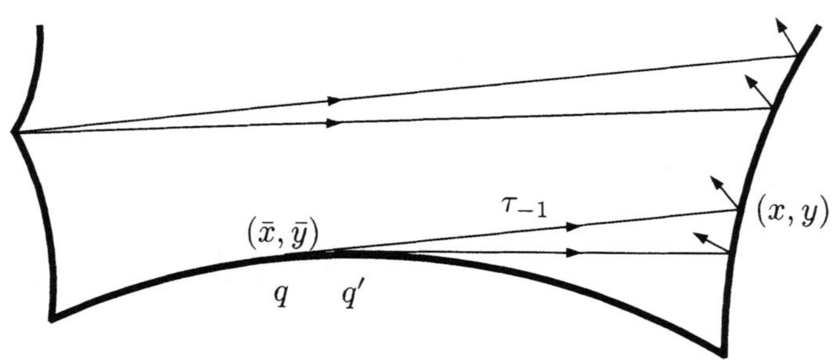

FIGURE 4.8. Trajectories making the set $\mathcal{S}_{-1}$.

While focusing at a corner point is obvious, we prove that focusing also occurs at a grazing collision. Consider two nearby points, q and $q' = q + dq$, on the wall where grazing collisions occur; see Fig. 4.8. Then the respective velocity vectors (which are, in this case, tangent to $\partial\mathcal{D}$) – we call them v and $v' = v + dv$ – differ by $\|dv\| \sim \mathcal{K}\|dq\|$. On the other hand, the distance from the point q to the line $q' + tv'$, $t \in \mathbb{R}$, is of order $\mathcal{O}(\|dq\|^2)$; more precisely it is $\frac{1}{2}\mathcal{K}\|dq\|^2 + o(\|dq\|^2)$. Hence the respective distance between the lines $q + tv$ and $q' + tv'$ along the orthogonal cross-section of this family of grazing trajectories is also $\|dq\|_\perp = \mathcal{O}(\|dq\|^2)$. Thus

$$\mathcal{B} = \lim_{\|dq\|\to 0} \|dv\|/\|dq\|_\perp = \infty.$$

Thus, every curve $S \subset \mathcal{S}_{-1} \setminus \mathcal{S}_0$ corresponds to a wave front whose preimage at the previous collision focuses. Therefore, when the above wave front approaches the collision at a point $x = (r, \varphi) \in S \subset \mathcal{S}_{-1} \setminus \mathcal{S}_0$, it has slope $\mathcal{B} = 1/\tau_{-1}$, so the slope of the curve S is, due to (3.39)

$$\mathcal{V} = d\varphi/dr = \mathcal{K} + \cos\varphi/\tau_{-1}, \tag{4.50}$$

where $\mathcal{K} = \mathcal{K}(x)$ and $\tau_{-1} = \tau_{-1}(x)$. Proposition 4.41 is now proved. □

We will sharpen the above proposition by showing that the slope of singularity curves $S \subset \mathcal{S}_{-n} \setminus \mathcal{S}_{-n+1}$ approaches, as n increases, the slope of the unstable space E^u_x at nearby points x.

PROPOSITION 4.42. *Let $\mathcal{D}$ be a billiard of category A, B, C, or D. Then there is a sequence $\beta_n \to 0$ as $n \to \infty$ such that for any curve $S \subset \mathcal{S}_{-n} \setminus \mathcal{S}_{-(n-1)}$ and any point $y \in S$ there is an open neighborhood $U_y \subset \mathcal{M}$ such that the slope $\mathcal{V}_y(S)$ of the curve S at y satisfies*

$$\sup_{x \in U_y} |\mathcal{V}^u_x - \mathcal{V}_y(S)| < \beta_n.$$

(Of course, the supremum here is taken over $x \in U_y$ where $\mathcal{V}^u_x$ is defined. Such points are described in Section 4.7.)

PROOF. Let t_{-n} denote the time of the nth collision in the past. It is easy to check that $|t_{-n}| > an - b$, where $a, b > 0$ are constants (indeed, for billiards of categories A and B, we can set $a = b = \tau_{\min}$, and for categories C and D see Exercise 2.18). Now we set $\beta_n = 1/(an - b)$, and then our proposition follows from Corollary 4.37. □

Loosely speaking, Proposition 4.42 says that the singularity curves $S \subset \mathcal{S}_{-n} \setminus \mathcal{S}_{-(n-1)}$ align with the unstable subspace E^u as n increases. This property is called the *alignment of singularity lines.*

PROPOSITION 4.43. *In billiards of categories A and B, every curve $S \subset \mathcal{S}_{-n} \setminus \mathcal{S}_0$, $n \geq 1$, is $C^{\ell-1}$ smooth with bounded derivatives (up to its endpoints). Moreover, for each $\nu = 1, \dots, \ell - 1$, the ν-th derivative is uniformly bounded by a constant $C_\nu(\mathcal{D}) > 0$ independent of n and the curve $S \subset \mathcal{S}_{-n} \setminus \mathcal{S}_0$.*

PROOF. It is enough to prove this for $n = 1$ and then use the results of Section 4.5. In (4.50), the function $\mathcal{K} = \mathcal{K}(r)$ is $C^{\ell-2}$ smooth with uniformly bounded derivatives. Hence it is enough to prove that $\tau_{-1} = \tau_{-1}(r)$ is $C^{\ell-2}$ smooth with uniformly bounded derivatives; here $\tau_{-1}(r) = \tau_{-1}(r, \varphi_S(r))$ is the function τ_{-1} restricted to the curve S. In fact, we prove a little more.

LEMMA 4.44. *The function $\tau_{-1}(r)$ along every curve $S \subset \mathcal{S}_{-1} \setminus \mathcal{S}_0$ is $C^{\ell-1}$ smooth, with uniformly bounded derivatives, up to the endpoints of S.*

PROOF. For every point $(r,\varphi) \in S$ we denote by $(\bar{r},\bar{\varphi}) = \mathcal{F}^{-1}(r,\varphi)$ its preimage, where, of course, $\bar{\varphi} = \pm\pi/2$. This makes $\bar{r}$ a function of r. Let (x,y) denote the Cartesian coordinates of the point $\pi_q(r,\varphi)$ and $(\bar{x},\bar{y})$ those of the point $\pi_q(\bar{r},\bar{\varphi})$; see Fig. 4.8. Observe that

$$\tau_{-1}^2 = (x-\bar{x})^2 + (y-\bar{y})^2 \tag{4.51}$$

and recall that x and y are C^ℓ functions of r (Section 2.1) and, likewise, that $\bar{x}$ and $\bar{y}$ are C^ℓ functions of $\bar{r}$. Thus it is enough to prove that $\bar{r}$ is a $C^{\ell-1}$ function of r. Let $(\dot{\bar{x}},\dot{\bar{y}}) = (d\bar{x}/d\bar{r}, d\bar{y}/d\bar{r})$ denote the unit tangent vector to $\partial\mathcal{D}$ at the point $(\bar{x},\bar{y})$. Observe that the vector $(x-\bar{x}, y-\bar{y})$ is parallel to $(\dot{\bar{x}},\dot{\bar{y}})$; hence

$$(x-\bar{x})\dot{\bar{y}} = (y-\bar{y})\dot{\bar{x}}.$$

Direct differentiation gives

$$\frac{d\bar{r}}{dr} = \frac{\dot{\bar{x}}\dot{y} - \dot{x}\dot{\bar{y}}}{(x-\bar{x})\,\ddot{\bar{y}} - (y-\bar{y})\,\ddot{\bar{x}}}, \tag{4.52}$$

which can also be derived from (2.26). Observe that $\bar{n} = (\ddot{\bar{x}},\ddot{\bar{y}})$ is a normal vector at the point $(\bar{x},\bar{y})$, and $\|\bar{n}\| = \bar{\mathcal{K}}$ is the curvature of $\partial\mathcal{D}$ at that same point. Therefore the absolute value of the denominator in (4.52) equals $\bar{\mathcal{K}}\tau_{-1} \geq \mathcal{K}_{\min}\tau_{\min} > 0$, so it is bounded away from zero. Now Lemma 4.44 follows from the Implicit Function Theorem. We note that our argument works for billiards with and without horizon. □

This completes the proof of Lemma 4.44 and Proposition 4.43. □

4.9. Singularities (global analysis)

PROPOSITION 4.45. *For dispersing billiards with finite horizon (categories A, C, and E), the set $\mathcal{S}_{-1} \setminus \mathcal{S}_0$ consists of finitely many compact*[2] *smooth unstable (increasing) curves, and the set $\mathcal{S}_1 \setminus \mathcal{S}_0$ consists of finitely many compact smooth stable (decreasing) curves.*

For billiards without horizon (categories B, D, and F), the same holds, except the number of the above curves may be infinite.

PROOF. Curves $\mathcal{S}_1 \setminus \mathcal{S}_0$ are traces left on $\mathcal{M}$ by finitely many dispersing wave fronts (those emanating from corner points and those formed by grazing collisions with walls). Now the results of Section 4.5, in particular Exercise 4.25, apply. □

EXERCISE 4.46. Show that Proposition 4.45 extends, without change, to the sets $\mathcal{S}_{-n} \setminus \mathcal{S}_0$ and $\mathcal{S}_n \setminus \mathcal{S}_0$ for all $n \geq 1$. Hint: Use induction on n and the last sentence of Remark 4.27.

It is easy to check that each smooth curve $S \subset \mathcal{S}_n \setminus \mathcal{S}_0$ terminates on $\mathcal{S}_0 = \partial\mathcal{M}$ or on another smooth curve $\tilde{S} \subset \mathcal{S}_n \setminus \mathcal{S}_0$. For example, Fig. 4.9 shows a curve $S' \subset \mathcal{S}_{-1} \setminus \mathcal{S}_0$ of type (i) (induced by grazing collisions) terminating on a curve $S'' \subset \mathcal{S}_{-1} \setminus \mathcal{S}_0$ of type (ii) (coming from a corner point). Observe that the curves S' and S'' have a common endpoint x (and, furthermore, a common tangent line with slope $\mathcal{V} = 1/\tau_{-1}(x)$ at the point x, but their curvatures at x may differ).

Fig. 4.10 shows a curve $S' \subset \mathcal{S}_{-1} \setminus \mathcal{S}_0$ of type (i), which is induced by grazing collisions with scatterer 1, terminating on another, longer curve $S'' \subset \mathcal{S}_{-1} \setminus \mathcal{S}_0$ of

[2]More precisely, their closures are compact; the removal of $\mathcal{S}_0$ deprives some curves in $\mathcal{S}_{-1}$ of their endpoints.

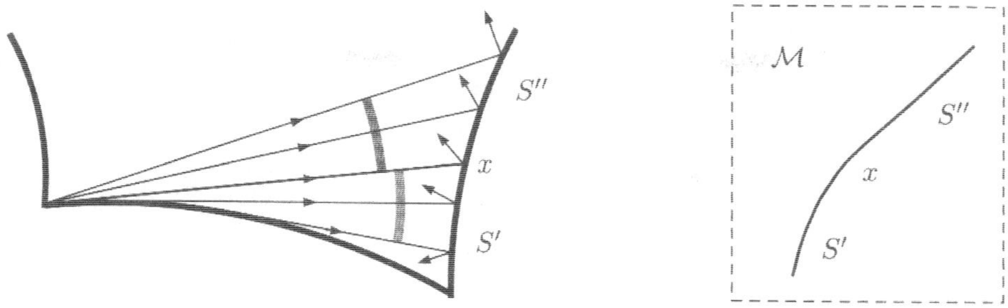

FIGURE 4.9. A junction of two singularity curves.

type (i), which is induced by grazing collisions with scatterer 2. Observe that S' terminates at an inner point x of S'' and they intersect transversally.

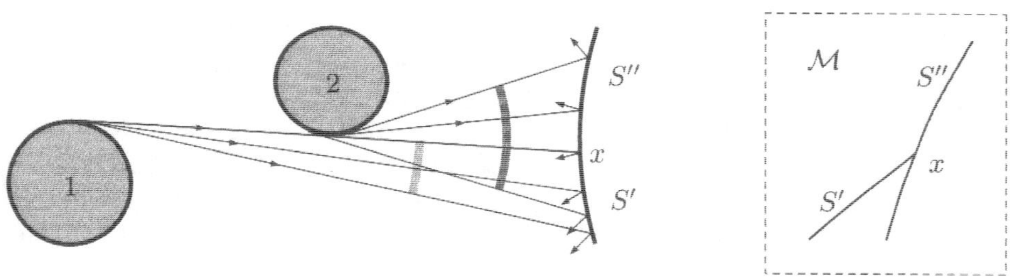

FIGURE 4.10. Another junction of two singularity curves.

PROPOSITION 4.47. *For each $n \neq 0$ every curve $S \subset \mathcal{S}_n \setminus \mathcal{S}_0$ is a part of some monotonic continuous (and piece-wise smooth) curve $\tilde{S} \subset \mathcal{S}_n \setminus \mathcal{S}_0$ which terminates on $\mathcal{S}_0 = \partial\mathcal{M}$ (the curve $\tilde{S}$ is monotonically increasing for $n < 0$ and monotonically decreasing for $n > 0$).*

Thus, one can continue every smooth curve $S \subset \mathcal{S}_n \setminus \mathcal{S}_0$ monotonically up to the natural boundary of $\mathcal{M}$. This property is often referred to as *continuation of singularity lines.*

EXERCISE 4.48. Prove Proposition 4.47. Hint: Check it for $n = \pm 1$ by direct inspection (two possible cases are shown in Figs. 4.9 and 4.10; all the others are similar) and then use induction on n.

EXERCISE 4.49. Show that for each $n', n'' \geq 0$ the set $\mathcal{M} \setminus (\mathcal{S}_{-n'} \cup \mathcal{S}_{n''})$ is a finite or countable union of open domains with piecewise smooth boundaries (curvilinear polygons) such that the interior angles made by their boundary components do not exceed π (i.e. those polygons are 'convex' as far as the interior angles are concerned). Some interior angles may be equal to zero; see below.

A typical structure of singularity curves is shown in Fig. 4.11. Note the symmetry established in Exercise 4.40.

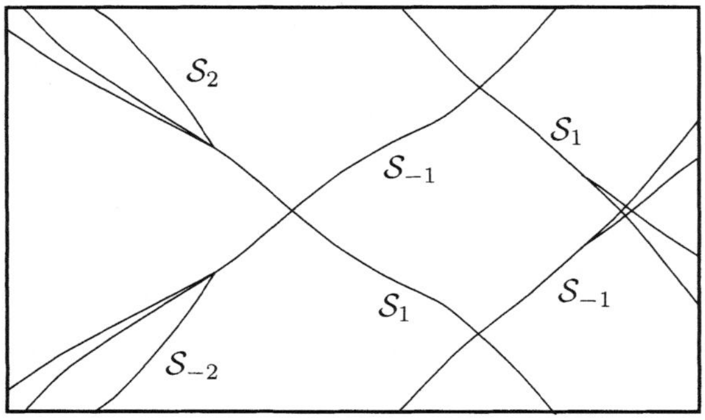

FIGURE 4.11. Singularity curves.

For billiards of categories A and B every curve $S \subset \mathcal{S}_1 \setminus \mathcal{S}_0$ consists of trajectories that start on a wall Γ_i and make a grazing collision with a wall Γ_j; see Fig. 4.12. (It is possible that $j = i$ (this happens if $\mathcal{D} \subsetneq \mathrm{Tor}^2$) and that the trajectories coming from Γ_i run around the torus and come back to Γ_i.)

The curve S then separates phase points $x \in \mathcal{M}$ whose trajectories hit Γ_j (almost tangentially) from those whose trajectories miss Γ_j (so they pass nearby). Accordingly, one side of S is mapped by $\mathcal{F}$ into a vicinity of $\mathcal{F}(S) \subset \mathcal{S}_0$, and then into a one-sided vicinity of the curve $S' = \mathcal{F}^2(S) \subset \mathcal{S}_{-1} \setminus \mathcal{S}_0$. Points on the other side of S are mapped directly into *the other* one-sided vicinity of S'; see the diagram in Fig. 4.13.

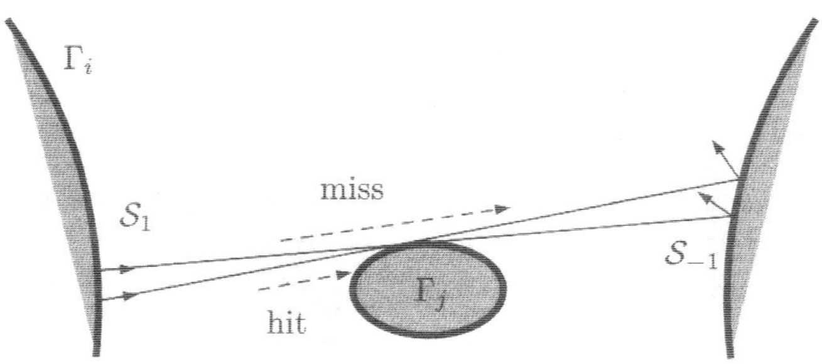

FIGURE 4.12. Singularities and nearby trajectories.

The first side of S experiences enormous expansion. Unstable curves of length ε terminating on S (dark grey dashes in Fig. 4.13) are mapped to unstable curves of length $\sim \sqrt{\varepsilon}$ terminating on $\mathcal{S}_0$ and then to unstable curves of length $\sim \sqrt{\varepsilon}$ terminating on S'. The latter are actually tangent to S' and lie within a distance $\sim \varepsilon$ of S'. All this can be verified by direct inspection, which we leave to the reader.

EXERCISE 4.50. Let $W \subset \mathcal{M}$ be an unstable curve. Show that $|\mathcal{F}(W)| \leq C\sqrt{|W|}$, where $C = C(\mathcal{D}) > 0$ is a constant.

On the other side of S, expansion is moderate: unstable curves of length ε terminating on S (light grey dashes in Fig. 4.13) are mapped to unstable curves of length $\sim \varepsilon$ terminating on S'. The latter make positive angles with S' at their terminal points.

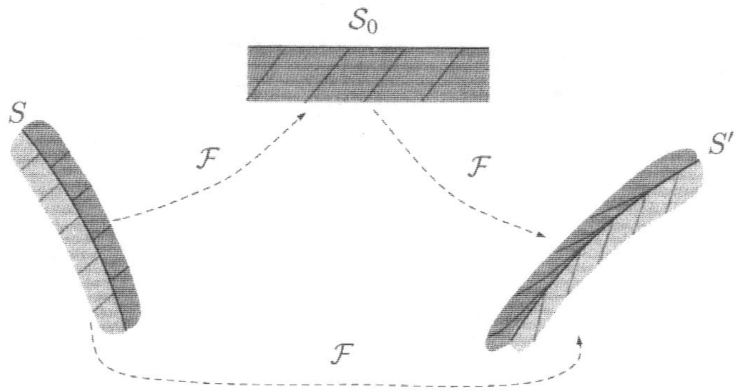

FIGURE 4.13. Action of $\mathcal{F}$ near singularities.

4.10. Singularities for type B billiard tables

In billiards of category B (with no horizon), the singularity set $\mathcal{S}_{-1}$ consists of an infinite number of smooth curves, which we describe briefly. They accumulate at points $x \in \mathcal{M}$ near which the function $\tau_{-1}(x)$ is unbounded.

EXERCISE 4.51. Show that there are only finitely many points $x_k = (r_k, \varphi_k) \in \mathcal{M}$, $1 \le k \le k_{\max}$, such that

$$\limsup_{x \to x_k} \tau_{-1}(x) = \infty.$$

These are points whose trajectories are periodic (closed geodesic lines) on the torus; cf. Exercise 2.15. Note that $\varphi_k = \pm\pi/2$ at every such point.

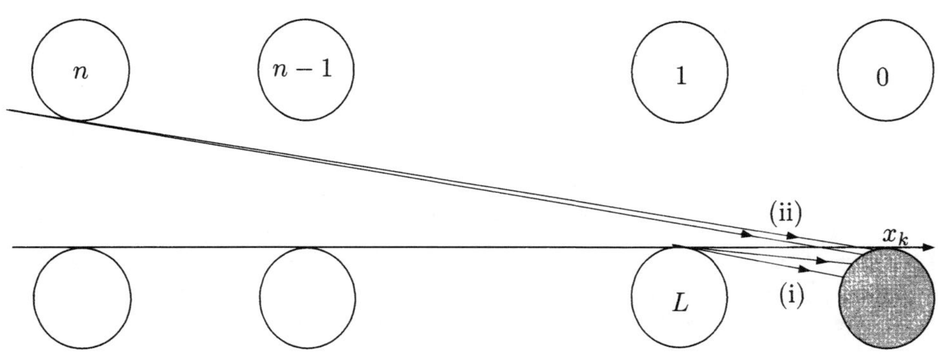

FIGURE 4.14. A point x_k where singularity curves accumulate.

For the particular billiard table shown in Fig. 2.4, let us consider a point $x_k = (r_k, \pi/2) \in \mathcal{M}$ whose trajectory is a horizontal line; see Fig. 4.14. In the neighborhood of this point, $\mathcal{S}_{-1}$ is made of curves of two types: (i) induced by grazing collisions with the very next scatterer, marked by L, to the left in the same row and (ii) coming from far away scatterers to the left in the upper row (Fig. 4.14).

There is only one curve of type (i), which we call $S^-_{k,0}$, running from x_k down into $\mathcal{M}$; see Fig. 4.15 (right). There are infinitely many curves of type (ii) running almost parallel to the curve $S^-_{k,0}$ on which they terminate (the other endpoint lies on the upper edge $\varphi = \pi/2$). If we number the scatterers in the upper row as shown in Fig. 4.14, then the curves of type (ii) can be labelled accordingly: the curve induced by grazing collisions with the nth scatterer is denoted by $S^-_{k,n}$.

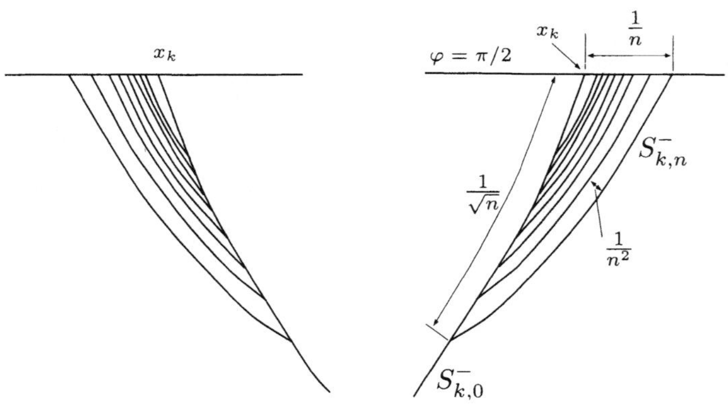

FIGURE 4.15. Singularity curves (unbounded horizon): $\mathcal{S}_1$ is on the left, $\mathcal{S}_{-1}$ on the right.

EXERCISE 4.52. Show that the length of $S^-_{k,n}$ is $|S^-_{k,n}| \sim 1/\sqrt{n}$, the distance between the upper endpoint of $S^-_{k,n}$ and the limit point x_k is $\sim 1/n$, and the distance between $S^-_{k,n}$ and $S^-_{k,n+1}$ is $\sim 1/n^2$, as indicated in Fig. 4.15 (right). Here the notation $\sim 1/n^a$, $a > 0$, means $= c/n^a + o(1/n^a)$ as $n \to \infty$ for some constant $c = c(k, \mathcal{D}) > 0$.

A similar analysis shows that the set $\mathcal{S}_1$ consists of infinitely many curves accumulating near the same points $x_k \in \mathcal{M}$, $1 \le k \le k_{\max}$. The curves of $\mathcal{S}_1$ near x_k can be obtained by reflecting those of $\mathcal{S}_{-1}$ in the vertical line $r = r_k$ (which passes through x_k), see Fig. 4.15 (left). (Note that the left and right parts of Fig. 4.15 should be superimposed for a proper view.) We denote the resulting curves by $S^+_{k,0}$ and $S^+_{k,n}$, respectively.

Next, let $D^-_{k,n}$ denote the region bounded by the curves $S^-_{k,n}$, $S^-_{k,n-1}$, $S^-_{k,0}$ and the upper edge $\varphi = \pi/2$. (This region is sometimes called *n-cell* in the literature. It is filled by all trajectories coming directly from the nth scatterer; see Fig. 4.14.) Similarly, the n-cell $D^+_{k,n}$ is defined to be the region bounded by the curves $S^+_{k,n}$, $S^+_{k,n-1}$, $S^+_{k,0}$ and the edge $\varphi = \pi/2$.

EXERCISE 4.53. Show that $\mathcal{F}^{-1}(D^-_{k,n}) = D^+_{k',n}$ for some $1 \le k' \le k_{\max}$; i.e. "positive" n-cells are transformed into "negative" n-cells by $\mathcal{F}$; see Fig. 4.16.

EXERCISE 4.54. Let $dx \in \mathcal{T}_x\mathcal{M}$ be an unstable tangent vector at a point $x \in D^+_{k,n}$. Prove that $\|D_x\mathcal{F}(dx)\|_p/\|dx\|_p \ge \text{const}\, n^{3/2}$, as well as $\|D_x\mathcal{F}(dx)\|/\|dx\| \ge$

const $n^{3/2}$. That is, the expansion factor in the n-cells (in both metrics) grows (explodes) as $n \to \infty$. Hint: Use (4.16) and (4.20) and note that $\tau(x) \geq \text{const}\, n$.

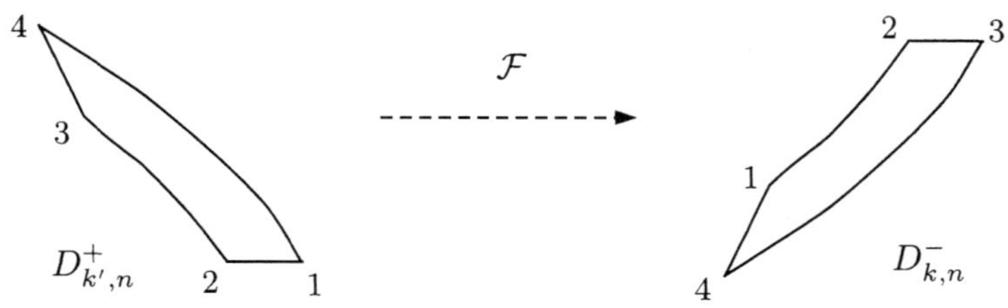

FIGURE 4.16. The map $\mathcal{F}$ transforms n-cells. The corners of $D^+_{k',n}$ and their respective images are numbered to indicate the action of $\mathcal{F}$.

This completes our short description of singularity curves for billiards of category B; for further details the interested reader is referred to [**BSC90, BSC91, Ch99**].

4.11. Stable and unstable manifolds

Consider the sets

$$\mathcal{S}_\infty = \cup_{n=0}^\infty \mathcal{S}_n \qquad \text{and} \qquad \mathcal{S}_{-\infty} = \cup_{n=0}^\infty \mathcal{S}_{-n} \tag{4.53}$$

of points where *some future* and, respectively, *some past* iterate of $\mathcal{F}$ is singular.

LEMMA 4.55. *Both sets $\mathcal{S}_\infty$ and $\mathcal{S}_{-\infty}$ are dense in $\mathcal{M}$.*

PROOF. Assume, for example, that S_∞ is not dense. Then an unstable curve $W \subset \mathcal{M} \setminus \mathcal{S}_\infty$ would exist and its images $\mathcal{F}^n(W)$ would grow exponentially fast with n without ever running into singularities and breaking up, so sooner or later they would have to extend beyond the natural boundaries of $\mathcal{M}$ due to (4.14), an impossibility. □

Observe that $\mathcal{S}_\infty$ is a countable union of smooth stable curves and the boundary $\partial\mathcal{M}$, while $\mathcal{S}_{-\infty}$ is a countable union of smooth unstable curves, plus $\partial\mathcal{M}$. For each $\mathcal{M}_i \subset \mathcal{M}$ (cf. (2.17)), the sets $\mathcal{M}_i \cap \mathcal{S}_\infty$ and $\mathcal{M}_i \cap \mathcal{S}_{-\infty}$ are pathwise connected; cf. Proposition 4.47.

On the other hand, the set $\mathcal{M} \setminus \mathcal{S}_{-\infty}$ has full μ measure, but it is badly disconnected. We analyze its connected components next. Let $x \in \mathcal{M} \setminus \mathcal{S}_{-\infty}$ and for any $n \geq 1$ denote by $\mathcal{Q}_{-n}(x)$ the connected component of the open set $\mathcal{M} \setminus \mathcal{S}_{-n}$ that contains x. Recall that $\mathcal{Q}_{-n}(x)$ is a curvilinear polygon with interior angles $\leq \pi$ (Exercise 4.49).

EXERCISE 4.56. Show that $\partial\mathcal{Q}_{-n}(x)$ consists of two monotonically increasing (and piecewise smooth) curves whose endpoints are the top and bottom vertices of $\mathcal{Q}_{-n}(x)$; see Fig. 4.17. We call those curves left and right sides and denote them by $\partial^L\mathcal{Q}_{-n}(x)$ and $\partial^R\mathcal{Q}_{-n}(x)$, respectively.

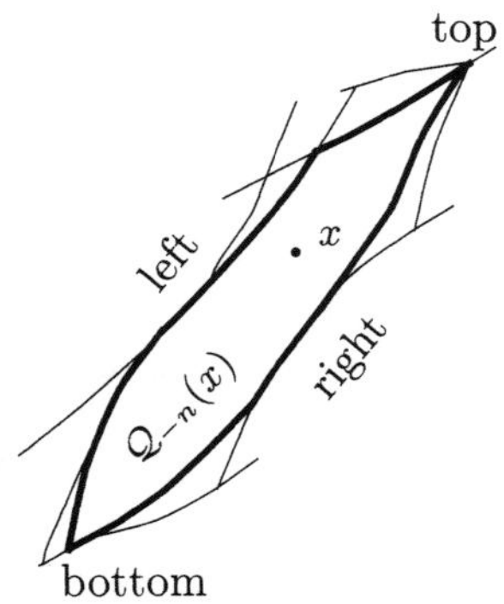

FIGURE 4.17. The connected component $\mathcal{Q}_{-n}(x)$.

Obviously, $\mathcal{Q}_{-n}(x) \supseteq \mathcal{Q}_{-(n+1)}(x)$ for all $n \geq 1$ and the intersection of their closures

$$\tilde{W}^u(x) \colon = \cap_{n=1}^{\infty} \overline{\mathcal{Q}_{-n}}(x)$$

is a closed continuous monotonically increasing curve. We denote by $W^u(x)$ the curve $\tilde{W}^u(x)$ without its endpoints.

LEMMA 4.57. *We have* $W^u(x) \subset \cap_{n\geq 1}\mathcal{Q}_{-n}(x)$.

PROOF. Suppose that some point $y \in W^u(x)$ belongs to a singularity curve $S \subset \partial\mathcal{Q}_{-n}(x)$ for a finite $n \geq 1$. Without loss of generality, we assume $S \subset \partial^L\mathcal{Q}_{-n}(x)$. Then y is also a limit point for a sequence of curves $S'_{-m} \subset \partial^R\mathcal{Q}_{-m}(x)$. Since $S \subset \mathcal{S}_{-n}$ for a finite n, the curve S has a slope at y different from the limit slope of the curves S'_{-m} (recall Proposition 4.42). This leads to a contradiction. See also Proposition 4.47 and Exercise 4.49. □

EXERCISE 4.58. Prove that at every point $y \in W^u(x)$ the slope of the curve $W^u(x)$ equals $\mathcal{V}^u(y)$. Hint: Use the continuity of $\mathcal{V}^u(y)$ (cf. (4.46)) and Proposition 4.42.

Observe that $\mathcal{F}^{-n}(W^u(x)) \subset \mathcal{M} \setminus \mathcal{S}_{-\infty}$ is an unstable curve for every $n \geq 1$. Let $\mathcal{L}_n \subset \mathcal{M}$ be the linear segment (in the r, φ coordinates) joining the endpoints of $\mathcal{F}^{-n}(W^u(x))$.

EXERCISE 4.59. Show that $\mathcal{L}_n \cap \mathcal{S}_n = \emptyset$. (Hint: Observe that $\mathcal{F}^{-n}(W^u(x)) \cap \mathcal{S}_n = \emptyset$ and use Proposition 4.47.) Then show that the unstable curves $\mathcal{F}^n(\mathcal{L}_n)$ converge, as $n \to \infty$, to $W^u(x)$ in the C^0 metric.

Observe that the unstable curves $\mathcal{F}^n(\mathcal{L}_n)$, $n \geq 1$, are $C^{\ell-1}$ smooth and have all their $\ell - 1$ derivatives uniformly bounded; see Proposition 4.29.

LEMMA 4.60. *Let $f_n(t)$, $n \geq 1$, be C^q smooth functions on an interval (a, b) such that $|f_n^{(\nu)}(t)| \leq C =$ const for all $n \geq 1$, $1 \leq \nu \leq q$, and $t \in (a, b)$. Suppose $\lim_{n\to\infty} f_n(t) = f(t)$ for all $a < t < b$. Then $f(t)$ is at least C^{q-1} smooth, $f_n^{(\nu)}(t) \to f^{(\nu)}(t)$ and $|f^{(\nu)}(t)| \leq C$ for all $1 \leq \nu \leq q-1$ and $t \in (a, b)$. Moreover, $f^{(q-1)}(t)$ is a Lipschitz continuous function with constant C; i.e. $|f^{(q-1)}(t) - f^{(q-1)}(t')| \leq C|t - t'|$.*

The proof is left as an exercise.

COROLLARY 4.61. *The curve $W^u(x)$ is $C^{\ell-2}$ smooth, all its $\ell-2$ derivatives are bounded by a constant independent of x, and its $(\ell-2)$nd derivative is Lipschitz continuous with a Lipschitz constant independent of x.*

REMARK 4.62. It is possible to show that for almost every $x \in \mathcal{M}$ the curve $W^u(x)$ is actually $C^{\ell-1}$ smooth, but this requires a separate argument; see e.g. Theorem 6.1 in [**KS86**, Part I].

Recall that $\mathcal{F}^{-n}(W^u(x)) \subset \mathcal{M} \setminus \mathcal{S}_{-\infty}$ is an unstable curve for every $n \geq 1$. The uniform hyperbolicity of $\mathcal{F}$ (see Section 4.4) implies

$$\left|\mathcal{F}^{-n}(W^u(x))\right| \leq C\Lambda^{-n}\left|W^u(x)\right|, \qquad \forall n \geq 1;$$

hence the preimages of $W^u(x)$ contract exponentially fast.

All our constructions have their time-reversals; see Remark 4.16. In particular, for every point $x \in \mathcal{M} \setminus S_\infty$ we have a stable curve $W^s(x)$ which satisfies

$$\left|\mathcal{F}^{n}(W^s(x))\right| \leq C\Lambda^{-n}\left|W^s(x)\right|, \qquad \forall n \geq 1.$$

DEFINITION 4.63. A smooth curve $W^u \subset \mathcal{M}$ is called an *unstable manifold* if the map $\mathcal{F}^{-n}$ is smooth on W^u for every $n \geq 1$ and

$$\lim_{n\to\infty} |\mathcal{F}^{-n}(W^u)| = 0.$$

Similarly, a smooth curve $W^s \subset \mathcal{M}$ is called a *stable manifold* if the map $\mathcal{F}^n$ is smooth on W^s for every $n \geq 1$ and

$$\lim_{n\to\infty} |\mathcal{F}^{n}(W^s)| = 0.$$

Our analysis shows that $W^u(x)$ and $W^s(x)$ are the maximal unstable and, respectively, stable manifolds passing through x, and the convergence in the above definition is actually exponentially fast. They also have the following invariance property:

EXERCISE 4.64. Prove that $W^u(x) \supseteq \mathcal{F}^{-k}\big[W^u(\mathcal{F}^k x)\big]$ for all $k \geq 1$. Accordingly, $\mathcal{F}^k\big(W^u(x)\big)$ is a union of some unstable manifolds.

REMARK 4.65. Stable and unstable manifolds were first constructed for smooth uniformly hyperbolic diffeomorphisms, where their graphs were obtained by successive approximations and by using the contracting map principle (this method goes back to Hadamard [**Had01**]). This construction was extended to smooth nonuniformly hyperbolic maps by Pesin [**Pes76, Pes77a**] and to hyperbolic maps with singularities by Katok and Strelcyn [**KS86**, Theorem 6.1]. We have presented here an alternative construction in which singularity curves play a key role. Our approach has two advantages: it gives uniform bounds on derivatives, and it allows us to directly estimate the size of unstable manifolds; see the next section.

4.12. Size of unstable manifolds

Let $W \subset \mathcal{M}$ be a smooth unstable (or stable) curve. Any point $x \in W$ divides W into two segments, and we denote by $r_W(x)$ the length (in the Euclidean metric) of the shorter one. For brevity, we put $r^u(x) := r_{W^u(x)}(x)$. If it happens that $W^u(x) = \emptyset$, then we set $r^u(x) = 0$. Clearly, $r^u(x)$ characterizes the size of $W^u(x)$.

THEOREM 4.66. *$W^u(x)$ exists (i.e. $r^u(x) > 0$) for almost every $x \in \mathcal{M}$. Furthermore,*

$$\mu\{x\colon r^u(x) < \varepsilon\} \le C\varepsilon \tag{4.54}$$

for some constant $C = C(\mathcal{D}) > 0$ and all $\varepsilon > 0$.

The existence of $W^u(x)$ follows from the general theory of hyperbolic maps with singularities [**KS86**, Theorem 6.1], but the estimate (4.54) is billiard specific. The theory guarantees only that $\mu\{x\colon r^u(x) < \varepsilon\} \le C\varepsilon^a$ for some $a > 0$.

PROOF. For every point $x \in \mathcal{M}$ denote by $d^u(x, \mathcal{S}_1)$ the length of the shortest unstable curve that connects x with the set $\mathcal{S}_1$.

LEMMA 4.67. *For any $x \in \mathcal{M}$*

$$r^u(x) \ge \min_{n\ge1} \hat{c}\,\Lambda^n\, d^u(\mathcal{F}^{-n}x, \mathcal{S}_1), \tag{4.55}$$

where $\hat{c} = \hat{c}(\mathcal{D}) > 0$ is a constant from (4.19).

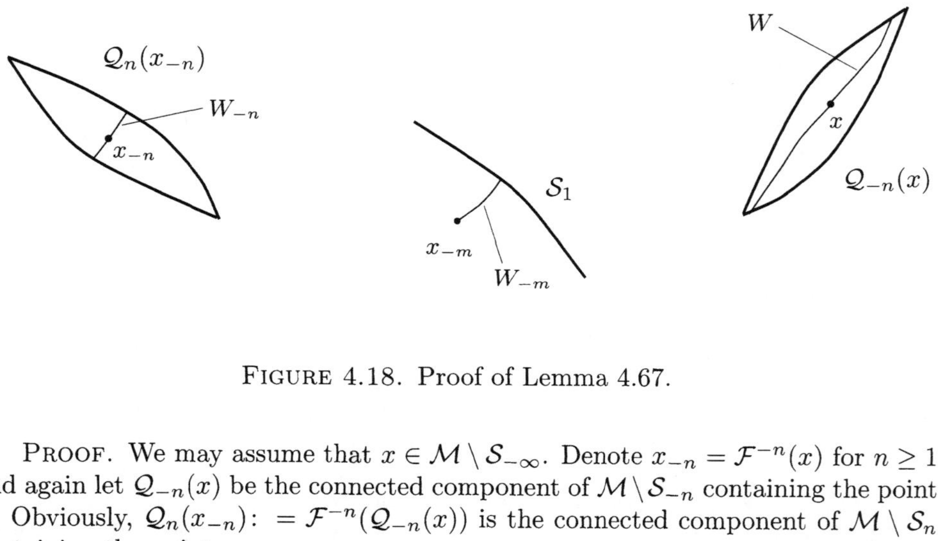

FIGURE 4.18. Proof of Lemma 4.67.

PROOF. We may assume that $x \in \mathcal{M} \setminus \mathcal{S}_{-\infty}$. Denote $x_{-n} = \mathcal{F}^{-n}(x)$ for $n \ge 1$ and again let $\mathcal{Q}_{-n}(x)$ be the connected component of $\mathcal{M}\setminus\mathcal{S}_{-n}$ containing the point x. Obviously, $\mathcal{Q}_n(x_{-n})\colon = \mathcal{F}^{-n}(\mathcal{Q}_{-n}(x))$ is the connected component of $\mathcal{M} \setminus \mathcal{S}_n$ containing the point x_{-n}.

Let W'_{-n} be an arbitrary unstable curve passing through x_{-n} and terminating on the opposite sides of $\mathcal{Q}_n(x_{-n})$; see Fig. 4.18. Then $W' = \mathcal{F}^n(W'_{-n})$ is an unstable curve passing through x and terminating on $\partial\mathcal{Q}_{-n}(x)$. It is divided by the point x into two segments, and we denote by W the shorter one. Put $W_{-m} = \mathcal{F}^{-m}(W)$ for all $m \ge 1$.

Since W_{-n} terminates on $\mathcal{S}_n$, there is an $m \in [1, n]$ such that W_{-m} joins x_{-m} with $\mathcal{S}_1$. Due to the uniform hyperbolicity of $\mathcal{F}$ (see (4.19)), we have

$$r_{W'}(x) = |W| \ge \hat{c}\,\Lambda^m |W_{-m}| \ge \hat{c}\,\Lambda^m d^u(\mathcal{F}^{-m}x, \mathcal{S}_1).$$

Since this holds for all $n \ge 1$, the lemma follows. □

Next, we pick $\hat{\Lambda} \in (1, \Lambda)$ and observe that

$$r^u(x) \ge r^u_*(x)\colon = \min_{n\ge1} \hat{c}\,\hat{\Lambda}^n\, d^u\big(\mathcal{F}^{-n}x, \mathcal{S}_1\big)$$

(we need to assume $\hat{\Lambda} < \Lambda$ for a later use).

So if $r^u(x) = 0$, then $r^u_*(x) = 0$; i.e. the past semitrajectory of x approaches the singularity set $\mathcal{S}_1$ faster than the exponential function $\hat{\Lambda}^{-n}$. We employ the Borel-Cantelli lemma to show that this happens with probability zero.

For any $\varepsilon > 0$ denote by $\mathcal{U}_\varepsilon(\mathcal{S}_1)$ the ε-neighborhood of the set $\mathcal{S}_1$ and let

$$\mathcal{U}^u_\varepsilon(\mathcal{S}_1) = \{x\colon d^u(x, \mathcal{S}_1) < \varepsilon\}$$

(this is sort of an ε-neighborhood of $\mathcal{S}_1$, measured along unstable curves). Due to (4.14), $\mathcal{U}^u_\varepsilon(\mathcal{S}_1) \subset \mathcal{U}_{D\varepsilon}(\mathcal{S}_1)$ for some constant $D > 0$ (because the set $\mathcal{S}_1$ consists of monotonically decreasing curves and horizontal lines); hence

$$\mu\big(\mathcal{U}^u_\varepsilon(\mathcal{S}_1)\big) < C\varepsilon \tag{4.56}$$

for some $C > 0$ and all $\varepsilon > 0$.

Next we show that $r^u_*(x) > 0$ for μ-almost every point $x \in \mathcal{M}$. Let

$$B_n\colon = \mathcal{F}^n\big(\mathcal{U}^u_{\hat{\Lambda}^{-n}}(\mathcal{S}_1)\big) = \big\{x\colon d^u\big(\mathcal{F}^{-n}(x), \mathcal{S}_1\big) \le \hat{\Lambda}^{-n}\big\}.$$

By the invariance of the measure μ we have

$$\sum_{n=1}^{\infty} \mu(B_n) = \sum_{n=1}^{\infty} \mu\big(\mathcal{U}^u_{\hat{\Lambda}^{-n}}(\mathcal{S}_1)\big) < \infty.$$

Denote by $B = \cap_{m\ge 1} \cup_{n\ge m} B_n$ the set of points that belong to infinitely many B_n's. The Borel-Cantelli lemma implies that $\mu(B) = 0$, and it is easy to check that $r^u_*(x) > 0$ for every $x \in \mathcal{M} \setminus B$.

LEMMA 4.68. *We have* $\mu\{x\colon r^u_*(x) < \varepsilon\} \le C\varepsilon$ *for some constant* $C = C(\mathcal{D}) > 0$ *and all* $\varepsilon > 0$.

PROOF. Observe that $r^u_*(x) < \varepsilon$ iff $x \in \mathcal{F}^n\big(\mathcal{U}^u_{\varepsilon\hat{c}^{-1}\hat{\Lambda}^{-n}}(\mathcal{S}_1)\big)$ for some $n \ge 1$. Now (4.56) implies the result. □

This completes the proof of Theorem 4.66. □

Observe that $x \in \mathcal{F}^n\big(\mathcal{U}^u_{\varepsilon\hat{c}^{-1}\hat{\Lambda}^{-n}}(\mathcal{S}_1)\big)$ means that x lies in a vicinity of S_{-n}. Hence, points x where $r^u_*(x) < \varepsilon$ (i.e., where unstable manifolds may be short) are concentrated in the vicinities of singularity curves $\mathcal{S}_{-n}$, primarily with small n (because the measure of points corresponding to large n's is exponentially small).

4.13. Additional facts about unstable manifolds

In the previous section, we used the Euclidean metric, but similar results hold in the p-metric as well, as we describe below.

Any smooth curve $W \subset \mathcal{M}$ is divided by any point $x \in W$ into two segments, and we denote by $p_W(x)$ the p-length (see Section 3.10) of the shorter one. We put $p^u(x)\colon = p_{W^u(x)}(x)$.

THEOREM 4.69. *We have* $\mu\{x\colon p^u(x) < \varepsilon\} \le C\varepsilon$ *for some constant* $C = C(\mathcal{D}) > 0$ *and all* $\varepsilon > 0$.

PROOF. Denote by $d^u_p(y, \mathcal{S}_1)$ the p-length of the shortest unstable curve that connects a point y with the set $\mathcal{S}_1$. An argument similar to the proof of Lemma 4.67 yields

$$p^u(x) \ge \min_{n\ge 1} \Lambda^n d^u_p(\mathcal{F}^{-n}x, \mathcal{S}_1). \tag{4.57}$$

We pick $\hat{\Lambda} \in (1, \Lambda)$ and observe that

$$p^u(x) \geq p^u_*(x) \colon = \min_{n\geq 1} \hat{\Lambda}^n \, d^u\big(\mathcal{F}^{-n}x, \mathcal{S}_1\big).$$

For any $\varepsilon > 0$ let

$$\mathcal{U}^u_{p,\varepsilon}(\mathcal{S}_1) = \{x \colon d^u_p(x, \mathcal{S}_1) < \varepsilon\}.$$

Next we prove an analogue of (4.56).

LEMMA 4.70. *We have*

$$\mu\big(\mathcal{U}^u_{p,\varepsilon}(\mathcal{S}_1)\big) < C\varepsilon, \tag{4.58}$$

where $C = C(\mathcal{D}) > 0$ is a constant.

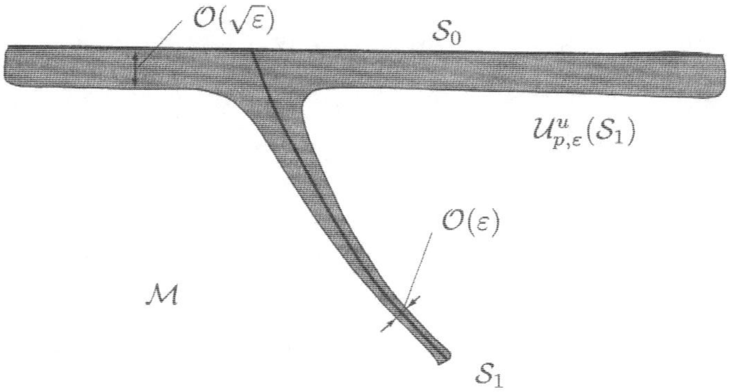

FIGURE 4.19. The set $\mathcal{U}^u_{p,\varepsilon}(\mathcal{S}_1)$.

PROOF. Recall that $\mathcal{S}_1$ consists of $\mathcal{S}_0 = \partial\mathcal{M}$ and finitely many smooth stable curves. The set $\mathcal{U}^u_{p,\varepsilon}(\mathcal{S}_1)$ is shown in Fig. 4.19; it gets thicker as $\cos\varphi \to 0$ and has width $\sim \sqrt{\varepsilon}$ near $\mathcal{S}_0$. We can foliate it by a one-parameter family of smooth unstable curves W_α, where $\alpha \in \mathfrak{A}$ is a parameter, such that $|W_\alpha|_p \leq 2\varepsilon$. Then

$$\mu\big(\mathcal{U}^u_{p,\varepsilon}(\mathcal{S}_1)\big) = \int_{\mathfrak{A}} \bigg(\int_{W_\alpha} \cos\varphi \, dm_\alpha\bigg) d\mu^f$$

where dm_α is the one-dimensional Lebesgue measure on W_α and $d\mu^f$ is the factor measure (the reader unfamiliar with measurable partitions and factor measures is referred to Appendix A). The inner integral is $\leq \text{const}\, \varepsilon$ for every W_α, since $\cos\varphi\, dm \asymp \cos\varphi\, dr = dp$ (the element of the p-length), and $|W_\alpha|_p \leq 2\varepsilon$. The factor measure is obviously finite, and its mass is bounded by a constant independent of ε. □

We now observe that $p^u_*(x) < \varepsilon$ iff $x \in \mathcal{F}^n\big(\mathcal{U}^u_{p,\varepsilon\hat{\Lambda}^{-n}}(\mathcal{S}_1)\big)$ for some $n \geq 1$. Now (4.58) implies that

$$\mu\{x \colon p^u_*(x) < \varepsilon\} \leq C\varepsilon$$

for some constant $C = C(\mathcal{D}) > 0$ and all $\varepsilon > 0$. This completes the proof of Theorem 4.69. □

Next we present a few more facts about unstable manifolds.

Let $\{x_m\}$ be a sequence of points in $\mathcal{M}$ converging to a point $y \in \mathcal{S}_{-\infty}$, as $m \to \infty$. It is easy to see that $r^u(x_m) \to 0$. Hence, arbitrarily short unstable manifolds are dense in $\mathcal{M}$: for any $\varepsilon > 0$ and any open set $\mathcal{U} \subset \mathcal{M}$ we have $\mu\{x \in \mathcal{U} \colon r^u(x) < \varepsilon\} > 0$.

EXERCISE 4.71. Show that if a sequence of unstable manifolds $\{W_m^u\}$ converges, as $m \to \infty$, to a curve W (in the C^0 metric), then W (taken without its endpoints) is an unstable manifold itself. Hint: Observe that $W \cap S_{-1} = \emptyset$, then use induction on n to prove that $\mathcal{F}^{-n}(W_m^u) \to \mathcal{F}^{-n}(W)$; hence $\mathcal{F}^{-n}(W) \cap S_{-1} = \emptyset$ for all $n \geq 1$.

Next, we consider the endpoints of the unstable manifold $W^u(x)$.

EXERCISE 4.72. Show that if $r_*^u(x) > 0$, then $r_*^u(y) > 0$ for all $y \in W^u(x)$. (Hint: Observe that $|\mathcal{F}^{-n}(W^u(x))| \leq \hat{c}\Lambda^{-n}$ and use the fact that $\hat{\Lambda} < \Lambda$.) Then argue that if y is an endpoint of $W^u(x)$, then either $r_*^u(y) > 0$ or $y \in \mathcal{S}_{-\infty}$. However, $r_*^u(y) > 0$ is impossible, since in that case $W^u(x)$ could be extended beyond the point y. Therefore, $y \in \mathcal{S}_{-\infty}$.

EXERCISE 4.73. Show that for μ-almost every $x \in \mathcal{M}$ both endpoints of $W^u(x)$ belong to $\mathcal{S}_{-\infty}$. (Use the previous exercise.)

Observe that no two unstable manifolds can intersect each other, but they can have a common endpoint (which lies on a singularity curve).

DEFINITION 4.74. For any $x \in \mathcal{M} \setminus \mathcal{S}_{-\infty}$, the *global* unstable manifold is

$$GW^u(x) = \cup_{n \geq 0} \mathcal{F}^n\big(W^u(\mathcal{F}^{-n}x)\big).$$

The global unstable manifold for a typical point x is a countable union of local unstable manifolds, some of which may have common endpoints.

4.14. Extension to type B billiard tables

In dispersing billiards of category B, the singularity set $\mathcal{S}_1$ consists of infinitely many curves (cf. Section 4.10) whose total length is infinite. Despite this, the estimates of Theorems 4.66 and 4.69 hold true:

THEOREM 4.75. *For billiards of category B, we have*

$$\mu\{x \colon r^u(x) < \varepsilon\} \leq C\varepsilon \tag{4.59}$$

$$\mu\{x \colon p^u(x) < \varepsilon\} \leq C\varepsilon \tag{4.60}$$

for some constant $C = C(\mathcal{D}) > 0$ and all $\varepsilon > 0$.

PROOF. We need to refine the arguments of the previous section, since the estimates (4.56) and (4.58), as stated, are not valid for the category B tables.

EXERCISE 4.76. Check that $\mu\big(\mathcal{U}_\varepsilon^u(\mathcal{S}_1)\big) \sim \varepsilon \ln(1/\varepsilon)$ and $\mu\big(\mathcal{U}_{p,\varepsilon}^u(\mathcal{S}_1)\big) \sim \varepsilon^{4/5}$. Hence both measures are $\gg \varepsilon$.

We will use the notation and results of Section 4.10. Let $E_u(x)$ be a piecewise constant function on $\mathcal{M}$ defined as follows: $E_u(x) = n^{3/2}$ for all $x \in D_{k,n}^+$ with $1 \leq k \leq k_{\max}$ and $n \geq 1$, and $E_u(x) = 1$ for all other points in $\mathcal{M}$.

EXERCISE 4.77. Prove that there is a small constant $\tilde{c} = \tilde{c}(\mathcal{D}) > 0$ such that $\tilde{c}E_u(x)$ is the lower bound on the factor of expansion of unstable vectors $dx \in \mathcal{T}_x\mathcal{M}$ for all $x \in \mathcal{M}$, i.e. $\|D_x\mathcal{F}(dx)\| \geq \tilde{c}E_u(x)\,\|dx\|$. Hint: Use (4.19) and Exercise 4.54.

Now (4.55) can be modified as follows:

$$r^u(x) \geq \min_{n\geq 1} \tilde{c}E_u(\mathcal{F}^{-n}x)\,\hat{c}\,\Lambda^{n-1}\,d^u(\mathcal{F}^{-n}x,\mathcal{S}_1). \tag{4.61}$$

Indeed, one can repeat the proof of Lemma 4.67 with the following modification:

$$|W| = |W_{-m}|\,\frac{|W_{-m+1}|}{|W_{-m}|}\,\frac{|W|}{|W_{-m+1}|} \geq \tilde{c}E_u(x_{-m})\,\hat{c}\,\Lambda^{m-1}|W_{-m}|;$$

then one gets (4.61). As before, we pick $\hat{\Lambda} \in (1,\Lambda)$ and observe that

$$r^u(x) \geq \tilde{r}^u_*(x) \colon = \min_{n\geq 1} \tilde{c}\,\hat{c}\,E_u(\mathcal{F}^{-n}x)\,\hat{\Lambda}^{n-1}\,d^u\big(\mathcal{F}^{-n}x,\mathcal{S}_1\big).$$

For any $\varepsilon > 0$ let

$$\tilde{\mathcal{U}}^u_\varepsilon(\mathcal{S}_1) = \{x\colon\, E_u(x)\,d^u(x,\mathcal{S}_1) < \varepsilon\}.$$

Observe that $\tilde{r}^u_*(x) < \varepsilon$ iff $x \in \mathcal{F}^n\big(\tilde{\mathcal{U}}^u_{\varepsilon\tilde{c}^{-1}\hat{c}^{-1}\hat{\Lambda}^{-n+1}}(\mathcal{S}_1)\big)$ for some $n \geq 1$. Thus, to prove (4.59) it is enough to derive the following analogue of (4.56):

$$\mu\big(\tilde{\mathcal{U}}^u_\varepsilon(\mathcal{S}_1)\big) < C\varepsilon \tag{4.62}$$

for some constant $C > 0$ and all $\varepsilon > 0$. For every $n \geq 1$ and $1 \leq k \leq k_{\max}$, the set $D^+_{k,n} \cap \tilde{\mathcal{U}}^u_\varepsilon(\mathcal{S}_1)$ lies within the $\mathcal{O}\big(n^{-3/2}\varepsilon\big)$-neighborhood of the boundary $\partial D^+_{n,k}$. Since the length of the latter is $\mathcal{O}\big(n^{-1/2}\big)$ (see Exercise 4.52), we get

$$\mu\big(D^+_{k,n} \cap \tilde{\mathcal{U}}^u_\varepsilon(\mathcal{S}_1)\big) < \text{const}\, n^{-5/2}\varepsilon \tag{4.63}$$

(the additional factor of $n^{-1/2}$ results from the density $\cos\varphi = \mathcal{O}\big(n^{-1/2}\big)$ of the measure μ on $D^+_{k,n}$). Denote

$$\mathcal{M}_0 = \mathcal{M} \setminus \cup_{k,n} D^+_{k,n}.$$

Since the total length of the singularity lines $\mathcal{S}_1$ within the set $\mathcal{M}_0$ is finite, it follows that

$$\mu\big(\mathcal{M}_0 \cap \tilde{\mathcal{U}}^u_\varepsilon(\mathcal{S}_1)\big) < \text{const}\,\varepsilon.$$

Adding these estimates up proves (4.62) and hence (4.59).

The proof of (4.60) is similar, but certain changes are noteworthy. The function $E_u(x)$ is defined in the same way as above, and $\tilde{c}E_u(x)$ will again be a lower bound on the factor of expansion of unstable vectors due to Exercise 4.54. Now (4.57) can be modified as follows:

$$p^u(x) \geq \min_{n\geq 1} \tilde{c}E_u(\mathcal{F}^{-n}x)\,\Lambda^{n-1}\,d^u_p(\mathcal{F}^{-n}x,\mathcal{S}_1).$$

We leave the verification to the reader. As before, we pick $\hat{\Lambda} \in (1,\Lambda)$ and observe that

$$p^u(x) \geq \tilde{p}^u_*(x) \colon = \min_{n\geq 1} \tilde{c}E_u(\mathcal{F}^{-n}x)\,\hat{\Lambda}^{n-1}\,d^u_p\big(\mathcal{F}^{-n}x,\mathcal{S}_1\big).$$

For any $\varepsilon > 0$ let

$$\tilde{\mathcal{U}}^u_{p,\varepsilon}(\mathcal{S}_1) = \{x\colon\, E_u(x)\,d^u_p(x,\mathcal{S}_1) < \varepsilon\}.$$

Observe that $\tilde{p}^u_*(x) < \varepsilon$ iff $x \in \mathcal{F}^n\big(\tilde{\mathcal{U}}^u_{p,\varepsilon\tilde{c}^{-1}\hat{\Lambda}^{-n+1}}(\mathcal{S}_1)\big)$ for some $n \geq 1$. Thus, to prove (4.60) it is enough to derive the following analogue of (4.58):

$$\mu\big(\tilde{\mathcal{U}}^u_{p,\varepsilon}(\mathcal{S}_1)\big) < C\varepsilon \tag{4.64}$$

for some constant $C > 0$ and all $\varepsilon > 0$.

EXERCISE 4.78. Show that for all $1 \le k \le k_{\max}$ and $n \ge 1$

$$\mu\big(D^+_{k,n} \cap \tilde{\mathcal{U}}^u_{p,\varepsilon}(\mathcal{S}_1)\big) < \text{const}\, n^{-2}\varepsilon.$$

Hint: Use the method of the proof of Lemma 4.70.

On the other hand,

$$\mu\big(\mathcal{M}_0 \cap \tilde{\mathcal{U}}^u_{p,\varepsilon}(\mathcal{S}_1)\big) < \text{const}\, \varepsilon,$$

which can be obtained as in the proof of Lemma 4.70. Adding these estimates up yields (4.64) and hence (4.60). □

CHAPTER 5

Dynamics of unstable manifolds

This chapter is devoted to three fundamental properties of stable and unstable manifolds in dispersing billiards. They are known by the names *distortion bounds*, *absolute continuity*, and *growth lemma*. These properties appear, in one way or another, in all advanced studies of billiard dynamics (most notably, in the proof of ergodicity and estimates on correlations; the latter are essential for the central limit theorem and many applications in physics).

These properties, in a basic form, were proved in Sinai's original paper [**Sin70**], but in subsequent studies they were substantially enhanced and refined. In this chapter we present their up-to-date versions in the most complete and versatile form.

We note that similar properties have been proved for many other classes of uniformly hyperbolic systems (with and without singularities); see references below. In billiards, however, these properties require a special construction ("homogeneous" unstable manifolds; see Section 5.4), and their proofs are rather difficult. They might constitute the most technical part of our theory, indeed, but they provide powerful tools to obtain far-reaching results.

Lastly, our discussion is restricted (as in the previous chapter) to dispersing billiards of main category A. Separate remarks are devoted to category B. The extension to other categories requires substantial technical work, and we omit it.

5.1. Measurable partition into unstable manifolds

We will define two important partitions of the collision space $\mathcal{M}$. As usual, a partition ξ is a collection $\{C_\alpha\}$ of disjoint subsets $C_\alpha \subset \mathcal{M}$ such that $\cup_\alpha C_\alpha = \mathcal{M}$; we denote by $\xi(x)$ the unique element of ξ containing the point $x \in \mathcal{M}$.

First, let ξ^u be the partition of $\mathcal{M}$ into unstable manifolds. Precisely, for every point $x \in \mathcal{M}$ we put $\xi^u(x) = W^u(x)$ if the latter is not empty, and $\xi^u(x) = \{x\}$ otherwise. Note that if y is an endpoint of $W^u(x)$, then $W^u(y) = \emptyset$; hence $\xi^u(y) = \{y\}$.

It is important that ξ^u is a *measurable* partition of $\mathcal{M}$. Measurable partitions have many useful properties; in particular, there are conditional measures on their elements. See Appendix A for the relevant definitions and facts.

To prove that ξ^u is a measurable partition, we need to construct a countable generator $\{B_m\}_{m=1}^\infty$ for it; see again Appendix A. Consider the countable collection of sets $\mathcal{Q}_{-n}(x)$ defined in Section 4.11 for all $x \in \mathcal{M}$ and $n \geq 1$, and their closures $\overline{\mathcal{Q}_{-n}(x)}$. Let $\{B_m\}$ denote this collection.

EXERCISE 5.1. Verify that $\{B_m\}$ is indeed a generator for the partition ξ^u. You need to verify two properties. First, for every element $C \in \xi^u$ and every B_m show that either $C \subset B_m$ or $C \cap B_m = \emptyset$. Second, for every two distinct elements

$C_1 \neq C_2$ of the partition ξ^u find a B_m such that $C_1 \subset B_m$ and $C_2 \cap B_m = \emptyset$, or vice versa.

As the partition ξ^u is measurable, we can apply operations on partitions described in Appendices A and C. First, the image $\xi^u_n = \{\mathcal{F}^n(C) \colon C \in \xi^u\}$ is a measurable partition of $\mathcal{M}$ for every $n \in \mathbb{Z}$. Second, the partition $\xi^u_{n_1}$ is finer than $\xi^u_{n_2}$ for any $n_1 < n_2$ (recall Exercise 4.64); i.e. the sequence of partitions $\{\xi^u_n\}$ is decreasing:

$$\cdots \succcurlyeq \xi^u_{-n} \succcurlyeq \cdots \succcurlyeq \xi^u_{-1} \succcurlyeq \xi^u_0 \succcurlyeq \xi^u_1 \succcurlyeq \cdots \succcurlyeq \xi^u_n \succcurlyeq \cdots . \tag{5.1}$$

Next we describe their join $\vee_{n=-\infty}^{\infty} \xi^u_n$ and their meet $\wedge_{n=-\infty}^{\infty} \xi^u_n$; see Appendix A. As the diameter of $\mathcal{F}^{-n}(W^u)$ converges to zero, as $n \to \infty$, we easily obtain that

$$\vee_{n=-\infty}^{\infty} \xi^u_n = \varepsilon, \tag{5.2}$$

where ε denotes the partition of $\mathcal{M}$ into individual points.

The meet $\pi^u = \wedge_{n=-\infty}^{\infty} \xi^u_n$ is more difficult to handle, because it cannot be constructed 'elementwise'. Indeed, the images $\mathcal{F}^n(W^u)$ of unstable manifolds grow, as $n \to \infty$, and converge to global unstable manifolds GW^u (see Definition 4.74), but the partition of $\mathcal{M}$ into global unstable manifolds is not measurable.

Since we will not use the partition π^u in our book, we discuss it only briefly for the curious reader. It is related to the so called Pinsker partition $\pi(\mathcal{F})$, which is the join of all finite partitions with zero entropy; see Appendix C. Now (5.1) and (5.2) imply that $\pi^u \succcurlyeq \pi(\mathcal{F})$; see Appendix C. Moreover, for certain classes of hyperbolic maps, including hyperbolic billiards, we have the identity $\pi^u = \pi(\mathcal{F})$ (mod 0) (see [**Sin66, Pes77b**]), so π^u coincides with the Pinsker partition. Next, it is a general fact in ergodic theory that an automorphism is K-mixing if and only if its Pinsker partition is trivial; i.e. if its elements have measure 0 or 1; see Appendix C. We will prove in Chapter 6 that our map $\mathcal{F}$ is indeed K-mixing; hence its Pinsker partition is trivial. Thus, π^u is actually a trivial partition.

Due to the time reversibility (see Remark 4.16), the partition ξ^s of $\mathcal{M}$ into stable manifolds has similar properties. In particular, its images $\xi^s_n = \mathcal{F}^n \xi^s$ make an increasing sequence

$$\cdots \preccurlyeq \xi^s_{-n} \preccurlyeq \cdots \preccurlyeq \xi^s_{-1} \preccurlyeq \xi^s_0 \preccurlyeq \xi^s_1 \preccurlyeq \cdots \preccurlyeq \xi^s_n \preccurlyeq \cdots ,$$

their join is $\vee_{n=-\infty}^{\infty} \xi^s_n = \varepsilon$, and their meet $\pi^s = \wedge_{n=-\infty}^{\infty} \xi^s_n$ has properties similar to those of π^u.

5.2. u-SRB densities

The measurability of the partition ξ^u implies the existence of conditional measures on its elements induced by the measure μ; see Appendix A. Here we describe those measures.

We introduce a handy notation first. For any point $x \in W$ on any unstable (or stable) curve $W \subset \mathcal{M}$ we denote by

$$\mathcal{J}_W \mathcal{F}^n(x) = \frac{\|D_x \mathcal{F}^n(dx)\|}{\|dx\|} \tag{5.3}$$

the Jacobian of the restriction of the map $\mathcal{F}^n$ to the one-dimensional curve W, at the point x, in the Euclidean metric (here dx denotes a nonzero tangent vector to W at x).

THEOREM 5.2. *For almost every $x \in \mathcal{M}$, the conditional measure $\nu_{W^u(x)}$ on the unstable manifold $W = W^u(x)$ is absolutely continuous with respect to the Lebesgue measure on W^u, and its density $\rho_W(y)$ is a $C^{\ell-2}$ smooth function satisfying*

$$\frac{\rho_W(y)}{\rho_W(z)} = \lim_{n\to\infty} \frac{\mathcal{J}_W\mathcal{F}^{-n}(y)}{\mathcal{J}_W\mathcal{F}^{-n}(z)} \tag{5.4}$$

for every $y, z \in W$ (note that this relation determines the density ρ_W on W up to a constant multiple, but since ν_W has to be a probability measure, ρ_W is specified completely by the condition $\int_W \rho_W(x)\,dx = 1$).

Smooth conditional measures on unstable manifolds with the above properties were first constructed by Sinai in 1968 [**Sin68a**] (for Anosov diffeomorphisms), and now all the above facts have been proven for large classes of hyperbolic maps and have become standard in ergodic theory. Their full proofs, however, are rather complicated; we refer the reader to [**Sin68a**] and [**PeS82**, Theorem 3] for a complete exposition (a similar argument is also given in [**Led84**, Proposition 3.1]). The smoothness of the conditional densities (for Anosov diffeomorphisms) was established in [**Ll92**, Corollary 4.4].

We sketch here only a semi-heuristic plausible argument explaining (but not proving) the above theorem. The existence of the limit (5.4) and the $C^{\ell-2}$ smoothness of ρ_W will be proven immediately after Proposition 5.4.

If unstable manifolds were parallel lines (like in an example in Appendix A illustrated in Fig. A.1), then the density of $\nu_{W^u(x)}$ would be proportional to that of μ when the latter is restricted to $W^u(x)$. To make this observation work for us, we apply $\mathcal{F}^{-n}$ and use the invariance of the measure.

Pick a large $n \gg 1$ and consider all unstable manifolds $W^u \subset \mathcal{Q}_{-2n}(x)$. Of course, the curves $W^u \subset \mathcal{Q}_{-2n}(x)$ are neither parallel nor rectilinear, since the function $\mathcal{V}^u(x)$ is not constant. However, the curves $W^u_n\colon = \mathcal{F}^{-n}(W^u)$ will be close enough to being parallel and rectilinear in the following sense.

Note that

$$W^u_n \subset \mathcal{Q}_{-n,n}\colon = \mathcal{Q}_{-n}(\mathcal{F}^{-n}x) \cap \mathcal{Q}_n(\mathcal{F}^{-n}x),$$

where $\mathcal{Q}_n(\mathcal{F}^{-n}x)$ denotes the connected components of $\mathcal{M}\setminus\mathcal{S}_n$ containing the point $\mathcal{F}^{-n}(x)$. The diameter of the domain $\mathcal{Q}_{-n,n}$ is less than $C\Lambda^{-n}$, where $C > 0$ is the maximal size of stable and unstable curves, so each curve will be nearly rectilinear (because the curvature of unstable manifolds is uniformly bounded). Also, since $\mathcal{F}^{-n}$ is smooth on $\mathcal{Q}_{-n,n}$, the oscillation of the function $\mathcal{B}^u$ on $\mathcal{Q}_{-n,n}$ can be made arbitrarily small by choosing n large enough; see Section 4.7. In this sense all the curves $W^u_n \subset \mathcal{Q}_{-n,n}$ are almost parallel.

Now the measure μ on $\mathcal{Q}_{-n,n}$ can be well approximated by a measure whose conditional distributions on W^u_n are uniform. Pushing this approximate measure forward under $\mathcal{F}^n$, we obtain an approximation to the measure μ on $\mathcal{Q}_{-2n}(x)$, whose conditional distributions on $W^u \subset \mathcal{Q}_{-2n}(x)$ have densities $\rho_{W^u,n}$ satisfying

$$\frac{\rho_{W^u,n}(y)}{\rho_{W^u,n}(z)} = \frac{\mathcal{J}_{W^u}\mathcal{F}^{-n}(y)}{\mathcal{J}_{W^u}\mathcal{F}^{-n}(z)}$$

for any $y, z \in W^u$. Taking the limit $n \to \infty$ gives the result.

DEFINITION 5.3. For any unstable manifold $W \subset \mathcal{M}$, the unique probability density ρ_W satisfying (5.4) is called the *u-SRB density*, and the corresponding probability measure ν_W is called the *u-SRB* measure (or, sometimes, the *Gibbs u-measure*).

The abbreviation SRB stands for the names of Sinai, Ruelle, and Bowen, who have studied these measures; see [**Sin68a, Ru76, Bow75**]. The u-SRB measures have the following invariance property:

PROPOSITION 5.4. *Let $W \subset \mathcal{M}$ be an unstable manifold and $\mathcal{F}(W) = \cup_i W_i$ its image, which is a finite or countable union of unstable manifolds W_i. Then the image $\mathcal{F}(\nu_W)$ of the measure ν_W, conditioned on each W_i, coincides with ν_{W_i}.*

The proof is left as an exercise. Hint: Use (5.4).

We now verify that the limit (5.4) exists, at least for typical unstable manifolds. Taking the logarithm and using the chain rule reduces this problem to the convergence of the series

$$\sum_{n=0}^{\infty}\Big(\ln \mathcal{J}_{W_n}\mathcal{F}^{-1}(y_n) - \ln \mathcal{J}_{W_n}\mathcal{F}^{-1}(z_n)\Big), \tag{5.5}$$

where $W_n = \mathcal{F}^{-n}(W)$ and $y_n = \mathcal{F}^{-n}(y)$ (the same for z_n).

To describe the contraction of unstable manifolds under the inverse map $\mathcal{F}^{-1}$, we pick $x \in W$ and denote $x_n = \mathcal{F}^{-n}(x)$. Then, using the results and notation of (4.18) and (3.40), we have

$$\begin{aligned}\mathcal{J}_{W_n}\mathcal{F}^{-1}(x_n) &= \frac{1}{1+\tau_{n+1}\,\mathcal{B}^+_{n+1}}\,\frac{\cos\varphi_n}{\cos\varphi_{n+1}}\,\frac{\sqrt{1+\mathcal{V}^2_{n+1}}}{\sqrt{1+\mathcal{V}^2_n}}\\ &= \frac{\cos\varphi_n}{2\mathcal{K}_{n+1}\tau_{n+1}+\cos\varphi_{n+1}\big(1+\tau_{n+1}\mathcal{B}^-_{n+1}\big)}\,\frac{\sqrt{1+\mathcal{V}^2_{n+1}}}{\sqrt{1+\mathcal{V}^2_n}},\end{aligned} \tag{5.6}$$

where $\tau_n = \tau(x_n)$, and $\mathcal{B}_n$, $\mathcal{K}_n$, etc., are taken at the point x_n; note that we used the mirror equation (3.33). Therefore

$$\begin{aligned}\ln \mathcal{J}_{W_n}\mathcal{F}^{-1}(x_n) &= \ln\cos\varphi_n + \tfrac{1}{2}\ln(1+\mathcal{V}^2_{n+1}) - \tfrac{1}{2}\ln(1+\mathcal{V}^2_n)\\ &\quad - \ln\big[2\mathcal{K}_{n+1}\tau_{n+1}+\cos\varphi_{n+1}\big(1+\tau_{n+1}\mathcal{B}^-_{n+1}\big)\big].\end{aligned} \tag{5.7}$$

Due to the uniform hyperbolicity of $\mathcal{F}$ (cf. (4.19)), $\operatorname{dist}(y_n, z_n) \le \operatorname{diam}(W_n) \le \hat{C}\Lambda^{-n}$, where $\hat{C} = 1/\hat{c}$. Thus,

$$\big|\ln \mathcal{J}_{W_n}\mathcal{F}^{-1}(y_n) - \ln \mathcal{J}_{W_n}\mathcal{F}^{-1}(z_n)\big| \le \hat{C}\Lambda^{-n}\max\Big|\frac{d}{dx_n}\ln \mathcal{J}_{W_n}\mathcal{F}^{-1}(x_n)\Big|$$

where the maximum is taken over all $x_n \in W_n$ and d/dx_n denotes the derivative with respect to the Euclidean length on W_n (we do not fix an orientation of W_n, since we need only the absolute value of this derivative).

Next we differentiate (5.7) with respect to x_n. Observe, however, that (5.7) contains functions that depend on x_{n+1}, but those can be handled by the chain rule

$$\frac{d}{dx_n} = \frac{dx_{n+1}}{dx_n}\,\frac{d}{dx_{n+1}} = \mathcal{J}_{W_n}\mathcal{F}^{-1}(x_n)\,\frac{d}{dx_{n+1}}$$

(note also that the factor $\mathcal{J}_{W_n}\mathcal{F}^{-1}(x_n)$ is uniformly bounded).

Now, we recall that the quantities φ, $\mathcal{K}$, $\mathcal{B}^-$, and $\mathcal{V}$ are $C^{\ell-2}$ smooth functions on unstable manifolds, with uniformly bounded derivatives (cf. Corollary 4.61, Remark 4.62, and Lemma 4.30), while τ_{n+1} is a C^ℓ smooth function of x_{n+1} and x_n, which easily follows from (4.51). Also, all the expressions on the right-hand side of (5.7), except the first one, are uniformly bounded. Thus,

$$\frac{d}{dx_n} \ln \mathcal{J}_{W_n} \mathcal{F}^{-1}(x_n) = \frac{Q_n(x_n)}{\cos \varphi_n} \tag{5.8}$$

where Q_n is a $C^{\ell-3}$ smooth uniformly bounded function of x_n with uniformly bounded derivatives.

REMARK 5.5. In billiards of category B, the free path τ is bounded below (by $\tau_{\min} > 0$), but not above; hence the last logarithmic term in (5.7) is bounded below, but not above. This does not lead to serious complications, however, since the derivative of that term is

$$\frac{\tilde{Q}(x_n)}{2\mathcal{K}_{n+1}\tau_{n+1} + \cos\varphi_{n+1}\left(1 + \tau_{n+1}\mathcal{B}^-_{n+1}\right)},$$

where $\tilde{Q}(x_n)$ is a smooth uniformly bounded function of x_n with uniformly bounded derivatives. The denominator here is bounded away from zero; hence (5.8) also holds for billiards of category B.

In billiards of categories C–F, the free path τ may be arbitrarily close to zero. This causes more serious complications in our estimates, but they can be adapted to these cases, too [**BSC91**, Appendix 1].

We continue our analysis. Due to (5.8)

$$\left|\ln \mathcal{J}_{W_n} \mathcal{F}^{-1}(y_n) - \ln \mathcal{J}_{W_n} \mathcal{F}^{-1}(z_n)\right| \leq \frac{\text{const} \cdot \Lambda^{-n}}{\min_x \cos\varphi_n},$$

where the minimum is taken over all $x \in W$. To deal with the potentially small denominator here, we need to assume that $r^u_*(x) > 0$ for some point $x \in W$ (and hence, for all points $x \in W$; see Exercise 4.72), which is true for typical W's; see Section 4.12. We also restrict our analysis to the segment $W[y,z] \subset W$ of the curve W with endpoints y and z. Then we have $d^u(\mathcal{F}^{-n}x, \mathcal{S}_1) \geq c\hat{\Lambda}^{-n}$; hence $\cos\varphi_n > c\hat{\Lambda}^{-n}$, for all $n \geq 1$ and all $x \in W[y,z]$ with some $c = c(y,z) > 0$. We also recall that $\hat{\Lambda} < \Lambda$; hence $\hat{\Lambda}^{-n} \gg \Lambda^{-n}$, and so

$$\min_x \cos\varphi_n > \tfrac{1}{2}\, c\hat{\Lambda}^{-n}.$$

Thus the series (5.5) converges exponentially, and the u-SRB density ρ_W exists.

REMARK 5.6. In the above proof we had to assume that $r^u_* > 0$ on W; i.e. we needed $\mathcal{F}^{-n}(W[y,z])$ to be at least a certain distance ($\geq$const $\hat{\Lambda}^{-n}$) from the singularities $\mathcal{S}_1$, and this distance must be much larger than the size of $\mathcal{F}^{-n}(W[y,z])$. Moreover, we believe that if this assumption fails; i.e. $\mathcal{F}^{-n}(W[y,z])$ comes too close to singularities too often, then the u-SRB density may not exist.

5.3. Distortion control and homogeneity strips

The u-SRB densities ρ_W exist on almost all unstable manifolds $W \subset \mathcal{M}$, but they may be very nonuniform. The fluctuations of ρ_W on W result from uneven contraction of the preimages $\mathcal{F}^{-n}(W)$ under the inverse map $\mathcal{F}^{-1}$ (see the previous section), and the greatest differences in the contraction factor (strongest distortions of unstable manifolds by $\mathcal{F}^{-1}$) occur near the singularities $\mathcal{S}_0 = \{\cos\varphi = 0\}$.

EXERCISE 5.7. Use (5.8) to show that if one endpoint of $\mathcal{F}^{-n}(W)$ lies on $\mathcal{S}_0$ for some $n \geq 0$, then the density ρ_W vanishes at the corresponding endpoint of W.

For our further studies, we will need to control fluctuations of the u-SRB densities ρ_W. Precisely, we need to divide every unstable manifold into pieces, on each of which the u-SRB density is almost uniform ($\rho_W \approx \text{const}$).

In the case of uniformly hyperbolic maps without singularities (Anosov and Axiom A diffeomorphisms), the u-SRB density satisfies a uniform bound

$$\left|\frac{d}{dx} \ln \rho_W(x)\right| \leq C = \text{const}; \tag{5.9}$$

see [**LMM86**, Section 2] and [**L192**]. Therefore for all $x, y \in W$

$$e^{-C|W|} \leq \frac{\rho_W(x)}{\rho_W(y)} \leq e^{C|W|}. \tag{5.10}$$

Thus, for any $\varepsilon > 0$ there is $\delta > 0$ such that the fluctuations of the u-SRB densities on δ-small unstable manifolds are ε-small.

In billiards, we do not have this luxury. As indicated by (5.8) and Exercise 5.7, the derivative $\frac{d}{dx} \ln \rho_W(x)$ may be unbounded, and fluctuations of the u-SRB density may be arbitrarily large on arbitrarily short unstable manifolds. To deal with this problem, we will subdivide unstable manifolds into pieces on which the fluctuations of the u-SRB measure will be bounded.

Since the greatest fluctuations occur in the vicinity of $\mathcal{S}_0$, Ya. Sinai proposed [**BSC91**] to partition unstable manifolds by countably many lines that are parallel to $\mathcal{S}_0$ and accumulate on $\mathcal{S}_0$, as defined below and illustrated by Fig. 5.1:

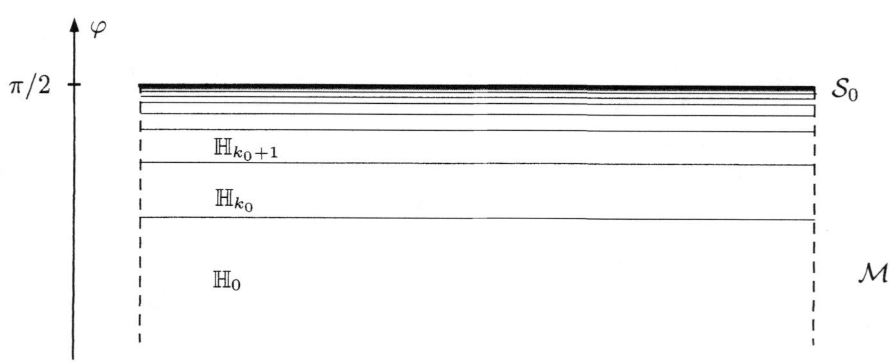

FIGURE 5.1. Homogeneity strips.

DEFINITION 5.8. Let $k_0 \geq 1$ be a large constant. For each $k \geq k_0$ we define two *homogeneity strips* $\mathbb{H}_{\pm k} \subset \mathcal{M}$ by

$$\mathbb{H}_k = \{(r, \varphi) \colon \pi/2 - k^{-2} < \varphi < \pi/2 - (k+1)^{-2}\}$$

and

$$\mathbb{H}_{-k} = \{(r,\varphi)\colon\ -\pi/2 + (k+1)^{-2} < \varphi < -\pi/2 + k^{-2}\}.$$

We also put

$$\mathbb{H}_0 = \{(r,\varphi)\colon\ -\pi/2 + k_0^{-2} < \varphi < \pi/2 - k_0^{-2}\}. \tag{5.11}$$

Now $\mathcal{M}$ is divided into homogeneity strips $\mathbb{H}_k$. Denote by

$$\mathbb{S}_k = \{(r,\varphi)\colon |\varphi| = \pi/2 - k^{-2}\}$$

for $|k| \geq k_0$ the boundaries of the homogeneity strips and put

$$\mathbb{S} = \cup_{|k|\geq k_0}\mathbb{S}_k. \tag{5.12}$$

The large constant $k_0 \gg 1$ will be selected in Section 5.10 (for the curious reader it is in the proof of Lemma 5.56).

DEFINITION 5.9. A stable or unstable curve $W \subset \mathcal{M}$ is said to be *weakly homogeneous* if W belongs to one strip $\mathbb{H}_k$.

We record an important estimate:

LEMMA 5.10. *For any point $x = (r,\varphi)$ on a weakly homogeneous unstable or stable curve $W \subset \mathbb{H}_k$ we have*

$$|W| \leq \text{const}\,(|k|+1)^{-3} \leq \text{const}\,\cos^{3/2}\varphi. \tag{5.13}$$

PROOF. This follows from the bounds (4.14) and Definition 5.8. □

5.4. Homogeneous unstable manifolds

DEFINITION 5.11. An unstable manifold $W \subset \mathcal{M}$ is said to be *homogeneous* if $\mathcal{F}^{-n}(W)$ is weakly homogeneous for every $n \geq 0$. Similarly, a stable manifold $W \subset \mathcal{M}$ is said to be homogeneous if $\mathcal{F}^n(W)$ is weakly homogeneous for every $n \geq 0$.

For brevity, we call homogeneous manifolds *H-manifolds.*

EXERCISE 5.12. Prove that given an unstable manifold $W \subset \mathcal{M}$, the connected components of $W \setminus \cup_{n\geq 0}\mathcal{F}^n(\mathbb{S})$ will be H-manifolds. Similarly, for any stable manifold $W \subset \mathcal{M}$ the connected components of $W \setminus \cup_{n\geq 0}\mathcal{F}^{-n}(\mathbb{S})$ will be stable H-manifolds.

Thus, the lines $\mathbb{S}_k \subset \mathbb{S}$ and their images act as singularities. It is natural then to combine them with the existing singularity sets $\mathcal{S}_n$, $-\infty < n < \infty$, and treat the union of the two as new ('extended') singularities. To make this formal, it is proposed in [**Ch99**] to redefine the collision space $\mathcal{M}$ as follows:

DEFINITION 5.13 (**New collision space**). Each connected component of the the old collision space $\mathcal{M}_i \subset \mathcal{M}$ (recall (2.17)) is divided into countably many connected homogeneity strips $\mathbb{H}_{i,k}\colon = \mathcal{M}_i \cap \mathbb{H}_k$. The new collision space $\mathcal{M}_{\mathbb{H}}$ is defined to be a disjoint union of the closures of the $\mathbb{H}_{i,k}$'s.

Observe that the new space $\mathcal{M}_{\mathbb{H}}$ is closed but no longer compact (it has countably many connected components). Also note that $\partial\mathcal{M}_{\mathbb{H}} = \mathbb{S}$.

The map $\mathcal{F}\colon \mathcal{M} \to \mathcal{M}$ naturally acts on the new collision space $\mathcal{M}_{\mathbb{H}}$, and we denote it by $\mathcal{F}_{\mathbb{H}}\colon \mathcal{M}_{\mathbb{H}} \to \mathcal{M}_{\mathbb{H}}$. Note that $\mathcal{F}_{\mathbb{H}}$ is not defined on $\mathcal{F}^{-1}(\mathcal{S}_0)$, since the line $\mathcal{S}_0$ does not belong in the new collision space. On the other hand, every line of $\mathbb{S}_k \subset \mathbb{S}$ is transformed into two lines in $\mathcal{M}_{\mathbb{H}}$; thus the map $\mathcal{F}_{\mathbb{H}}$ is two-valued on $\mathcal{F}^{-1}(\mathbb{S})$ (this should come as no surprise to us, since generic billiard maps and flows are multivalued at singular points; cf. Chapter 3).

EXERCISE 5.14. Show that the map $\mathcal{F}_{\mathbb{H}}$ lacks smoothness on the 'extended' singularity set $\mathcal{S}_1 \cup \mathbb{S} \cup \mathcal{F}^{-1}(\mathbb{S})$. Similarly, the map $\mathcal{F}_{\mathbb{H}}^n$ is not smooth on the 'extended' singularity set

$$\mathcal{S}_n^{\mathbb{H}} := \mathcal{S}_n \cup \left(\cup_{m=0}^n \mathcal{F}^{-m}(\mathbb{S})\right). \tag{5.14}$$

The inverse map $\mathcal{F}_{\mathbb{H}}^{-n}$ is not smooth on the 'extended' singularity set

$$\mathcal{S}_{-n}^{\mathbb{H}} := \mathcal{S}_{-n} \cup \left(\cup_{m=0}^n \mathcal{F}^{m}(\mathbb{S})\right). \tag{5.15}$$

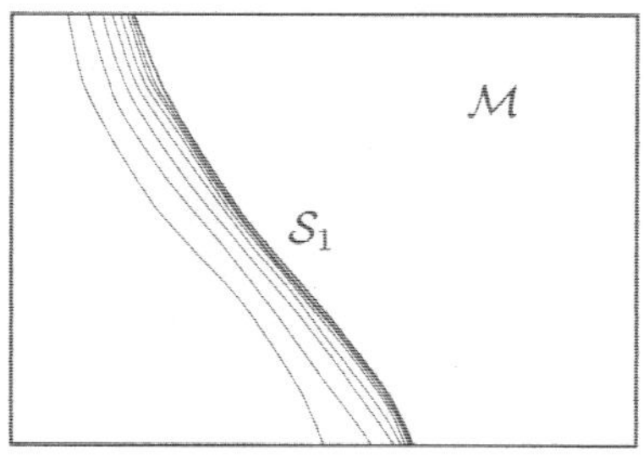

FIGURE 5.2. Curves of $\mathcal{F}^{-1}(\mathbb{S})$ accumulating on $\mathcal{S}_1$.

Observe that $\mathcal{F}^{-1}(\mathbb{S})$ is a countable union of stable curves that are almost parallel to each other and accumulate on the old singular curves $\mathcal{S}_1 \setminus \mathcal{S}_0$; see Fig. 5.2. (Note that $\mathcal{F}^{-1}(\mathbb{S})$ accumulate on one side of each curve $\mathcal{S}_1 \setminus \mathcal{S}_0$, the side which is mapped onto a vicinity of $\mathcal{S}_0$; cf. Section 4.9.) Thus, for every $n \geq 1$ the set $\mathcal{F}^{-n}(\mathbb{S})$ is a countable union of compact $C^{\ell-1}$ smooth stable curves. Similarly, $\mathcal{F}^n(\mathbb{S})$ is a countable union of compact $C^{\ell-1}$ smooth unstable curves.

EXERCISE 5.15. Based on the results of Section 4.5 show that all $\ell-1$ derivatives of those curves are uniformly bounded.

EXERCISE 5.16. Verify that the alignment property (Section 4.8), the continuation property (Section 4.9) and the 'convexity' property (Exercise 4.49) hold for the 'extended' singularity set $\mathcal{S}_n^{\mathbb{H}}$.

Let $\mathcal{Q}_{-n}^{\mathbb{H}}(x)$ denote the connected component of the set $\mathcal{M} \setminus \mathcal{S}_{-n}^{\mathbb{H}}$ containing x. These new domains $\mathcal{Q}_{-n}^{\mathbb{H}}(x)$ are nested in the old domains $\mathcal{Q}_{-n}(x)$ used in Section 4.11 and have a similar shape. The arguments of that section carry over to these new domains and imply that the intersection $\cap_{n=1}^{\infty} \overline{\mathcal{Q}_{-n}^{\mathbb{H}}(x)}$ is a closed unstable curve. Let $W_{\mathbb{H}}^u(x)$ denote that curve without its endpoints. Then $W_{\mathbb{H}}^u(x)$ is the maximal unstable H-manifold passing through x.

All the facts mentioned in Section 5.2 hold for unstable H-manifolds. First, the partition $\xi_{\mathbb{H}}^u$ of $\mathcal{M}$ into (maximal) unstable H-manifolds is measurable. The corresponding conditional measures on the curves $W_{\mathbb{H}}^u$ are u-SRB measures whose density is determined by (5.4).

Due to the time reversibility (see Remark 4.16), the definition and construction of stable H-manifolds $W_{\mathbb{H}}^s(x)$ are very similar. The partition $\xi_{\mathbb{H}}^s$ of $\mathcal{M}$ into (maximal) stable H-manifolds is also measurable.

5.5. Size of H-manifolds

Our analysis in the previous section clearly shows that unstable (and stable) H-manifolds for the map $\mathcal{F}\colon \mathcal{M} \to \mathcal{M}$ are unstable (respectively, stable) manifolds for the map $\mathcal{F}_{\mathbb{H}}\colon \mathcal{M}_{\mathbb{H}} \to \mathcal{M}_{\mathbb{H}}$ on the new collision space $\mathcal{M}_{\mathbb{H}}$; the map $\mathcal{F}_{\mathbb{H}}$ has "more singularities" than $\mathcal{F}$ does.

This fact allows us to use the techniques of Sections 4.11–4.14 to estimate the length of the H-manifold $W^u_{\mathbb{H}}(x)$. Denote by $r^u_{\mathbb{H}}(x) = r_{W^u_{\mathbb{H}}(x)}(x)$ the distance, measured along $W^u_{\mathbb{H}}(x)$, from x to the nearest endpoint of $W^u_{\mathbb{H}}(x)$.

THEOREM 5.17. *$W^u_{\mathbb{H}}(x)$ exists (i.e. $r^u_{\mathbb{H}}(x) > 0$) for almost every $x \in \mathcal{M}$. Furthermore,*

$$\mu\{x\colon r^u_{\mathbb{H}}(x) < \varepsilon\} \leq C\varepsilon \tag{5.16}$$

for some constant $C = C(\mathcal{D}) > 0$ and all $\varepsilon > 0$.

PROOF. Our argument closely follows the proof of (4.59) in Section 4.14, and we leave certain steps to the reader as exercises. Let $E_u(x)$ be a piecewise constant function on $\mathcal{M}$ defined by $E_u(x) = (|k|+1)^2$ for all $x \in \mathcal{F}^{-1}(\mathbb{H}_k)$.

EXERCISE 5.18. Show that there is a small constant $\tilde{c} = \tilde{c}(\mathcal{D}) > 0$ such that the factor of expansion of unstable vectors $dx \in \mathcal{T}_x\mathcal{M}$ for all $x \in \mathcal{M}$ will be between $\tilde{c}E_u(x)$ and $\tilde{c}^{-1}E_u(x)$, i.e.

$$\tilde{c}E_u(x) \leq \frac{\|D_x\mathcal{F}(dx)\|}{\|dx\|} \leq \tilde{c}^{-1}E_u(x).$$

Hint: Note that if $x \in \mathcal{F}^{-1}(\mathbb{H}_k)$, then $\cos\varphi \approx (|k|+1)^{-2}$ at the point $\mathcal{F}(x) \in \mathbb{H}_k$ and use the relation (4.20) to conclude that $\|D_x\mathcal{F}(dx)\|/\|dx\| \asymp (|k|+1)^2$.

For every point $x \in \mathcal{M}$ denote by $d^u(x, \mathcal{S}^{\mathbb{H}}_1)$ the length of the shortest unstable curve that connects x with the set $\mathcal{S}^{\mathbb{H}}_1$.

EXERCISE 5.19. Adapt the proof of (4.61) to show that

$$r^u_{\mathbb{H}}(x) \geq \min_{n\geq 1} \tilde{c}E_u(\mathcal{F}^{-n}x)\,\hat{c}\,\Lambda^{n-1}\,d^u(\mathcal{F}^{-n}x, \mathcal{S}^{\mathbb{H}}_1).$$

We pick $\hat{\Lambda} \in (1, \Lambda)$ and observe that

$$r^u_{\mathbb{H}}(x) \geq r^u_{\mathbb{H}*}(x)\colon = \min_{n\geq 1} \tilde{c}\,\hat{c}\,E_u(\mathcal{F}^{-n}x)\hat{\Lambda}^{n-1}\,d^u\big(\mathcal{F}^{-n}x, \mathcal{S}^{\mathbb{H}}_1\big).$$

For any $\varepsilon > 0$ let

$$\mathcal{U}^u_\varepsilon(\mathcal{S}^{\mathbb{H}}_1) = \{x\colon E_u(x)\,d^u(x, \mathcal{S}^{\mathbb{H}}_1) < \varepsilon\}.$$

Observe that $r^u_{\mathbb{H}*}(x) < \varepsilon$ iff $x \in \mathcal{F}^n\big(\mathcal{U}^u_{\varepsilon\tilde{c}^{-1}\hat{c}^{-1}\hat{\Lambda}^{-n+1}}(\mathcal{S}^{\mathbb{H}}_1)\big)$ for some $n \geq 1$.

EXERCISE 5.20. Adapt the proof of (4.62) to show that $\mu\big(\mathcal{U}^u_\varepsilon(\mathcal{S}^{\mathbb{H}}_1)\big) < C\varepsilon$ for some constant $C > 0$ and all $\varepsilon > 0$. Hint: For every k the set $\mathcal{F}^{-1}(\mathbb{H}_k) \cap \mathcal{U}^u_\varepsilon(\mathcal{S}^{\mathbb{H}}_1)$ lies within the $\mathcal{O}\big((|k|+1)^{-2}\varepsilon\big)$-neighborhood of the boundary $\partial\mathcal{F}^{-1}(\mathbb{H}_k)$; hence

$$\mu\big(\mathcal{F}^{-1}(\mathbb{H}_k) \cap \mathcal{U}^u_\varepsilon(\mathcal{S}^{\mathbb{H}}_1)\big) < \text{const}\,(|k|+1)^{-2}\varepsilon.$$

Also check that the ε-neighborhood of the set $\mathbb{S}$ (which is a part of $\mathcal{S}^{\mathbb{H}}_1$) has μ-measure $\mathcal{O}(\varepsilon)$.

It now follows that $\mu\{x\colon r^u_{\mathbb{H}*}(x) < \varepsilon\} \leq C\varepsilon$; hence the theorem. □

Theorem 5.17 shows that unstable H-manifolds, despite being much shorter than ordinary unstable manifolds, satisfy the same linear 'tail bound' (compare (4.54) and (5.16)!). Note, however, that there is no linear tail bound on the length of H-manifolds in the p-metric, as the following exercise shows.

EXERCISE 5.21. Let $p^u_{\mathbb{H}}(x)$ denote the distance from x to the nearest endpoint of the H-manifold $W^u_{\mathbb{H}}(x)$ in the p-metric. Show that $\mu\{x\colon p^u_{\mathbb{H}}(x) < \varepsilon\} \geq C\varepsilon^{4/5}$ for some constant $C > 0$ and all sufficiently small $\varepsilon > 0$. Hint: For any $x \in \mathbb{H}_k$ we have $|W^u_{\mathbb{H}}(x)|_p \leq \text{const}\, k^{-5}$.

EXERCISE 5.22. Show that $\mu\{x\colon p^u_{\mathbb{H}}(x) < \varepsilon\} \leq C\varepsilon^{4/5}$ for some constant $C > 0$ and all sufficiently small $\varepsilon > 0$.

Next we present a few additional facts about H-manifolds.

EXERCISE 5.23. Show that for almost every point $x \in \mathcal{M}$ the set $W^u(x) \cap \mathcal{F}^n(\mathbb{S})$ is empty for all but finitely many $n \geq 0$. Hint: Note that if $W^u(x) \cap \mathcal{F}^n(\mathbb{S}) \neq \emptyset$, then $\mathcal{F}^{-n}(W^u(x)) \subset \mathcal{U}^u_{C\hat{\Lambda}^{-n+1}}(\mathbb{S})$ for some constant $C > 0$. Then use the estimate in Exercise 5.20 and Borel-Cantelli lemma.

Nonetheless, for every $n \geq 0$ the set $W^u(x) \cap \mathcal{F}^n(\mathbb{S})$ is either finite or countable, but in the latter case its points converge to an endpoint of $W^u(x)$. Thus, the set $W^u(x) \cap \left(\cup_{n\geq 0}\mathcal{F}^n(\mathbb{S})\right)$ of the endpoints of H-manifolds $W^u_{\mathbb{H}} \subset W(x)$ is either finite or countable, and in the latter case it accumulates on the endpoints of $W^u(x)$ (but there are no accumulation points inside $W^u(x)$); see Fig. 5.3.

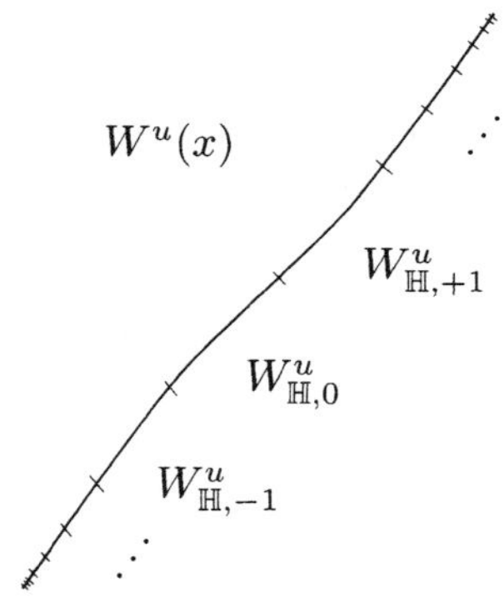

FIGURE 5.3. Partition of a typical unstable manifold into H-manifolds.

EXERCISE 5.24. Use the method of Exercise 4.71 to show that if a sequence of unstable H-manifolds $\{W^u_m\}$ converges, as $m \to \infty$, to a curve W (in the C^0 metric), then W (taken without its endpoints) is an unstable H-manifold itself.

EXERCISE 5.25. Use the method of Exercise 4.72 to show that if $r^u_{\mathbb{H}*}(x) > 0$, then $r^u_{\mathbb{H}*}(y) > 0$ for all $y \in W^u_{\mathbb{H}}(x)$. Then show that for μ-almost every $x \in \mathcal{M}$ both endpoints of $W^u_{\mathbb{H}}(x)$ belong to $\cup_{n\geq 0}\mathcal{S}^{\mathbb{H}}_{-n}$.

Lastly, Theorem 5.17 can be extended to billiards of category B, and we indicate necessary modifications in its proof for this case. Some details are left to the reader.

First, the function $E_u(x)$ must be redefined so that for each $x \in \mathcal{F}^{-1}(\mathbb{H}_k \cap D^-_{l,n})$ (equivalently, for each $x \in \mathcal{F}^{-1}(\mathbb{H}_k) \cap D^+_{l,n}$) (see the definition of the n-cells $D^{\pm}_{l,n}$ in Section 4.10), we have $E_u(x) = n(|k|+1)^2$.

EXERCISE 5.26. Show that there is a small constant $\tilde{c} = \tilde{c}(\mathcal{D}) > 0$ such that the factor of expansion of unstable vectors $dx \in \mathcal{T}_x\mathcal{M}$ for all $x \in \mathcal{M}$ will be between $\tilde{c}E_u(x)$ and $\tilde{c}^{-1}E_u(x)$, i.e.

$$\tilde{c}E_u(x) \leq \frac{\|D_x\mathcal{F}(dx)\|}{\|dx\|} \leq \tilde{c}^{-1}E_u(x).$$

Hint: Argue as in Exercise 5.18 and use the fact that $\tau(x) \asymp n$ for $x \in D^+_{l,n}$.

The last modification is required in the proof of the crucial estimate $\mu\big(\mathcal{U}^u_\varepsilon(\mathcal{S}^{\mathbb{H}}_1)\big) < C\varepsilon$. One can check directly that for each n-cell $D^+_{l,n}$ the set $\mathcal{F}^{-1}(\mathbb{H}_k) \cap D^+_{l,n}$ is a narrow strip stretching from top to bottom of this cell; see Fig. 5.4. The length of this cell is $\mathcal{O}\big(n^{-1/2}\big)$ (see Section 4.10); hence

$$\mu\big(\mathcal{F}^{-1}(\mathbb{H}_k) \cap D^+_{l,n} \cap \mathcal{U}^u_\varepsilon(\mathcal{S}^{\mathbb{H}}_1)\big) < \text{const}\, n^{-2}(|k|+1)^{-2}\varepsilon$$

(the additional factor of $n^{-1/2}$ results from the density $\cos\varphi = \mathcal{O}\big(n^{-1/2}\big)$ of the measure μ on $D^+_{l,n}$, just as in our early estimate (4.63)). Now adding these estimates over all k and n completes the proof.

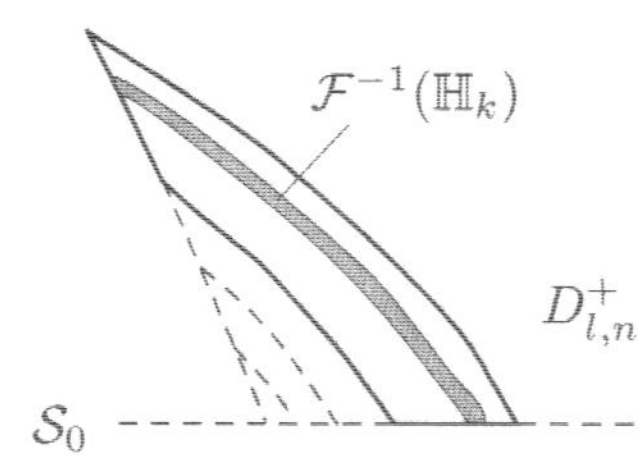

FIGURE 5.4. The set $\mathcal{F}^{-1}(\mathbb{H}_k) \cap D^+_{l,n}$ within the cell $D^+_{l,n}$ for billiards of category B.

5.6. Distortion bounds

As we noted, uniform bounds similar to (5.9) and (5.10) do not hold for billiards. Here we derive nonuniform bounds on the fluctuations of the u-SRB densities on unstable H-manifolds that do hold for billiards.

Our first lemma will apply to rather generic unstable curves (provided they are homogeneous in the appropriate sense), but further theorems will be restricted to unstable H-manifolds.

Let W be an unstable curve such that $W_n = \mathcal{F}^{-n}(W)$ is a weakly homogeneous unstable curve for all $0 \leq n \leq N-1$. We assume that the curves W_n are $C^{\ell-1}$ smooth and all their $\ell - 1$ derivatives are uniformly bounded (say, by constants C_ν, $1 \leq \nu \leq \ell - 1$, involved in Proposition 4.29).

We use the notation of Section 5.4. In particular for every point $x \in W$ we put $x_n = \mathcal{F}^{-n}(x)$ and denote by $\mathcal{J}_W\mathcal{F}^{-n}(x)$ the factor of contraction of W under the map $\mathcal{F}^{-n}$ at x. The following is an analogue of (5.10):

LEMMA 5.27. *For every $y, z \in W$ and every $1 \le n \le N$ we have*

$$C_{\mathrm{d}}^{-1} \le e^{-C|W|^{1/3}} \le \frac{\mathcal{J}_W \mathcal{F}^{-n}(y)}{\mathcal{J}_W \mathcal{F}^{-n}(z)} \le e^{C|W|^{1/3}} \le C_{\mathrm{d}},$$

where $C = C(\mathcal{D}) > 0$ and $C_{\mathrm{d}} = C_{\mathrm{d}}(\mathcal{D}) > 0$ are constants (the subscript d *in C_{d} stands for 'distortions').*

Note that, obviously, $|W|$ can be replaced here by the distance between the points y and z.

PROOF. Taking the logarithm and using the chain rule gives

$$\begin{aligned}\left|\ln \mathcal{J}_W \mathcal{F}^{-n}(y) - \ln \mathcal{J}_W \mathcal{F}^{-n}(z)\right| &\le \sum_{m=0}^{n-1} \left|\ln \mathcal{J}_{W_m} \mathcal{F}^{-1}(y_m) - \ln \mathcal{J}_{W_m} \mathcal{F}^{-1}(z_m)\right| \\ &\le \sum_{m=0}^{n-1} |W_m| \max\left|\frac{d}{dx_m} \ln \mathcal{J}_{W_m} \mathcal{F}^{-1}(x_m)\right| \\ &\le \text{const} \sum_{m=0}^{n-1} |W_m| / \cos\varphi_m \\ &\le \text{const} \sum_{m=0}^{n-1} |W_m|^{1/3}.\end{aligned}$$

In the last two steps we used (5.8) and (5.13). Note that we need uniform bounds on the derivatives of W_n's to use (5.8). Now, due to the uniform hyperbolicity of $\mathcal{F}$, we have $\sum_{m=0}^{n-1} |W_n|^{1/3} \le \text{const}\,|W|^{1/3}$. □

COROLLARY 5.28. *For every $x \in W$ and every $1 \le n \le N$*

$$C_{\mathrm{d}}^{-1} \le \frac{\mathcal{J}_W \mathcal{F}^{-n}(x)}{|W_n|/|W|} \le C_{\mathrm{d}}.$$

The following is an analogue of (5.9):

THEOREM 5.29 (Distortion bounds). *For every unstable H-manifold $W \subset \mathcal{M}$*

$$\left|\frac{d}{dx} \ln \rho_W(x)\right| \le \frac{C}{|W|^{2/3}},$$

where $C = C(\mathcal{D}) > 0$ is a constant.

PROOF. Fix a point $\bar{x} \in W$. Then due to (5.4) and the chain rule

$$\ln \rho_W(x) = \ln \rho_W(\bar{x}) + \sum_{n=0}^{\infty} \big(\ln \mathcal{J}_{W_n} \mathcal{F}^{-1}(x_n) - \ln \mathcal{J}_{W_n} \mathcal{F}^{-1}(\bar{x}_n)\big),$$

(the exponential convergence of this series was established in Section 5.2). Hence

$$\begin{aligned}\frac{d}{dx} \ln \rho_W(x) &= \sum_{n=0}^{\infty} \frac{d}{dx} \ln \mathcal{J}_{W_n} \mathcal{F}^{-1}(x_n) \\ &= \sum_{n=0}^{\infty} \frac{dx_n}{dx} \frac{d}{dx_n} \ln \mathcal{J}_{W_n} \mathcal{F}^{-1}(x_n) \\ &= \sum_{n=0}^{\infty} \mathcal{J}_W \mathcal{F}^{-n}(x) \frac{Q_n(x_n)}{\cos\varphi_n} \qquad (5.17)\end{aligned}$$

(provided this series converges uniformly in x, which will follow from the subsequent analysis). Here Q_n is a $C^{\ell-3}$ smooth uniformly bounded function of x_n with uniformly bounded derivatives; cf. (5.8). Therefore,

$$
\begin{aligned}
\left|\frac{d}{dx}\ln\rho_W(x)\right| &\le \text{const}\sum_{n=0}^{\infty}\frac{\mathcal{J}_W\mathcal{F}^{-n}(x)}{\cos\varphi_n}\\
&\le \text{const}\sum_{n=0}^{\infty}\frac{|W_n|/|W|}{|W_n|^{2/3}}\\
&\le \text{const}\sum_{n=0}^{\infty}\frac{\Lambda^{-n/3}}{|W|^{2/3}} \qquad (5.18)
\end{aligned}
$$

where we used (5.13), Corollary 5.28, and the uniform hyperbolicity of $\mathcal{F}$; cf. (4.19). Theorem 5.29 is proved. □

COROLLARY 5.30. *For every $x,y\in W$*

$$C_{\mathrm{d}}^{-1}\le e^{-C|W|^{1/3}}\le\frac{\rho_W(x)}{\rho_W(y)}\le e^{C|W|^{1/3}}\le C_{\mathrm{d}},$$

where $C=C(\mathcal{D})>0$ is a constant.

Theorem 5.29 can be generalized to higher order derivatives:

THEOREM 5.31 (Higher order distortion bounds). *For every $\nu=1,\dots,\ell-2$ there is a constant $C_\nu=C_\nu(\mathcal{D})>0$ such that*

$$\left|\frac{d^\nu}{dx^\nu}\ln\rho_W(x)\right|\le\frac{C_\nu}{|W|^{2\nu/3}}.$$

PROOF. For $\nu=1$ see the previous theorem. For $\nu\ge2$ we need to subsequently differentiate (5.17). First observe that

$$
\begin{aligned}
\frac{d}{dx_n}\mathcal{J}_{W_n}\mathcal{F}^{-1}(x_n) &= \mathcal{J}_{W_n}\mathcal{F}^{-1}(x_n)\,\frac{d}{dx_n}\ln\mathcal{J}_{W_n}\mathcal{F}^{-1}(x_n)\\
&= \mathcal{J}_{W_n}\mathcal{F}^{-1}(x_n)\,\frac{Q_n(x_n)}{\cos\varphi_n};
\end{aligned}
$$

hence

$$
\begin{aligned}
\frac{d}{dx}\mathcal{J}_W\mathcal{F}^{-n}(x) &= \frac{d}{dx}\prod_{m=0}^{n-1}\mathcal{J}_{W_m}\mathcal{F}^{-1}(x_m)\\
&= \sum_{m=0}^{n-1}\frac{\mathcal{J}_W\mathcal{F}^{-n}(x)}{\mathcal{J}_{W_m}\mathcal{F}^{-1}(x_m)}\,\frac{dx_m}{dx}\,\frac{d}{dx_m}\mathcal{J}_{W_m}\mathcal{F}^{-1}(x_m)\\
&= \mathcal{J}_W\mathcal{F}^{-n}(x)\sum_{m=0}^{n-1}\mathcal{J}_W\mathcal{F}^{-m}(x)\,\frac{Q_m(x_m)}{\cos\varphi_m}.
\end{aligned}
$$

Also,

$$
\begin{aligned}
\frac{d}{dx}\frac{Q_n(x_n)}{\cos\varphi_n} &= \frac{dx_n}{dx}\,\frac{d}{dx_n}\,\frac{Q_n(x_n)}{\cos\varphi_n}\\
&= \mathcal{J}_W\mathcal{F}^{-n}(x)\left[\frac{Q_n'(x_n)}{\cos\varphi_n}+\frac{Q_n(x_n)\sin\varphi_n}{\cos^2\varphi_n}\,\frac{d\varphi_n}{dx_n}\right],
\end{aligned}
$$

where

$$\frac{d\varphi_n}{dx_n} = \frac{d\varphi_n}{\sqrt{dr_n^2 + d\varphi_n^2}} = \frac{\mathcal{V}_n}{\sqrt{1+\mathcal{V}_n^2}}$$

is a $C^{\ell-2}$ smooth uniformly bounded function with uniformly bounded derivatives along unstable manifolds; see the notation of (5.6) and (5.7). Hence

$$\frac{d}{dx}\frac{Q_n(x_n)}{\cos\varphi_n} = \mathcal{J}_W\mathcal{F}^{-n}(x)\,\frac{Q_n^{(1)}(x_n)}{\cos^2\varphi_n},$$

where $Q_n^{(1)}$ is a $C^{\ell-4}$ smooth uniformly bounded function of x_n with uniformly bounded derivatives along unstable manifolds. We note that the smoothness of the functions involved in our analysis decreases as we estimate the derivatives of $\ln \rho_W(x)$. Thus, bounds on higher derivatives require higher smoothness of the billiard table $\mathcal{D}$.

Combining the above estimates gives

$$(5.19)\qquad \begin{aligned}\frac{d^2}{dx^2}\ln\rho_W(x) &= \sum_{n=0}^{\infty}\sum_{m=0}^{n-1}\mathcal{J}_W\mathcal{F}^{-n}(x)\,\mathcal{J}_W\mathcal{F}^{-m}(x)\,\frac{Q_n(x_n)\,Q_m(x_m)}{\cos\varphi_n\,\cos\varphi_m}\\ &\quad+\sum_{n=0}^{\infty}\bigl[\mathcal{J}_W\mathcal{F}^{-n}(x)\bigr]^2\,\frac{Q_n^{(1)}(x_n)}{\cos^2\varphi_n}.\end{aligned}$$

Repeating the estimates in (5.18) we obtain

$$\left|\frac{d^2}{dx^2}\ln\rho_W(x)\right| \le \text{const}\left(\sum_{n=0}^{\infty}\frac{\Lambda^{-n/3}}{|W|^{2/3}}\right)^2 \le \frac{\text{const}}{|W|^{4/3}},$$

thus proving the theorem for $\nu = 2$. The proof for $\nu \ge 3$ is a straightforward extension of the above arguments, which we leave to the reader. $\square$

EXERCISE 5.32. Prove Theorem 5.31 for $\nu = 3$. You'll need to express the third derivative in a way similar to (5.19). Hint: The most important term of the respective formula will look like

$$\sum_{n=0}^{\infty}\bigl[\mathcal{J}_W\mathcal{F}^{-n}(x)\bigr]^3\,\frac{Q_n^{(2)}(x_n)}{\cos^3\varphi_n},$$

where $Q_n^{(2)}$ will be a $C^{\ell-5}$ smooth uniformly bounded function of x_n with uniformly bounded derivatives along unstable manifolds.

REMARK 5.33. Theorem 5.29 was first proved in [**BSC91**, Appendix 1] (though in a slightly weaker form). Theorem 5.31 was proved in [**CD06**, Appendix B], but only for $\nu = 1$ and $\nu = 2$. Its extension to $\nu > 2$ was just published in [**Ch06a**].

The above estimates can be extended to unstable curves under the conditions described right before Lemma 5.27. Namely, let W be an unstable curve such that $W_n = \mathcal{F}^{-n}(W)$ is a weakly homogeneous unstable curve for all $0 \le n \le N-1$. We assume that the curves W_n are $C^{\ell-1}$ smooth and all their $\ell-1$ derivatives are uniformly bounded (in accordance with Proposition 4.29).

PROPOSITION 5.34. *For every* $\nu = 1, \dots, \ell-2$ *there is a constant* $C_\nu = C_\nu(\mathcal{D}) > 0$ *such that*

$$\left|\frac{d^\nu}{dx^\nu}\ln\mathcal{J}_W\mathcal{F}^{-n}(x)\right| \le \frac{C_\nu}{|W|^{2\nu/3}}.$$

PROOF. Our argument is based on Lemma 5.27. It differs only slightly from the proofs of Theorems 5.29 and 5.31, mainly because the above derivatives are represented by finite, rather than infinite, series. We recommend that the reader verify this. □

In ergodic theory, one often has to iterate measures. The classical Krylov-Bogolubov theorem [**KH95**, Theorem 4.1.1] states that any continuous transformation $F\colon X \to X$ of a compact metric space X has at least one invariant measure μ. To prove this, one chooses an arbitrary probability measure μ_0 on X and argues that any limit point of Cesaro averages

$$\mu_n^{\mathrm{a}} = \frac{1}{n}\sum_{i=0}^{n-1} F^i \mu_0 \tag{5.20}$$

is an F-invariant measure.

Even though the Krylov-Bogolubov theorem does not apply to billiards (the map $\mathcal{F}$ is not continuous), iterations of certain measures under $\mathcal{F}$ play an important role in the study of the ergodic properties of the collision map $\mathcal{F}$.

Let μ_0 be a probability measure concentrated on an unstable curve W with density $\rho_W^{(0)}$ with respect to the Lebesgue measure on W. Then for every $n \geq 1$ the measure $\mu_n = \mathcal{F}^n \mu_0$ will have density on the set $\mathcal{F}^n(W)$ given by

$$\rho^{(n)}(x) = \mathcal{J}_{\mathcal{F}^n W} \mathcal{F}^{-n}(x)\, \rho_W^{(0)}\big(\mathcal{F}^{-n}(x)\big).$$

The Jacobian $\mathcal{J}_{\mathcal{F}^n W} \mathcal{F}^{-n}(x)$ can be controlled by Lemma 5.27 and Proposition 5.34, but this requires partitioning of W and its images into weakly homogeneous unstable curves, which we call *homogeneous components*, or H-components.

DEFINITION 5.35. Given an unstable curve $W \subset \mathcal{M}$ and $n \geq 0$ we define *H-components* of $\mathcal{F}^n(W)$ to be maximal subcurves $W' \subset \mathcal{F}^n(W)$ such that $\mathcal{F}^{-i}(W')$ is a weakly homogeneous unstable curve for every $0 \leq i \leq n$.

Observe that the H-components of $\mathcal{F}^n(W)$ can also be defined as smooth pieces of $\mathcal{F}_{\mathbb{H}}^n(W) \subset \mathcal{M}_{\mathbb{H}}$, where $\mathcal{F}_{\mathbb{H}}\colon \mathcal{M}_{\mathbb{H}} \to \mathcal{M}_{\mathbb{H}}$ is the map on the "new collision space" introduced in Section 5.4.

We assume that the curve W is $C^{\ell-1}$ smooth and all its $\ell - 1$ derivatives are uniformly bounded, in accordance with Proposition 4.29. Then the same will hold for every H-component of $\mathcal{F}^n(W)$, $n \geq 1$. Now Proposition 5.34 applies and we obtain

PROPOSITION 5.36. *If the initial density $\rho_W^{(0)}$ on W is $C^{\ell-2}$ smooth, then $\rho^{(n)}$ will be $C^{\ell-2}$ smooth on every H-component W' of $\mathcal{F}^n(W)$. Moreover, for every $\nu = 1, \dots, \ell - 2$ there will be a constant $C_\nu = C_\nu(\mathcal{D}) > 0$ such that*

$$\left|\frac{d^\nu}{dx^\nu} \ln \rho^{(n)}(x)\right| \leq \frac{C_\nu}{|W'|^{2\nu/3}}$$

at every point $x \in W'$ and for all j and $n \geq n_W$, where n_W depends on the maximal value of the derivatives of the initial density $\rho_W^{(0)}$ (if they are small enough, then $n_W = 0$; i.e. the above estimate holds for all $n \geq 0$).

5.7. Holonomy map

Let us consider again an unstable curve W and a smooth probability measure μ_0 concentrated on W. We will see later in Chapter 7 that the measure $\mu_n = \mathcal{F}^n \mu_0$ weakly converges, as $n \to \infty$, to the same $\mathcal{F}$-invariant measure μ, which is defined in Section 2.12 (Definition 2.36). That is, the limit measure will be independent of W and μ_0. The proof of this fact uses the following construction that provides means of control over the evolution of the exponentially growing number of H-components of $\mathcal{F}^n(W)$.

Let $W^1, W^2 \subset \mathcal{F}^n(W)$ be two H-components of $\mathcal{F}^n(W)$ sufficiently close to each other. Some points $x \in W^1$ may be joined with some other points $\bar{x} \in W^2$ by pieces of stable manifolds. The images of the points x and $\bar{x}$ will get closer together exponentially fast in the future, so their asymptotic behavior will be almost identical.

This allows us to identify the images of all points on all H-components of $\mathcal{F}^n(W)$ that lie on the same stable manifold and respectively lump together all the measures (induced by μ_n) they carry.

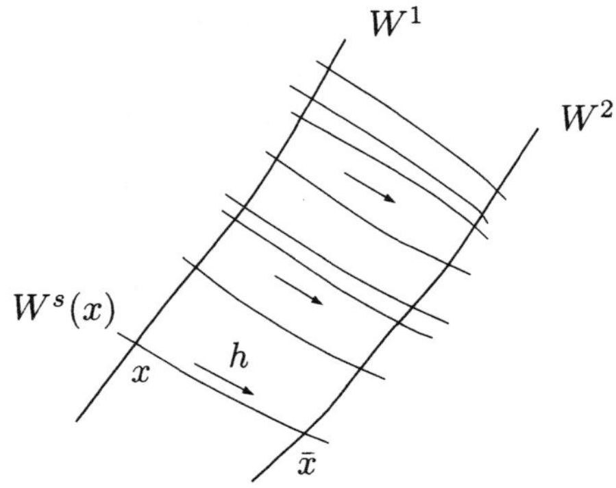

FIGURE 5.5. Holonomy map.

DEFINITION 5.37 ('Classical' holonomy map). Let $W^1, W^2 \subset \mathcal{M}$ be two unstable curves. Denote

$$W^i_* = \{x \in W^i \colon W^s(x) \cap W^{3-i} \neq \emptyset\}$$

for $i = 1, 2$. The map $\mathbf{h}\colon W^1_* \to W^2_*$ taking every point $x \in W^1_*$ to $\bar{x} = W^s(x) \cap W^2$ is called the *holonomy map* (see Fig. 5.5).

This map is often described as *sliding along stable manifolds.* Earlier publications, especially by the Russian school, used the term *canonical isomorphism.* Note that the map $\mathbf{h}$ is an order-preserving homeomorphism between two subsets of the curves W^1 and W^2 (indeed, recall that stable manifolds cannot cross each other).

The complementary sets $W^i \setminus W^i_*$ consist of intervals around singularity curves $S \subset \mathcal{S}_\infty$ that cross W^i where the stable manifolds are not long enough to reach the other curve W^{3-i}.

EXERCISE 5.38. Recall that arbitrarily short stable manifolds are dense in $\mathcal{M}$; cf. Section 4.13. Argue that the sets $W^i_* \subset W^i$ are nowhere dense in W^i (their closures are Cantor sets). We will see, however, that the sets W_*^i, despite being nowhere dense, usually have positive Lebesgue measure on W^i.

The future trajectories of the points $x \in W_*^1$ and $\bar{x} = \mathbf{h}(x)$ exponentially converge to each other (indeed, $\operatorname{dist}(\mathcal{F}^m(x), \mathcal{F}^m(\bar{x})) \leq C\Lambda^{-m}$, due to the uniform hyperbolicity of $\mathcal{F}$). Thus, the future images of the sets W_*^1 and W_*^2 may be identified, at least in the topological sense. We will develop these ideas further in Chapter 7.

However, in order to identify (lump together) the measures they carry, as we suggested above, we need the Lebesgue measure $\mathbf{m}_1$ on W_*^1 to be transformed by $\mathbf{h}$ into the Lebesgue measure $\mathbf{m}_2$ on W_*^2 (i.e., the holonomy map $\mathbf{h}$ does not 'distort' the Lebesgue measure). Of course, for nonlinear maps certain distortions do occur, so that the measure $\mathbf{m}_1^* = \mathbf{h}^{-1}(\mathbf{m}_2)$ on W_*^1 differs from $\mathbf{m}_1$. But we will show that the distortions are small – i.e. the Jacobian of the holonomy map $\mathbf{h}$

$$J\mathbf{h}(x) = \frac{d\mathbf{m}_1^*}{d\mathbf{m}_1}(x)$$

is bounded – and under certain conditions it is close to one. First, we derive a general formula for the Jacobian $J\mathbf{h}$:

THEOREM 5.39 ([**AS67, Sin70**]). *At $\mathbf{m}_1$-almost every point $x \in W_*^1$ the Jacobian of the holonomy map satisfies*

$$J\mathbf{h}(x) = \lim_{n\to\infty} \frac{\mathcal{J}_{W^1}\mathcal{F}^n(x)}{\mathcal{J}_{W^2}\mathcal{F}^n(\mathbf{h}(x))}. \tag{5.21}$$

This limit is finite and positive at $\mathbf{m}_1$-almost every point $x \in W_*^1$; i.e. the holonomy map $\mathbf{h}$ is absolutely continuous.

This is yet another standard fact in ergodic theory (like the earlier Theorem 5.2) whose meaning is rather transparent but whose full proof is fairly lengthy, and for this reason we will omit it. We sketch here only a plausible argument justifying (but not proving) the above theorem.

Let $x \in W_*^1$. We may assume that x is a Lebesgue density point of the set W_*^1 on the curve W^1 in order to ensure abundance of other points of W_*^1 in arbitrarily small vicinities of $x \in W^1$. More precisely, for any $\varepsilon > 0$ there is a $\delta > 0$ such that for any segment $V^1 \subset W^1$ such that $x \in V^1$ and $\mathbf{m}_1(V_1) = |V^1| < \delta$ we have

$$\mathbf{m}_1(V^1 \cap W_*^1) > (1-\varepsilon)\mathbf{m}_1(V^1).$$

Pick a large n and consider the connected component $\mathcal{Q}_n(x)$ of the set $\mathcal{M} \setminus \mathcal{S}_n$ containing the point x (see Section 4.11). Take two points $y^1, y^2 \subset W_*^1 \cap \mathcal{Q}_n(x)$ so that x lies between y^1 and y^2, and denote $\bar{y}^i = \mathbf{h}(y^i)$ for $i = 1, 2$. It is intuitively clear that

$$J\mathbf{h}(x) \approx \frac{\operatorname{dist}(\bar{y}_1, \bar{y}_2)}{\operatorname{dist}(y_1, y_2)}$$

and this approximation improves as n grows (since the y_i's get closer to x).

Denote by V^1 the segment of the curve W^1 between the points y_1 and y_2 and by V^2 the segment of W^2 between the corresponding points $\bar{y}_1$ and $\bar{y}_2$. Obviously, $V^1, V^2 \subset \mathcal{Q}_n(x)$, so the map $\mathcal{F}^n$ is smooth on both V^1 and V^2. Their images $V_n^i = \mathcal{F}^n(V^i)$, $i = 1, 2$, are two almost parallel unstable curves that run very closely to each other; see Fig. 5.6. In fact, due to the uniform hyperbolicity (4.19), the distance between V_n^1 and V_n^2 is $\leq \text{const}\,\Lambda^{-n}$.

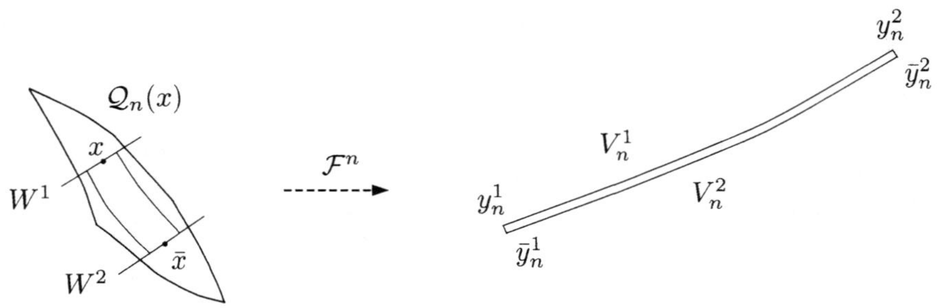

FIGURE 5.6. 'Proof' of Theorem 5.39.

Thus, $|V_n^1| \approx |V_n^2|$, and this approximation improves as n grows. The abundance of points $y \in W_*^1$ around x allows us to select the points y_1 and y_2 so that the segments V_n^1 and V_n^2 will be as short as we wish in order to minimize their distortions under the map $\mathcal{F}^{-n}$. Then, using the distortion bounds we obtain

$$\frac{\operatorname{dist}(\bar{y}_1, \bar{y}_2)}{\operatorname{dist}(y_1, y_2)} \approx \frac{\mathcal{J}_{V_n^2}\mathcal{F}^{-n}(\bar{x}_n)}{\mathcal{J}_{V_n^1}\mathcal{F}^{-n}(x_n)} = \frac{\mathcal{J}_{W^1}\mathcal{F}^n(x)}{\mathcal{J}_{W^2}\mathcal{F}^n(\bar{x})},$$

where $x_n = \mathcal{F}^n(x)$ and $\bar{x}_n = \mathcal{F}^n(\bar{x})$. This approximation improves as n grows and as long as we make appropriate selection of y_1 and y_2. Taking the limit $n \to \infty$ completes our 'proof'.

REMARK 5.40. Absolute continuity for holonomy maps, along with the formula (5.21), was first established by Anosov and Sinai in 1967 [**AS67**, Equation (5.3)] for smooth uniformly hyperbolic systems and later extended to large classes of hyperbolic maps. We refer the reader to [**AS67**] and [**BP01**, Theorem 4.4.1] for full proofs of these facts.

5.8. Absolute continuity

Theorem 5.39 ensures absolute continuity of the holonomy map but does not provide any specific estimates on its Jacobian. In fact, only very crude estimates can be achieved if we use stable manifolds to connect points of two given unstable curves, since there is little control on the distortion of stable manifolds under the billiard collision map $\mathcal{F}$; see Section 5.3. Certain "crude" estimates on the Jacobian $\mathcal{J}\mathbf{h}$ are proved in [**Sin70**, Section 5] and [**Ga74**, Section 4].

However, if we use stable H-manifolds instead of stable manifolds, then much finer estimates on the Jacobian of the holonomy map can be obtained. Since such estimates are essential for our further studies, we redefine the holonomy map as follows:

DEFINITION 5.41 (Modified holonomy map). Let $W^1, W^2 \subset \mathcal{M}$ be two unstable curves. Denote

$$W_*^i = \{x \in W^i \colon W_{\mathbb{H}}^s(x) \cap W^{3-i} \neq \emptyset\}$$

for $i = 1, 2$, where $W_{\mathbb{H}}^s(x)$ is the stable H-manifold passing through x. The map $\mathbf{h}\colon W_*^1 \to W_*^2$ taking every point $x \in W_*^1$ to $\bar{x} = W_{\mathbb{H}}^s(x) \cap W^2$ is called the *(modified) holonomy map*.

We assume that W^1 and W^2 are $C^{\ell-1}$ smooth curves and at least their first and second derivatives are uniformly bounded (then their future images will have uniformly bounded derivatives as well, due to Proposition 4.29).

Let $x \in W_*^1$ and $\bar{x} = \mathbf{h}(x) \in W_*^2$. We put $\delta = \operatorname{dist}(x, \bar{x})$ and denote by γ the angle between the tangent vectors to the curves W^1 and W^2 at the points x and $\bar{x}$, respectively.

THEOREM 5.42. *The Jacobian of the holonomy map* $\mathbf{h}$ *is uniformly bounded*

$$C^{-1} \leq J\mathbf{h}(x) \leq C,$$

for all $x \in W_*^1$*; here* $C = C(\mathcal{D}) > 1$ *is a constant. Moreover,*

$$A^{-\gamma-\delta^{1/3}} \leq J\mathbf{h}(x) \leq A^{\gamma+\delta^{1/3}}, \tag{5.22}$$

where $A = A(\mathcal{D}) > 1$ *is a constant.*

PROOF. We denote $W_n^i = \mathcal{F}^n(W^i)$ for $i = 1, 2$ and $n \geq 1$. We also denote $x_n = \mathcal{F}^n(x)$ and $\bar{x}_n = \mathcal{F}^n(\bar{x})$ and put $\delta_n = \operatorname{dist}(x_n, \bar{x}_n)$. Due to the uniform hyperbolicity of $\mathcal{F}$ (cf. (4.19)), $\delta_n \leq \hat{C}\delta\Lambda^{-n}$, where $\hat{C} = 1/\hat{c}$.

Taking the logarithm of (5.21) and using the chain rule give

$$\ln J\mathbf{h}(x) = \sum_{n=0}^{\infty} \Big(\ln \mathcal{J}_{W_n^1}\mathcal{F}(x_n) - \ln \mathcal{J}_{W_n^2}\mathcal{F}(\bar{x}_n) \Big). \tag{5.23}$$

Due to (4.18), (3.40), and (3.39)

$$\begin{aligned} \mathcal{J}_{W_n^1}\mathcal{F}(x_n) &= (1 + \tau_n \mathcal{B}_n^+) \frac{\cos\varphi_n}{\cos\varphi_{n+1}} \frac{\sqrt{1+\mathcal{V}_{n+1}^2}}{\sqrt{1+\mathcal{V}_n^2}} \\ &= \frac{\cos\varphi_n + \tau_n(\mathcal{K}_n + \mathcal{V}_n)}{\cos\varphi_{n+1}} \frac{\sqrt{1+\mathcal{V}_{n+1}^2}}{\sqrt{1+\mathcal{V}_n^2}}, \end{aligned}$$

where, as usual, $\tau_n = \tau(x_n)$, and $\mathcal{B}_n$, $\mathcal{K}_n$, etc., are taken at the point x_n. Therefore

$$\begin{aligned} \ln \mathcal{J}_{W_n^1}\mathcal{F}(x_n) &= -\ln\cos\varphi_{n+1} + \tfrac{1}{2}\ln(1+\mathcal{V}_{n+1}^2) - \tfrac{1}{2}\ln(1+\mathcal{V}_n^2) \\ &\quad + \ln\big[\cos\varphi_n + \tau_n(\mathcal{K}_n + \mathcal{V}_n)\big]. \end{aligned} \tag{5.24}$$

Using similar notation at $\bar{x}_n$ we get

$$\begin{aligned} \ln \mathcal{J}_{W_n^2}\mathcal{F}(\bar{x}_n) &= -\ln\cos\bar{\varphi}_{n+1} + \tfrac{1}{2}\ln(1+\bar{\mathcal{V}}_{n+1}^2) - \tfrac{1}{2}\ln(1+\bar{\mathcal{V}}_n^2) \\ &\quad + \ln\big[\cos\bar{\varphi}_n + \bar{\tau}_n(\bar{\mathcal{K}}_n + \bar{\mathcal{V}}_n)\big]. \end{aligned} \tag{5.25}$$

Comparing the first terms of the above expressions gives

$$\begin{aligned} \big|\ln\cos\varphi_{n+1} - \ln\cos\bar{\varphi}_{n+1}\big| &\leq \frac{\text{const}\,|\varphi_{n+1} - \bar{\varphi}_{n+1}|}{\cos\varphi_{n+1}} \\ &\leq \frac{\text{const}\,\delta_{n+1}}{\delta_{n+1}^{2/3}} = \text{const}\,\delta_{n+1}^{1/3} \end{aligned}$$

where we used (5.13). All the other terms in (5.24) and (5.25) are uniformly bounded and differentiable; thus it easily follows that

$$\big|\ln \mathcal{J}_{W_n^1}\mathcal{F}(x_n) - \ln \mathcal{J}_{W_n^2}\mathcal{F}(\bar{x}_n)\big| \leq C\big(\delta_{n+1}^{1/3} + \gamma_n + \gamma_{n+1} + \delta_n\big) \tag{5.26}$$

where $C > 0$ is a constant and $\gamma_n = |\Delta\mathcal{V}_n| = |\mathcal{V}_n - \bar{\mathcal{V}}_n|$ stands for the angle between the tangent vectors to the curves W_n^1 and W_n^2 at the points x_n and $\bar{x}_n$, respectively (note that $\gamma_0 = \gamma$).

Due to Exercise 4.50 (more precisely, due to its 'time reversal' counterpart; see explanation in Section 4.3), we have $\delta_n \leq C\delta_{n+1}^{1/2}$; thus the term δ_n in (5.26) may be dropped. It remains to estimate γ_n for all $n \geq 0$. For brevity, we denote $\Delta\mathcal{B}_n^- = \mathcal{B}_n^- - \bar{\mathcal{B}}_n^-$, $\Delta\tau_n = \tau_n - \bar{\tau}_n$, etc. First we estimate $\Delta\mathcal{B}_{n+1}^-$ by using (3.31):

$$\begin{aligned}
\Delta\mathcal{B}_{n+1}^- &= \frac{1}{\tau_n + 1/\mathcal{B}_n^+} - \frac{1}{\bar{\tau}_n + 1/\bar{\mathcal{B}}_n^+} \\
&= -\frac{1}{\tau_n + 1/\mathcal{B}_n^+}\,\frac{1}{\bar{\tau}_n + 1/\bar{\mathcal{B}}_n^+}\left(\Delta\tau_n + \frac{1}{\mathcal{B}_n^+} - \frac{1}{\bar{\mathcal{B}}_n^+}\right) \\
&= -\frac{\Delta\tau_n}{\left(\tau_n + 1/\mathcal{B}_n^+\right)\left(\bar{\tau}_n + 1/\bar{\mathcal{B}}_n^+\right)} + \frac{\Delta\mathcal{R}_n + \Delta\mathcal{B}_n^-}{\left(1 + \tau_n\mathcal{B}_n^+\right)\left(1 + \bar{\tau}_n\bar{\mathcal{B}}_n^+\right)}
\end{aligned} \tag{5.27}$$

(recall that $\mathcal{B}_n^+ = \mathcal{R}_n + \mathcal{B}_n^-$). The first term in (5.27) is bounded by

$$\text{const}\,\Delta\tau_n \leq \text{const}(\delta_n + \delta_{n+1}) \leq \text{const}\,\delta_n. \tag{5.28}$$

Next,

$$\frac{\Delta\mathcal{R}_n}{\left(1 + \tau_n\mathcal{B}_n^+\right)\left(1 + \bar{\tau}_n\bar{\mathcal{B}}_n^+\right)} \leq \frac{2\mathcal{K}_n\cos\bar{\varphi}_n - 2\bar{\mathcal{K}}_n\cos\varphi_n}{\left(\cos\varphi_n + \tau_n(\mathcal{K}_n + \mathcal{V}_n)\right)\left(\cos\bar{\varphi}_n + \bar{\tau}_n(\bar{\mathcal{K}}_n + \bar{\mathcal{V}}_n)\right)}.$$

Since the denominator is bounded away from zero, the fraction does not exceed $\text{const}\,\delta_n$; thus (5.27) reduces to

$$\Delta\mathcal{B}_{n+1}^- = \tilde{Q}_n^{(1)} + \frac{\Delta\mathcal{B}_n^-}{\left(1 + \tau_n\mathcal{B}_n^+\right)\left(1 + \bar{\tau}_n\bar{\mathcal{B}}_n^+\right)},$$

where $|\tilde{Q}_n^{(1)}| \leq \text{const}\,\delta_n$. Next, due to (3.39)

$$\begin{aligned}
\Delta\mathcal{V}_{n+1} &= \Delta\mathcal{K}_{n+1} + \mathcal{B}_{n+1}^-\cos\varphi_{n+1} - \bar{\mathcal{B}}_{n+1}^-\cos\bar{\varphi}_{n+1} \\
&= \tilde{Q}_{n+1}^{(2)} + \cos\varphi_{n+1}\,\Delta\mathcal{B}_{n+1}^-,
\end{aligned}$$

where $|\tilde{Q}_{n+1}^{(2)}| \leq \text{const}\,\delta_{n+1}$. Combining the last two estimates gives

$$\Delta\mathcal{V}_{n+1} = \tilde{Q}_n^{(3)} + \frac{\cos\varphi_{n+1}}{\cos\varphi_n}\,\frac{\Delta\mathcal{V}_n}{\left(1 + \tau_n\mathcal{B}_n^+\right)\left(1 + \bar{\tau}_n\bar{\mathcal{B}}_n^+\right)},$$

where $|\tilde{Q}_n^{(3)}| \leq \text{const}\,\delta_n$. Now consider the fraction

$$u_n := \frac{\cos\varphi_{n+1}}{\cos\varphi_n\left(1 + \tau_n\mathcal{B}_n^+\right)\left(1 + \bar{\tau}_n\bar{\mathcal{B}}_n^+\right)}.$$

EXERCISE 5.43. Verify that for all $0 \leq k \leq n$

$$u_n u_{n-1} \cdots u_k \leq C/\Lambda^{n-k}$$

where $C = C(\mathcal{D}) > 0$ is a constant. Hints:

$$\mathcal{B}_n^+\cos\varphi_n = 2\mathcal{K}_n + \mathcal{B}_n^-\cos\varphi_n \geq 2\mathcal{K}_{\min} > 0$$

and $1 + \bar{\tau}_n\bar{\mathcal{B}}_n^+ \geq \Lambda$.

EXERCISE 5.44. Use the result of the previous exercise to prove that

$$\gamma_n = |\Delta \mathcal{V}_n| \le \text{const}\Big(\sum_{k=0}^{n} \delta_k/\Lambda^{n-k} + \gamma/\Lambda^n\Big)$$
$$\le \text{const}\Big(\delta n/\Lambda^n + \gamma/\Lambda^n\Big) \tag{5.29}$$

(recall that $\delta_k \le \hat{C}\delta/\Lambda^k$ due to uniform hyperbolicity).

Combining (5.29) with (5.26) yields

$$\Big|\ln \mathcal{J}_{W_n^1}\mathcal{F}(x_n) - \ln \mathcal{J}_{W_n^2}\mathcal{F}(\bar{x}_n)\Big| \le C\Big(\frac{\delta^{1/3}}{\Lambda^{n/3}} + \frac{\delta n}{\Lambda^n} + \frac{\gamma}{\Lambda^n}\Big), \tag{5.30}$$

where $C = C(\mathcal{D}) > 0$ is a constant. Summing up over n gives

$$\big|\ln J\mathbf{h}(x)\big| \le \text{const}\big(\gamma + \delta^{1/3}\big),$$

which completes the proof of the theorem. □

REMARK 5.45. The above theorem and its proof extends to billiards of category B without changes. Even though the intercollision times τ_n and $\bar{\tau}_n$ will no longer be bounded above, the crucial estimate (5.28) will obviously hold true.

The main estimate (5.22) involves two parameters, δ and γ. In most applications, though, the curves W^1 and W^2 are disjoint, and then γ is unnecessary, as we show next. Let

$$W_\diamond^1 = \{x \in W_*^1 \colon r_{W^1}(x) > D\delta^{1/2},\ r_{W^2}(\bar{x}) > D\delta^{1/2}\},$$

where $D > 0$ is a sufficiently large constant. This is the set of points of W_*^1 that are at least $D\delta^{1/2}$-away from the endpoints of both curves.

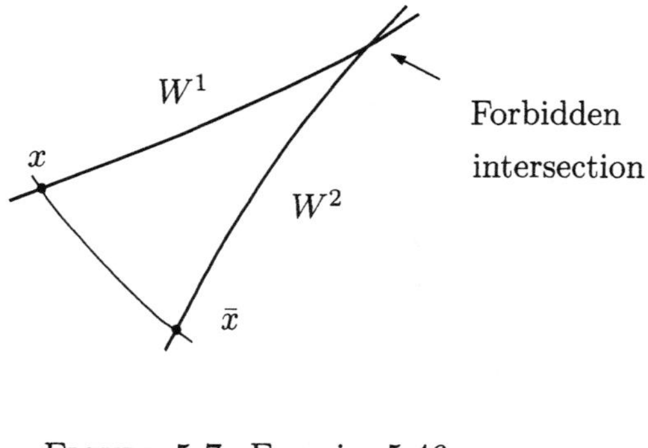

FIGURE 5.7. Exercise 5.46.

EXERCISE 5.46. Show that if $W^1 \cap W^2 \ne \emptyset$, then for every $x \in W_\diamond^1$ we have $\gamma \le C\delta^{1/2}$, where $C = C(\mathcal{D}) > 0$ is a constant. Hint: Use uniform bounds on the first and second derivatives of the curves W^1 and W^2 to prove that if the above inequality failed, then W^1 and W^2 would have to cross each other (see Fig. 5.7), thus contradicting our assumption.

So for all the points $x \in W_\diamond^1$ we have $\gamma \ll \delta^{1/3}$; hence the estimate (5.22) takes a simpler form: $A^{-\delta^{1/3}} \le J\mathbf{h}(x) \le A^{\delta^{1/3}}$.

EXERCISE 5.47. Prove that the Jacobian $J\mathbf{h}(x)$ is a continuous function on W_*^1. Hint: Note that the series (5.23) converges uniformly in x due to (5.30) and every term depends on x continuously.

The Jacobian $J\mathbf{h}(x)$ is not necessarily Hölder continuous (unlike its counterpart in Anosov and Axiom A systems), but it has a similar property sometimes called 'dynamically defined Hölder continuity' [**Yo98**, p. 597], which we describe below.

Let $\mathcal{Q}_n^{\mathbb{H}}(x)$ denote the open connected component of the set $\mathcal{M} \setminus \mathcal{S}_n^{\mathbb{H}}$ containing the point x (see Section 5.4). For two points $x, y \in \mathcal{M}$ denote by

$$\mathbf{s}_+(x,y) = \min\{n \geq 0\colon y \notin \mathcal{Q}_n^{\mathbb{H}}(x)\} \tag{5.31}$$

the 'separation time' (the images $\mathcal{F}^n(x)$ and $\mathcal{F}^n(y)$ get separated at time $n = s_+(x,y)$ as they lie in different connected components of the new collision space $\mathcal{M}_{\mathbb{H}}$). Clearly, this function is symmetric: $\mathbf{s}_+(x,y) = \mathbf{s}_+(y,x)$. If $y \in \mathcal{Q}_n^{\mathbb{H}}(x)$ for all $n \geq 0$, then $y \in W_{\mathbb{H}}^s(x)$; in that case we set $\mathbf{s}_+(x,y) = \infty$. Observe that if x and y lie on one unstable curve $W \subset \mathcal{M}$, then

$$\operatorname{dist}(x,y) \leq C\Lambda^{-\mathbf{s}_+(x,y)} \tag{5.32}$$

for some constant $C = C(\mathcal{D}) > 0$.

PROPOSITION 5.48. *There are constants $C > 0$ and $\boldsymbol{\theta} \in (0,1)$ such that*

$$|\ln J\mathbf{h}(x) - \ln J\mathbf{h}(y)| \leq C\boldsymbol{\theta}^{\mathbf{s}_+(x,y)}.$$

PROOF. Denote $\bar{x} = \mathbf{h}(x)$ and $\bar{y} = \mathbf{h}(y)$. Observe that $\mathbf{s}_+(\bar{x},\bar{y}) = \mathbf{s}_+(x,y)$; this follows from Proposition 4.47. Using the notation of (5.23) and the triangle inequality gives

$$\begin{aligned}\Delta\colon &= |\ln J\mathbf{h}(x) - \ln J\mathbf{h}(y)| \\ &\leq \sum_{n=0}^{\infty} |\ln \mathcal{J}_{W_n^1}\mathcal{F}(x_n) - \ln \mathcal{J}_{W_n^2}\mathcal{F}(\bar{x}_n) - \ln \mathcal{J}_{W_n^1}\mathcal{F}(y_n) + \ln \mathcal{J}_{W_n^2}\mathcal{F}(\bar{x}_n)|.\end{aligned}$$

Let $m = \mathbf{s}_+(x,y)/2$. Then we apply the estimate (5.30) to all $n > m$ and Lemma 5.27 (on distortion bounds) to all $n \leq m$. This proves our claim with $\boldsymbol{\theta} = \Lambda^{-1/6}$. □

5.9. Two growth lemmas

The map $\mathcal{F}$ is uniformly hyperbolic; i.e. $\mathcal{F}^n$ expands unstable curves $W \subset \mathcal{M}$ by a factor $\geq \hat{c}\Lambda^n$ according to (4.19). At the same time, the images $\mathcal{F}^n(W)$ get broken by singularities into pieces, some of which may be long and some others short. In this process, arbitrarily short pieces may appear, and the total number of pieces grows exponentially with n (it may even be infinite, if we deal with billiards of category B, see Section 4.10).

The hyperbolicity of $\mathcal{F}$ guarantees exponential growth of unstable curves only in a *local sense*, i.e. in terms of a uniform expansion of their tangent vectors. It *does not* guarantee that typical pieces of $\mathcal{F}^n(W)$ will grow (get longer) with n. In fact, they may conceivably get shorter as n increases, so there will be no growth in a *global sense*; see the next paragraph.

Imagine the following scenario. A very short unstable curve of length $\delta > 0$ gets expanded under $\mathcal{F}$ by a factor of 1.8, but a singularity curve breaks it into two equal pieces. Then the image $\mathcal{F}(W)$ would be the union of two unstable curves, each of length $0.9\,\delta$. If the same happens to each piece at every iteration of the map (see Fig. 5.8), then $\mathcal{F}^n(W)$ would consist of 2^n unstable curves, each of length $(0.9)^n\delta$. Then unstable curves, being expanded locally, would get shorter "globally".

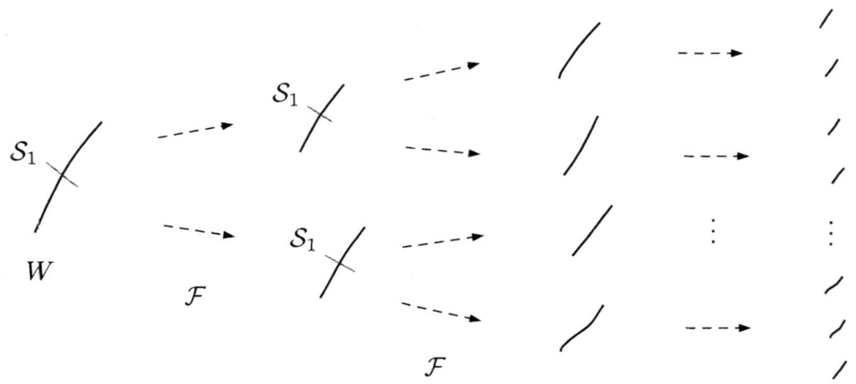

FIGURE 5.8. A paradox: unstable curves get shorter under the map $\mathcal{F}$.

Such a behavior of unstable curves would be catastrophic; it would effectively undo the uniform hyperbolicity of the map $\mathcal{F}$. This problem is common for all hyperbolic maps with discontinuities, but in billiards it is aggravated by the necessity to subdivide unstable curves further into H-components, which we defined in Section 5.6 in order to ensure proper bounds on distortions, probability densities, and the Jacobian of the holonomy map.

The H-components of the set $\mathcal{F}^n(W)$ will generally be shorter than its smooth components. For example, every unstable curve terminating on $\mathcal{S}_0$ must be broken into an infinite number of tiny H-components. While the situation with unstable curves may look bad enough (Fig. 5.8), it apparently gets even worse when we partition them into smaller H-components.

It is then remarkable that, despite odds, *typical* H-components of $\mathcal{F}^n(W)$ for *every* small unstable curve W grow monotonically and exponentially in n, until they reach the size of order one, see below. This is a fundamental fact in the theory of chaotic billiards. It may be aptly described as a victory of the hyperbolicity over singularities (and inhomogeneities): expansion always prevails over fragmentation! If this was not the case, the theory of chaotic billiards might have been quite different.

We now turn to precise definitions. Let W be a weakly homogeneous unstable curve. Denote by $\mathbf{m}_W$ the Lebesgue measure[1] on W. For every $x \in W$ and $n \geq 0$ denote by $W_n(x)$ the H-component of $\mathcal{F}^n(W)$ containing the point $\mathcal{F}^n(x)$ and by

$$r_n(x) = r_{W_n(x)}(\mathcal{F}^n x) \tag{5.33}$$

the distance from the point $\mathcal{F}^n(x)$ to the nearest endpoint of $W_n(x)$. Clearly, $r_n(x)$ is a function on W that characterizes the size of the H-components of $\mathcal{F}^n(W)$.

[1] We will denote by $\mathbf{m}$ various one-dimensional Lebesgue measures.

EXERCISE 5.49. Verify that $\mathbf{m}_W\big(r_0(x) < \varepsilon\big) = \min\{2\varepsilon, \mathbf{m}_W(W)\}$, where $\mathbf{m}_W(W) = |W|$ is simply the length of W.

EXERCISE 5.50. Assume that the map $\mathcal{F}^n$ expands W uniformly (as a linear map) by a factor $\Lambda_n > 1$ and $\mathcal{F}^n(W)$ constitutes a single H-component. Show that

$$\mathbf{m}_W\big(r_n(x) < \varepsilon\big) = \min\{2\varepsilon/\Lambda_n, \mathbf{m}_W(W)\}.$$

EXERCISE 5.51. Consider an almost ideal situation where the map $\mathcal{F}^n$ expands W uniformly (as a linear map) by a factor Λ^n (where $\Lambda > 1$), the set $\mathcal{F}^n(W)$ remains a single H-component as long as $|\mathcal{F}^n(W)| < 1$, after which $\mathcal{F}^n(W)$ is divided into H-components of length between 0.5 and 1 (note that the first division occurs when $n \approx \big|\ln|W|\big|/\ln\Lambda$); see Fig. 5.9. Prove that

$$\begin{aligned}\mathbf{m}_W\big(r_n(x) < \varepsilon\big) &\leq \mathbf{m}_W\big(r_0 < \varepsilon/\Lambda^n\big) + 4\varepsilon\,\mathbf{m}_W(W)\\ &\leq 2\varepsilon/\Lambda^n + 4\varepsilon\,\mathbf{m}_W(W).\end{aligned}$$

Thus, if $|W|$ is very small, then the measure $\mathbf{m}_W\big(r_n(x) < \varepsilon\big)$ decreases exponentially in n until $\mathcal{F}^n(W)$ reaches the size > 1, and after that (i.e., for all $n \geq \big|\ln|W|\big|/\ln\Lambda$) this measure will be of order ε:

$$\mathbf{m}_W\big(r_n(x) < \varepsilon\big) \leq 5\varepsilon\,\mathbf{m}_W(W).$$

Clearly, these estimates are quite sharp for all sufficiently small $\varepsilon > 0$.

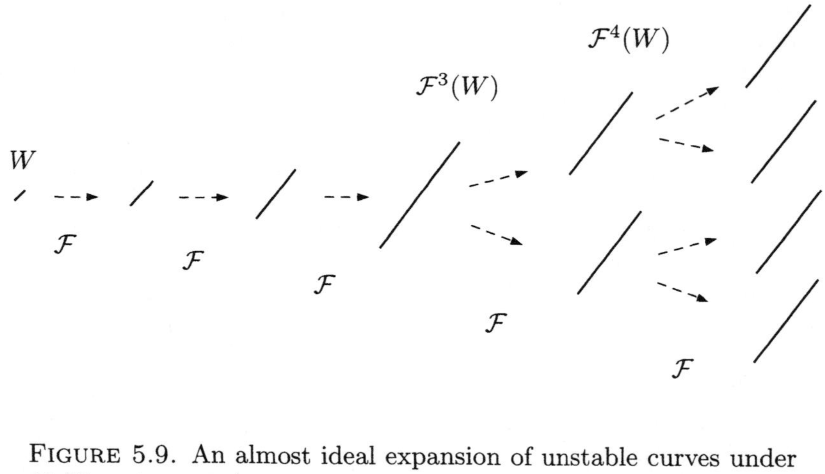

FIGURE 5.9. An almost ideal expansion of unstable curves under $\mathcal{F}$ (Exercises 5.51).

Now consider a sufficiently short unstable curve $W \subset \mathcal{M}$.

THEOREM 5.52 (First growth lemma). *There are constants $\hat{\Lambda} > 1$, $\vartheta_1 \in (0,1)$, and $c_1, c_2 > 0$, such that for all $n \geq 0$ and $\varepsilon > 0$*

$$\mathbf{m}_W\big(r_n(x) < \varepsilon\big) \leq c_1(\vartheta_1\hat{\Lambda})^n\,\mathbf{m}_W\big(r_0(x) < \varepsilon/\hat{\Lambda}^n\big) + c_2\varepsilon\,\mathbf{m}_W(W).$$

Due to Exercise 5.49, the first term on the right-hand side equals $c_1\vartheta_1^n\varepsilon$ for all $\varepsilon < \hat{\Lambda}^n|W|/2$; hence it decreases exponentially in n.

THEOREM 5.53 (Second growth lemma). *There are constants* $\varkappa > 0$ *and* $c_3 > 0$, *such that for all* $n \geq \varkappa \big|\ln |W|\big|$ *and* $\varepsilon > 0$

$$\mathbf{m}_W\big(r_n(x) < \varepsilon\big) \leq c_3 \varepsilon\, \mathbf{m}_W(W).$$

The reader should note a striking similarity between the estimates in the above two theorems stated for billiards and those obtained in Exercise 5.51 for the nearly ideal situation where the components of $\mathcal{F}^n(W)$ are broken only into sufficiently long pieces. Therefore, the appearance of arbitrarily short H-components in the real billiard dynamics does not prevent *typical* H-components from evolving in the same general manner as described in Exercise 5.51.

Clearly there are two regimes in the evolution of the H-components of $\mathcal{F}^n(W)$ if the initial curve $W \subset \mathcal{M}$ is short:

(i) during the first $\sim \big|\ln |W|\big|$ iterations of $\mathcal{F}$, typical H-components of $\mathcal{F}^n(W)$ grow monotonically and exponentially fast in n, until they reach the size of order one (the first growth lemma);

(ii) after that, typical H-components of $\mathcal{F}^n(W)$ remain large at all times (the second growth lemma).

This concludes our analysis of the global expansion of unstable curves.

5.10. Proofs of two growth lemmas

Here we prove Theorem 5.52 and 5.53 (the first and second growth lemmas).

First observe that we can replace the Lebesgue measure $\mathbf{m}_W$ on W with any uniformly equivalent measure $\mathbf{m}_W^*$, i.e. such that

$$0 < C^{-1} \leq d\mathbf{m}_W / d\mathbf{m}_W^* \leq C < \infty, \tag{5.34}$$

where $C = C(\mathcal{D}) > 1$ is a constant, and prove the growth lemmas for the new measure $\mathbf{m}_W^*$ (the constant C can be incorporated into the factors $c_1 \div c_3$ in the growth lemmas).

The Lebesgue measure $\mathbf{m}_W$ (induced by the Euclidean metric on $\mathcal{M}$) is slightly inconvenient since the expansion of unstable curves is not always monotone (Exercise 4.23). In the ergodic theory of uniformly hyperbolic maps on Riemannian manifolds, the given Riemannian metric can always be replaced by a uniformly equivalent one in which the expansion of unstable tangent vectors is monotone; see the original proof in [**HP70**] and a recent book [**BrS02**, Section 5.2]. That metric is called an *adapted metric* or *Lyapunov metric*.

In dispersing billiards an adapted metric can be defined in many ways. One of them (perhaps, the simplest one) is

$$\|dx\|_* = \frac{\mathcal{K} + |\mathcal{V}|}{\sqrt{1+\mathcal{V}^2}}\, \|dx\|,$$

where, as usual, $dx = (dr, d\varphi)$ denotes a tangent vector, $\mathcal{K}$ denotes the curvature of $\partial\mathcal{D}$ at the given point and $\mathcal{V} = d\varphi/dr$ (in the case $dr = 0$ we define $\|dx\|_* = \|dx\|$ by continuity).

EXERCISE 5.54. Verify that the metrics $\|\cdot\|_*$ and $\|\cdot\|$ are uniformly equivalent.

EXERCISE 5.55. Verify that the expansion of unstable vectors $dx_0 \in C_x^u$ in the metric $\|\cdot\|_*$ satisfies

$$\frac{\|dx_1\|_*}{\|dx_0\|_*} \geq \Lambda_* > 1, \tag{5.35}$$

where $dx_1 = D_x\mathcal{F}(dx_0)$ and $\Lambda_* > 1$ is a constant. Hint: In the notation of (4.18) and (4.15)

$$\frac{\|dx_1\|_*}{\|dx_0\|_*} = (1 + \tau_0 \mathcal{B}_0^+)\,\frac{\cos\varphi_0}{\cos\varphi_1}\,\frac{\mathcal{K}_1 + \mathcal{V}_1}{\mathcal{K}_0 + \mathcal{V}_0}.$$

Next, use (3.31), (3.33), and (3.39) to obtain

$$1 + \tau_0 \mathcal{B}_0^+ = \frac{\mathcal{B}_0^+}{\mathcal{B}_1^-} = \frac{\frac{2\mathcal{K}_0}{\cos\varphi_0} + \mathcal{B}_0^-}{\mathcal{B}_1^-} = \frac{\mathcal{K}_0 + \mathcal{V}_0}{\mathcal{B}_1^- \cos\varphi_0}.$$

This gives

$$\frac{\|dx_1\|_*}{\|dx_0\|_*} = \frac{\mathcal{K}_1 + \mathcal{V}_1}{\mathcal{B}_1^- \cos\varphi_1} = 1 + \frac{2\mathcal{K}_1}{\mathcal{B}_1^- \cos\varphi_1}. \tag{5.36}$$

Now argue that the last fraction is bounded away from zero.

We define $\mathbf{m}_W^*$ to be the Lebesgue measure on W induced by the metric $\|\cdot\|_*$. We redefine (5.3) accordingly:

$$\mathcal{J}_W^* \mathcal{F}^n(x) = \frac{\|D_x\mathcal{F}^n(dx)\|_*}{\|dx\|_*}, \tag{5.37}$$

where dx denotes a nonzero tangent vector to the curve W at x.

LEMMA 5.56. *Given an unstable curve $W \subset \mathcal{M}$, we denote by W_i all the H-components of $\mathcal{F}(W)$ and put*

$$\lambda(W_i) = \max_{y \in W_i} \mathcal{J}_{W_i}^* \mathcal{F}^{-1}(y).$$

Then

$$\liminf_{\delta\to 0} \sup_{W\colon |W|_* < \delta} \sum_i \lambda(W_i) < 1, \tag{5.38}$$

where $|W|_* = \mathbf{m}_W^*(W)$ denotes the length of W in the $\|\cdot\|_*$-metric.

Note that the inequality (5.38) prevents the disastrous scenario shown in Fig. 5.8, because there $\lambda(W_i) = 1.8^{-1}$ for $i = 1, 2$; hence $\sum_i \lambda(W_i) = 2/1.8 > 1$.

PROOF. Let W be a small unstable curve. It gives rise to a family of billiard trajectories (directed lines on the table $\mathcal{D}$) starting at points $x \in W$. Observe that $x \in W \cap \mathcal{S}_1$ iff the directed line originating at x is tangent to a scatterer; i.e. the orbit of x experiences a grazing (tangential) collision (see Fig. 5.10).

At a grazing collision, W is sliced into two parts: one lands on $\partial\mathcal{D}$ and gets reflected (almost tangentially, because W is small), and the other misses the collision (passes by). The reflecting part is further subdivided into countably many H-components by the lines $\mathbb{S}_k$ bounding the homogeneity strips $\mathbb{H}_k$.

EXERCISE 5.57. Denote by W_k' the H-component of $\mathcal{F}(W)$ landing in the homogeneity strip $\mathbb{H}_k$. Prove that

$$\sum_k \lambda(W_k') < \text{const}\, |W|_*^{1/4}.$$

Hints: Observe that the length of the part $\mathcal{F}(W)$ landing on $\partial\mathcal{D}$ is $\le \text{const}\,|W|_*^{1/2}$ (cf. Exercise 4.50); hence it intersects $\mathbb{H}_k$ only for k satisfying $|k| \ge \text{const}\,|W|_*^{-1/4}$. Then recall that $\lambda(W_k') \asymp k^{-2}$ (Exercise 5.18).

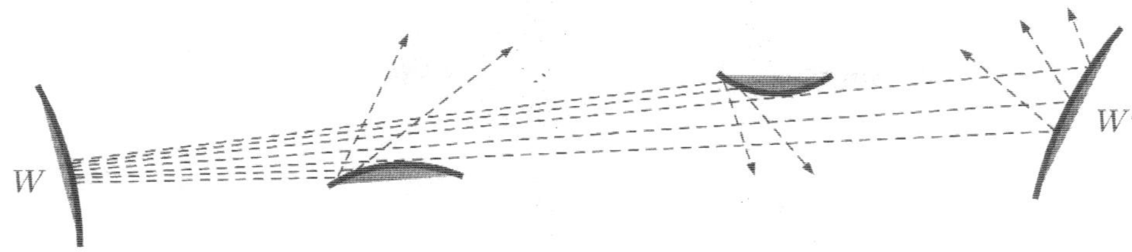

FIGURE 5.10. Proof of Lemma 5.56.

The part of W passing by without collision may be sliced again by a grazing collision with $\partial\mathcal{D}$ later on (see Fig. 5.10), thus creating another countable set of reflecting H-components. This can happen at most $\tau_{\max}/\tau_{\min}$ times, where $\tau_{\min}$ and $\tau_{\max}$ are the minimal and maximal free paths (Exercise 4.6).

In the end, we will have $N \leq \tau_{\max}/\tau_{\min} =$ const countable sets of H-components resulting from almost grazing collisions and at most one unstable curve $W' \subset \mathcal{F}(W)$ that misses all the grazing collisions and lands somewhere else on $\partial\mathcal{D}$; see Fig. 5.10.

Assume first that $W' \cap \mathbb{S} = \emptyset$. Then W' is a single H-component. It is guaranteed to contract under $\mathcal{F}^{-1}$ only by a moderate factor of Λ_*^{-1}; see (5.35). Thus, we arrive at

$$\sum_i \lambda(W_i) \leq \Lambda_*^{-1} + \text{const}\,|W|_*^{1/4}. \tag{5.39}$$

This is certainly enough to prove (5.38), since $\Lambda_* > 1$.

But it may happen that W' crosses some lines $\mathbb{S}_k$, $|k| \geq k_0$, and breaks into two or more H-components W'_j. This case requires some extra work.

EXERCISE 5.58. Show that in the last case $\liminf_{\delta\to 0} \sup_{W:\,|W|_*<\delta} \sum_j \lambda(W'_j) < C/k_0^2$, where $C = C(\mathcal{D}) > 0$ is a constant. Hint: If W' crosses only one line $\mathbb{S}_k$, then use the estimate of Exercise 5.18. If W' crosses several lines $\mathbb{S}_k \subset \mathbb{S}$, then the analysis is similar to Exercise 5.57.

Now we make the long-promised choice of k_0 introduced in Definition 5.8. We fix k_0 so that $C/k_0^2 < \Lambda_*^{-1}$; then our estimate in Exercise 5.58 shows that (5.39) holds in all cases. Now Lemma 5.56 easily follows. □

REMARK 5.59. Lemma 5.56 can be extended to billiards of category B, but its proof requires certain modifications, because it is now possible that the curve W intersects infinitely many n-cells $D_{l,n}^+$ (for some l and all sufficiently large n's, say all $n \geq n_W$); the n-cells were defined in Section 4.10. If this happens, the image $\mathcal{F}(W)$ will consist of *infinitely* many countable sets of H-components resulting from nearly grazing collisions. For $n \geq n_W$ and $|k| \geq k_0$, let $W_{k,n}$ denote the H-component of $\mathcal{F}(W)$ originating from $D_{l,n}^+$ and landing in the homogeneity strip $\mathbb{H}_k$. Then $\lambda(W_{k,n}) \leq Cn^{-1}k^{-2}$, where $C > 0$ is a constant; see Exercise 5.26. It is important to observe that $\mathcal{F}(D_{l,n}^+) = D_{l',n}^-$ is a "negative" n-cell and its size is $\mathcal{O}(n^{-1/2})$ (see Exercise 4.53); hence it can intersect only homogeneous strips $\mathbb{H}_k$ with $|k| \geq cn^{1/4}$, where $c > 0$ is a constant. Thus we obtain

$$\sum_{n=n_W}^{\infty} \sum_{|k|=cn^{1/4}}^{\infty} \lambda(W_{k,n}) \le \sum_{n=n_W}^{\infty} \sum_{|k|=cn^{1/4}}^{\infty} Cn^{-1}k^{-2}$$
$$\le \sum_{n=n_W}^{\infty} 2\,Cc^{-1}n^{-5/4}$$
$$\le 10\,Cc^{-1}n_W^{-1/4}.$$

Since $n_W \to \infty$ as $|W| \to 0$, we obtain Lemma 5.56 for billiards of type B.

We now continue our proofs of the growth lemmas. Due to (5.38) there exists $\delta_0 > 0$ such that

$$\vartheta_1 \colon = e^{C\delta_0^{1/3}} \max\Big\{ \Lambda_*^{-1}, \sup_{W\colon |W|_* < \delta_0} \sum_i \lambda(W_i)\Big\} < 1, \tag{5.40}$$

where $C > 0$ is a constant involved in Lemma 5.27. We also denote

$$\vartheta_0 \colon = e^{-C\delta_0^{1/3}} \vartheta_1 = \max\Big\{ \Lambda_*^{-1}, \sup_{W\colon |W|_* < \delta_0} \sum_i \lambda(W_i)\Big\}. \tag{5.41}$$

Note that $\vartheta_0 < \vartheta_1$. In addition, we assume that $\vartheta_0 \Lambda_* \ge 1$.

PROOF OF THEOREM 5.52. We use induction on n. First we set $n = 1$ and prove a slightly stronger bound

$$\mathbf{m}_W^*\big(r_1(x) < \varepsilon\big) \le \vartheta_0 \Lambda_* \, \mathbf{m}_W^*\big(r_0(x) < \varepsilon/\Lambda_*\big) + \zeta_0 \varepsilon \, \mathbf{m}_W^*(W) \tag{5.42}$$

for some constant $\zeta_0 > 0$ defined below. Observe that this implies the estimate in Theorem 5.52 for $n = 1$ with $c_1 = 1$ and $c_2 = \zeta_0$.

Assume that $|W|_* < \delta_0$. Let W_i denote all the H-components of $\mathcal{F}(W)$. For each i, the set $W_i \cap \mathcal{F}\big(\{r_1(x) < \varepsilon\}\big)$ is the union of two subintervals of the component W_i, each of length ε, adjacent to its endpoints (these intervals may overlap if $2\varepsilon > |W_i|_*$). Then the set $\mathcal{F}^{-1}(W_i) \cap \{r_1(x) < \varepsilon\}$ is a subset of the union of two subintervals in $\mathcal{F}^{-1}(W_i)$ of length $\lambda(W_i)\varepsilon$; therefore,

$$\mathbf{m}_W^*(r_1(x) < \varepsilon) \le \sum_i 2\varepsilon\lambda(W_i) \le 2\vartheta_0\varepsilon.$$

This implies (see the next exercise)

$$\mathbf{m}_W^*\big(r_1(x) < \varepsilon\big) \le \vartheta_0 \Lambda_* \, \mathbf{m}_W^*\big(r_0(x) < \varepsilon/\Lambda_*\big). \tag{5.43}$$

EXERCISE 5.60. Derive the bound (5.43). Hint: Use the result of Exercise 5.49, which takes care of the case $\varepsilon < \Lambda_*|W|/2$. Otherwise note that our choice of ϑ_1 guarantees that $\vartheta_0 \Lambda_* \ge 1$.

This bound appears to be even better than (5.42), but remember we assumed that $|W|_* < \delta_0$. So we cannot use this bound in our inductive scheme, since some H-components of $\mathcal{F}(W)$ may be longer than δ_0. To proceed with our induction, we will partition the H-components of $\mathcal{F}(W)$ into pieces that are shorter than δ_0. This will only enlarge the set $\{r_1(x) < \varepsilon\}$. Hence if we prove the desired bound on the measure of the larger set, it would be more than enough. It is this enlargement of the set $\{r_1(x) < \varepsilon\}$ that gives us the second term in (5.42).

More precisely, let us divide each H-component W_i of $\mathcal{F}(W)$ with length $|W_i|_* > \delta_0$ into $k_i + 1$ equal segments, where $k_i = [|W_i|_*/\delta_0]$, so that each segment will be shorter than δ_0. If $|W_i|_* < \delta_0$, then we set $k_i = 0$ and leave W_i unchanged. We call the obtained segments *shortened H-components* of $\mathcal{F}(W)$.

On each original (not shortened) H-component W_i there are exactly k_i points that divide it into the shortened H-components. The union of the preimages of the ε-neighborhoods of these partition points has measure bounded above by

$$\begin{aligned}\leq 2\varepsilon \sum_i k_i \lambda(W_i) &\leq 2\delta_0^{-1}\varepsilon \sum_i |W_i|_* \lambda(W_i)\\ &\leq 2e^{C\delta_0^{1/3}}\delta_0^{-1}\varepsilon \sum_i |\mathcal{F}^{-1}(W_i)|_*\\ &= 2e^{C\delta_0^{1/3}}\delta_0^{-1}\varepsilon\,|W|_*,\end{aligned}$$

where the factor $e^{C\delta_0^{1/3}}$ accounts for possible distortions of W_i under the map $\mathcal{F}^{-1}$; see Lemma 5.27. This completes the proof of (5.42) with $\zeta_0 = 2e^{C\delta_0^{1/3}}\delta_0^{-1}$.

Next we prove, by induction on $n \geq 1$, that

$$\begin{aligned}\mathbf{m}_W^*\big(r_n(x) < \varepsilon\big) &\leq (\vartheta_1\Lambda_*)^n\,\mathbf{m}_W^*\big(r_0(x) < \varepsilon/\Lambda_*^n\big)\\ &\quad + \zeta_1\big(1 + \vartheta_1 + \cdots + \vartheta_1^{n-1}\big)\,\varepsilon\,\mathbf{m}_W^*(W),\end{aligned} \tag{5.44}$$

where $\zeta_1 = e^{C\delta_0^{1/3}}\zeta_0$. The case $n = 1$ follows from (5.42), since $\vartheta_0 < \vartheta_1$ and $\zeta_0 < \zeta_1$. Assuming that (5.44) is proved for some $n \geq 1$, we apply (5.42) to each shortened H-component $W' \subset \mathcal{F}^n(W)$ and obtain

$$\mathbf{m}_{W'}^*\big(r_1(y) < \varepsilon\big) \leq \vartheta_0\Lambda_*\,\mathbf{m}_{W'}^*\big(r_0(y) < \varepsilon/\Lambda_*\big) + \zeta_0\varepsilon\,|W'|_*.$$

Hence on its preimage $W'' = \mathcal{F}^{-n}(W')$ we get

$$\mathbf{m}_{W''}^*\big(r_{n+1}(y) < \varepsilon\big) \leq \vartheta_1\Lambda_*\,\mathbf{m}_{W''}^*\big(r_n(y) < \varepsilon/\Lambda_*\big) + \zeta_1\varepsilon\,|W''|_*$$

(to take care of possible distortions of W' under the map $\mathcal{F}^{-n}$, we replace ϑ_0 by ϑ_1 and ζ_0 by ζ_1, in accordance with Lemma 5.27). Summing up over all $W'' \subset W$ and then using the inductive assumption yield

$$\begin{aligned}\mathbf{m}_W^*\big(r_{n+1}(x) < \varepsilon\big) &\leq (\vartheta_1\Lambda_*)\,\mathbf{m}_W^*\big(r_n(x) < \varepsilon/\Lambda_*\big) + \zeta_1\varepsilon\,|W|_*\\ &\leq (\vartheta_1\Lambda_*)^{n+1}\,\mathbf{m}_W^*\big(r_0(x) < \varepsilon/\Lambda_*^{n+1}\big)\\ &\quad + \zeta_1\big(1 + \vartheta_1 + \cdots + \vartheta_1^n\big)\,\varepsilon\,\mathbf{m}_W^*(W),\end{aligned}$$

which proves (5.44) for all $n \geq 1$ by induction. This implies Theorem 5.52, for the 'adapted' measure $\mathbf{m}_W^*$, with $c_1 = 1$ and $c_2 = \zeta_1/(1 - \vartheta_1)$. Lastly, the claim of Theorem 5.52 for the Lebesgue measure $\mathbf{m}_W$ follows, because the constant C from (5.34) can be incorporated into the values of c_1 and c_2. □

PROOF OF THEOREM 5.53. The second growth lemma follows from the first in a general manner (no specifics of the billiard dynamics are involved here). We set $\varkappa = 1/\min\{|\ln\vartheta_1|, |\ln\hat{\Lambda}|\}$. It is easy to check that for all $n \geq \varkappa\big|\ln|W|\big|$ we have $\hat{\Lambda}^n > |W|^{-1}$ and $\vartheta_1^n < |W|$. Observe also that $\mathbf{m}_W\big(r_0 < \varepsilon/\hat{\Lambda}^n\big) < 2\varepsilon/\hat{\Lambda}^n$. Therefore, due to Theorem 5.52

$$\begin{aligned}\mathbf{m}_W\big(r_n(x) < \varepsilon\big) &\leq 2\zeta_1\vartheta_1^n\varepsilon + c_2\varepsilon\,\mathbf{m}_W(W)\\ &\leq (2\zeta_1 + c_2)\,\varepsilon\,\mathbf{m}_W(W).\end{aligned}$$

□

5.11. Third growth lemma

Recall that the expansion of any small unstable curve $W \subset \mathcal{M}$ goes through two stages (regimes); see the end of Section 5.9. During the first stage, typical H-components of $\mathcal{F}^n(W)$ keep growing, while during the second stage typical H-components remain large enough (of length $\mathcal{O}(1)$) at all times. More precisely, according to the second growth lemma, the orbits of typical points $x \in W$ will stay, most of the time, in long enough H-components of $\mathcal{F}^n(W)$ and far enough from their endpoints; i.e. $r_n(x) = \mathcal{O}(1)$ for most values of n.

However, the images of some 'unfortunate' points $x \in W$ may be trapped in short H-components (or in small vicinities of the endpoints of long H-components, which is equally bad) for an extended period of time. That is, for some $x \in W$ we may have $r_n(x) < \varepsilon$ during an interval $n_1 \leq n < n_2$, where $\varepsilon > 0$ is small. Here we estimate the measure of such 'hapless' points.

First we argue heuristically. Let us choose $\varepsilon > 0$ so small that

$$\mathbf{m}_W\big(r_n(x) < \varepsilon\big)/\mathbf{m}_W(W) \leq \theta < 1 \qquad \forall n \geq \varkappa\big|\ln|W|\big|$$

(this is possible due to Theorem 5.53). Suppose, rather ideally, that at every iteration of $\mathcal{F}$ all the points $x \in \mathcal{F}^n(W)$ are "reshuffled randomly", so that the "events" $\{r_n(x) < \varepsilon\}$ are independent in a pure probabilistic sense. Then their intersection would have probability

$$\mathbf{m}_W\Big(\max_{n_1 \leq n < n_2} r_n(x) < \varepsilon\Big)/\mathbf{m}_W(W) \leq \theta^{n_2 - n_1}. \tag{5.45}$$

The real billiard dynamics is deterministic, so the above events cannot be truly independent, but an analogue of the estimate (5.45) holds!

THEOREM 5.61 (Third growth lemma). *There are constants $\varepsilon_0 > 0$ and $\vartheta_2 \in (0,1)$ such that for any $n_2 > n_1 > \varkappa\big|\ln|W|\big|$ we have*

$$\mathbf{m}_W\Big(\max_{n_1 \leq n < n_2} r_n(x) < \varepsilon_0\Big) \leq \vartheta_2^{n_2 - n_1}\mathbf{m}_W(W).$$

In other words, the probability that the orbit of $x \in W$ stays too close to the endpoints of H-components during $k = n_2 - n_1$ consecutive iterations is exponentially small in k.

We prove this theorem at the end of the section, but first we derive a corollary. Given $n_2 > n_1 > \varkappa\big|\ln|W|\big|$, for each point $x \in W$ we put

$$\bar{n}(x) = \min_{n \geq n_1}\{n\colon\ |W_n(x)| \geq 2\varepsilon_0\}$$

where $W_n(x)$ denotes the H-component of $\mathcal{F}^n(W)$ that contains the point $\mathcal{F}^n(x)$. In other words, $\bar{n}(x)$ is the first time, after the initial n_1 iterations of $\mathcal{F}$, when the image of the point x belongs in an H-component of length $\geq 2\varepsilon_0$ (if this never happens, we set $\bar{n}(x) = \infty$). Denote $W_{n_1,n_2} = \{x \in W \colon \bar{n}(x) < n_2\}$.

EXERCISE 5.62. Show that if $\bar{n}(x) = n$, then $\bar{n}(y) = n$ for all $y \in \mathcal{F}^{-1}(W_n(x))$. Conclude that the set $\{\mathcal{F}^{\bar{n}(x)}(x)\colon\ x \in W_{n_1,n_2}\}$ is a union of weakly homogeneous unstable curves of length $\geq 2\varepsilon_0$.

Suppose we apply the following 'stopping rule' in the evolution of the H-components of $\mathcal{F}^n(W)$. At time $n = n_1$ all the H-components of $\mathcal{F}^n(W)$ whose length is $\geq 2\varepsilon_0$ are stopped ('frozen'), and the map $\mathcal{F}$ will act only on the other (shorter) H-components. The images of the latter will consist of H-components, and at every iteration of $\mathcal{F}$ we will stop ('freeze') every H-component whose length reaches $2\varepsilon_0$ or exceeds it. At time $n = n_2 - 1$ only those H-components whose length is $< 2\varepsilon_0$ and whose preimages at all times $n_1 \leq n < n_2$ belonged in H-components shorter than $2\varepsilon_0$ will stay 'alive'. Obviously, the union of these 'live' H-components will be $\mathcal{F}^{n_2-1}(W \setminus W_{n_1,n_2})$.

Theorem 5.61 now implies

COROLLARY 5.63.

$$\mathbf{m}_W(W \setminus W_{n_1,n_2}) \leq \vartheta_2^{n_2-n_1} \mathbf{m}_W(W).$$

Thus, the $k = n_2 - n_1$ consecutive iterations of $\mathcal{F}$, with the above stopping rule, transform the curve W, up to a set of exponentially (in k) small measure, into 'frozen' H-components of length $\geq 2\varepsilon_0$.

PROOF OF THEOREM 5.61. This theorem was first proved in [**BSC91**, Appendix 3], and again in [**Yo98**, p. 639] by billiard-specific arguments. However, Theorem 5.61 can be derived from the second growth lemma in a very general manner without using specifics of billiard dynamics, and here we do just that.

First we fix some θ satisfying

$$e^{-1/\varkappa} < \theta < 1. \tag{5.46}$$

Then we choose $\varepsilon_0 > 0$ and $b > 0$ so small that

$$|\ln \varepsilon_0| \geq \frac{|\ln \theta| + |\ln b|}{1 - \varkappa |\ln \theta|} \tag{5.47}$$

and

$$\vartheta_2 \colon = \theta \Big(1 + bc_3 e^{C(2\varepsilon_0)^{1/3}} \Big) < 1 \tag{5.48}$$

where $C > 0$ is the constant in the distortion bound Lemma 5.27 (the reason for these choices will become clear below).

Next we construct three objects:

(a) a decreasing sequence of subsets

$$W = \tilde{W}_0 \supset \tilde{W}_1 \supset \tilde{W}_2 \supset \cdots \supset \tilde{W}_{n_2-n_1};$$

(b) a "characteristic" function $\mathbf{n}(x)$ on W that takes value $\mathbf{n}(x) = i$ at all points $x \in \tilde{W}_i \setminus \tilde{W}_{i+1}$;

(c) a sequence of additional functions: $\mathbf{k}_1(x)$ on $\tilde{W}_1$, $\mathbf{k}_2(x)$ on $\tilde{W}_2$, etc., whose values are natural numbers, see below.

For brevity, we will also use notation $\mathbf{t}_i(x) = n_1 + \mathbf{k}_1(x) + \cdots + \mathbf{k}_{i-1}(x)$ for all $x \in \tilde{W}_{i-1}$ and $i \geq 2$ (and $\mathbf{t}_1(x) \equiv n_1$ on the entire W). The functions $\mathbf{k}_i$ will be the lengths of certain time intervals, and $\mathbf{t}_i$'s will be time moments when the sets $\tilde{W}_i$ are defined. The values $\mathbf{k}_i$ will satisfy the constraint

$$\mathbf{k}_1(x) + \cdots + \mathbf{k}_{\mathbf{n}(x)}(x) = n_2 - n_1$$

at all points $x \in W$ where $\max_{n_1 \leq n < n_2} r_n(x) < \varepsilon_0$.

First, we define $\tilde{W}_0 = W$ and

$$\tilde{W}_1 = \{x \in \tilde{W}_0 \colon r_{n_1}(x) < \varepsilon_0\}.$$

We set $\mathbf{n}(x) = 0$ on $\tilde{W}_0 \setminus \tilde{W}_1$. On the set $\tilde{W}_1$, the function $\mathbf{n}(x)$ will take positive values, as defined below.

The set $\mathcal{F}^{n_1}(\tilde{W}_1)$ is a union of unstable curves that we denote by $\hat{W}_{1,j}$, $j \geq 1$. Observe that each curve $\hat{W}_{1,j}$ is either an H-component of $\mathcal{F}^{n_1}(W)$ of length $\leq 2\varepsilon_0$ or an interval of length ε_0 adjacent to one of the two endpoints of a longer H-component of $\mathcal{F}^{n_1}(W)$.

On every curve $\tilde{W}_{1,j} := \mathcal{F}^{-n_1}(\hat{W}_{1,j}) \subset \tilde{W}_1$ we define the function

$$\mathbf{k}_1(x) = \big[\varkappa|\ln|\hat{W}_{1,j}||\big] + 1. \tag{5.49}$$

If it happens that $n_1 + \mathbf{k}_1(x) \geq n_2$ we reset $\mathbf{k}_1(x) = n_2 - n_1$. Thus the function $\mathbf{k}_1(x)$ is defined on $\tilde{W}_1$. For brevity, we set $\mathbf{t}_1(x) \equiv n_1$ and $\mathbf{t}_2(x) = \mathbf{t}_1 + \mathbf{k}_1(x)$.

Next we define, inductively, for $i \geq 2$

$$\tilde{W}_i = \{x \in \tilde{W}_{i-1} : \mathbf{t}_i(x) < n_2 \ \&\ r_{\mathbf{t}_i(x)}(x) < \varepsilon_0\}.$$

The set $\{\mathcal{F}^{\mathbf{t}_i(x)}(x) : x \in \tilde{W}_i\}$ is a union of unstable curves that we denote by $\hat{W}_{i,j}$, $j \geq 1$. These curves belong to the images of W obtained at different iterations of $\mathcal{F}$, because the function $\mathbf{t}_i(x)$ is variable. However, each curve $\hat{W}_{i,j}$ is either an H-component (for some iteration of W) of length $\leq 2\varepsilon_0$ or a subinterval of an interval of length ε_0 adjacent to one of the two endpoints of a longer H-component.

Now on every curve $\tilde{W}_{i,j} := \mathcal{F}^{-\mathbf{t}_i}(\hat{W}_{i,j}) \subset \tilde{W}_i$ we define the function

$$\mathbf{k}_i(x) = \big[\varkappa|\ln|\hat{W}_{i,j}||\big] + 1. \tag{5.50}$$

If it happens that $\mathbf{t}_i(x) + \mathbf{k}_i(x) \geq n_2$ we reset $\mathbf{k}_i(x) = n_2 - \mathbf{t}_i(x)$. Thus the function $\mathbf{k}_i(x)$ is defined on $\tilde{W}_i$. For brevity, we set $\mathbf{t}_{i+1}(x) = \mathbf{t}_i(x) + \mathbf{k}_i(x)$. We set $\mathbf{n}(x) = i - 1$ at all points $x \in \tilde{W}_{i-1} \setminus \tilde{W}_i$.

Obviously, after $\leq n_2 - n_1$ steps this inductive construction will be completed.

EXERCISE 5.64. Verify that $\mathbf{k}_1(x) + \cdots + \mathbf{k}_{\mathbf{n}(x)}(x) = n_2 - n_1$ at all points $x \in W$ where $\max_{n_1 \leq n < n_2} r_n(x) < \varepsilon_0$.

Now we use Theorem 5.53 to make some measure estimates. First,

$$\mathbf{m}_W(\tilde{W}_1)/\mathbf{m}_W(W) \leq c_3\varepsilon_0.$$

Furthermore, for every $k_1 \geq 1$ let us consider the intervals $\tilde{W}_{1,j} \subset \tilde{W}_1$ on which $\mathbf{k}_1(x) = k_1$. Their images $\mathcal{F}^{n_1}(\tilde{W}_{1,j}) = \hat{W}_{1,j}$ have length $|\hat{W}_{1,j}| \leq e^{-(k_1-1)/\varkappa}$, due to (5.49); hence $r_{n_1}(x) < e^{-(k_1-1)/\varkappa}$ on $\tilde{W}_{1,j}$, and so

$$\mathbf{m}_W(\mathbf{k}_1(x) = k_1)/\mathbf{m}_W(W) \leq c_3 e^{-(k_1-1)/\varkappa}.$$

Combining the above estimates gives

$$\mathbf{m}_W(\mathbf{k}_1(x) = k_1)/\mathbf{m}_W(W) \leq c_3 \min\big\{\varepsilon_0, e^{-(k_1-1)/\varkappa}\big\}.$$

EXERCISE 5.65. Show that under our condition (5.47) we have

$$\min\big\{\varepsilon_0, e^{-(k_1-1)/\varkappa}\big\} \leq b\theta^{k_1}$$

for all $k_1 \geq 1$.

Thus we arrive at the bound

$$\frac{\mathbf{m}_W(\mathbf{k}_1(x) = k_1)}{\mathbf{m}_W(W)} \leq bc_3\theta^{k_1}. \tag{5.51}$$

Next, for any $i \geq 1$ consider an arbitrary interval $\hat{W}_{i,j}$. Recall that the functions $\mathbf{t}_i(x)$, $\mathbf{k}_i(x)$, and $\mathbf{t}_{i+1}(x) = \mathbf{t}_i(x) + \mathbf{k}_i(x)$ are all constant on $\tilde{W}_{i,j} = \mathcal{F}^{-\mathbf{t}_i}(\hat{W}_{i,j})$; we denote their values simply by $\mathbf{t}_i$, $\mathbf{k}_i$, and $\mathbf{t}_{i+1}$, respectively, dropping the argument x.

Assume that $\mathbf{t}_{i+1} < n_2$. The set $\mathcal{F}^{\mathbf{k}_i}(\hat{W}_{i,j})$ has its own H-components. It is easy to see that the set $\mathcal{F}^{\mathbf{t}_{i+1}}(\tilde{W}_{i,j} \cap \tilde{W}_{i+1})$ lies within the union of the ε_0-neighborhoods of the endpoints of these H-components. We apply Theorem 5.53 to the curve $\hat{W}_{i,j}$, which gives

$$\mathbf{m}_{\hat{W}_{i,j}}\big(\hat{W}_{i,j} \cap \mathcal{F}^{\mathbf{t}_i}(\tilde{W}_{i+1})\big) \le c_3\varepsilon_0 \mathbf{m}_{\hat{W}_{i,j}}(\hat{W}_{i,j})$$

(note that Theorem 5.53 can be used here due to our choice of $\mathbf{k}_i$ in (5.50), and this explains our definition of $\mathbf{k}_i$).

Furthermore, for every integer $k_{i+1} \ge 1$, consider the intervals $\tilde{W}_{i+1,l} \subset \tilde{W}_{i,j} \cap \tilde{W}_{i+1}$ on which $\mathbf{k}_{i+1}(x) = k_{i+1}$. Their images $\mathcal{F}^{\mathbf{t}_{i+1}}(\tilde{W}_{i+1,l}) = \hat{W}_{i+1,l}$ have length $|\hat{W}_{i+1,l}| \le e^{-(k_{i+1}-1)/\varkappa}$, due to (5.50); hence $r_{\mathbf{t}_{i+1}}(x) < e^{-(k_{i+1}-1)/\varkappa}$ on $\tilde{W}_{i+1,l}$, and so

$$\mathbf{m}_{\hat{W}_{i,j}}\big(\hat{W}_{i,j} \cap \mathcal{F}^{\mathbf{t}_i}(\mathbf{k}_{i+1}(x) = k_{i+1})\big) \le c_3 e^{-(k_{i+1}-1)/\varkappa} \mathbf{m}_{\hat{W}_{i,j}}(\hat{W}_{i,j}).$$

Combining the above estimates, with the help of Exercise 5.65, gives

$$\mathbf{m}_{\hat{W}_{i,j}}\big(\hat{W}_{i,j} \cap \mathcal{F}^{\mathbf{t}_i}(\mathbf{k}_{i+1}(x) = k_{i+1})\big) \le bc_3\theta^{k_{i+1}} \mathbf{m}_{\hat{W}_{i,j}}(\hat{W}_{i,j}).$$

Pulling this estimate back to W under $\mathcal{F}^{-\mathbf{t}_i}$ gives

$$\mathbf{m}_{\tilde{W}_{i,j}}\big(\tilde{W}_{i,j} \cap (\mathbf{k}_{i+1}(x) = k_{i+1})\big) \le \zeta\theta^{k_{i+1}} \mathbf{m}_{\tilde{W}_{i,j}}(\tilde{W}_{i,j}),$$

where $\zeta = bc_3 e^{C(2\varepsilon_0)^{1/3}}$, for brevity. The extra exponential factor $e^{C(2\varepsilon_0)^{1/3}}$ is included here to account for possible distortions, according to Lemma 5.27.

Now observe that the functions $\mathbf{k}_1, \dots, \mathbf{k}_i$ are constant on every curve $\tilde{W}_{i,j}$. Let $k_1, \dots, k_i$ be an arbitrary sequence of natural numbers, and denote by $\tilde{W}_{k_1,\dots,k_i}$ the union of all the intervals $\tilde{W}_{i,j}$ on which $\mathbf{k}_1(x) = k_1, \dots, \mathbf{k}_i(x) = k_i$. Adding up the previous estimate over all intervals $\tilde{W}_{i,j} \subset \tilde{W}_{k_1,\dots,k_i}$ gives

$$\frac{\mathbf{m}_W\big(\tilde{W}_{k_1,\dots,k_i} \cap \{\mathbf{k}_{i+1}(x) = k_{i+1}\}\big)}{\mathbf{m}_W(\tilde{W}_{k_1,\dots,k_i})} \le \zeta\theta^{k_{i+1}}. \tag{5.52}$$

Of course, the set in the numerator is nothing but $\tilde{W}_{k_1,\dots,k_{i+1}}$. The probabilistic-type estimates (5.51) and (5.52) are crucial in our analysis, as we will see next.

Next let $m \ge 1$ and $S = \{k_1, \dots, k_m\}$ be a sequence of natural numbers such that

$$k_1 + \dots + k_m = n_2 - n_1. \tag{5.53}$$

Denote

$$\tilde{W}_S = \tilde{W}_{k_1,\dots,k_m} = \big\{x \in W \colon \mathbf{n}(x) = m,\ \mathbf{k}_1(x) = k_1, \dots, \mathbf{k}_m(x) = k_m\big\}.$$

Multiplying the estimates (5.51) and (5.52) for all $i = 1, \dots, m-1$ gives

$$\mathbf{m}_W(\tilde{W}_S) \le \zeta^m \theta^{n_2 - n_1} \mathbf{m}_W(W) \tag{5.54}$$

(note here that $bc_3 < \zeta$).

Now, adding (5.54) over all possible m and sequences $S = \{k_1, \dots, k_m\}$ satisfying (5.53) and using Exercise 5.64 give

$$\begin{aligned} \mathbf{m}_W(\tilde{W}) &\le \sum_{m=1}^{n_2-n_1} \binom{n_2-n_1-1}{m-1} \zeta^m \theta^{n_2-n_1} \mathbf{m}_W(W) \\ &= \zeta(1+\zeta)^{n_2-n_1-1} \theta^{n_2-n_1} \mathbf{m}_W(W) \end{aligned}$$

where $\binom{n_2-n_1-1}{m-1}$ denote the binomial coefficients coming from counting the number of respective sequences $\{S\}$. Using (5.48) completes the proof of Theorem 5.61. □

5.12. Size of H-manifolds (a local estimate)

In Section 5.5 we obtained a linear 'tail bound' (5.16) on the size of unstable H-manifolds $W^u_{\mathbb{H}}(x) \subset \mathcal{M}$, which reads

$$\mu\{x \colon r^u_{\mathbb{H}}(x) < \varepsilon\} \le C\varepsilon.$$

Here $r^u_{\mathbb{H}}(x)$ denotes the distance in the Euclidean metric, measured along $W^u_{\mathbb{H}}(x)$, from x to the nearest endpoint of $W^u_{\mathbb{H}}(x)$ (in a way, $r^u_{\mathbb{H}}(x)$ characterizes the size of $W^u_{\mathbb{H}}(x)$). This is a global estimate on unstable H-manifolds in the entire collision space $\mathcal{M}$. Here we deduce a similar local estimate on H-manifolds crossing a given stable curve W in terms of the Lebesgue measure on W.

Let $W \subset \mathcal{M}$ be a stable curve and $\mathbf{m}_W$ denote the Lebesgue measure on it.

THEOREM 5.66. *For any stable curve $W \subset \mathcal{M}$, the H-manifold $W^u_{\mathbb{H}}(x)$ exists (i.e. $r^u_{\mathbb{H}}(x) > 0$) for $\mathbf{m}_W$-almost every $x \in W$. Furthermore,*

$$\mathbf{m}_W\{x \in W \colon r^u_{\mathbb{H}}(x) < \varepsilon\} \le C\varepsilon \tag{5.55}$$

for some constant $C = C(\mathcal{D}) > 0$ and all $\varepsilon > 0$.

This theorem has its 'time reversal' counterpart (the time reversal principle was explained in Section 4.3):

THEOREM 5.67. *For any unstable curve $W \subset \mathcal{M}$, the stable H-manifold $W^s_{\mathbb{H}}(x)$ exists (i.e. $r^s_{\mathbb{H}}(x) > 0$) for $\mathbf{m}_W$-almost every $x \in W$. Furthermore,*

$$\mathbf{m}_W\{x \in W \colon r^s_{\mathbb{H}}(x) < \varepsilon\} \le C\varepsilon, \tag{5.56}$$

where $C = C(\mathcal{D}) > 0$ is a constant.

Here $W^s_{\mathbb{H}}(x)$ denotes the maximal stable H-manifold passing through $x \in \mathcal{M}$, and $r^s_{\mathbb{H}}(x)$ the distance, measured along $W^s_{\mathbb{H}}(x)$, from x to the nearest endpoint of $W^s_{\mathbb{H}}(x)$. It is enough to prove either Theorem 5.66 or Theorem 5.67, and we prove the latter.

PROOF. We will derive our theorem from the first growth lemma. First we need to state a 'time reversal' counterpart of Exercise 5.19, which requires some new notation. Let $E_s(x)$ be a piecewise constant function on $\mathcal{M}$ defined by $E_s(x) = (|k|+1)^2$ for all $x \in \mathcal{F}(\mathbb{H}_k)$. Observe that $E_s(x) = E_u(\mathcal{F}^{-2}x)$.

Exercise 5.18 has its 'time reversal', too, and it says that

$$\check{c}E_s(x) \le \frac{\|D_x\mathcal{F}^{-1}(dx)\|}{\|dx\|} \le \check{c}^{-1}E_s(x) \tag{5.57}$$

for all stable vectors $dx \in \mathcal{T}_x\mathcal{M}$ and all $x \in \mathcal{M}$, where $\tilde{c} = \tilde{c}(\mathcal{D}) > 0$ is a constant[2].

For every point $x \in \mathcal{M}$ denote by $d^s(x, \mathcal{S}_{-1}^{\mathbb{H}})$ the length of the shortest unstable curve that connects x with the set $\mathcal{S}_{-1}^{\mathbb{H}}$. Now the 'time reversal' of Exercise 5.19 reads

$$r_{\mathbb{H}}^s(x) \geq \min_{n\geq 1} \tilde{c}\,\hat{c}\, E_s(\mathcal{F}^n x)\Lambda^{n-1}\, d^s(\mathcal{F}^n x, \mathcal{S}_{-1}^{\mathbb{H}}). \tag{5.58}$$

EXERCISE 5.68. Prove formulas (5.57) and (5.58).

Next, given an unstable curve $W \subset \mathcal{M}$, recall that $r_n(x)$ denotes the distance from the point $\mathcal{F}^n(x)$ to the nearest endpoint of the H-component of $\mathcal{F}^n(W)$ containing the point $\mathcal{F}^n(x)$.

EXERCISE 5.69. Prove that there is a large constant $\tilde{C} = \tilde{C}(\mathcal{D}) > 0$ such that

$$\tilde{C}\,\tilde{c}\,\hat{c}\, E_s(\mathcal{F}^n x)\, d^s(\mathcal{F}^n x, \mathcal{S}_{-1}^{\mathbb{H}}) \geq r_{n-1}(x)$$

for all $n \geq 1$. Hint: Use the second inequality in (5.57) to show that the length of the shortest stable curve that connects $\mathcal{F}^{n-1}(x)$ with the set $\mathbb{S}_0$ does not exceed $\tilde{c}^{-1} E_s(\mathcal{F}^n x)\, d^s(\mathcal{F}^n x, \mathcal{S}_{-1}^{\mathbb{H}})$. Next, due to (4.14), the length of the *longest* stable curve that connects $\mathcal{F}^{n-1}(x)$ with the set $\mathbb{S}_0$ does not exceed $C E_s(\mathcal{F}^n x)\, d^s(\mathcal{F}^n x, \mathcal{S}_{-1}^{\mathbb{H}})$ for some constant $C = C(\mathcal{D}) > 0$. Due to time reversibility, the same is true for the length of the longest *unstable* curve that connects $\mathcal{F}^{n-1}(x)$ with the set $\mathbb{S}_0$.

As a result, we arrive at the estimate

$$r_{\mathbb{H}}^s(x) \geq \min_{n\geq 0} \tilde{C}^{-1}\Lambda^n\, r_n(x).$$

Therefore

$$\mathbf{m}_W\{x \in W\colon r_{\mathbb{H}}^s(x) < \varepsilon\} \leq \sum_{n=0}^{\infty} \mathbf{m}_W\{r_n(x) < \tilde{C}\Lambda^{-n}\varepsilon\}.$$

Due to the first growth lemma (Theorem 5.52 and the remark that immediately follows) we have

$$\mathbf{m}_W\{r_n(x) < \tilde{C}\Lambda^{-n}\varepsilon\} \leq \tilde{C}c_1\vartheta_1^n\Lambda^{-n}\varepsilon + \tilde{C}c_2\Lambda^{-n}\varepsilon\, \mathbf{m}_W(W).$$

Now summation over $n \geq 0$ proves (5.56). □

5.13. Fundamental theorem

We conclude this chapter with the so-called "fundamental theorem of the theory of dispersing billiards". Generalizing (4.53), we define

$$\mathcal{S}_{\infty}^{\mathbb{H}} = \cup_{n=0}^{\infty} \mathcal{S}_n^{\mathbb{H}} \qquad \text{and} \qquad \mathcal{S}_{-\infty}^{\mathbb{H}} = \cup_{n=0}^{\infty} \mathcal{S}_{-n}^{\mathbb{H}}. \tag{5.59}$$

Recall that the sets $\mathcal{S}_{\pm n}^{\mathbb{H}}$ are the singularities for the maps $\mathcal{F}_{\mathbb{H}}^{\pm n}$; the sets $\mathcal{S}_{\pm n}^{\mathbb{H}}$ include the 'real singularities' $\mathcal{S}_{\pm n}$ and the images (preimages) of the boundaries of the homogeneity strips:

$$\mathcal{S}_{\pm n}^{\mathbb{H}} = \mathcal{S}_{\pm n} \cup \big(\cup_{m=0}^{n} \mathcal{F}^{\mp m}(\mathbb{S})\big). \tag{5.60}$$

[2]For billiards of category B (according to our analysis at the end of Section 5.5), the function E_s needs to be redefined in the n-cells $D_{l,n}^-$, which were described in Section 4.10. Namely, for each $x \in \mathcal{F}(\mathbb{H}_k) \cap D_{l,n}^-$, we need to set $E_s(x) = n(|k|+1)^2$. Then the 'time reversal' of Exercise 5.26 would provide us with (5.57) for the category B tables.

The following fact was called the *Fundamental theorem of the theory of dispersing billiards* by Sinai [**Sin70, BS73**]:

THEOREM 5.70. *Let $x \in \mathcal{M} \setminus \mathcal{S}_{\infty}^{\mathbb{H}}$. Then for any $q > 0$ and $A > 0$ there exists an open neighborhood $\mathcal{U}_x^u \subset \mathcal{M}$ of x such that for any unstable curve $W \subset \mathcal{U}_x^u$*

$$\mathbf{m}_W\big(y \in W\colon r_{\mathbb{H}}^s(y) > A|W|\big) \geq (1-q)\,\mathbf{m}_W(W).$$

Similarly, let $x \in \mathcal{M} \setminus \mathcal{S}_{-\infty}^{\mathbb{H}}$. Then for any $q > 0$ and $A > 0$ there exists an open neighborhood $\mathcal{U}_x^s \subset \mathcal{M}$ of x such that for any stable curve $W \subset \mathcal{U}_x^s$

$$\mathbf{m}_W\big(y \in W\colon r_{\mathbb{H}}^u(y) > A|W|\big) \geq (1-q)\,\mathbf{m}_W(W).$$

Note that the second part is just the time-reversal of the first.

The significance of this theorem becomes evident when we choose a large $A \gg 1$ and a small $q \approx 0$. Then, accordingly, a vast majority of points $y \in W$ would lie on stable (unstable) H-manifolds which are much longer than the curve W itself. Note that those H-manifolds stretch far beyond W on *both* sides of W; see Fig. 5.11 (left).

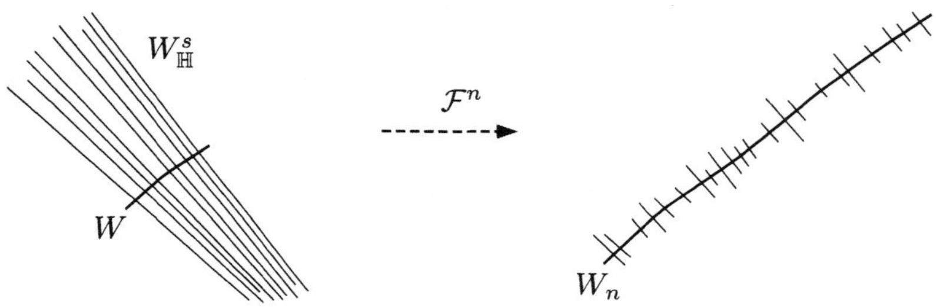

FIGURE 5.11. Fundamental theorem 5.70.

We note that in Sinai's original papers [**Sin70, BS73**] this theorem was given in a slightly weaker form: it involved unstable and stable manifolds rather than H-manifolds (and the former are, generally, much longer; see Section 5.5). The notion of H-manifolds was actually introduced much later in [**BSC91**].

PROOF OF THEOREM 5.70. It is enough to establish the first part. Our main idea is to apply Theorem 5.67 to the curve $W_n = \mathcal{F}^n(W)$ for a sufficiently large n, which ensures plenty of short stable H-manifolds crossing W_n. Then we pull them back under $\mathcal{F}^{-n}$ and obtain plenty of long stable H-manifolds crossing W; see Fig. 5.11.

Given $q > 0$, we set $p = C^{-1}C_{\rm d}^{-1}q$, where $C > 0$ is the constant in Theorem 5.67 and $C_{\rm d} > 0$ is the constant in the distortion bounds Lemma 5.27. Next, given $A > 0$, we choose $n \gg 1$ so large that $A \leq p\hat{c}^2\Lambda^{2n}$, where $\hat{c}$ and Λ appear in (4.19) (the reason for these choices will be clear below).

Now denote by $\mathcal{Q}_n^{\mathbb{H}}(x)$ the open connected component of the set $\mathcal{M}\setminus\mathcal{S}_n^{\mathbb{H}}$ containing the point x. Then for any unstable curve $W \subset \mathcal{Q}_n^{\mathbb{H}}(x)$ its image $W_n = \mathcal{F}^n(W)$ will consist of a single H-component. Choose $\mathcal{U}_x^u$ to be a small open neighborhood of x such that

$$\operatorname{dist}\big(\mathcal{U}_x^u, \partial\mathcal{Q}_n^{\mathbb{H}}(x)\big) > A \operatorname{diam}\mathcal{U}_x^u.$$

Now for any unstable curve $W \subset \mathcal{U}_x^u$ we apply Theorem 5.67 to the curve $W_n = \mathcal{F}^n(W)$ and obtain

$$\mathbf{m}_{W_n}\{z \in W_n \colon r^s_{\mathbb{H}}(z) < p|W_n|\} \le Cp|W_n|. \tag{5.61}$$

Let $z \in W_n$ be a point where $r^s_{\mathbb{H}}(z) \ge p|W_n|$. Then, due to the uniform hyperbolicity of $\mathcal{F}$ (cf. (4.19)), we have

$$\begin{aligned} r^s_{\mathbb{H}}(\mathcal{F}^{-n}z) &\ge \min\bigl\{\operatorname{dist}\bigl(W, \partial Q^{\mathbb{H}}_n(x)\bigr), p\hat{c}\Lambda^n|W_n|\bigr\} \\ &\ge \min\{A|W|, p\hat{c}^2\Lambda^{2n}|W|\} \\ &= A|W| \qquad\qquad \text{(due to our choice of } n). \end{aligned}$$

Since all the curves $\mathcal{F}^i(W)$, $0 \le i \le n$, are weakly homogeneous, the distortion bound Lemma 5.27 and the estimate (5.61) imply

$$\mathbf{m}_W\{y \in W \colon r^s_{\mathbb{H}}(y) < A|W|\} \le C_{\mathrm{d}}Cp|W| = q|W|$$

due to our choice of p. Theorem 5.70 is proved. □

CHAPTER 6

Ergodic properties

Here we prove that dispersing billiards of categories A and B, as dynamical systems, are ergodic, mixing and have Bernoulli property.

6.1. History

The ergodicity and mixing for dispersing billiards of categories A and B were first established by Sinai [**Sin70**] in 1970. He used a rather direct argument based on the 'fundamental theorem' (our Theorem 5.70) and Hopf's method (see below). Bernoulli property was obtained in 1974 by G. Gallavotti and D. Ornstein [**GO74**].

Later it turned out that Sinai's argument [**Sin70**] worked well for uniformly hyperbolic billiards, i.e. such that the unstable vectors expanded uniformly, in the sense of (4.19), but not for nonuniformly hyperbolic billiards. For the latter (most notably, for hard ball gases) the 'fundamental theorem', as it is stated in Section 5.13, does not hold.

To establish ergodicity for hard ball systems, a much more elaborate construction of 'regular coverings' was developed in subsequent works by Sinai and Chernov [**Sin79, SC87, Ch93**] and others [**KSS90, LW95**]. It was that construction, combined with Hopf's method, which was used to prove ergodicity for various systems of hard balls [**KSS91, KSS92, BLPS92, Sim99, Sim03, Sim04**]. Regular coverings were also employed in some proofs of ergodicity for other nonuniformly hyperbolic billiards [**Bu90, Sz92, Mar93b, DM01, DMM03**].

However, in our book we use Sinai's original argument; his method is simple and elegant, and it will actually apply to all major types of chaotic billiards covered in this book. For good tutorials on the more complicated method of regular coverings we refer the reader to [**KSS90, LW95, Va92**].

6.2. Hopf's method: heuristics

Nearly all the proofs of ergodicity for hyperbolic systems are based on Hopf's method [**Ho39, Ho40**], which we introduce here informally, leaving the formalities for the next two sections.

Suppose $F\colon M \to M$ is a smooth hyperbolic map preserving a probability measure μ. Suppose at every point $x \in M$ there is a stable manifold $W^s(x)$ and an unstable manifold $W^u(x)$.

Claim. *Almost every stable and unstable manifold belongs (mod 0) to one ergodic component of the map F.*

For the formal definition of ergodic components, conditional measures, generators, and other notions used below we refer the reader to Appendix C.

PROOF (HEURISTIC). Let $A \subset M$ be an ergodic component of the map F. Then $F\colon A \to A$ preserves the conditional measure μ_A, which is ergodic. The trajectory of μ_A-almost every point $x \in A$ is distributed in A according to the measure μ_A. Precisely, if $\{B_k\}$ is a countable generator of the σ-algebra of measurable subsets of A, then there is a subset $A_1 \subset A$ of full measure, $\mu_A(A \setminus A_1) = 0$, such that for every $x \in A_1$ and every k we have

$$\lim_{N\to\infty} \#\{n\colon 0 < n < N,\ F^n(x) \in B_k\}/N = \mu_A(B_k).$$

If a stable manifold W^s intersects two distinct ergodic components $A, B \subset M$, then the trajectories of typical points $x \in W^s \cap A$ and $y \in W^s \cap B$ are distributed differently (according to two different measures, μ_A and μ_B). On the other hand, the future semi-trajectories of the points x and y converge to each other; hence they must have identical distribution. This contradiction proves the claim for stable manifolds. The argument for unstable manifolds is similar, except one uses the past semi-trajectories of $x, y \in W^u$. □

Let $W_1, W_2, \dots, W_n$ be a finite sequence of stable and unstable manifolds such that $W_i \cap W_{i+1} \neq \emptyset$ for all $i = 1, \dots, n-1$. This is possible if stable and unstable manifolds in this sequence alternate. If these manifolds are *typical* and if they intersect each other in *typical points*, then according to the above claim their union $\cup W_i$ belongs (mod 0) to one ergodic component of the map F. Such a sequence of stable and unstable manifolds is called a *Hopf chain* or a *zig-zag line* or a *us-path* ('us' stays for 'unstable-stable'). It is instrumental in the construction of ergodic components for hyperbolic maps.

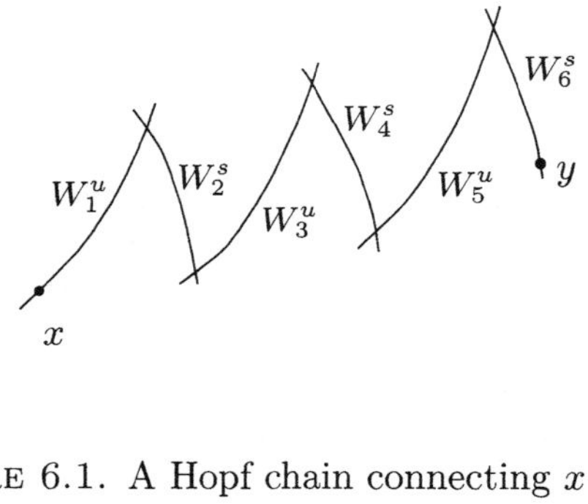

FIGURE 6.1. A Hopf chain connecting x and y.

Let $A \subset M$ be a set such that any two typical points $x, y \in A$ can be joined by a Hopf chain; see Fig. 6.1. Then A belongs (mod 0) to one ergodic component of the map F. If this holds for $A = M$, then the map F is ergodic.

This is the essence of Hopf's method: to prove ergodicity for a hyperbolic map F, it suffices to show that any two typical points $x, y \in M$ can be joined by a Hopf chain of typical stable and unstable manifolds intersecting one another in typical points. For nonergodic maps, Hopf's method can be used to identify their ergodic components.

The reader experienced in ergodic theory may find this argument already sufficient and proceed to Section 6.5, where Hopf chains are constructed for dispersing billiards. For a less experienced reader, we include a formal presentation of Hopf's method in the next two sections.

6.3. Hopf's method: preliminaries

Here we describe general conditions under which Hopf chains can be constructed (the reader will easily recognize that all of them have been established for dispersing billiards in Chapters 4 and 5). We emphasize that here we are constructing only Hopf chains. This construction by no means implies ergodicity, which will be proved only with the help of the Fundamental Theorem.

(H1) Smoothness with singularities. Suppose M is a compact Riemannian manifold (perhaps, with boundary and corners). Let F be a C^r smooth ($r \geq 2$) invertible map on M defined everywhere except singularities (a formal description was given in Section 3.1). We assume that F preserves a probability measure μ on M, which is equivalent to the Lebesgue measure (i.e. has an a.e. positive density). The Oseledets theorem 3.1 implies that Lyapunov exponents exist on a set of full μ-measure.

(H2) Hyperbolicity. Suppose F is hyperbolic; i.e. there exists a full measure set $H \subset M$, $\mu(H) = 1$, such that every point $x \in H$ is hyperbolic (has no zero Lyapunov exponents). Cf. the definitions given in Section 3.1. Then we have $\mathcal{T}_x M = E^u_x \oplus E^s_x$, where E^u_x and E^s_x denote unstable and stable subspaces, respectively. We also assume that $d^u = \dim E^u_x$ and $d^s = \dim E^s_x$ are constant on H, and the spaces E^u_x and E^s_x depend continuously on the point $x \in H$.

(H3) Unstable and stable manifolds. Suppose that at μ-almost every point $x \in H$ there exist a (local) stable manifold $W^s(x)$ and a (local) unstable manifold $W^s(x)$. More precisely, for any sufficiently small open set $\mathcal{U} \subset M$ and for μ-almost every point $x \in \mathcal{U}$ there is an unstable manifold $x \in W^u(x) \subset \mathcal{U}$ such that

$$\mathcal{T}_x W^u(x) = E^u(x), \qquad \operatorname{diam} F^{-n}\big(W^u(x)\big) \to 0 \quad \text{as } n \to \infty$$

and a stable manifold $x \in W^s(x) \subset \mathcal{U}$ such that

$$\mathcal{T}_x W^s(x) = E^s(x), \qquad \operatorname{diam} F^{n}\big(W^u(x)\big) \to 0 \quad \text{as } n \to \infty$$

(the convergence here is not necessarily exponential). Both manifolds are C^r smooth and have finite size. If $\mathcal{U}$ is small enough, then E^u_x and E^s_x will be uniformly transversal to each other in $\mathcal{U}$, and then for any $x, y \in \mathcal{U}$ the intersection $W^u(x) \cap W^s(y)$ will consist of at most one point.

Note: We simply assume here the existence of stable and unstable manifolds at almost every point. Katok and Strelcyn [**KS86**] (following Pesin's works [**Pes76, Pes77a**]) prove this existence under very mild assumptions on the singularities of the map F.

(H4) Abundance of unstable and stable manifolds. We assume also that for every unstable manifold $W^u(x)$ there is a stable manifold $W^s(y)$ through almost every point $y \in W^u(x)$ with respect to the internal Lebesgue measures on $W^u(x)$, and vice versa.

(H5) Measurability. Let $\mathcal{U} \subset M$ be a sufficiently small open set (e.g., a ball). Then the partition of $\mathcal{U}$ into local unstable manifolds $W^u(x) \cap \mathcal{U}$, $x \in \mathcal{U}$, is measurable (relevant definitions were given in Appendix A) and the conditional measures induced by μ on unstable manifolds are equivalent to the internal Lebesgue measures on them. The same is true for the partition of $\mathcal{U}$ into stable manifolds.

(H6) Absolute continuity. Let $V^1, V^2 \subset M$ be two small nearby d^u-dimensional submanifolds uniformly transversal to stable subspaces E_x^s. Denote

$$V_*^i = \{x \in V^i \colon W^s(x) \cap V^{3-i} \neq \emptyset\}$$

for $i = 1, 2$ (the above intersection is either empty or consists of a single point). We assume that the *holonomy* map defined by

$$\mathbf{h} \colon V_*^1 \to V_*^2, \quad x \mapsto W^s(x) \cap V^2$$

is absolutely continuous; i.e. it has a finite positive Jacobian with respect to the internal Lebesgue measures on V^1 and V^2. Similarly, the holonomy map defined by sliding along unstable manifolds is absolutely continuous.

Next we derive simple facts from our general assumptions (H1)–(H6). We emphasize again that (H1)–(H6) by no means guarantee the ergodicity of the map F; they only allow us to construct Hopf's chains and identify its ergodic components.

Let $V \subset \mathcal{U}$ be a d^u-dimensional submanifold uniformly transversal to stable subspaces E_x^s. Denote by $\mathbf{m}_V$ the internal Lebesgue measure on V. It may not be true that for $\mathbf{m}_V$-a.e. point $x \in V$ there is a stable manifold $W^s(x)$ (for example, V may be a singularity manifold for the map F^n, $n > 0$). We denote

$$V_* = \{x \in V \colon W^s(x) \ \text{ exists}\}.$$

LEMMA 6.1. *For any measurable subset $B \subset V_*$ we have*

$$\mathbf{m}_V(B) > 0 \Longleftrightarrow \mu\big(\cup_{x \in B} W^s(x)\big) > 0.$$

PROOF. We can foliate a small open neighborhood of V by some smooth d^u-dimensional manifolds (all transversal to unstable subspaces), so that V will be one of them. Then the absolute continuity (H6) and the Fubini theorem (cf. also (A.2)) imply the result. $\square$

Maps satisfying (H1)–(H6) are more general than collision maps for dispersing billiard tables: the dimensions of stable and unstable subspaces may be arbitrary and the hyperbolicity may be nonuniform.

6.4. Hopf's method: main construction

Here we turn to the ergodic properties of the map F. Recall a few standard facts from ergodic theory; cf. Appendix C.

Facts. A map F preserving the measure μ is ergodic iff every F-invariant (mod 0) measurable function $f \colon M \to \mathbb{R}$ (i.e. such that $f(Fx) = f(x)$ for μ-almost every x) is (mod 0) constant; i.e. $\mu\{f(x) = c\} = 1$ for some $c \in \mathbb{R}$. If the map F is not ergodic, then the space M can be partitioned into ergodic components of F (some of them may have positive measure, or all of them may have zero measure).

Furthermore, a set $A \subset M$ of positive measure belongs (mod 0) to one ergodic component of F iff every F-invariant function f is (mod 0) constant on A; i.e. $\mu\big(A \setminus \{f(x) = c\}\big) = 0$ for some $c \in \mathbb{R}$.

Now let $f \colon M \to \mathbb{R}$ be a continuous function. Denote by

$$\bar{f}_+(x) = \lim_{n \to \infty} \frac{1}{n} \sum_{i=0}^{n-1} f(F^i x), \quad \bar{f}_-(x) = \lim_{n \to \infty} \frac{1}{n} \sum_{i=0}^{n-1} f(F^{-i} x)$$

its *time averages* (forward and backward). By the Birkhoff ergodic theorem, these limits exist and coincide a.e.; i.e. the set

$$M_f = \{x \in M \colon \bar{f}_+(x) = \bar{f}_-(x)\}$$

has full measure, $\mu(M_f) = 1$. We define the function $\bar{f}(x) \colon = \bar{f}_+(x) = \bar{f}_-(x)$ on M_f.

PROPOSITION 6.2. *Suppose $\bar{f}(x)$ is (mod 0) constant for every continuous function f. Then the map F is ergodic. (The converse also holds.)*

PROOF. If F is not ergodic, then there is a nontrivial F-invariant set B, i.e. such that $F(B) = B$ and $0 < \mu(B) < 1$. Let g be the indicator function of B. The function g can be approximated in the $L^1(M, \mu)$ norm by continuous functions; i.e. for any $\varepsilon > 0$ there is a continuous function f such that $\|f - g\|_1 < \varepsilon$. Due to the invariance of the measure μ and the function g we have $\|f \circ F^n - g\|_1 = \|f - g\|_1$. Therefore

$$\begin{aligned}
\|\bar{f}_+ - g\|_1 &= \left\| \lim_{n\to\infty} \frac{1}{n} \sum_{i=0}^{n-1} (f(F^i x) - g) \right\|_1 \\
&= \lim_{n\to\infty} \frac{1}{n} \left\| \sum_{i=0}^{n-1} (f(F^i x) - g) \right\|_1 \\
&\le \lim_{n\to\infty} \frac{1}{n} \sum_{i=0}^{n-1} \left\| (f(F^i x) - g) \right\|_1 \\
&= \|f - g\|_1 < \varepsilon
\end{aligned}$$

where we used the Lebesgue Dominated Convergence Theorem and the triangle inequality. Since $\bar{f}_+$ is (mod 0) constant, we arrive at a contradiction, so F is ergodic. □

PROPOSITION 6.3. *Let $A \subset M$ be a subset of positive measure. Suppose $\bar{f}(x)$ is (mod 0) constant on A for every continuous function f. Then A belongs (mod 0) to one ergodic component of F.*

EXERCISE 6.4. Prove Proposition 6.3 (follow the lines of the previous argument).

The above propositions allow us to restrict our discussion to continuous functions $f \colon M \to \mathbb{R}$.

LEMMA 6.5. *For any stable or unstable manifold W the function $\bar{f}(x)$ is constant on the set $W \cap M_f$.*

PROOF. Without loss of generality we may consider a stable manifold $W = W^s$. Since the function f is continuous, it is also uniformly continuous on the compact set M. Then, for every $x, y \in W^s$ we have

$$\lim_{n\to\infty} \left| f(F^n x) - f(F^n y) \right| \to 0.$$

This implies $\bar{f}(x) = \bar{f}_+(x) = \bar{f}_+(y) = \bar{f}(y)$ for any $x, y \in W^s \cap M_f$. □

We say that a stable or unstable manifold $W \subset M$ is M_f-full if $\mathbf{m}_W(W \setminus M_f) = 0$ (recall that $\mathbf{m}_W$ denotes the internal Lebesgue measure on W).

LEMMA 6.6. *The union of all unstable and stable manifolds that are* not *M_f-full has zero μ-measure.*

EXERCISE 6.7. Prove Lemma 6.6. Hint: Use measurability (H5), as well as the equivalence of μ and the Lebesgue measure.

The above lemma is sometimes loosely stated as "almost every unstable and stable manifold is M_f-full".

Let $G_f \subset M_f$ denote the set of points such that for every $x \in G_f$ the manifolds $W^u(x)$ and $W^s(x)$ exist and are M_f-full. Obviously, $\mu(G_f) = 1$. We now say that a stable or unstable manifold $W \subset M$ is G_f-full if $\mathbf{m}_W(W \setminus G_f) = 0$.

LEMMA 6.8. *If a stable or unstable manifold is M_f-full, then it is also G_f-full.*

PROOF. Consider an unstable manifold W^u. First, through almost every point $y \in W^u$ (with respect to the internal Lebesgue measure $\mathbf{m}_{W^u}$) there is a stable manifold $W^s(y)$ due to Assumption (H3). Those stable manifolds which are not M_f-full make a set of measure zero (Lemma 6.6), so they correspond to a set of points $y \in W^u$ of zero $\mathbf{m}_{W^u}$-measure (Lemma 6.1). □

COROLLARY 6.9. *Let $x \in G_f$. Then both manifolds $W^u(x)$ and $W^s(x)$ exist and are G_f-full.*

The last corollary suggests the following construction of a certain increasing sequence of subsets (*Hopf subsets*):

$$H_1 \subset H_2 \subset H_3 \subset \cdots \subset M.$$

Pick a point $x_0 \in G_f$. Define the set

$$H_1(x_0) = W^u(x_0) \cap G_f$$

and

$$H_2(x_0) = \cup_{y \in H_1(x_0)} W^s(y) \cap G_f.$$

Then inductively, given a set $H_{2n}(x_0)$ define

$$H_{2n+1}(x_0) = \cup_{x \in H_{2n}(x_0)} W^u(x) \cap G_f$$

and

$$H_{2n+2}(x_0) = \cup_{y \in H_{2n+1}(x_0)} W^s(y) \cap G_f.$$

Plainly, we build an unstable manifold through x_0, then a stable manifold through every point $y \in G_f$ on it, then an unstable manifold through every point $x \in G_f$ on the previous stable manifolds, etc. Lastly, denote

$$H_\infty(x_0) = \cup_{n \geq 1} H_n(x_0).$$

Alternatively, one can start with a stable manifold through x_0, build unstable manifolds through every point $y \in G_f$ on it, etc., so that there are in fact two sequences of Hopf sets. They have similar properties, listed below:

(a) Every stable and unstable manifold in the above construction is G_f-full (by Corollary 6.9).
(b) $\mu(H_2) > 0$ (by Lemma 6.1); hence $\mu(H_n) > 0$ for all $n \geq 2$.
(c) The function $\bar{f}$ is constant on every H_n, hence on H_∞ (by Lemma 6.5).

There emerges a procedure, which can be used to verify that a given set $A \subset M$ of positive measure belongs (mod 0) to one ergodic component.

Let f be an arbitrary continuous function. Pick a point $x_0 \in A \cap G_f$ (i.e. any point in A except in the null set $A \setminus G_f$). Verify that the corresponding set $H_\infty(x_0)$ covers (mod 0) the set A, i.e. $\mu(A \setminus H_\infty(x_0)) = 0$. Then the function f is constant (mod 0) on A. Therefore A belongs (mod 0) to one ergodic component (by Proposition 6.3).

The crucial condition that $H_\infty(x_0)$ covers A (mod 0) can be described as follows. For almost every pair of points $x, y \in A \cap G_f$ there is a sequence of alternating unstable and stable manifolds $W_1, W_2, \dots, W_n$ such that $x \in W_1$, $y \in W_n$, and for every $i = 1, \dots, n-1$ the manifolds W_i and W_{i+1} intersect in exactly one point $z_i = W_i \cap W_{i+1} \in G_f$. This is a Hopf chain mentioned in Section 6.2 and shown in Fig. 6.1.

To prove the ergodicity of the map F, one may employ the following strategy:

(1) Find a set $M_1 \subset M$ such that for every point $x \in M_1$ there is a neighborhood $\mathcal{U}(x)$ that belongs (mod 0) to one ergodic component of F.
(2) Show that M_1 has full measure and is pathwise connected.

The first step is known as 'local' ergodicity and the second as 'global' ergodicity. We will employ this strategy in the next two sections.

6.5. Local ergodicity

Here we apply Hopf's method to dispersing billiards. Our presentation follows general lines of Sinai's original work [**Sin70**].

Recall that $\mathcal{S}_\infty^{\mathbb{H}}$ and $\mathcal{S}_{-\infty}^{\mathbb{H}}$ are sets (countable unions of smooth curves) on which positive and, respectively, negative iterations of the "new collision map" $\mathcal{F}_{\mathbb{H}}$ are discontinuous; see (5.59) and (5.60) and relevant definitions in Sections 5.13 and 5.4.

THEOREM 6.10 (Local ergodicity - I). *For any point $x \in \mathcal{M} \setminus (\mathcal{S}_\infty^{\mathbb{H}} \cup \mathcal{S}_{-\infty}^{\mathbb{H}})$ there is an open neighborhood $\mathcal{U}_x \subset \mathcal{M}$ that belongs (mod 0) to one ergodic component of the billiard map $\mathcal{F}$.*

PROOF. Choose a large $A \gg 1$ and a small $q > 0$ and consider the open sets $\mathcal{U}_x^u$ and $\mathcal{U}_x^s$ constructed in Theorem 5.70.

EXERCISE 6.11. Show that there are arbitrarily small 'rectangular-shaped' open neighborhoods $\mathcal{U}_x \subset \mathcal{U}_x^u \cap \mathcal{U}_x^s$ of the point x, each of which is bounded by two unstable H-manifolds and two stable H-manifolds. Hint: Let $W^u \ni x$ and $W^s \ni x$ be arbitrarily small unstable and stable curves contained in $\mathcal{U}_x^u \cap \mathcal{U}_x^s$. Theorem 5.70 provides plenty of long stable H-manifolds through W^u and plenty of long unstable H-manifolds through W^s.

EXERCISE 6.12. Show that if $q > 0$ is small enough, the domain $\mathcal{U}_x$ in the previous exercise can be made arbitrarily close to a rhombus, as shown in Fig. 6.2. Hint: Recall that stable lines E_y^s and unstable lines E_y^u (which are tangent to the stable and unstable H-manifolds $W_{\mathbb{H}}^u(x)$ and $W_{\mathbb{H}}^s(x)$, respectively) are continuous functions in their domains (cf. Section 4.7); hence they are both continuous at x. Thus all stable H-manifolds in $\mathcal{U}_x$ are almost parallel to each other, and so are all unstable H-manifolds in $\mathcal{U}_x$.

Let us fix $\mathcal{U}_x$ described in Exercises 6.11 and 6.12. We will prove that $\mathcal{U}_x$ belongs (mod 0) to one ergodic component of the map $\mathcal{F}$ by joining typical points $x_0, y_0 \in \mathcal{U}_x$ with Hopf chains.

Let $f\colon \mathcal{M} \to \mathbb{R}$ be a continuous function and $x_0 \in \mathcal{U}_x \cap G_f$ (in terms of the previous section). We will show that

$$\mathcal{U}_x \cap G_f \subset H_\infty(x_0),$$

which is more than enough to prove the theorem by Proposition 6.3.

Let $x_0 \in \mathcal{U}_x \cap G_f$. We start Hopf's construction at the point x_0 and define $W_1 \colon = W^u_{\mathbb{H}}(x_0) \cap \mathcal{U}_x$. Since the unstable H-manifold W_1 is G_f-full, we have by Theorem 5.70 that there is a stable H-manifold $W^s_{\mathbb{H}}(x_1)$ for some point $x_1 \in W_1 \cap G_f$ that stretches on both sides of W_1 by a distance $> A|W_1|$. We set $W_2 \colon = W^s_{\mathbb{H}}(x_1) \cap \mathcal{U}_x$.

EXERCISE 6.13. Show that either $|W_2| \geq A|W_1|$ or $\partial W_2 \subset \partial \mathcal{U}_x$ (in the latter case both endpoints of W_2 lie on the two unstable H-manifolds bounding $\mathcal{U}_x$).

Next, according to Corollary 6.9, the manifold W_2 is also G_f-full. By Theorem 5.70 there is an unstable H-manifold $W^u_{\mathbb{H}}(x_2)$ for some point $x_2 \in W_2 \cap G_f$ that stretches on both sides of W_2 by a distance $> A|W_2|$. We set $W_3 \colon = W^u_{\mathbb{H}}(x_2) \cap \mathcal{U}_x$. Again, either $|W_3| \geq A|W_2|$ or the manifold W_3 terminates on $\partial \mathcal{U}_x$ (more specifically, on the two stable H-manifolds bounding $\mathcal{U}_x$).

Obviously we can continue this alternating construction of stable and unstable H-manifolds until we get an unstable H-manifold W_n that terminates on $\partial \mathcal{U}_x$; see black lines in Fig. 6.2 (where $n = 5$). Note that W_n is G_f-full and $W_n \cap G_f \subset H_\infty(x_0)$.

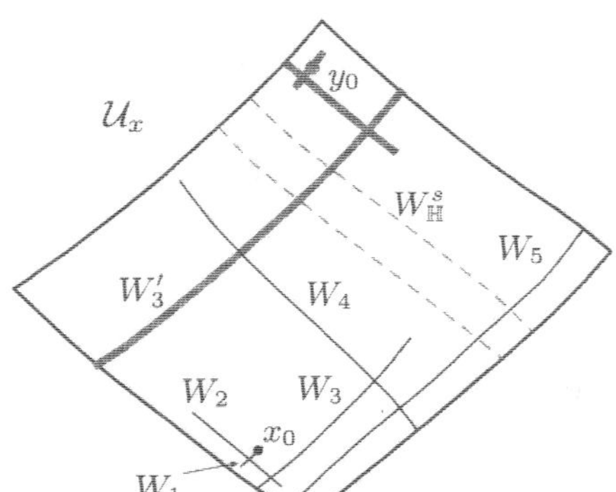

FIGURE 6.2. Proof of Theorem 6.10.

Starting from any other point $y_0 \in \mathcal{U}_x \cap G_f$ we build a similar sequence of alternating unstable and stable manifolds ending with an unstable H-manifold W'_m that terminates on $\partial \mathcal{U}_x$; see thick gray lines in Fig. 6.2 (where $m = 3$). Note that W'_m is G_f-full and $W'_m \cap G_f \subset H_\infty(y_0)$.

Due to Theorem 5.70 there are plenty of stable H-manifolds $W^s_{\mathbb{H}} \subset \mathcal{U}_x$ crossing both W_n and W'_m; see dashed lines in Fig. 6.2. Actually the union of such stable H-manifolds has positive measure. Due to Lemma 6.1, almost all of them intersect W_n and W'_m in points that belong to the set G_f.

This proves that $y_0 \in H_\infty(x_0)$ and vice versa. $\square$

THEOREM 6.14 (Local ergodicity - II). *Suppose that a point $x \in \mathcal{M}$ lies on exactly one smooth singularity curve $S \subset \mathcal{S}_\infty^{\mathbb{H}} \cup \mathcal{S}_{-\infty}^{\mathbb{H}}$. Then there is an open neighborhood $\mathcal{U}_x \subset \mathcal{M}$ that belongs (mod 0) to one ergodic component of $\mathcal{F}$.*

PROOF. There is exactly one $n \in \mathbb{Z}$ such that $S \subset \mathcal{F}^{-n}(\mathcal{S}_0 \cup \mathbb{S})$, according to Definitions (5.59)–(5.60) (recall relevant definitions in Sections 5.13, 5.3, 5.4, 2.10, and 2.11). The case $S \subset \mathcal{F}^{-n}(\mathbb{S})$ is easy. Indeed, our choice of homogeneity strips was somewhat arbitrary, so their boundary points must have the same properties (regarding local ergodicity) as the interior points. So we assume that $x \in S \subset \mathcal{F}^{-n}(\mathcal{S}_0)$.

Next, the case $n = 0$ (where $x \in \mathcal{S}_0 = \partial\mathcal{M}$) will easily follow from our analysis of the case $n \neq 0$. Without loss of generality, we assume $n > 0$. Since the map $\mathcal{F}^{n-1}$ is smooth at x and every ergodic component is $\mathcal{F}$-invariant, it is enough to prove the theorem for the point $\mathcal{F}^{n-1}(x)$. Thus we can assume that $n = 1$; i.e. $\mathcal{F}$ is discontinuous at x.

The curve S divides every open neighborhood V of x into two parts V_1 and V_2 on which $\mathcal{F}$ acts as described in the end of Section 4.9; see Fig. 6.3. Precisely, the first image $V_1' = \mathcal{F}(V_1)$ is an open neighborhood of $x' = \mathcal{F}(x) \in \mathcal{S}_0$ adjacent to the edge $\mathcal{S}_0$ and the second image $V_1'' = \mathcal{F}^2(V_1)$ is a semi-neighborhood of the point $x'' = \mathcal{F}^2(x)$ adjacent to the singularity curve $S' = \mathcal{F}^2(S) \subset \mathcal{S}_{-1}$. On the other hand, the first image $V_2' = \mathcal{F}(V_2)$ is a semi-neighborhood of the point x'' adjacent to the other side of the singularity curve S'. It can be checked directly that $V' = V_1'' \cup V_2'$ is an open neighborhood of the point x''; see Fig. 6.3.

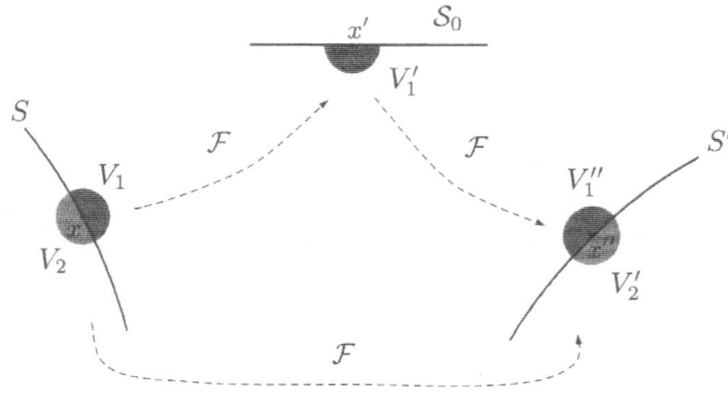

FIGURE 6.3. Action of $\mathcal{F}$ in a neighborhood of $x \in S$.

The second part of Theorem 5.70 applies to sufficiently small neighborhoods of the point x. On the other hand, the first part of Theorem 5.70 applies to sufficiently small neighborhoods of the point x'. Pulling it back to x we derive the following:

THEOREM 6.15. *For any $q > 0$ and $A > 0$ there exists an open neighborhood V of x such that for any unstable curve $W \subset V$ there is a subset $W' \subset W$ such that $\mathbf{m}_W(W') \geq (1-q)\,\mathbf{m}_W(W)$ and the stable H-manifold $W_{\mathbb{H}}^s(y)$ passing through every $y \in W'$ extends by a distance $> A|W|$ on both sides of y unless it terminates on the singularity curve S.*

The values of q and A here are certainly different from those used in Theorem 5.70. The possible termination on S is a new element here, compared to Theorem 5.70. Note that every stable H-manifold $W^s_{\mathbb{H}}(y)$ can terminate on S at only one of its endpoints (but not both).

Next we choose a large $A > 1$ and a small $q > 0$ and construct a neighborhood $\mathcal{U}_x$. In the vicinity of x, all unstable manifolds are almost parallel, but this is not true for stable manifolds. In fact, stable manifolds $W^s \subset V_1$ are almost tangent to S (they terminate on S tangentially) (see Sections 4.9 and 4.11), but stable manifolds $W^s \subset V_2$ are transversal to S

EXERCISE 6.16. Show that stable manifolds $W^s \subset V_2$ are almost parallel to each other. Hint: Use the results of Section 4.7 to show that the function $\mathcal{B}^{s+}(y)$, $y \in V_2$, has a continuous extension to the point x.

As in the previous theorem, we construct a neighborhood $\mathcal{U}_x$ bounded by two unstable and two stable H-manifolds. The part $\mathcal{U}_x \cap V_1$ may be made arbitrarily close to a rhombus provided $q > 0$ is small enough. But the part $\mathcal{U}_x \cap V_2$ will only be close to a trapezoid with almost parallel unstable sides but transversal stable sides; see Fig 6.4.

EXERCISE 6.17. Show that the part $\mathcal{U}_x \cap V_2$ can be made arbitrarily close to a trapezoid with almost parallel unstable sides. It can also be constructed so that its smaller unstable side is almost equal to its smaller stable side, so that the sides of this trapezoid will be comparable (none will be too short).

Next, for an arbitrary point $x_0 \in \mathcal{U}_x \cap G_f$ we construct a sequence of stable and unstable H-manifolds $x \in W_1, W_2, \dots, W_n$ as in the previous theorem, with the following two modifications.

First, we should stay away from the singularity curve S. This can be enforced in the following way. For every curve W_i we find its endpoint e_i that is farther from S than the other endpoint. Then we choose the next curve W_{i+1} so that it intersects W_i at a point x_i close enough to e_i (we can pick x_i so that $\text{dist}(x_i, e_i) < 2q|W_i|$).

Second, it is more convenient now to terminate the construction by a stable manifold W_n whose endpoints both lie on the two unstable sides of $\mathcal{U}_x$; see black lines in Fig. 6.4 (where $n = 4$). As in the previous theorem, W_n will be G_f-full and $W_n \cap G_f \subset H_\infty(x_0)$.

Similarly, for any other point $y_0 \in \mathcal{U}_x \cap G_f$ we build a sequence of alternating unstable and stable manifolds ending with a stable H-manifold W'_m that terminates on the two unstable sides of $\mathcal{U}_x$; see thick gray lines in Fig. 6.4 (where $m = 4$). Note that W'_m is G_f-full and $W'_m \cap G_f \subset H_\infty(y_0)$.

The stable H-manifolds W_n and W'_m may lie on one side of S (then they will be almost parallel to each other) or on the opposite sides, as is shown in Fig. 6.4.

In any case, due to Theorem 5.70 there are plenty of unstable H-manifolds $W^u_{\mathbb{H}} \subset \mathcal{U}_x$ crossing both W_n and W'_m; see dashed lines in Fig. 6.4. The union of such unstable H-manifolds has positive measure. Due to Lemma 6.1, almost all of them intersect W_n and W'_m in points that belong to the set G_f.

This proves that $y_0 \in H_\infty(x_0)$ and vice versa. □

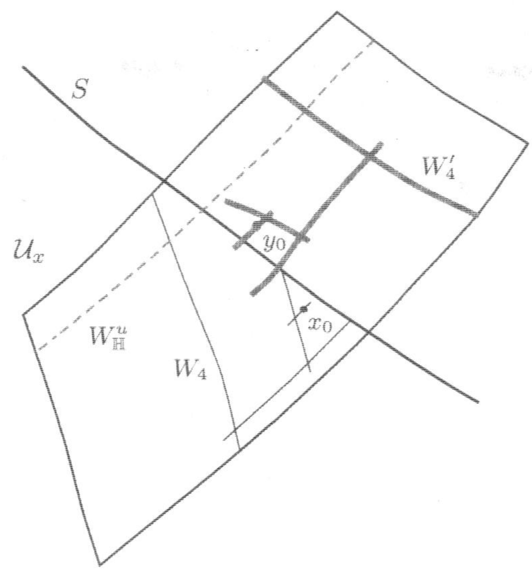

FIGURE 6.4. Proof of Theorem 6.14.

6.6. Global ergodicity

The 'local ergodic theorems' 6.10 and 6.14 apply to every point $x \in \mathcal{M}$ except those which belong to more than one smooth curve of the singularity set $\mathcal{S}_{\infty}^{\mathbb{H}} \cup \mathcal{S}_{-\infty}^{\mathbb{H}}$. We denote that set by $\mathcal{M}^{\#}$. The following lemma holds for all dispersing billiards:

LEMMA 6.18. *The set $\mathcal{M}^{\#}$ is countable.*

PROOF. It is enough to show that any two distinct curves $S, S' \subset \mathcal{S}_{\infty}^{\mathbb{H}} \cup \mathcal{S}_{-\infty}^{\mathbb{H}}$ intersect each other in at most one point. Let $S \subset \mathcal{F}_{\mathbb{H}}^{n}(\mathcal{S}_0 \cup \mathbb{S})$ and $S' \subset \mathcal{F}_{\mathbb{H}}^{n'}(\mathcal{S}_0 \cup \mathbb{S})$. If $n > 0$ and $n' < 0$, then the curve S is unstable and the curve S' is stable, so our claim is obvious.

Assume that $n' \geq n \geq 0$. Then $\mathcal{F}^{-n}(S) \subset \mathcal{S}_0 \cup \mathbb{S}$ is a horizontal curve (defined by an equation $\varphi = \text{const}$). At the same time $\mathcal{F}^{-n}(S')$ is either unstable (if $n' > n$) or also horizontal (if $n' = n$). In both cases their intersection is at most one point, as we claimed. The remaining case $n, n' < 0$ is treated similarly. $\square$

Now we need a general fact that easily follows from Proposition 6.3:

Fact: If sets A and B of positive measure belong (mod 0) in one ergodic component and $\mu(A \cap B) > 0$, then the union $A \cup B$ belongs (mod 0) to one ergodic component.

Recall that the collision space $\mathcal{M}$ consists of connected components $\mathcal{M}_i$; see Section 2.9.

PROPOSITION 6.19. *Every connected component $\mathcal{M}_i$ of the collision space $\mathcal{M}$ belongs (mod 0) in one ergodic component of the map $\mathcal{F}$.*

PROOF. Since the set $\mathcal{M}^{\#}$ is countable, its complement $\mathcal{M}_i \setminus \mathcal{M}^{\#}$ is arcwise connected. Suppose $x, y \in \mathcal{M}_i \setminus \mathcal{M}^{\#}$ are connected by a compact continuous curve

$C \subset \mathcal{M}_i \setminus \mathcal{M}^\#$. Every point $z \in C$ satisfies the 'local ergodic theorem' 6.10 or 6.14; thus an open neighborhood $\mathcal{U}_z$ belongs (mod 0) in one ergodic component. Due to the compactness of C it can be covered by a finite number of such open (and overlapping) neighborhoods; see Fig. 6.5. Now the result follows from the above general fact. □

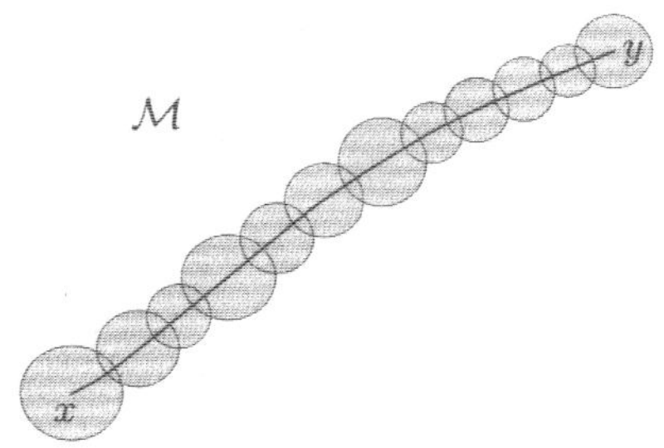

FIGURE 6.5. Proof of Proposition 6.19.

THEOREM 6.20. *The collision map $\mathcal{F}\colon \mathcal{M} \to \mathcal{M}$ is ergodic.*

PROOF. Due to the previous proposition, every ergodic component of $\mathcal{F}$ is a union of some $\mathcal{M}_i$'s. Each $\mathcal{M}_i$ consists of all collisions on one scatterer. It is obvious that from any scatterer, billiard trajectories run to nearby scatterers (and the set of such trajectories has positive measure), from those to their neighbors, etc. Thus if the billiard table $\mathcal{D}$ is connected, the entire collision space $\mathcal{M}$ belongs in one ergodic component. □

6.7. Mixing properties

Mixing properties for the collision map $\mathcal{F}\colon \mathcal{M} \to \mathcal{M}$ can be established relatively easily based on general results due to Pesin [**Pes77a, Pes92**], which we describe below.

Let $F\colon M \to M$ be a hyperbolic map preserving a probability measure μ.

DEFINITION 6.21. We say that F satisfies *ergodic decomposition* (or spectral decomposition) if there is a finite or countable partition

$$M = M_0 \cup M_1 \cup M_2 \cdots$$

such that
(a) each M_i is F-invariant, $\mu(M_0) = 0$ and $\mu(M_i) > 0$ for $i \geq 1$;
(b) the restriction $F\colon M_i \to M_i$ is ergodic for every $i \geq 1$.
Furthermore, for every $i \geq 1$ there is a finite partition

$$M_i = M_i^1 \cup \cdots \cup M_i^{k_i}$$

such that
(c) $F(M_i^j) = M_i^{j+1}$ and $F(M_i^{k_i}) = M_i^1$ (cyclic permutation);
(d) the map $F^{k_i}\colon M_i^j \to M_i^j$ is mixing, K-mixing and Bernoulli.

Pesin [**Pes92**] established the ergodic decomposition for very general classes of hyperbolic maps and their invariant measures that include all dispersing billiards of categories A and B.

We have already proved ergodicity for the collision map; hence its ergodic decomposition has a single ergodic component M_1. Next we want to show that $k_1 = 1$ in the clause (c); i.e. the cycle is trivial.

Indeed, if $k_1 \geq 2$, then the map $\mathcal{F}^{k_1}$ would not be ergodic (its ergodic components would be the subsets $M_1^1, \dots, M_1^{k_1}$). But this is impossible due to the following:

THEOREM 6.22. *The map $\mathcal{F}^n \colon \mathcal{M} \to \mathcal{M}$ is ergodic for every $n \geq 2$.*

PROOF. The map $\mathcal{F}^n$ satisfies the assumptions (H1)–(H6) of Section 6.3; hence Hopf's method will work for it. The map $\mathcal{F}$ and the maps $\mathcal{F}^n$ have the same stable and unstable manifolds, and they have the same singularity sets $\mathcal{S}_\infty^{\mathbb{H}} \cup \mathcal{S}_{-\infty}^{\mathbb{H}}$. Then the fundamental theorem is valid for $\mathcal{F}^n$. All these facts can be checked by direct inspection, which we leave to the reader.

Thus Local Ergodic Theorems 6.10 and 6.14, Lemma 6.18 and Proposition 6.19 carry over to the map $\mathcal{F}^n$ without change. Hence every connected component $\mathcal{M}_i \subset \mathcal{M}$ (built on a curve $\Gamma_i \subset \Gamma$) belongs (mod 0) to one ergodic component of $\mathcal{F}^n$ for every $n \geq 2$. It remains to verify the following:

LEMMA 6.23. *Let the boundary $\Gamma = \partial\mathcal{D}$ of the billiard table consist of r smooth components, $\Gamma = \Gamma_1 \cup \cdots \cup \Gamma_r$ (they are closed curves for tables of categories A and B). Then it is impossible to divide the set $\{1, \dots, r\}$ into groups $G_1, \dots, G_k$ with $k \geq 2$, so that billiard trajectories originating on a curve Γ_i, $i \in G_p$, $1 \leq p \leq k$, will necessarily land on a curve Γ_j, $j \in G_{p+1}$ (with the identification $G_{k+1} = G_1$).*

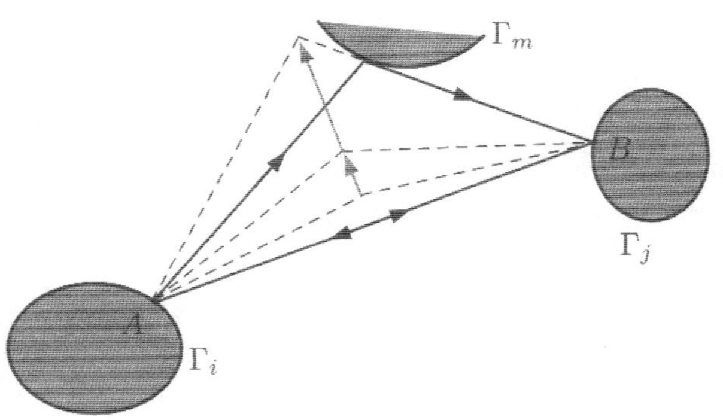

FIGURE 6.6. Proof of Lemma 6.23.

PROOF. Consider an arbitrary billiard trajectory originating at some point $A \in \Gamma_i$ and landing at some other point $B \in \Gamma_j$; see Fig. 6.6. Reversing the time we obtain a trajectory running from B back to A. This shows that if $\Gamma_i \in G_p$, then $\Gamma_j \in G_{p+1}$ and $\Gamma_i \in G_{p+2}$; hence $G_p = G_{p+2}$, which is only possible if $k = 1$ or $k = 2$. Suppose $k = 2$. Without loss of generality, let $\Gamma_i \in G_1$ and $\Gamma_j \in G_2$. Consider an isosceles triangle with base AB and let its height grow; see Fig. 6.6. The first time this triangle touches another curve $\Gamma_m \subset \Gamma$ gives us billiard trajectories running from Γ_i to Γ_m; hence $\Gamma_m \in G_2$, and some other billiard trajectories running from Γ_m to Γ_j, hence $\Gamma_j \in G_1$. It follows that $G_1 = G_2$, so $k = 1$. □

This proves Theorem 6.22. □

COROLLARY 6.24. *For dispersing billiards of categories A and B the map* $\mathcal{F}\colon \mathcal{M} \to \mathcal{M}$ *is mixing, K-mixing and Bernoulli.*

REMARK 6.25. Ergodicity has also been established for dispersing billiards of categories C, D, F, and E; see [**Re95**].

REMARK 6.26. The Bernoulli property holds for K-mixing hyperbolic maps (and flows) under very general conditions [**CH96, OW98**].

6.8. Ergodicity and invariant manifolds for billiard flows

Here we discuss the ergodic properties of the billiard flow $\Phi^t\colon \Omega \to \Omega$, which preserves the measure μ_Ω; see the notation in Section 2.12. For an extensive studies of ergodic properties of more general hyperbolic flows we refer the reader to [**KB94**].

Recall that Φ^t is a suspension flow whose base transformation is $\mathcal{F}\colon \mathcal{M} \to \mathcal{M}$ and whose ceiling function is $\tau(x)$; see Section 2.9.

EXERCISE 6.27. Show that a suspension flow is ergodic if and only if its base transformation is ergodic. Hint: If the flow Φ^t is not ergodic, then there exists a Φ^t-invariant subset $B \subset \Omega$ with measure $0 < \mu_\Omega(B) < 1$. In that case the set $A\colon = B \cap \mathcal{M}$ would be invariant under the map $\mathcal{F}$, and one can check using (2.30) that $0 < \mu(A) < 1$.

As we have established ergodicity for $\mathcal{F}$, it follows that

PROPOSITION 6.28. *For billiards of categories A and B, the flow* $\Phi^t\colon \Omega \to \Omega$ *is ergodic.*

It would be tempting to derive mixing properties for a suspension flow from those of a base transformation, but this is impossible. Whatever the base transformation, the flow may fail to be mixing.

EXERCISE 6.29. Show that if the ceiling function of a suspension flow is constant, then the flow cannot be mixing. Extend this conclusion to suspension flows whose ceiling function takes two values, p and q, such that p/q is rational.

To derive mixing properties for the flow, we need to construct and use its stable and unstable manifolds.

The billiard flow Φ^t is hyperbolic; i.e. μ_Ω-almost every point $X \in \Omega$ has one positive and one negative Lyapunov exponent (the third exponent is necessarily zero). This follows from the hyperbolicity of the map $\mathcal{F}$ proved in Section 4.4 and the general results of Section 3.3. The corresponding characteristic subspaces E^u_X and E^s_X are one-dimensional, and we have $E^u_X, E^s_X \subset \mathcal{T}^\perp_X\Omega$ (i.e. they are 'orthogonal' to the velocity vector); see Section 4.7.

We will show that for μ_Ω-almost every point $X \in \Omega$ there exist a one-dimensional local stable manifold $\mathcal{W}^s(X) \subset \Omega$ and a one-dimensional local unstable manifold

$\mathcal{W}^u(X) \subset \Omega$, both passing through X, that satisfy[1]

$$\left|\Phi^t\left(W^s(X)\right)\right| \to 0, \qquad \left|\Phi^{-t}\left(W^u(X)\right)\right| \to 0$$

as $t \to \infty$. The manifolds $\mathcal{W}^u(X)$ and $\mathcal{W}^s(X)$ are tangent to the subspaces E^u_X and E^s_X, respectively.

PROPOSITION 6.30. *For μ_Ω-almost every point $X \in \Omega$ there is a stable manifold $\mathcal{W}^s(X)$ and an unstable manifold $\mathcal{W}^u(X)$. The stable manifold $W^s(X)$ is a focusing wave front, and the unstable manifold $W^u(X)$ is a dispersing wave front.*

PROOF. Let $X = (q, v) \in \Omega$ be a phase point such that $q \in \operatorname{int} \mathcal{D}$ and $v \in S^1$, in the notation of Section 2.5. Denote by $x = \mathbf{P}^-(X) \in \mathcal{M}$ the natural projection of X defined in Section 3.3 and by $W^u(x) \subset \mathcal{M}$ the (local) unstable manifold for the map $\mathcal{F}$ (one exists for μ-almost every x). The outgoing trajectories of points $y \in W^u(x)$ make a one-parameter family of directed lines in the domain $\mathcal{D}$ (see Fig. 6.7); one of these lines passes through the point q and contains the vector v. Let σ be the orthogonal cross-section of this family (cf. Section 3.7) passing through the point q. Denote by $\mathcal{W}^u(X) \subset \Omega$ the curve σ equipped with unit normal vectors pointing along our directed lines. Then $\mathcal{W}^u(X)$ is a dispersing wave front; see Sections 4.3 and 4.5.

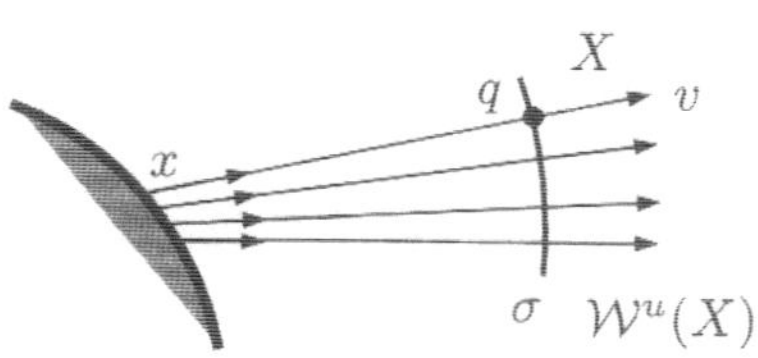

FIGURE 6.7. Proof of Proposition 6.30.

Since $\mathcal{F}^{-n}W^u(x)$ is an unstable curve for every $n \geq 1$, it follows that $\Phi^{-t}\mathcal{W}^u(X)$ is a dispersing wave front for all $t > 0$. Also, the length of $\mathcal{F}^{-n}W^u(x)$ converges to zero as $n \to \infty$; hence so does the length of $\Phi^{-t}\mathcal{W}^u(X)$ as $t \to \infty$. Thus we constructed the unstable manifold.

The construction of the stable manifold $\mathcal{W}^s(X)$ is the time reversal of the above. $\square$

EXERCISE 6.31. Show that for dispersing billiards of category A, $\left|\Phi^{-t}\left(\mathcal{W}^u(X)\right)\right|$ converges to zero exponentially fast in t. Hint: Use the exponential convergence $\left|\mathcal{F}^{-n}\left(W^u(x)\right)\right| \to 0$ and the bounds $0 < \tau_{\min} \leq \tau(x) \leq \tau_{\max} < \infty$.

In billiards of category B, there is no upper bound on $\tau(x)$, and so the convergence $\left|\Phi^{-t}\left(\mathcal{W}^u(X)\right)\right| \to 0$ may be arbitrarily slow.

For a small $\varepsilon > 0$, the sets

$$\mathcal{W}^u_\varepsilon(X) = \cup_{|t|<\varepsilon}\Phi^t\mathcal{W}^u(X), \qquad \mathcal{W}^s_\varepsilon(X) = \cup_{|t|<\varepsilon}\Phi^t\mathcal{W}^s(X),$$

[1]Here $|\Phi^t(W)|$ denotes the total length of the one-dimensional manifold $\Phi^t(W)$. Since $\Phi^t(W^s)$ may consist of several disconnected components, we cannot claim that its diameter converges to zero.

are two-dimensional surfaces in Ω. They are called (local) *weakly unstable* and *weakly stable* manifolds, respectively. They intersect in a piece of the trajectory of X

$$\mathcal{W}^u_\varepsilon(X) \cap \mathcal{W}^s_\varepsilon(X) = \cup_{|t|<\varepsilon} \Phi^t(X);$$

see Fig. 6.8. Note that the $\mathbf{P}^-$-projections of $\mathcal{W}^u_\varepsilon(X)$ and $\mathcal{W}^s_\varepsilon(X)$ onto $\mathcal{M}$ are the manifolds $W^u(x)$ and $W^s(x)$, respectively, where $x = \mathbf{P}^-(X)$.

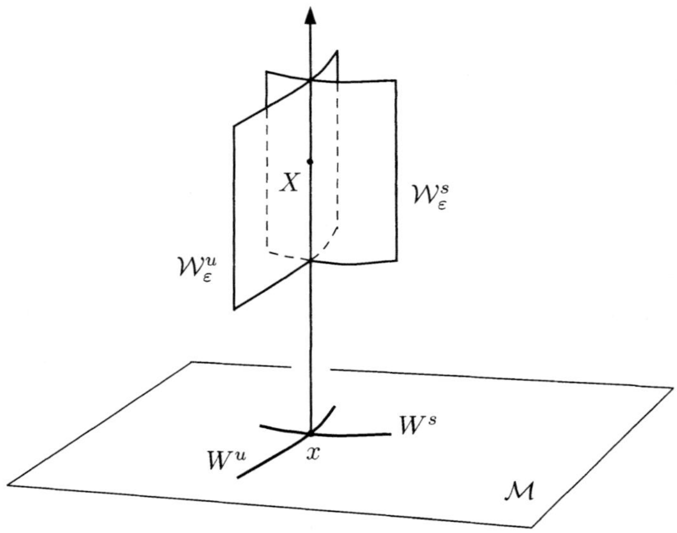

FIGURE 6.8. Weakly stable and unstable manifolds.

6.9. Mixing properties of the flow and 4-loops

Now we use stable and unstable manifolds of the billiard flow $\Phi^t\colon \Omega \to \Omega$ to investigate its mixing properties.

We begin with an instructive (counter) example. Let $F\colon M \to M$ be a hyperbolic map on a 2D manifold M preserving a measure μ, such that for μ-almost every point $x \in M$ there is an unstable manifold $W^u(x)$ and a stable manifold $W^s(x)$. Consider a suspension flow Φ^t constructed over the map $T\colon M \to M$ under a *constant* ceiling function $L(x) = L_0 > 0$. Note that it acts on the 3D manifold $\Omega = M \times [0, L_0]$, and it is not mixing according to Exercise 6.29.

EXERCISE 6.32. Show that if $W^u(x)$ and $W^s(x)$ exist, then for any point $X = (x,t) \in \Omega$ the unstable manifold of the flow is $\mathcal{W}^u(X) = W^u(x) \times \{t\}$ and its stable manifold is $\mathcal{W}^s(X) = W^s(x) \times \{t\}$.

Hence both manifolds $\mathcal{W}^u(X)$ and $\mathcal{W}^s(X)$ belong to the same 'horizontal section' $M \times \{t\} \subset \Omega$. Furthermore, any Hopf chain[2] starting at X will belong to the same 'horizontal section'. Observe that every such section $M \times \{t\}$ is a 2D surface in Ω. In that case the families of stable and unstable manifolds are said to

[2]As in Section 6.2, we call a Hopf chain any finite sequence of alternating stable and unstable manifolds $\mathcal{W}_1, \dots, \mathcal{W}_n$ such that $\mathcal{W}_i \cap \mathcal{W}_{i+1} \neq \emptyset$ for all $i = 1, \dots, n-1$.

be (locally) *integrable.* More precisely, local integrability at X means that all short enough Hopf chains starting at X lie in one hypersurface of Ω.

Now suppose that Φ^t is a suspension flow over a hyperbolic base map $T\colon M \to M$ with an arbitrary (not necessarily constant) ceiling function $\tau(x)$. We assume, for convenience, that $\tau(x)$ is continuous. Local integrability of the families of unstable and stable manifolds of Φ^t can be verified as follows.

Let $x_1, x_3 \in M$ be two nearby points such that

$$x_2 \colon = W^u(x_1) \cap W^s(x_3) \neq \emptyset, \quad x_4 \colon = W^s(x_1) \cap W^u(x_3) \neq \emptyset; \tag{6.1}$$

see Fig. 6.9. Let $X_1 = (x_1, t)$ and $X_3 = (x_3, t)$. Then the weakly stable and unstable manifolds at the points X_1 and X_3 intersect each other in pieces of trajectories of the points x_1, x_2, x_3, x_4; see Fig. 6.9. Now let $Y_1 = X_1$:

$$Y_2 = \mathcal{W}^u(Y_1) \cap \mathcal{W}^s_\varepsilon(X_3), \qquad Y_3 = \mathcal{W}^s(Y_2) \cap \mathcal{W}^u_\varepsilon(X_3),$$

$$Y_4 = \mathcal{W}^u(Y_3) \cap \mathcal{W}^s_\varepsilon(X_1), \qquad Y_5 = \mathcal{W}^s(Y_4) \cap \mathcal{W}^u_\varepsilon(X_1);$$

see Fig. 6.9 (Here ε should be large enough, compared to the distance between X_1 and X_3. Also, due to the continuity of $\tau(x)$ we can assume that the entire construction stays well under the ceiling.)

Observe that all the points Y_1, Y_2, Y_3, Y_4, Y_5 belong to one Hopf chain which starts at Y_1 and terminates at Y_5. Furthermore, its endpoints Y_1 and Y_5 lie on the same trajectory of the flow. Our Hopf chain 'loops' around the flow line $\Phi^t(X_1)$.

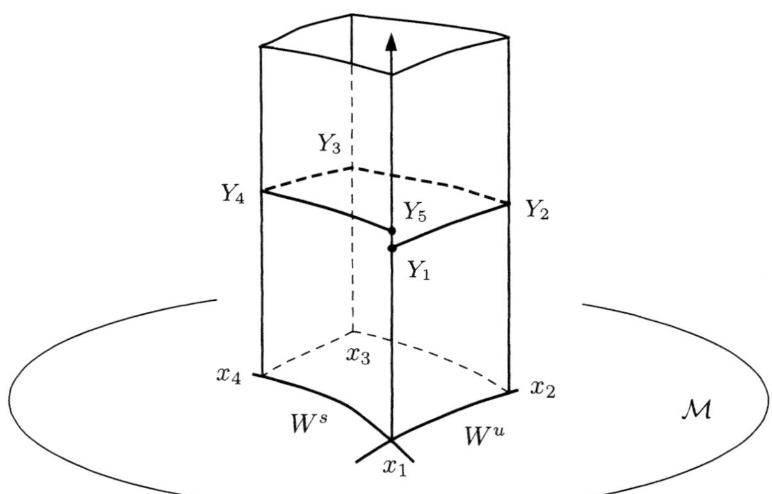

FIGURE 6.9. An open 4-loop.

We call a *4-loop* a Hopf chain of four (sufficiently short) alternating stable and unstable manifolds that begins and terminates on the same flow trajectory. We say that a 4-loop is *open* if its endpoints are distinct and *closed* otherwise (in the latter case the 4-loop is a closed curve in Ω).

It is clear that the families of stable and unstable manifolds of a hyperbolic flow Φ^t are locally integrable at X if every 4-loop in a vicinity of X is closed.

On the other hand, if every 4-loop is open and the flow Φ^t is ergodic, then it is automatically mixing, K-mixing, and Bernoulli (actually, it is enough to assume that the 4-loop is open for pairs of points $x_1, x_3 \in M$ satisfying (6.1) that make a set of positive measure). This is a fairly standard fact in smooth ergodic theory; it is proven, for example, in [**KM81**, Theorems 4 and 5]. We sketch the argument in the next section for the sake of completeness.

6.10. Using 4-loops to prove K-mixing

Here we assume that every 4-loop is open and prove the K-mixing of the billiard flow. The openness of the 4-loops will be verified in the next section. Our argument will heavily use the notion of measurable partitions, already mentioned in Section 5.1 and fully presented in Appendix A.

Let ξ^u denote the partition of Ω into (the smooth components of) unstable manifolds.

EXERCISE 6.33. Verify that ξ^u is a measurable partition. This requires constructing a countable generator; see Appendix A. Use the one constructed in Section 5.1 for the 'namesake' partition ξ^u of the collision space $\mathcal{M}$ and 'lift' it to Ω.

The properties of the partition of ξ^u of the collision space $\mathcal{M}$ discussed in Section 5.1 easily extend to the partition ξ^u of Ω.

First, the partition $\xi^u_t = \{\Phi^t(C)\colon C \in \xi^u\}$ is measurable for every $t \in \mathbb{R}$ and the family of partitions $\{\xi^u_t\}$, $t \in \mathbb{R}$, is decreasing; i.e. $\xi^u_{t_1} \succcurlyeq \xi^u_{t_2}$ for any $t_1 < t_2$. Clearly, $\vee_{t=-\infty}^{\infty}\xi^u_t = \varepsilon$, which denotes the partition of Ω into individual points.

Let ξ^s denote the partition of Ω into (the smooth components of) stable manifolds. It has similar properties, in particular, $\xi^s_{t_1} \preccurlyeq \xi^s_{t_2}$ for any $t_1 < t_2$ and $\vee_{t=-\infty}^{\infty}\xi^s_t = \varepsilon$. According to a general fact in ergodic theory (see Appendix C), the partitions

$$\pi^u = \wedge_{t=-\infty}^{\infty}\xi^u_t \quad \text{and} \quad \pi^s = \wedge_{t=-\infty}^{\infty}\xi^s_t$$

are finer than the Pinsker partition $\pi = \pi(\{\Phi^t\})$ of the flow Φ^t, i.e. $\pi^u \succcurlyeq \pi$ and $\pi^s \succcurlyeq \pi$. Therefore, $\pi^u \wedge \pi^s \succcurlyeq \pi$.

In fact, one can prove that $\pi^u = \pi^s = \pi$ (see [**KM81**, Theorem 4]), but we will not need this.

Let $\eta_t = \xi^u_t \wedge \xi^s_{-t}$ for every $t \in \mathbb{R}$ and $\eta_\infty = \wedge_{t=-\infty}^{\infty}\eta_t$.

EXERCISE 6.34. Show that $\eta_\infty = \pi^u \wedge \pi^s$; hence $\eta_\infty \succcurlyeq \pi$. Also, verify that the partition η_∞ is invariant under the flow Φ^t. This means that the corresponding complete sub-σ-algebra $\mathfrak{B}(\eta_\infty)$ is Φ^t-invariant for every $t \in \mathbb{R}$. Hint: Note that $A \in \mathfrak{B}(\eta_\infty)$ if and only if $A \in \mathfrak{B}(\xi^u_t)$ and $A \in \mathfrak{B}(\xi^s_t)$ for every $t \in \mathbb{R}$.

It is a standard fact in ergodic theory that a flow is K-mixing if and only if its Pinsker partition π is trivial, i.e. $\pi = \nu$ (which denotes the partition whose elements have measure 0 or 1); see Appendix C. Thus it is enough to verify that the partition η_∞ is trivial.

We heavily use the notion and notation of Appendix A, especially those related to measurable partitions. Let $D \subset \Omega$ be a full measure set and ξ^u_D and ξ^s_D denote the restrictions of the partitions ξ^u and ξ^s to D. According to our discussion at the end of Appendix A, the set D can be chosen so that $\xi^u_D \curlywedge \xi^s_D \succcurlyeq \xi^u_D \wedge \xi^s_D$; see the definition of $\xi^u_D \curlywedge \xi^s_D$ in Appendix A. We fix such a set D.

For any point $x \in \mathcal{M}$, consider its trajectory segment $\mathcal{L}_x = \{\Phi^t x : 0 < t < \tau(x)\}$. Clearly, the partition of Ω into segments $\{\mathcal{L}_x : x \in \mathcal{M}\}$ is measurable and the conditional measures on $\mathcal{L}_x$ are uniform. Let $E_1 = \{x \in \mathcal{M} : \mathbf{m}(\mathcal{L}_x \setminus D) = 0\}$, where $\mathbf{m}$ denotes the Lebesgue measure on $\mathcal{L}_x$. Obviously, $\mu(E_1) = 1$.

We call a point $x_1 \in E_1$ *rich* if for any $\delta > 0$ there is another point $x_3 \in E_1$ such that $0 < \mathrm{dist}(x_1, x_3) < \delta$ and (6.1) holds with $x_2, x_4 \in E_1$. Denote by $E_2 \subset E_1$ the set of rich points.

EXERCISE 6.35. Verify that $\mu(E_2) = 1$. In fact, this can be derived from our general assumptions (H1)–(H6).

Next, let $x_1 \in E_2$. Fix a sequence of points $x_3^{(i)} \in E_1$, $i \geq 1$, such that $0 < \mathrm{dist}(x_1, x_3^{(i)}) < 1/i$ and (6.1) holds with $x_2^{(i)}, x_4^{(i)} \in E_1$.

Let $x_3 = x_3^{(i)} \in E_1$ be one of the above points and $0 < t < \tau(x_1)$. Consider the 4-loop starting at $Y_1 = (x_1, t)$ and terminating at Y_5, as shown in Fig. 6.9. Our main assumption is that all 4-loops are open, i.e. $Y_5 \neq Y_1$; hence $Y_5 = \Phi^{\sigma_i}(Y_1)$ for some small $\sigma_i \neq 0$.

EXERCISE 6.36. Show that σ_i does not depend on the choice of t and that $\sigma_i \to 0$ as $i \to \infty$.

Note that the transformation $Y_1 \mapsto Y_5 = \Phi^{\sigma_i}(Y_1)$ is just a shift along the line segment $\mathcal{L}_{x_1}$; hence it preserves the Lebesgue measure $\mathbf{m}$ on it. Similarly, the transformation $\mathcal{L}_{x_j} \to \mathcal{L}_{x_{j+1}}$ defined by $Y_j \mapsto Y_{j+1}$ (with the identification $x_5 = x_1$, of course) is affine; hence it preserves the Lebesgue measures on those segments. (These transformations are similar to the holonomy map.)

Recall that for each i and each $j = 2, 3, 4$ we have $\mathbf{m}\big(\mathcal{L}_{x_j^{(i)}} \setminus D\big) = 0$; hence for $\mathbf{m}$-almost every point $Y_1 \in \mathcal{L}_{x_1}$ all the five points Y_1, Y_2, Y_3, Y_4, Y_5 belong in D.

Now let $A \in \mathfrak{G}(\eta_\infty)$ be any element of the complete sub-σ-algebra generated by the partition η_∞. Since $\eta_\infty \preccurlyeq \xi^u \wedge \xi^s \preccurlyeq \xi^u_D \curlywedge \xi^s_D$, our set A is a $(\xi^u_D \curlywedge \xi^s_D)$-set (mod 0); i.e. there exists a set $A' = A$ (mod 0) that is a union of some elements of the partition $\xi^u_D \curlywedge \xi^s_D$. In particular, if $Y_1 \in A'$ and $Y_1, Y_2, Y_3, Y_4, Y_5 \in D$ (as above), then all these points belong in A' as well. Thus the set $A' \cap \mathcal{L}_{x_1}$ is invariant (mod 0) under the map Φ^{σ_i} for every $i \geq 1$ in the following sense:

$$\mathcal{L}_{x_1} \cap \Phi^{\sigma_i}(A' \cap \mathcal{L}_{x_1}) = A' \cap \Phi^{\sigma_i}(\mathcal{L}_{x_1}) \pmod 0$$

(i.e. the symmetric difference of these two sets has $\mathbf{m}$-measure zero).

Since $\sigma_i \to 0$ as $i \to \infty$, it follows that either $\mathbf{m}\big(A' \cap \mathcal{L}_{x_1}\big) = 0$ or $\mathbf{m}\big(\mathcal{L}_{x_1} \setminus A'\big) = 0$, according to a general fact mentioned the subsection on Lebesgue density points in Appendix A. Thus every set $A \in \mathfrak{G}(\eta_\infty)$ is (mod 0) a union of some segments $\mathcal{L}_x$, $x \in \mathcal{M}$. We denote by B_A the set of points $x \in \mathcal{M}$ for which the segment $\mathcal{L}_x$ belongs (mod 0) to A (we think of B_A as the 'base' of A).

Suppose the partition η_∞ is not trivial. Then there is a set $A \in \mathfrak{G}(\eta_\infty)$ such that $0 < \mu_\Omega(A) < 1$ (hence $0 < \mu(B_A) < 1$). Since the σ-algebra $\mathfrak{G}(\eta_\infty)$ is Φ^t-invariant for every $t \in \mathbb{R}$ (Exercise 6.34), it follows that $\Phi^t(A) \in \mathfrak{G}(\eta_\infty)$; hence $\Phi^t(A)$ is also a (mod 0) union of some segments $\mathcal{L}_x$, $x \in \mathcal{M}$.

EXERCISE 6.37. Show that the set A must be Φ^t-invariant (mod 0), i.e. $A = \Phi^t(A)$ (mod 0) for every $t > 0$. Hint: It is enough to prove that B_A is $\mathcal{F}$-invariant (mod 0). If this is not the case, then $\mu\big(B_A \setminus \mathcal{F}(B_A)\big) > 0$. Argue that if $t > 0$ is small enough, then the set $A \cap \Phi^t(A)$ contains only parts of the segments $\mathcal{L}_x$ for $x \in B_A \setminus \mathcal{F}(B_A)$, which is impossible since $A \cap \Phi^t(A) \in \mathfrak{G}(\eta_\infty)$.

Now the ergodicity of the flow Φ^t implies that $\mu_\Omega(A) = 1$; hence the σ-algebra $\mathfrak{G}(\eta_\infty)$ is trivial. Thus the Pinsker partition π is trivial, too, and the flow Φ^t is K-mixing.

6.11. Mixing properties for dispersing billiard flows

Here we verify that all the 4-loops are open. Our argument uses the fact that unstable and stable manifolds are wave fronts whose curvature has the opposite sign.

Following our early notation of Sections 3.11 and 4.7, we denote by $\mathcal{B}^u(X) > 0$ and $\mathcal{B}^s(X) < 0$ the curvatures of the wave fronts $\mathcal{W}^u(X)$ and $\mathcal{W}^s(X)$, respectively, at the point X. It easily follows from the results of Sections 4.3 and 4.7 that

$$0 < \mathcal{B}_{\min} \leq \mathcal{B}^u(X) < 1/t^-(X)$$

and similarly

$$-1/t^+(X) < \mathcal{B}^s(X) \leq -\mathcal{B}_{\min} < 0.$$

Also, the function $\mathcal{B}^s(X)$ is continuous on the set of points whose future semitrajectories avoid grazing collisions. Similarly, the function $\mathcal{B}^u(X)$ is continuous on the set of points whose past semitrajectories avoid grazing collisions.

Now let $X_1 \in \Omega$ be a point whose entire trajectory avoids grazing collisions at all times. Then in a vicinity of X_1 the functions $\mathcal{B}^u$ and $\mathcal{B}^s$ are approximately constant. Hence stable and unstable manifolds are wave fronts whose curves on the billiard table are approximately circles.

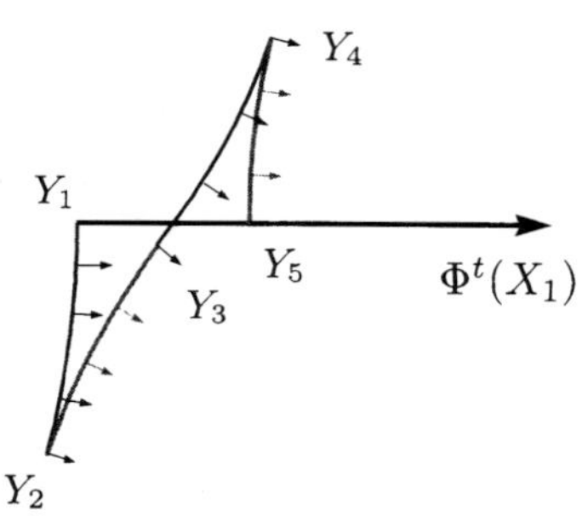

FIGURE 6.10. An open 4-loop on a billiard table. Unstable manifolds are black, stable manifolds are grey.

Now we draw the 4-loop depicted in Figure 6.9 as a sequence of four wave fronts on the billiard table; see Figure 6.10. It is evident that $Y_5 \neq Y_1$.

COROLLARY 6.38. *For dispersing billiards of categories A and B the flow* $\Phi^t \colon \Omega \to \Omega$ *is mixing, K-mixing and Bernoulli.*

EXERCISE 6.39. Show that if the lengths of the wave fronts in Fig. 6.10 are $\sim \delta$, then $|\sigma| = \operatorname{dist}(Y_1, Y_5) \sim \delta^2$. Recall that σ is defined by $Y_5 = \Phi^\sigma(Y_1)$. Hint: Approximate the wave fronts by circles.

Interestingly, there is an exact formula for σ for any 4-loop in a dispersing billiard:

LEMMA 6.40. *Let $Y_1, \ldots, Y_5$ be the endpoints of a 4-loop and $Y_5 = \Phi^\sigma(Y_1)$. Denote by K the projection of the 4-loop onto the base $\mathcal{M}$ (note that K is a domain bounded by two stable and two unstable manifolds of the map $\mathcal{F}$). Then $|\sigma| = \mu(K)$.*

PROOF. Our argument follows that of [**KM81**, Lemma 5.1].

EXERCISE 6.41. Let $D \subset \mathcal{M}$ be any domain with a piecewise smooth boundary. Use Green's theorem to verify that

$$\mu(D) = -\int_{\partial D} \sin\varphi \, dr$$

where the boundary ∂D is traversed counterclockwise in the $r\varphi$ coordinate plane.

Now, given a 4-loop whose projection onto $\mathcal{M}$ is K, let $s(y)$ denote the function on the boundary ∂K that represents the (temporal) height of our 4-loop over the base $\mathcal{M}$. That is, for any $y \in \partial K$ we set $s(y)$ to the unique moment of time $s \in (0, \tau(y))$ such that the point $\Phi^s(y)$ lies on the 4-loop.

Observe that $s(y)$ is smooth on each stable and unstable manifold bounding K; it is continuous at the points x_2, x_3 and x_4; and it has a jump at x_1 equal to σ. Therefore,

$$\sigma = \pm \int_{\partial K} ds(y)$$

where the sign depends on the direction in which we traverse ∂K.

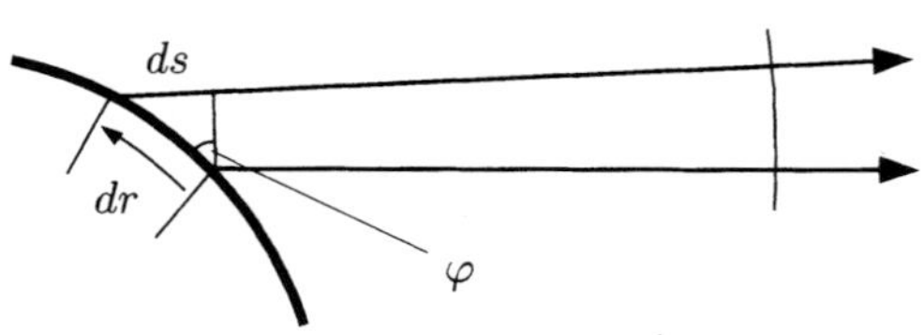

FIGURE 6.11. Proof of the relation $ds = \sin\varphi \, dr$.

It remains to notice that $ds(y) = \pm \sin\varphi \, dr$, where the sign depends on the direction of motion along ∂K (see Figure 6.11) to complete the proof of Lemma 6.40.

□

CHAPTER 7

Statistical properties

7.1. Introduction

In ergodic theory, which studies measure preserving transformations, a hierarchy of properties exists that characterize increasingly stronger degrees of chaos:[1] ergodicity, weak mixing, strong mixing, multiple mixing, K-mixing, and Bernoulli. Each property in this list implies all the previous (weaker) properties. The highest degree of chaos (Bernoulliness) implies all the others. All Bernoulli automorphisms with the same entropy are isomorphic. Thus, Bernoulli property is indeed the highest possible degree of chaos, in terms of pure ergodic theory.

However, when we deal with smooth maps (with or without singularities), other issues come into play, which are strongly motivated by applications in physics. Given a smooth map $F\colon M \to M$ preserving a measure μ and a smooth function $f\colon M \to \mathbb{R}$ ('observable'), one can study the speed of the decay of correlations, a central limit theorem, and other limit theorems; see definitions below. All these are collectively known as *statistical properties*; see definitions in the next section.

While ergodic properties listed above characterize chaos in measure-theoretic terms, statistical properties 'speak' in the language of probability theory: they pronounce a close similarity between the given dynamical system (M, F, μ, f) and sequences of i.i.d. (independent identically distributed) random variables.

A question arises how to compare ergodic properties and statistical ones. Strictly speaking, the former are neither stronger nor weaker than the latter; they simply characterize chaos in different terms. For example, there are smooth Bernoulli maps for which statistical properties are very poor for generic smooth observables (correlations decay slowly, the central limit theorem fails, etc.). On the other hand, there are systems with poor ergodic characteristics (not even mixing) for which the Central Limit Theorem surprisingly holds for certain observables (this happens for irrational circle rotations [**BD87**] and cycles [**Ch95**, Proposition 3.6]).

For typical models, however, strong statistical properties require at least mixing and hardly can be imagined without K-mixing and Bernoulli. In the study of billiards, ergodic properties (including Bernoulli) definitely constitute a more basic level of our knowledge than, say, estimates on correlations and limit theorems. This is why we presented the ergodic properties of billiards earlier and now are turning to their statistical properties.

7.2. Definitions

Let $F\colon M \to M$ be an invertible map preserving a probability measure μ, and $f\colon M \to \mathbb{R}$ a measurable function (in physics, f is normally called an *observable*).

[1]For brevity, we restrict this discussion to invertible transformations (automorphisms).

LEMMA 7.1. *The sequence*

$$\mathcal{X}_n = f \circ F^n, \qquad n \in \mathbb{Z}, \tag{7.1}$$

is a stationary stochastic process, with (M, μ) *being the underlying probability space.*

PROOF. The stationarity of the sequence $\{\mathcal{X}_n\}$ means that its finite-dimensional distributions are translationally invariant; see Appendix B. Precisely, for any distinct $k_1, \dots, k_m \in \mathbb{Z}$ the joint distribution of $\mathcal{X}_{k+k_1}, \dots, \mathcal{X}_{k+k_m}$ is independent of $k \in \mathbb{Z}$; i.e. for any Borel set $B \subset \mathbb{R}^m$ the probability

$$\mu\big((\mathcal{X}_{k+k_1}, \dots, \mathcal{X}_{k+k_m}) \in B\big)$$

does not depend on k. This fact follows from the invariance of the measure μ. We leave its verification to the reader as an exercise. □

Thus, given a function $f\colon M \to \mathbb{R}$, its values $f \circ F^n$ at all times $n \in \mathbb{Z}$ make a sequence of (identically distributed) random variables. We can consider them as a stationary stochastic process (with discrete time) and invoke notions of probability theory to describe its behavior.

For example, let $f = \mathbf{1}_A$ be the indicator of a set $A \subset M$ (i.e., $f(x) = 1$ for $x \in A$ and 0 for $x \notin A$); then $\mathcal{X}_n = \mathbf{1}_{F^{-n}A}$ is the indicator of the preimage $F^{-n}(A)$.

We say that $\mathcal{X}_n$ and $\mathcal{X}_m$ as defined by (7.1) are independent if for any Borel sets $E, G \subset \mathbb{R}$

$$\mu(\mathcal{X}_n \in E,\, \mathcal{X}_m \in G) = \mu(\mathcal{X}_n \in E)\, \mu(\mathcal{X}_m \in G).$$

For the indicator $f = \mathbf{1}_A$, this amounts to the independence of the sets $F^{-n}(A)$ and $F^{-m}(A)$.

Of particular interest to us are the partial sums

$$S_n = f + f \circ F + f \circ F^2 + \cdots + f \circ F^{n-1}. \tag{7.2}$$

We say that the sequence (7.1) satisfies the Strong Law of Large Numbers (SLLN) if

$$\frac{S_n}{n} \to \langle f \rangle \colon = \int_M f \, d\mu \tag{7.3}$$

almost everywhere (cf. Appendix B). It is easy to see that the SLLN is guaranteed by Birkhoff's ergodic theorem for all ergodic maps and integrable observables.

Further probabilistic laws that characterize stochastic processes are the Central Limit Theorem (CLT), the Invariance Principle (see Section 7.9), etc. These laws are established in classical probability theory for i.i.d. (independent identically distributed) random variables and have been extended to various classes of weakly dependent stationary processes in advanced studies; see survey [**DPS86**].

We must realize, though, that the random variables $\mathcal{X}_n = f \circ F^n$ resulting from a dynamical system $F\colon M \to M$ are *not* independent (apart from trivial examples). Thus advanced tools of probability theory must be used to establish analogues of the CLT, Invariance Principle, etc., for dynamically generated random processes.

The dependence between random variables $\mathcal{X}_n$ and $\mathcal{X}_m$ is characterized by their covariance

$$\operatorname{Cov}(\mathcal{X}_n, \mathcal{X}_m) = \langle \mathcal{X}_n \mathcal{X}_m \rangle - \langle \mathcal{X}_n \rangle \langle \mathcal{X}_m \rangle$$

where $\langle \cdot \rangle$ denotes expectation; cf. (7.3). For stationary processes, $\langle \mathcal{X}_n \rangle = \langle \mathcal{X}_m \rangle$, and $\operatorname{Cov}(\mathcal{X}_n, \mathcal{X}_m)$ depends only on $|n - m|$. Accordingly, we define

$$\mathbf{C}_f(n) = \operatorname{Cov}(f, f \circ F^n) = \langle f \cdot (f \circ F^n) \rangle - \langle f \rangle^2. \tag{7.4}$$

EXERCISE 7.2. Show that $\mathbf{C}_f(-n) = \mathbf{C}_f(n)$ and $\mathbf{C}_f(n-m) = \operatorname{Cov}(\mathcal{X}_n, \mathcal{X}_m)$.

In the studies of dynamical systems, following a tradition in physics literature, the quantity (7.4) is called *correlation* rather than covariance. It is actually common to study more general correlation functions, namely

$$\mathbf{C}_{f,g}(n) = \langle f \cdot (g \circ F^n) \rangle - \langle f \rangle \langle g \rangle \tag{7.5}$$

where $f, g \colon M \to \mathbb{R}$ are two observables. To distinguish $\mathbf{C}_f(n)$ from $\mathbf{C}_{f,g}(n)$, one calls the former an *autocorrelation function.*

If the variables $\mathcal{X}_n$ and $\mathcal{X}_m$ were independent for $n \neq m$, then we would have $\langle \mathcal{X}_n \mathcal{X}_m \rangle = \langle \mathcal{X}_n \rangle \langle \mathcal{X}_m \rangle$; hence $\mathbf{C}_f(n-m) = 0$. But as we said, one can hardly expect independence in a deterministic system; after all, its present state $x \in M$ completely determines its state $x_n = F^n(x)$ at any time $n \in \mathbb{Z}$. This seems to prevent any independence (or even slightest uncertainty about the future!).

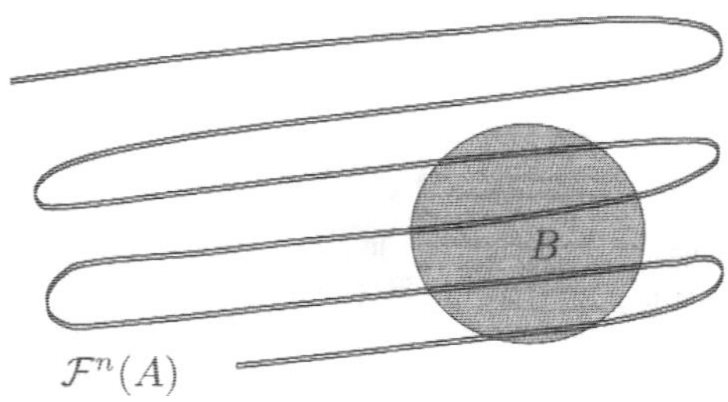

FIGURE 7.1. Asymptotic independence in mixing maps.

However, suppose we know the state of the system only approximately. Now the picture changes. Say we know that $x \in A$ and want to determine if $x_n = F^n(x) \in B$, where $A, B \subset M$ are certain subsets (possibly very small). Then, assuming that F is mixing, the chance that $x_n \in B$ given that $x \in A$ will be

$$\mu\big(F^n(x) \in B / x \in A\big) = \mu(B \cap F^n A)/\mu(A) \to \mu(B)$$

as $n \to \infty$, so the events $\{x \in A\}$ and $\{x_n \in B\}$ become asymptotically independent!

The asymptotic independence of x and $x_n = F^n(x)$, in this sense, is equivalent to the asymptotic vanishing of correlations (*decay of correlations*), which is a standard fact in ergodic theory.

Fact. The map F is mixing iff for any two observables $f, g \in L^2_\mu(M)$ we have $\mathbf{C}_{f,g}(n) \to 0$ as $n \to \infty$ (Appendix C).

So the convergence $\mathbf{C}_{f,g}(n) \to 0$ is guaranteed by mixing, but to prove the CLT and other limit laws, we need a fast convergence in the sense defined below. And here is an unpleasant surprise: for typical observables f, g, the convergence is very slow, too slow for any limit theorems to hold.

Precisely, for every aperiodic measure-preserving transformation and any sequence $c_n \to 0$ (however slowly convergent) we have $|\mathbf{C}_{f,g}(n)| \geq |c_n|$ for typical integrable (and even for typical bounded) functions f and g. Even for continuous maps on compact metric spaces, the convergence is arbitrarily slow for generic continuous functions; cf. a discussion in [**Ch95**]. To ensure a fast convergence, we need to restrict the class of maps and observables to smooth or nearly smooth ones.

Fortunately, this is exactly the type of maps and observables we encounter in physical models, where the phase space M is commonly supposed to be a smooth (often compact) manifold and the map F is smooth or piecewise smooth. Physically interesting observables are usually smooth (or piecewise smooth) functions on phase space. Within this class of maps and observables, the speed of convergence $\mathbf{C}_{f,g}(n) \to 0$ becomes tractable.

The following two types of convergence are common in applications:

DEFINITION 7.3. We say that correlations decay exponentially fast if

$$|\mathbf{C}_{f,g}(n)| \leq \text{const}\, e^{-a|n|} \tag{7.6}$$

for some $a > 0$ and all $n \neq 0$. We say that correlations decay algebraically (or polynomially) if

$$|\mathbf{C}_{f,g}(n)| \leq \text{const}\, |n|^{-a} \tag{7.7}$$

for some $a > 0$ and all $n \neq 0$.

Exponential decay of correlations is usually sufficient for the proofs of limit theorems. Algebraic decay of correlations may be sufficient if a is high enough, but the proofs become more difficult. If $a \leq 1$, then limit theorems usually fail.

For smooth uniformly hyperbolic maps (Anosov and Axiom A diffeomorphisms), correlations decay exponentially for all smooth (even all Hölder continuous) observables [**Bow75, Ru78**]. For nonuniformly hyperbolic maps, correlations usually decay algebraically [**Hu01**].

We will see that for dispersing billiards of categories A and B (where the hyperbolicity is uniform), correlations decay exponentially, just like for Anosov and Axiom A maps. For dispersing billiards of category E and for many nondispersing chaotic billiards, the hyperbolicity is nonuniform and correlations decay polynomially [**Mac83, Mar04, CZ05**].

Next we formulate the Central Limit Theorem:

DEFINITION 7.4. We say that the sequence (7.1) satisfies the Central Limit Theorem (CLT) if

$$\lim_{n\to\infty} \mu\left\{\frac{S_n - n\langle f\rangle}{\sqrt{n}} \leq z\right\} = \frac{1}{\sqrt{2\pi}\sigma}\int_{-\infty}^{z} e^{-\frac{s^2}{2\sigma^2}}\, ds \tag{7.8}$$

for all $-\infty < z < \infty$. Here $\sigma = \sigma_f \geq 0$ is a constant.[2] Equivalently, $(S_n - n\langle f\rangle)/\sqrt{n}$ converges in distribution to the normal law $\mathcal{N}(0, \sigma_f^2)$.

The variance σ_f^2 in the CLT is related to correlations (7.4) by

$$\sigma_f^2 = \sum_{n=-\infty}^{\infty} \mathbf{C}_f(n) = \mathbf{C}_f(0) + 2\sum_{n=1}^{\infty} \mathbf{C}_f(n). \tag{7.9}$$

It follows that a prerequisite for the Central Limit Theorem is the summability of the correlation function $\mathbf{C}_f(n)$. One can regard this condition as a minimal requirement on the speed of convergence $\mathbf{C}_f(n) \to 0$.

[2]In the case $\sigma_f = 0$, the right-hand side of (7.8) is to be read as 0 for $z < 0$ and 1 for $z > 0$.

REMARK 7.5. It is interesting to examine the degenerate case $\sigma_f^2 = 0$. This occurs if and only if the function f is cohomologous to a constant, which means that $f = K + g - g \circ F$ for some $g \in L^2(M)$ and a constant K.

EXERCISE 7.6. Show that if $f = K + g - g \circ F$ for some $g \in L^2(M)$, then $\langle f \rangle = K$ and the CLT holds with $\sigma_f^2 = 0$. (The converse is true, but its proof is much harder; see Section 7.8.)

We will introduce some other limit theorems in Section 7.9. For further general discussion of statistical properties for dynamical systems we refer the reader to [**Ch95, CY00, Den89**].

7.3. Historic overview

Various approaches have been used in the study of statistical properties of smooth hyperbolic dynamical systems.

It all began in the late 1960s, when Markov partitions were constructed for several classes of smooth hyperbolic maps. First it was done for linear toral automorphisms by Adler and Weiss [**AW67**], then for Anosov diffeomorphisms by Sinai [**Sin68a, Sin68b**], and for Axiom A diffeomorphisms by Bowen [**Bow70**]. A finite Markov partition can be used to code the trajectory of each point by an infinite sequence of symbols from a finite alphabet. This gives a representation of the dynamics by a symbolic system of a certain type called 'topological Markov chain' or 'subshift of finite type'.

On the other hand, an infinite sequence of symbols can be identified with a configuration of a one-dimensional lattice gas, a classical object of study in statistical mechanics. Thus the study of hyperbolic maps became closely and intimately connected with the theory of lattice gases, a link that would have a great impact on both disciplines.

As it happened, at about the same time (the late 1960s), physicists invented Gibbs states for one-dimensional lattice gases; this was done by Dobrushin [**Dob68a, Dob68b, Dob68c**] and Lanford and Ruelle [**LR69**]. In the early 1970s, Sinai [**Sin72**] and Ruelle [**Ru76**] employed the theory of Gibbs states to construct Gibbs invariant measures for Anosov and Axiom A diffeomorphisms. Thus a large class of physically relevant invariant measures was found. The theory of Gibbs states in physics (which is known under the name 'thermodynamic formalism') also contained facts that 'translated' into exponential decay of correlations and the Central Limit Theorem for Gibbs measures and Hölder continuous observables; see [**Bow75**].

A key element in the study of Gibbs states was a spectral gap for the Perron-Frobenius operator acting on Hölder continuous densities. The Perron-Frobenius operator is a very efficient tool, but its use involves highly abstract elements of functional analysis, which in a sense obscure the dynamical content of the problem. Anyway, such was the dominant approach throughout the 1970s. For a comprehensive exposition of the respective methods and results, we refer the reader to surveys [**Bow75, Ru78, PP90, Ba00, Ch02**].

In the early 1980s, an attempt was made to extend the thermodynamic formalism to dispersing billiards. It started when Bunimovich and Sinai [**BS80**] described Markov partitions for dispersing billiards of category A; for a more formal construction of Markov partitions for tables of categories A, B and C see [**BSC90**]. However it was realized that such partitions can never be finite (they are always countable).

This fact rendered the symbolic representation inefficient and the Perron-Frobenius operator unhandy.

Instead, Bunimovich and Sinai [**BS81**] approximated the collision map $\mathcal{F}\colon \mathcal{M} \to \mathcal{M}$ by a probabilistic Markov chain with countably many states. Then they derived statistical properties of the map $\mathcal{F}$ from a fast mixing in the approximating chain. The key ingredient in their analysis was the verification of Doeblin's condition. Alternatively, the map $\mathcal{F}$ can be approximated by a countable family of finite Markov chains, and Doeblin's condition can be replaced by better mixing conditions, which leads to sharper results; see [**BSC91**].

In any case, approximations by Markov chains yielded the Central Limit Theorem and some other limit laws but less than perfect estimates on the correlation function. It was proved in [**BS81**] that

$$|\mathbf{C}_{f,g}(n)| \leq \text{const}\, e^{-a|n|^{\gamma}} \tag{7.10}$$

for some $a > 0$ and $\gamma \in (0,1)$; compare this to (7.6). The above mode of decay was termed *stretched exponential.* The estimate (7.10) was improved in [**BSC91**], where it was shown that $\gamma \geq 1/2$, but that was still weaker than (7.6). We refer the reader to [**BSC91, Ch95**] for a detailed presentation of these ideas.

The lack of exponential estimates on correlation functions for dispersing billiards was bothersome, and in 1998 L.-S. Young derived one! She developed [**Yo98**] (see also [**BY00**]) a new approach to hyperbolic maps with singularities, based on *tower construction* (which, in a sense, is a more tractable version of Markov partition), and found a way to reemploy the Perron-Frobenius operator. This produced an exponential bound on correlations (7.6) for dispersing billiards of category A and Hölder continuous observables. Chernov extended [**Ch99**] Young's results to billiards of categories B and C.

Thus Young revitalized the power of the Perron-Frobenius operator, but only a year later she abandoned it in favor of a better tool. She rederived [**Yo99**] the exponential bound on correlations (7.6) for hyperbolic maps by the coupling method (borrowed from probability theory), thus bypassing symbolic formalism and the Perron-Frobenius operator altogether. The underlying idea of this new approach is that images of different smooth measures on $\mathcal{M}$ are getting closer together and thus converge to a common limit; the degree of 'closeness' and the speed of convergence are controlled by coupling. Her entire argument was intrinsically dynamical, unlike earlier operator-based techniques.

The elegance of Young's new method was recently demonstrated by Bressaud and Liverani [**BL02**], who applied it to Anosov maps and recovered Bowen's results of the 1970s. Dolgopyat further simplified the coupling method by replacing smooth measures on $\mathcal{M}$ with one-dimensional measures on unstable curves, which made the argument even more transparent. Here we present the most recent version of the coupling method following [**CD06**, Appendix A] and [**Ch06b**].

As in the previous chapters, our discussion is restricted to billiards of category A, but with some extra effort the main conclusions can be extended to the categories B and C. Actually, many of our arguments are quite general and can be applied to other hyperbolic maps with singularities.

7.4. Standard pairs and families

We start with an observation that motivates the use of one-dimensional measures on unstable curves, as proposed by Dolgopyat.

Let us partition the collision space $\mathcal{M}$ into small subdomains (cells) $D_i \subset \mathcal{M}$ and represent a smooth measure μ_0 on $\mathcal{M}$ as a weighted sum of its restrictions to those cells. Now the image of a small domain $D \subset \mathcal{M}$ under the map $\mathcal{F}^n$ gets strongly expanded in the unstable direction, strongly contracted in the stable direction, and possibly cut by singularities into many pieces. Thus, $\mathcal{F}^n(D)$ will soon look like a union of one-dimensional curves, each of which resembles an unstable manifold of the collision map $\mathcal{F}$. Henceforth the measure $\mathcal{F}^n(\mu_0)$ will evolve as a weighted sum of smooth measures on unstable curves. We note that, in addition, one has to 'cut' $\mathcal{F}^n(D)$ along the homogeneity lines $\mathbb{S}_k$, $|k| \geq k_0$, introduced in Section 5.3 to ensure distortion control.

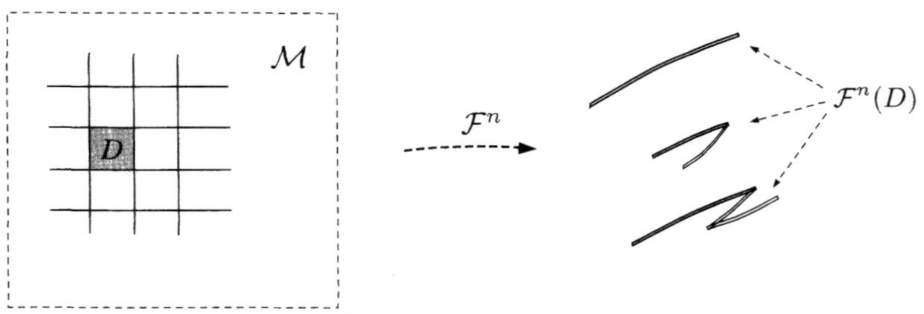

FIGURE 7.2. Evolution of a cell $D \subset \mathcal{M}$.

Accordingly, we define a class of probability measures supported on unstable curves (we remind the reader that unstable curves introduced in Section 4.5 are more general than unstable manifolds):

DEFINITION 7.7. A *standard pair* (W, ν) is a homogeneous unstable curve $W \subset \mathcal{M}$ with a probability measure ν on it whose density ρ with respect to the Lebesgue measure on W is regular; see below.

The regularity of the density $\rho(x)$ should be comparable to the regularity of the map $\mathcal{F}$; the latter is expressed by two key estimates: distortion bounds (Section 5.6) and absolute continuity (Section 5.8). While distortions are fairly smooth (Theorems 5.29 and 5.31), the Jacobian of the holonomy map satisfies only 'dynamically defined Hölder continuity' (Proposition 5.48). Accordingly, we require the density $\rho(x)$ be dynamically Hölder continuous, in the sense formulated below.

Let $\mathbf{s}_+(x, y)$ be the 'separation time' function defined in (5.31) and $\boldsymbol{\theta} = \Lambda^{-1/6} < 1$ as introduced in the proof of Proposition 5.48. We also fix a sufficiently large constant $C_{\mathrm{r}} > 0$.

DEFINITION 7.8. A density $\rho(x)$ on a homogeneous unstable curve $W \subset \mathcal{M}$ is said to be *regular* if

$$|\ln \rho(x) - \ln \rho(y)| \leq C_{\mathrm{r}}\, \boldsymbol{\theta}^{\mathbf{s}_+(x,y)} \tag{7.11}$$

(the subscript r in C_{r} stands for 'regular').

EXERCISE 7.9. Verify that the density satisfying (7.11) may have a discontinuity at every point of the countable and dense set $W \cap \mathcal{S}_\infty^{\mathbb{H}}$, but it must be continuous on its complement $W \setminus \mathcal{S}_\infty^{\mathbb{H}}$. Also, show that

$$\max_{x\in W} \rho(x) / \min_{x\in W} \rho(x) \le \text{const} = e^{C_{\mathrm{r}}}. \tag{7.12}$$

REMARK 7.10. The condition (7.11) will not be altered if we multiply the density $\rho(x)$ by a constant. Therefore, given a standard pair (W,ν), any subcurve $W' \subset W$ with the conditional measure induced by ν on it will make a standard pair.

Recall that the partition $\xi_{\mathbb{H}}^u$ of $\mathcal{M}$ into (maximal) unstable H-manifolds $\{W_{\mathbb{H}}^u\}$ is measurable; the conditional measure on each $W_{\mathbb{H}}^u$ induced by μ is the u-SRB measure $\nu_{W_{\mathbb{H}}^u}$ whose density is determined by (5.4); cf. Section 5.4.

EXERCISE 7.11. Verify that any unstable H-manifold $W_{\mathbb{H}}^u$ with the u-SRB measure $\nu_{W_{\mathbb{H}}^u}$ on it makes a standard pair. Hint: Use (5.32) and Corollary 5.30.

The class of standard pairs is invariant under $\mathcal{F}$ in the following sense:

PROPOSITION 7.12. *Let (W,ν) be a standard pair. For each $n \ge 0$, denote by $W_{i,n}$ the H-components of $\mathcal{F}^n(W)$. Then $\mathcal{F}^n(\nu) = \sum_i c_{i,n}\nu_{i,n}$ where $\sum_i c_{i,n} = 1$ and each $(W_{i,n},\nu_{i,n})$ is a standard pair.*

PROOF. By induction, it is enough to prove this for $n = 1$. Let $W_{i,1}$ be an H-component of $\mathcal{F}(W)$ and $x, y \in W_{i,1}$. Denote $x_1 = \mathcal{F}^{-1}(x)$ and $y_1 = \mathcal{F}^{-1}(y)$. Observe that $\mathbf{s}_+(x,y) = \mathbf{s}_+(x_1,y_1) - 1$. Now using Lemma 5.27 on distortion bounds and (5.32) gives

$$\begin{aligned} |\ln\rho(x) - \ln\rho(y)| &\le |\ln\rho_{i,1}(x_1) - \ln\rho_{i,1}(y_1)| \\ &\quad + |\ln \mathcal{J}_W\mathcal{F}^{-1}(x) - \ln\mathcal{J}_W\mathcal{F}^{-1}(y)| \\ &\le C_{\mathrm{r}}\boldsymbol{\theta}^{\mathbf{s}_+(x_1,y_1)} + C|W_{i,1}|^{1/3} \\ &\le C_{\mathrm{r}}\boldsymbol{\theta}\,\boldsymbol{\theta}^{\mathbf{s}_+(x,y)} + C'\boldsymbol{\theta}^{\mathbf{s}_+(x,y)} \end{aligned}$$

for some constant $C' = C'(\mathcal{D}) > 0$. Thus it is enough to assume that C_{r} is so large that $C_{\mathrm{r}}\boldsymbol{\theta} + C' \le C_{\mathrm{r}}$. □

We see that $\mathcal{F}^n$ transforms the measure ν from a standard pair into a weighted sum of measures on finitely or countably many standard pairs. Motivated by this observation, we introduce even more general families of standard pairs:

DEFINITION 7.13. A *standard family* is an arbitrary (countable or uncountable) family $\mathcal{G} = \{(W_\alpha,\nu_\alpha)\}$, $\alpha \in \mathfrak{A}$, of standard pairs with a probability factor measure $\lambda_{\mathcal{G}}$ on the index set $\mathfrak{A}$. Such a family induces a probability measure $\mu_{\mathcal{G}}$ on the union $\cup_\alpha W_\alpha$ (and thus on $\mathcal{M}$) defined by

$$\mu_{\mathcal{G}}(B) = \int \nu_\alpha(B\cap W_\alpha)\, d\lambda_{\mathcal{G}}(\alpha) \qquad \forall B \subset \mathcal{M}.$$

Proposition 7.12 now simply says that $\mathcal{F}^n$ transforms a standard pair into a countable standard family (whose factor measure is defined by the sequence of the coefficients $\{c_{i,n}\}$). Similarly, any standard family $\mathcal{G}$ is mapped by $\mathcal{F}^n$ into another standard family $\mathcal{G}_n = \mathcal{F}^n(\mathcal{G})$. It is easy to see that $\mu_{\mathcal{G}_n} = \mathcal{F}^n(\mu_{\mathcal{G}})$.

It will be important to control the size of curves W_α in a standard family $\mathcal{G} = \{(W_\alpha, \nu_\alpha)\}$. For every $\alpha \in \mathfrak{A}$ and $x \in W_\alpha$, the point x divides the curve W_α into two parts, and we denote by $r_\mathcal{G}(x)$ the length of the shorter one (in the Euclidean metric). This defines a function $r_\mathcal{G}$ on[3] $\cup_\alpha W_\alpha$. Now we denote

$$\mathcal{Z}_\mathcal{G} : = \sup_{\varepsilon>0} \frac{\mu_\mathcal{G}(r_\mathcal{G} < \varepsilon)}{\varepsilon} = \sup_{\varepsilon>0} \frac{\int \nu_\alpha\big(x \in W_\alpha \colon r_\mathcal{G}(x) < \varepsilon\big)\, d\lambda_\mathcal{G}(\alpha)}{\varepsilon}.$$

EXERCISE 7.14. Let $\mathcal{G}$ consist of a single standard pair (W, ν). Suppose that ν is the normalized Lebesgue measure on W; show that $\mathcal{Z}_\mathcal{G} = 2/|W|$. Next, let ν have an arbitrary regular density; then show that $\mathcal{Z}_\mathcal{G} \asymp 1/|W|$; i.e. $C_1 < \mathcal{Z}_\mathcal{G}|W| < C_2$, where $C_1 = C_1(\mathcal{D}) > 0$ and $C_2 = C_2(\mathcal{D}) > 0$ are constants (the symbol $\asymp$ was introduced in Section 4.3). Hint: Use the last result of Exercise 7.9.

The next exercise generalizes the previous one:

EXERCISE 7.15. Let $\mathcal{G} = \{(W_\alpha, \nu_\alpha)\}$ be a standard family. Show that

$$\mathcal{Z}_\mathcal{G} \asymp \int \frac{d\lambda_\mathcal{G}(\alpha)}{|W_\alpha|}$$

(in this formula, either both quantities are finite or both are infinite).

We will consider only standard families $\mathcal{G}$ with $\mathcal{Z}_\mathcal{G} < \infty$. Next we investigate how the quantity $\mathcal{Z}_{\mathcal{G}_n}$, where $\mathcal{G}_n = \mathcal{F}^n(\mathcal{G})$, changes with n.

EXERCISE 7.16. Let $\mathcal{G}$ be a standard family consisting of a single standard pair (W, ν). Let $\mathcal{G}_n = \mathcal{F}^n(\mathcal{G})$. Use the first growth lemma (Theorem 5.52) and (7.12) to show that for all $n \geq 0$ and $\varepsilon > 0$

$$\begin{aligned} \nu\big(r_{\mathcal{G}_n}(\mathcal{F}^n x) < \varepsilon\big) &\leq \text{const}\big[(\vartheta_1 \hat{\Lambda})^n\, \nu\big(r_\mathcal{G}(x) < \varepsilon/\hat{\Lambda}^n\big) + \varepsilon\big] \\ &\leq \text{const}\big[\vartheta_1^n \varepsilon/|W| + \varepsilon\big] \end{aligned} \tag{7.13}$$

where the constants depend on the table $\mathcal{D}$ only. (Recall that $0 < \vartheta_1 < 1$.)

PROPOSITION 7.17. *Let $\mathcal{G} = \{(W_\alpha, \nu_\alpha)\}$, $\alpha \in \mathfrak{A}$, be a standard family and $\mathcal{G}_n = \mathcal{F}^n(\mathcal{G})$. Then for all $n \geq 0$ and $\varepsilon > 0$*

$$\mathcal{Z}_{\mathcal{G}_n} \leq c_1 \vartheta_1^n \mathcal{Z}_\mathcal{G} + c_2$$

for some constants $c_i = c_i(\mathcal{D}) > 0$, $i = 1, 2$.

PROOF. It is enough to integrate (7.13) with respect to the factor measure $\lambda_\mathcal{G}$ and use the result of Exercise 7.15. □

We see that if $\mathcal{Z}_\mathcal{G}$ is very large, the sequence $\mathcal{Z}_{\mathcal{G}_n}$ will decrease exponentially fast until it goes under a certain threshold, say $c_1 + c_2$.

COROLLARY 7.18. *For all $n \geq \varkappa \ln \mathcal{Z}_\mathcal{G}$ we have $\mathcal{Z}_{\mathcal{G}_n} \leq c_3$ for some constants $\varkappa, c_3 > 0$.*

The reader should notice similarities between the statements 7.17–7.18 and two growth lemmas (Theorems 5.52–5.53).

Motivated by these facts, we introduce the notion of a proper family. Let $C_\text{p} > 0$ be a sufficiently large constant (the subscript p in C_p stands for 'proper').

[3] In all our subsequent proofs, the curves W_α of every standard family will be disjoint; however, in this general definition we need not assume this: the function $r_\mathcal{G}$ is simply defined on every curve W_α separately.

DEFINITION 7.19. A standard family $\mathcal{G}$ is said to be *proper* if $\mathcal{Z}_{\mathcal{G}} \leq C_{\rm p}$.

EXERCISE 7.20. Check that a family consisting of a single standard pair (W, ν) is proper iff $|W| \geq c_0 = \text{const} > 0$, i.e. iff the curve W is not too short (c_0 can be made arbitrarily small by choosing $C_{\rm p}$ sufficiently large). For brevity, we call such (W, ν) *proper standard pairs.*

EXAMPLE 7.21. In Exercise 7.11 we verified that every unstable H-manifold $W_{\mathbb{H}}^u$ with the u-SRB measure $\nu_{W_{\mathbb{H}}^u}$ on it was a standard pair. Thus the partition of $\mathcal{M}$ into maximal unstable H-manifolds $\{W_{\mathbb{H}}^u\}$ with u-SRB measures $\{\nu_{W_{\mathbb{H}}^u}\}$ on them and the factor measure induced by μ makes a (special) standard family $\mathcal{E}$. For this family $\mu_{\mathcal{E}} = \mu$, of course. Note also that $\mathcal{E}$ is mapped by $\mathcal{F}$ into itself (cf. Proposition 5.4).

EXERCISE 7.22. Verify that the family $\mathcal{E}$ is proper. Hint: Use Theorem 5.17.

More generally, let $\{W\}$ be a smooth foliation of $\mathcal{M}$ into unstable curves that stretch from $\varphi = -\pi/2$ to $\varphi = \pi/2$. Dividing them by the lines $\mathbb{S}_k$, $|k| \geq k_0$, introduced in Section 5.3 gives a smooth foliation of $\mathcal{M}$ into homogeneous unstable curves $\{W_\alpha\}$. The measure μ induces smooth conditional measures $\{\nu_\alpha\}$ on $\{W_\alpha\}$ and a factor measure on the index set. It is easy to check that the so defined standard family $\mathcal{G}$ is proper and $\mu_{\mathcal{G}} = \mu$.

EXERCISE 7.23. Let $D \subset \mathcal{M}$ be an arbitrary disk. Consider an arbitrary smooth foliation of D by homogeneous unstable curves $\{W_\alpha\}$. The measure μ induces conditional measures $\{\nu_\alpha\}$ on them and a factor measure λ on the index set $\{\alpha\}$. Show that for the resulting family $\mathcal{G} = \{(W_\alpha, \nu_\alpha)\}$ we have $\mathcal{Z}_{\mathcal{G}} < \infty$, and if D is not too small, then $\mathcal{G}$ is proper.

7.5. Coupling lemma

In this section we obtain the main tool for our studies of statistical properties of dispersing billiards.

Let (W_1, ν_1) and (W_2, ν_2) be two standard pairs. For a large n, their images $\mathcal{F}^n W_1$ and $\mathcal{F}^n W_2$ consist of many H-components scattered all over $\mathcal{M}$. Some H-components $W' \subset \mathcal{F}^n W_1$ of the first image may lie close to some H-components $W'' \subset \mathcal{F}^n W_2$ of the other image. Then certain points $x' \in W'$ can be joined by stable manifolds with points $x'' \in W''$, so that their further iterations will get closer together exponentially fast. In this way we can 'link' the measures they carry and eventually show that the asymptotic behavior of the two measures $\mathcal{F}^n(\nu_1)$ and $\mathcal{F}^n(\nu_2)$ becomes identical.

This argument may run into an obvious problem, though: the H-components W' and W'' may carry different amount (mass) of the corresponding measures. To resolve this problem we may, so to say, couple a heavy piece with several light ones. This can be done by splitting a heavy piece into several 'thinner' curves, each coupled to a different partner.

To implement this idea, it is convenient to split each original curve W_1 and W_2 into uncountably many 'fibers'. To this end, given a standard pair (W, ν), we consider $\hat{W} \colon = W \times [0, 1]$ and equip $\hat{W}$ with a probability measure $\hat{\nu}$ defined by

$$d\hat{\nu}(x, t) = d\nu(x)\, dt = \rho(x)\, dx\, dt \tag{7.14}$$

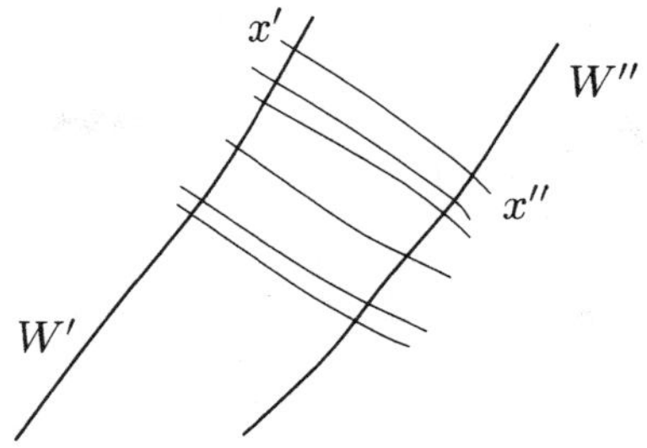

FIGURE 7.3. Coupling principle.

where $\rho(x)$ is the density of ν and $0 \le t \le 1$. We call $\hat{W}$ a *rectangle* with *base* W; see Fig. 7.4. The map $\mathcal{F}^n$ can be naturally defined on $\hat{W}$ by $\mathcal{F}^n(x,t) = (\mathcal{F}^n x, t)$, and any function f initially defined on W can also be extended to $\hat{W}$ by $f(x,t) = f(x)$.

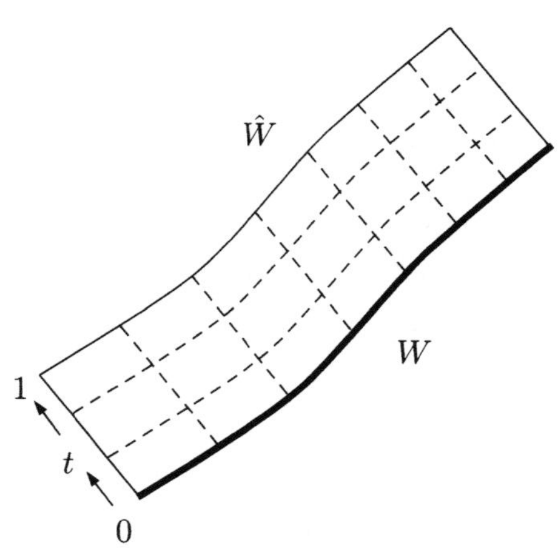

FIGURE 7.4. A rectangle $\hat{W} = W \times [0,1]$ over a 'base' W.

Next, recall that $\mathcal{F}^n(W_1)$ and $\mathcal{F}^n(W_2)$ are, generally, countable standard families. Thus, in the above construction we may start with two standard families, rather than two standard pairs, and couple the images of measures initially defined on the two families. In that case we need to split *every* unstable curve W in *each* original family into uncountably many 'fibers' by constructing a rectangle over W.

Given a standard family $\mathcal{G} = (W_\alpha, \nu_\alpha)$, $\alpha \in \mathfrak{A}$, with a factor measure $\lambda_{\mathcal{G}}$, we denote by $\hat{\mathcal{G}} = (\hat{W}_\alpha, \hat{\nu}_\alpha)$ the family of the corresponding rectangles, with $\hat{W}_\alpha$ being the rectangle with base W_α, equip $\hat{\mathcal{G}}$ with the same factor measure $\lambda_{\mathcal{G}}$, and denote by $\hat{\mu}_{\mathcal{G}}$ the induced measure on the union $\cup_\alpha \hat{W}_\alpha$.

The following lemma is the key instrument of the coupling method:

LEMMA 7.24 (Coupling Lemma). *Let $\mathcal{G} = (W_\alpha, \nu_\alpha)$, $\alpha \in \mathfrak{A}$, and $\mathcal{E} = (W_\beta, \nu_\beta)$, $\beta \in \mathfrak{B}$, be two proper standard families. Then there exist a bijection (a coupling map) $\Theta\colon \cup_\alpha \hat{W}_\alpha \to \cup_\beta \hat{W}_\beta$ that preserves measure; i.e. $\Theta(\hat{\mu}_{\mathcal{G}}) = \hat{\mu}_{\mathcal{E}}$, and a (coupling time) function $\Upsilon : \cup_\alpha \hat{W}_\alpha \to \mathbb{N}$ such that*

A. *Let* $(x,t) \in \hat{W}_\alpha$, $\alpha \in \mathfrak{A}$, *and* $\Theta(x,t) = (y,s) \in \hat{W}_\beta$, $\beta \in \mathfrak{B}$. *Denote* $m = \Upsilon(x,t) \in \mathbb{N}$. *Then the points* $\mathcal{F}^m(x)$ *and* $\mathcal{F}^m(y)$ *lie on the same stable H-manifold* $W^s_{\mathbb{H}} \subset \mathcal{M}$.

B. *There is a uniform exponential tail bound on the function* Υ*:*

$$\hat{\mu}_{\mathcal{G}_1}(\Upsilon > n) \le C_\Upsilon \vartheta_\Upsilon^n, \tag{7.15}$$

for some constants $C_\Upsilon = C_\Upsilon(\mathcal{D}) > 0$ *and* $\vartheta_\Upsilon = \vartheta_\Upsilon(\mathcal{D}) < 1$.

We postpone, until Sections 7.12–7.15, the proof of the Coupling Lemma and turn to statistical properties for dispersing billiards. First we introduce a class of appropriate functions on $\mathcal{M}$ (observables). They will be characterized by dynamically defined Hölder continuity.

Recall the separation time $\mathbf{s}_+(x,y)$ introduced in (5.31). Similarly, we introduce the *past separation time*:

$$\mathbf{s}_-(x,y) = \min\{n \ge 0\colon y \notin \mathcal{Q}^{\mathbb{H}}_{-n}(x)\}, \tag{7.16}$$

where $\mathcal{Q}^{\mathbb{H}}_{-n}(x)$ denotes the open connected component of the set $\mathcal{M}\setminus\mathcal{S}^{\mathbb{H}}_{-n}$ containing the point x (clearly, this function is symmetric: $\mathbf{s}_-(x,y) = \mathbf{s}_-(y,x)$). Observe that the preimages $\mathcal{F}^{-n}(x)$ and $\mathcal{F}^{-n}(y)$ get separated at time $n = \mathbf{s}_-(x,y)$ as they lie in different connected components of the 'new' collision space $\mathcal{M}_{\mathbb{H}}$. If $y \in \mathcal{Q}^{\mathbb{H}}_{-n}(x)$ for all $n \ge 0$, then $y \in W^u_{\mathbb{H}}(x)$; in that case we set $\mathbf{s}_-(x,y) = \infty$.

EXERCISE 7.25. Verify that if x and y lie on one stable curve $W \subset \mathcal{M}$, then

$$\operatorname{dist}(x,y) \le C\Lambda^{-\mathbf{s}_-(x,y)} \tag{7.17}$$

for some constant $C = C(\mathcal{D}) > 0$. Hint: This is similar to (5.32).

DEFINITION 7.26. A function $f\colon \mathcal{M} \to \mathbb{R}$ is said to be *dynamically Hölder continuous* if there are $\vartheta_f \in (0,1)$ and $K_f > 0$ such that for any x and y lying on one unstable curve

$$|f(x) - f(y)| \le K_f \vartheta_f^{\mathbf{s}_+(x,y)} \tag{7.18}$$

and for any x and y lying on one stable curve

$$|f(x) - f(y)| \le K_f \vartheta_f^{\mathbf{s}_-(x,y)}. \tag{7.19}$$

We denote the space of such functions by $\mathcal{H}$.

The class of dynamically Hölder continuous functions is larger than the class of (ordinary) Hölder continuous functions $f\colon \mathcal{M} \to \mathbb{R}$, which are characterized by

$$|f(x) - f(y)| \le C_f[\operatorname{dist}(x,y)]^{\alpha_f} \qquad \forall x,y \in \mathcal{M}. \tag{7.20}$$

Here $\alpha_f \in (0,1]$ is the Hölder exponent, and the minimal $C_f > 0$ satisfying (7.20) is the Hölder norm of f.

EXERCISE 7.27. Prove that any Hölder continuous function $f\colon \mathcal{M} \to \mathbb{R}$ is dynamically Hölder continuous, with $\vartheta_f = \Lambda^{-\alpha_f}$. Hint: Use (5.31) and (7.16).

EXERCISE 7.28. Prove that if there are $n_1, n_2 \ge 0$ such that f is Hölder continuous on every connected component of the set $\mathcal{M} \setminus (\mathcal{S}^{\mathbb{H}}_{n_1} \cup \mathcal{S}^{\mathbb{H}}_{-n_2})$; i.e. (7.20) holds, with the same C_f and α_f whenever x and y belong in the same component of the set $\mathcal{M} \setminus (\mathcal{S}^{\mathbb{H}}_{n_1} \cup \mathcal{S}^{\mathbb{H}}_{-n_2})$, then f is dynamically Hölder continuous.

EXERCISE 7.29. Verify that in billiards of category A the return time function $\tau(x)$ is Hölder continuous on the connected components of the set $\mathcal{M} \setminus \mathcal{S}_1^{\mathbb{H}}$. What is its Hölder exponent? Conclude that $\tau(x)$ is dynamically Hölder continuous.

REMARK 7.30. Occasionally we will deal with functions satisfying only one of the conditions (7.18) and (7.19). We denote the space of functions satisfying (7.18) by $\mathcal{H}^+$, and those satisfying (7.19) by $\mathcal{H}^-$.

7.6. Equidistribution property

Here we derive the first statistical property from the Coupling Lemma. It says that for any proper standard family $\mathcal{G}$ the images $\mathcal{F}^n(\mu_\mathcal{G})$ of its measure $\mu_\mathcal{G}$ weakly converge, as $n \to \infty$, to the $\mathcal{F}$-invariant measure μ. Furthermore, in a certain sense the speed of convergence is exponential. We call this property *equidistribution.*

We note that a proper family $\mathcal{G}$ may be quite 'localized' in $\mathcal{M}$. It can consist of a single standard pair, or its measure $\mu_\mathcal{G}$ may be supported on some disk in $\mathcal{M}$; see Exercise 7.23. In that case the equidistribution property effectively pronounces asymptotic independence between the present and the future of the system.

THEOREM 7.31 (Equidistribution). *Let $\mathcal{G}$ be a proper standard family. For any dynamically Hölder continuous function $f \in \mathcal{H}$ and $n \geq 0$*

$$\left| \int_\mathcal{M} f \circ \mathcal{F}^n \, d\mu_\mathcal{G} - \int_\mathcal{M} f \, d\mu \right| \leq B_f \theta_f^n \tag{7.21}$$

where

$$B_f = 2C_\Upsilon \bigl(K_f + \|f\|_\infty\bigr)$$

and

$$\theta_f = \bigl[\max\{\vartheta_\Upsilon, \vartheta_f\}\bigr]^{1/2} < 1. \tag{7.22}$$

PROOF. Recall that there is a proper $\mathcal{F}$-invariant family $\mathcal{E}$ such that $\mu = \mu_\mathcal{E}$; see Example 7.21. The Coupling Lemma 7.24 gives us a coupling map Θ between the families $\mathcal{G}$ and $\mathcal{E}$ and the corresponding coupling time function Υ. Then

$$\begin{aligned} \Delta\colon &= \int_\mathcal{M} f \circ \mathcal{F}^n \, d\mu_\mathcal{G} - \int_\mathcal{M} f \circ \mathcal{F}^n \, d\mu_\mathcal{E} \\ &= \int_{\hat{\mathcal{G}}} f(\mathcal{F}^n(x,t)) \, d\hat{\mu}_\mathcal{G} - \int_{\hat{\mathcal{E}}} f(\mathcal{F}^n(y,s)) \, d\hat{\mu}_\mathcal{E} \\ &= \int_{\hat{\mathcal{G}}} \bigl[f(\mathcal{F}^n(x,t)) - f(\mathcal{F}^n(\Theta(x,t)))\bigr] \, d\hat{\mu}_\mathcal{G}. \end{aligned} \tag{7.23}$$

Note that if $\Theta(x,t) = (y,s)$ and $m\colon = \Upsilon(x,t) \leq n$, then $\mathbf{s}_-(\mathcal{F}^n x, \mathcal{F}^n y) > n - m$, hence

$$\bigl|f(\mathcal{F}^n(x,t) - f(\mathcal{F}^n(\Theta(x,t)))\bigr| \leq K_f \vartheta_f^{n-m}. \tag{7.24}$$

Now the last integral in (7.23) can be decomposed as

$$\int_{\hat{\mathcal{G}}} [\dots] = \int_{\Upsilon \leq n/2} [\dots] + \int_{\Upsilon > n/2} [\dots] = I + I\!I. \tag{7.25}$$

Clause A of Lemma 7.24 and (7.24) imply that $|I| \leq K_f \vartheta_f^{n/2}$, and clause B implies that $|I\!I| \leq 2\|f\|_\infty C_\Upsilon \vartheta_\Upsilon^{n/2}$. □

EXERCISE 7.32. The estimate (7.21) can be slightly improved if we refine the decomposition (7.25) by using smaller sets $\{\Upsilon = m\}$ for all $0 \le m \le n$ and the remainder set $\{\Upsilon > n\}$. Work out the details and obtain the improvement (we will not use it later, though).

Theorem 7.31 can be extended to 'multiple' observables, i.e. observations made at multiple times. Let $f_0, f_1, \dots, f_k \in \mathcal{H}$ be dynamically Hölder continuous functions with the same $\vartheta_f = \vartheta_{f_i}$, the same $K_f = K_{f_i}$, and the same $\|f\|_\infty = \|f_i\|_\infty$, $0 \le i \le k$. (For example, we can take $f_0 = f_1 = \dots = f_k = f$.) Consider the product

$$\tilde{f} = f_0 \cdot (f_1 \circ \mathcal{F}) \cdot (f_2 \circ \mathcal{F}^2) \cdots (f_k \circ \mathcal{F}^k).$$

THEOREM 7.33. *Let $\mathcal{G}$ be a proper standard family. Then for $n \ge 0$*

$$\left| \int_{\mathcal{M}} \tilde{f} \circ \mathcal{F}^n \, d\mu_{\mathcal{G}} - \int_{\mathcal{M}} \tilde{f} \, d\mu \right| \le B_{\tilde{f}} \, \theta_f^n \tag{7.26}$$

where $\theta_f < 1$ is as in (7.22) and

$$B_{\tilde{f}} = 2C_\Upsilon \|f\|_\infty^k (1 - \vartheta_f)^{-1} \big(K_f + \|f\|_\infty\big).$$

PROOF. The argument is almost identical to the proof of Theorem 7.31, except (7.24) must be replaced by

$$\begin{aligned} \big|\tilde{f}(\mathcal{F}^n(x,t)) - \tilde{f}(\mathcal{F}^n(\Theta(x,t)))\big| &\le K_f \|f\|_\infty^k \big(\vartheta_f^{n-m} + \dots + \vartheta_f^{n-m+k}\big) \\ &\le K_f \|f\|_\infty^k (1 - \vartheta_f)^{-1} \vartheta_f^{n-m}. \end{aligned}$$

□

REMARK 7.34. We used consecutive time moments $0, 1, \dots, k$ for simplicity. The statement (and the proof) will be the same for any increasing sequence of time moments $0 < t_1 < t_2 < \dots < t_k$.

REMARK 7.35. In Theorems 7.31 and 7.33 we assumed that the initial standard family $\mathcal{G}$ was proper. If it is not, then we will have to wait until its image $\mathcal{F}^m\mathcal{G}$ becomes proper for some $m \ge 1$ and then apply these theorems with n replaced by $n - m$. If $\mathcal{Z}_{\mathcal{G}} < \infty$, then the waiting time is $m = \varkappa \ln \mathcal{Z}_{\mathcal{G}}$ iterations, according to Corollary 7.18. If $\mathcal{G}$ consists of a single standard pair (W, ν), the waiting time is $m = \varkappa \big|\ln |W|\big|$.

REMARK 7.36. In Theorems 7.31 and 7.33 it is enough to assume that $f \in \mathcal{H}^-$ (respectively, $f_0, \dots, f_k \in \mathcal{H}^-$); cf. Remark 7.30.

7.7. Exponential decay of correlations

Here we establish the bound (7.6) for dynamically Hölder continuous observables. Our results are similar (but somewhat sharper) than those in [**Yo98, Ch99**]. For brevity, we use the notation $\langle f \rangle = \int_{\mathcal{M}} f \, d\mu$.

THEOREM 7.37 (Exponential decay of correlations). *For every pair of dynamically Hölder continuous functions $f, g \in \mathcal{H}$ and $n \ge 0$*

$$\big|\langle f \cdot (g \circ \mathcal{F}^n) \rangle - \langle f \rangle \langle g \rangle\big| \le B_{f,g} \, \theta_{f,g}^n \tag{7.27}$$

where

$$\theta_{f,g} = \big[\max\{\vartheta_\Upsilon, \vartheta_f, \vartheta_g, e^{-1/\varkappa}\}\big]^{1/4} < 1, \tag{7.28}$$

where $\varkappa > 0$ is the constant of Theorem 5.53,

$$B_{f,g} = C_0\bigl(K_f\|g\|_\infty + K_g\|f\|_\infty + \|f\|_\infty\|g\|_\infty\bigr), \tag{7.29}$$

and $C_0 = C_0(\mathcal{D}) > 0$ is a constant.

PROOF. Let $\mathcal{E} = \{W_\alpha, \nu_\alpha\}$, $\alpha \in \mathfrak{A}$, be the standard family consisting of all maximal unstable H-manifolds and u-SRB measures on them; see Example 7.21. Recall that $\mu_{\mathcal{E}} = \mu$ and $\mathcal{E}$ is a proper family (Exercise 7.22). We can divide the time interval $[0, n]$ in the ratio 1:3 and write

$$\langle f \cdot (g \circ \mathcal{F}^n)\rangle = \int \bigl(f \circ \mathcal{F}^{-n/4}\bigr)\bigl(g \circ \mathcal{F}^{3n/4}\bigr)\, d\mu_{\mathcal{E}}$$

(here $n/4$ means an integer closest to $n/4$, etc.). Let $\bar{f}$ denote the conditional expectation of $f \circ \mathcal{F}^{-n/4}$ on the unstable H-manifolds W_α with respect to the u-SRB measure ν_α, i.e.

$$\bar{f}(x) = \int_{W_\alpha} f \circ \mathcal{F}^{-n/4}\, d\nu_\alpha \qquad \forall x \in W_\alpha \quad \forall \alpha \in \mathfrak{A}.$$

Clearly the function $f \circ \mathcal{F}^{-n/4}$ is almost constant on each H-manifold W_α. Precisely, due to the dynamical Hölder continuity of f

$$\sup_{W_\alpha} f \circ \mathcal{F}^{-n/4} - \inf_{W_\alpha} f \circ \mathcal{F}^{-n/4} \leq K_f \vartheta_f^{n/4}; \tag{7.30}$$

hence

$$\sup_{x \in \mathcal{M}} |\bar{f}(x) - f \circ \mathcal{F}^{-n/4}(x)| \leq K_f \vartheta_f^{n/4}.$$

Note also that $\langle f\rangle = \langle \bar{f}\rangle$; thus

$$\begin{aligned}\Delta\colon &= \langle f \cdot (g \circ \mathcal{F}^n)\rangle - \langle f\rangle\langle g\rangle \\ &= \langle \bar{f} \cdot (g \circ \mathcal{F}^{3n/4})\rangle - \langle \bar{f}\rangle\langle g\rangle + \delta_1,\end{aligned}$$

where $|\delta_1| \leq K_f\|g\|_\infty \vartheta_f^{n/4}$. Since the function $\bar{f}$ is constant on every H-manifold W_α, we denote its value by $\bar{f}_\alpha$ and use Theorem 7.31 and Remark 7.35 to obtain

$$\int_{W_\alpha} \bar{f} \cdot (g \circ \mathcal{F}^{3n/4})\, d\nu_\alpha = \bar{f}_\alpha \langle g\rangle + \delta_\alpha \tag{7.31}$$

where

$$\begin{aligned}|\delta_\alpha| \leq 2\min\Bigl\{&\|f\|_\infty\|g\|_\infty, \\ &\|f\|_\infty C_\Upsilon\bigl(K_g + \|g\|_\infty\bigr)\theta_g^{3n/4 - \varkappa|\ln|W_\alpha||}\Bigr\}\end{aligned}$$

and $\theta_g = \bigl[\max\{\vartheta_\Upsilon, \vartheta_g\}\bigr]^{1/2}$. Integrating (7.31) with respect to the factor measure $\lambda_{\mathcal{E}}$ of the family $\mathcal{E}$ gives

$$\langle \bar{f} \cdot (g \circ \mathcal{F}^{3n/4})\rangle = \langle \bar{f}\rangle\langle g\rangle + \int_{\mathfrak{A}} \delta_\alpha\, d\lambda_{\mathcal{E}}(\alpha).$$

It remains to estimate the last term here. Since $\mathcal{E}$ is a proper family

$$\mu\bigl(\cup W_\alpha\colon |W_\alpha| < e^{-\frac{n}{4\varkappa}}\bigr) \leq C_{\mathrm{p}} e^{-\frac{n}{4\varkappa}}.$$

For H-manifolds W_α satisfying $|W_\alpha| \geq e^{-\frac{n}{4\varkappa}}$ we have $3n/4 - \varkappa\big|\ln|W_\alpha|\big| \geq n/2$. Therefore

$$\left|\int \delta_\alpha \, d\lambda_{\mathcal{E}}(\alpha)\right| \leq 2C_{\mathrm{p}}\|f\|_\infty\|g\|_\infty e^{-\frac{n}{4\varkappa}} + 2\|f\|_\infty C_\Upsilon\big(K_g + \|g\|_\infty\big)\theta_g^{n/2}.$$

Theorem 7.37 is now proved. □

REMARK 7.38. It is enough to assume that $f \in \mathcal{H}^+$ and $g \in \mathcal{H}^-$; cf. Remarks 7.30 and 7.36.

We derive an important corollary from the above remark. Suppose f is constant on every unstable H-manifold $W^u_{\mathbb{H}}$ and g is constant on every stable H-manifold $W^s_{\mathbb{H}}$. We can also redefine unstable and stable cones so that they will degenerate to lines tangent to the unstable and stable manifolds, respectively.[4] In that case we may assume that $K_f = K_g = 0$ and $\theta_f = \theta_g = 0$, and therefore

$$\big|\langle f\cdot(g\circ\mathcal{F}^n)\rangle - \langle f\rangle\langle g\rangle\big| \leq C_0\|f\|_\infty\|g\|_\infty\theta_0^n \tag{7.32}$$

for all $n \geq 0$, where $\theta_0 = \big[\max\{\vartheta_\Upsilon, e^{-1/\varkappa}\}\big]^{1/4} < 1$.

COROLLARY 7.39. *Let $A, B \subset \mathcal{M}$ be two measurable sets such that $A = \cup W^u_{\mathbb{H}}$ is the union of* some *unstable H-manifolds and $B = \cup W^s_{\mathbb{H}}$ is the union of* some *stable H-manifolds (here we mean* maximal *H-manifolds; cf. Sections* 4.11 *and* 5.4*). Then*

$$\big|\mu(A\cap\mathcal{F}^{-n}(B)) - \mu(A)\mu(B)\big| \leq C_0\theta_0^n. \tag{7.33}$$

EXERCISE 7.40. Verify that (7.32) follows from (7.27)–(7.29) and (7.33) follows from (7.32). Hint: Use the characteristic functions (indicators) of the sets A and B.

Theorem 7.37 can be extended to the so-called 'multiple' correlations, i.e. correlations between observations made at multiple times. Let $f_0, f_1, \dots, f_r \in \mathcal{H}$ and $g_0, g_1, \dots, g_k \in \mathcal{H}$ be two sets of dynamically Hölder continuous functions. We suppose f's have identical parameters $\vartheta_f = \vartheta_{f_i}$, $K_f = K_{f_i}$, and $\|f\|_\infty = \|f_i\|_\infty$ for all $0 \leq i \leq r$. Similarly, let g's have identical parameters $\vartheta_g = \vartheta_{g_i}$, $K_g = K_{g_i}$, and $\|g\|_\infty = \|g_i\|_\infty$ for all $0 \leq i \leq k$. Consider two products

$$\tilde f = f_0\cdot(f_1\circ\mathcal{F}^{-1})\cdot(f_2\circ\mathcal{F}^{-2})\cdots(f_r\circ\mathcal{F}^{-r})$$

and

$$\tilde g = g_0\cdot(g_1\circ\mathcal{F})\cdot(g_2\circ\mathcal{F}^2)\cdots(g_k\circ\mathcal{F}^k).$$

THEOREM 7.41 (Exponential decay of multiple correlations). *For all $n > 0$*

$$\big|\langle\tilde f\cdot(\tilde g\circ\mathcal{F}^n)\rangle - \langle\tilde f\rangle\langle\tilde g\rangle\big| \leq B_{\tilde f,\tilde g}\,\theta_{f,g}^{|n|} \tag{7.34}$$

where $\theta_{f,g}$ is as in (7.28) and

$$B_{\tilde f,\tilde g} = C_0\|f\|_\infty^r\|g\|_\infty^k\left[\frac{K_f\|g\|_\infty}{1-\vartheta_f} + \frac{K_g\|f\|_\infty}{1-\vartheta_g} + \|f\|_\infty\|g\|_\infty\right] \tag{7.35}$$

---

[4]As usual in the study of hyperbolic maps, the construction of invariant cones is rather loose. For example, given a family of unstable cones C^u_x we can replace it with more narrow cones $C^{u,n}_x := D\mathcal{F}^n(C^u_{\mathcal{F}^{-n}x})$. This of course reduces the class of unstable curves. Ultimately, we can set $n=\infty$; then the cones become lines tangent to unstable manifolds, and unstable curves reduce to unstable manifolds.

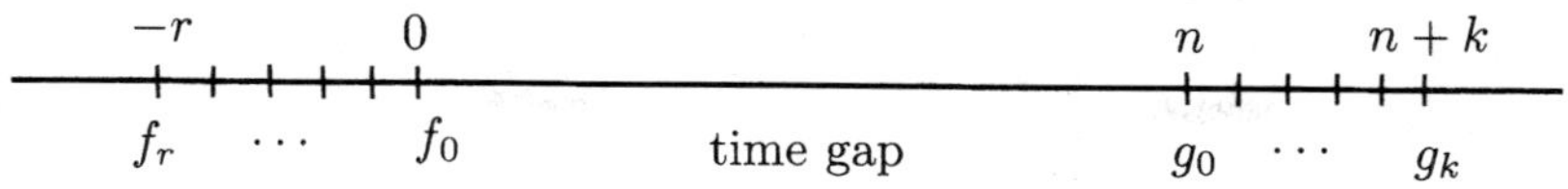

FIGURE 7.5. Multiple correlations.

and $C_0 = C_0(\mathcal{D}) > 0$ is a constant.

PROOF. The argument is almost identical to the proof of Theorem 7.37, with a few modifications. First, we note obvious bounds:

$$\|\tilde{f}\|_\infty \leq \|f\|_\infty^{r+1}, \qquad \|\tilde{g}\|_\infty \leq \|g\|_\infty^{k+1}.$$

Second, (7.30) is replaced with

$$\begin{aligned} \sup_{W_\alpha} \tilde{f} \circ \mathcal{F}^{-n/4} - \inf_{W_\alpha} \tilde{f} \circ \mathcal{F}^{-n/4} &\leq K_f \|f\|_\infty^r (\vartheta_f^{n/4} + \cdots + \vartheta_f^{n/4+r}) \\ &\leq K_f \|f\|_\infty^r (1-\vartheta_f)^{-1} \vartheta_f^{n/4}. \end{aligned}$$

Lastly, Theorem 7.33 must be used instead of 7.31. □

REMARK 7.42. We used consecutive time moments $-r, \dots, -1, 0$ and $n, n+1, \dots, n+k$ for simplicity. The statement (and the proof) will be the same for any two increasing sequences of time moments $t_{-r} < \cdots < t_{-1} < 0$ and $n < t_1 < \cdots < t_k$ separated by a 'time gap' of length n.

7.8. Central Limit Theorem

Here we prove that the sequence $\mathcal{X}_n = f \circ F^n$ introduced in (7.1) satisfies the CLT (7.8), along with formula (7.9), for every dynamically Hölder continuous observable $f \in \mathcal{H}$. Our arguments here and in the next section use certain advanced facts of probability theory, but the reader need not worry about it; we include full statements and/or references to standard textbooks here and in Appendix B.

First we make a methodological remark. In the theory of chaotic dynamical systems, the CLT is usually established after obtaining good estimates on correlation functions, and the proof of the CLT heavily uses bounds on correlations. One may wonder if, generally, a fast decay of correlations implies the CLT in a formal manner. Standard proofs of the CLT do not answer this question, though, since they all use some other (stronger) mixing properties of the dynamics in addition to explicit bounds on correlations.

We believe that fast decay of the two-correlation function $\mathbf{C}_f(n)$ alone would not be sufficient for the CLT, but good bounds on multiple correlations can do the job. Here we present a proof of the CLT that relies solely on the exponential decay of multiple correlations (our Theorem 7.41). So we demonstrate that the CLT is, indeed, a formal consequence of a fast decay of multiple correlations.

THEOREM 7.43. *Let $f \in \mathcal{H}$ be a dynamically Hölder continuous observable. Then the CLT (7.8) holds, along with formula (7.9) for σ_f^2 and the statement of Remark 7.5.*

PROOF. Our argument follows the lines of [**BSC91**, Section 6] and [**Ch95**, Section 3]. This proof uses certain standard facts of probability theory; the reader unfamiliar with probabilities is advised to consult our Appendix B and [**BSC91, Ch95**].

First, we can assume that $\langle f\rangle = 0$ (otherwise we replace f with $f - \langle f\rangle$). Now we have $\mathbf{C}_f(n) = \langle f\cdot (f\circ \mathcal{F}^n)\rangle$.

EXERCISE 7.44. Verify the following formula:

$$\operatorname{Var} S_n = \langle S_n^2\rangle = n\,\mathbf{C}_f(0) + 2\sum_{i=1}^{n-1}(n-i)\,\mathbf{C}_f(i). \tag{7.36}$$

Hint: Recall Exercise 7.2.

Since $\mathbf{C}_f(n)$ converges to zero exponentially fast, it follows that

$$\operatorname{Var} S_n = n\sigma_f^2 + O(1), \tag{7.37}$$

where σ_f^2 is given by (7.9).

EXERCISE 7.45. Prove that there exists a finite limit

$$\lim_{n\to\infty}(\operatorname{Var} S_n - n\sigma_f^2) = -2\sum_{n=1}^{\infty} n\mathbf{C}_f(n) < \infty. \tag{7.38}$$

There are two cases now. First, it may happen that $\sigma_f^2 = 0$. Then, due to (7.38), the sequence $\{\operatorname{Var} S_n\}$ is bounded and has a finite limit. Now we can use a general result by Leonov [**Leo61**] (see also [**IL71**, Theorem 18.2.2]), according to which there is another function, $g\in L^2_\mu(\mathcal{M})$, such that $f(x) = g(\mathcal{F}x) - g(x)$ a.e. This proves our claim in Remark 7.5.

From now on we assume $\sigma_f^2 > 0$. Then, due to (7.38), the main expression (7.8) is equivalent to

$$\lim_{n\to\infty}\mu\left\{\frac{S_n}{\sqrt{\operatorname{Var} S_n}}\le z\right\} = \frac{1}{\sqrt{2\pi}}\int_{-\infty}^{z} e^{-s^2/2}\,ds; \tag{7.39}$$

i.e. $S_n/\sqrt{\operatorname{Var} S_n}$ converges to the standard normal distribution $\mathcal{N}(0,1)$. It is actually (7.39), rather than (7.8), which is called the Central Limit Theorem in some probability textbooks.

The proof of (7.39) is based on Bernstein's technique of approximating sums of weakly dependent random variables by sums of *independent* variables. This method is also known as "the big small block technique".

We partition the time interval $[0,n]$ into a sequence of alternating big intervals (blocks) of length $p = [n^a]$ and small blocks of length $q = [n^b]$ for some $0 < b < a < 1$ (later we will also assume that $a < 1/2$). The number of big blocks is $k = [n/(p+q)] \sim n^{1-a}$. The last remaining block has length $l = n - kp - (k-1)q < p+q$. We remark that the lengths of both big and small blocks grow as $n\to\infty$.

We denote by $\Delta_r, 1\le r\le k$, the big blocks and set

$$S_p^{(r)}(x) = \sum_{i\in\Delta_r} f(\mathcal{F}^i x), \qquad S_n' = \sum_{r=1}^{k} S_p^{(r)}$$

and $S_n'' = S_n - S_n' = \sum_{i\in[0,n-1]\setminus\cup\Delta_r} f(\mathcal{F}^i x)$. The second sum S_n'' contains no more than $n'' = kq + p \le 2n^h$ terms, where $h = \max\{a, 1-a+b\} < 1$.

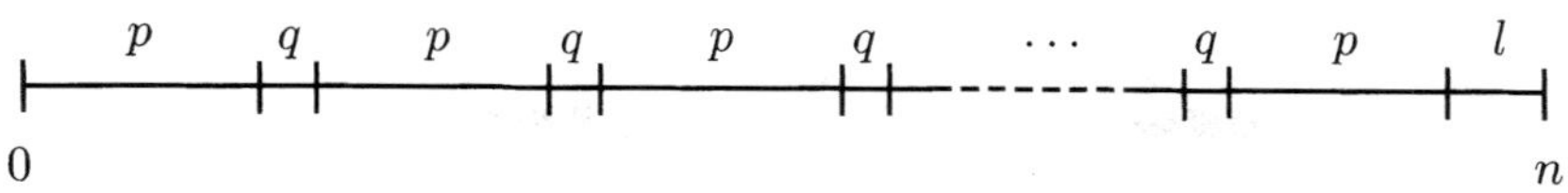

FIGURE 7.6. Big and small blocks alternate.

EXERCISE 7.46. Verify that

$$\operatorname{Var} S_n'' \leq n''\Big(|\mathbf{C}_f(0)| + 2\sum_{n=1}^{\infty} |\mathbf{C}_f(n)|\Big) \leq \text{const} \cdot n^h.$$

Hint: Follow the lines of Exercise 7.44.

Using Chebyshev's inequality gives

$$\mu\Big\{x : \frac{|S_n''|}{\sigma_f \sqrt{n}} > \varepsilon\Big\} \leq \frac{\operatorname{Var} S_n''}{\varepsilon^2 \sigma_f^2 n} \leq \text{const}\, n^{h-1} \to 0 \tag{7.40}$$

as $n \to \infty$. Hence $S_n''/\sqrt{\operatorname{Var} S_n}$ converges to zero in probability. The following fact is standard in probability theory (see, e.g., [**IL71**, Lemma 18.4.1]), and we invite the reader to verify it directly:

LEMMA 7.47. *If a sequence of random variables $\mathcal{X}_n$ converges to $\mathcal{X}$ in distribution and η_n converges to zero in probability, then $\mathcal{X}_n + \eta_n$ converges to $\mathcal{X}$ in distribution.*

Thus we can replace S_n by $S_n' = \sum S_p^{(r)}$ in (7.39).

EXERCISE 7.48. Verify that the observables $S_p^{(1)}, \dots, S_p^{(k)}$ have the same distribution. In particular, $\operatorname{Var} S_p^{(r)} = \operatorname{Var} S_p$ for all $r = 1, \dots, k$.

EXERCISE 7.49. Verify that

$$|\operatorname{Var} S_n - k \operatorname{Var} S_p| \leq n''|\mathbf{C}_f(0)| + 2\sum_{i=1}^{p-1}(n'' + ik)|\,\mathbf{C}_f(i)| + 2\sum_{i=p}^{n-1}(n-i)|\mathbf{C}_f(i)|.$$

Hint: Follow the lines of Exercises 7.44 and 7.46.

As a result, $|\operatorname{Var} S_n - k \operatorname{Var} S_p| = \mathrm{o}(n) = \mathrm{o}(\operatorname{Var} S_n)$, so we can replace $\operatorname{Var} S_n$ in (7.39) by $k \operatorname{Var} S_p$. Thus (7.39) becomes equivalent to

$$\lim_{n\to\infty} \mu\Big\{\frac{S_n'}{\sqrt{k \operatorname{Var} S_p}} \leq z\Big\} = \frac{1}{\sqrt{2\pi}} \int_{-\infty}^{z} e^{-\frac{s^2}{2}}\, ds. \tag{7.41}$$

Denote

$$U_n = \frac{S_n'}{\sqrt{k \operatorname{Var} S_p}} = \frac{S_p^{(1)} + \cdots + S_p^{(k)}}{\sqrt{k \operatorname{Var} S_p}} = u_1 + \cdots + u_k \tag{7.42}$$

where $u_r = S_p^{(r)}/\sqrt{k \operatorname{Var} S_p}$. Observe that U_n is the sum of identically distributed but not necessarily independent random variables $u_1, \dots, u_k$.

Now the big blocks $\Delta_1, \dots, \Delta_k$ are separated by small blocks whose length $q \to \infty$ as $n \to \infty$. Based on this we will show that $S_p^{(1)}, \dots, S_p^{(k)}$ (and thus $u_1, \dots, u_k$) become almost independent:

LEMMA 7.50. *U_n has the same limit distribution as the sum*

$$V_n = v_1 + \cdots + v_k \tag{7.43}$$

*of k **independent** random variables $v_1, \ldots, v_k$, each of which has the same distribution as u_1.*

PROOF. The Lévy continuity theorem (see Appendix B) says that a sequence of random variables $\{W_n\}$ converges in distribution to a random variable W if and only if their characteristic functions $\varphi_{W_n}(t) = \langle \exp(\mathbf{i}tW_n)\rangle$ converge pointwise to the characteristic function of W, i.e. to $\varphi_W(t) = \langle \exp(\mathbf{i}tW)\rangle$; here $\mathbf{i} = \sqrt{-1}$. Hence it suffices to show that $\varphi_{U_n}(t) \to \varphi_{V_n}(t)$ for any $t \in \mathbb{R}$. Denote $w_r = \exp(\mathbf{i}tu_r)$ for $1 \le r \le k$; then

$$\varphi_{U_n}(t) = \langle w_1 \cdots w_k \rangle.$$

Due to the independence of $v_1, \ldots, v_k$ we have

$$\langle e^{\mathbf{i}tv_1} \cdots e^{\mathbf{i}tv_k}\rangle = \langle e^{\mathbf{i}tv_1}\rangle \cdots \langle e^{\mathbf{i}tv_k}\rangle = \langle e^{\mathbf{i}tu_1}\rangle^k;$$

hence

$$\varphi_{V_n}(t) = \langle w_1\rangle^k = \langle w_1\rangle \cdots \langle w_k\rangle.$$

Next, denote $g(x) = \exp\bigl(\mathbf{i}tf(x)/\sqrt{k \operatorname{Var} S_p}\bigr)$. Then

$$w_r = \bigl(g \circ \mathcal{F}^{(p+q)(r-1)}\bigr) \cdot \bigl(g \circ \mathcal{F}^{(p+q)(r-1)+1}\bigr) \cdots \bigl(g \circ \mathcal{F}^{(p+q)(r-1)+p-1}\bigr).$$

EXERCISE 7.51. Show that if $f \in \mathcal{H}$, then $g \in \mathcal{H}$. Furthermore, verify that $\vartheta_g = \vartheta_f$, $K_g = K_f t/\sqrt{k \operatorname{Var} S_p}$, and $\|g\|_\infty \le 1$.

Now Theorem 7.41 gives us, for every $r = 1, \ldots, k-1$,

$$|\langle w_r w_{r+1} \cdots w_k\rangle - \langle w_r\rangle\langle w_{r+1} \cdots w_k\rangle| \le B\theta^q \tag{7.44}$$

where

$$\theta = \max\{\vartheta_\Upsilon, \vartheta_f, e^{-1/\varkappa}\} < 1$$

and

$$B = C_0\left[\frac{2K_f|t|}{(1-\vartheta_f)\sqrt{k \operatorname{Var} S_p}} + 1\right].$$

Combining (7.44) for all $r = 1, \ldots, k-1$ gives

$$|\langle w_1 \cdots w_k\rangle - \langle w_1\rangle \cdots \langle w_k\rangle| \le Bk\,\theta^q.$$

This implies Lemma 7.50. □

It remains to show that V_n converges, as $n \to \infty$, to the standard normal law in distribution. We use the following fact of advanced probability theory (see, e.g., the proof of Theorem 18.4.1 in [**IL71**]):

PROPOSITION 7.52. *Assume that for every $n \ge 1$ a random variable V_n is the sum of $k = k_n$ independent identically distributed random variables $v_1, \ldots, v_k$ normalized so that $E(V_n) = 0$ and $\operatorname{Var} V_n = 1$. Assume also that $k_n \to \infty$ as $n \to \infty$. Then V_n converges to the standard normal distribution if and only if the following analogue of Lindeberg's condition holds*[5] *for every $\varepsilon > 0$:*

$$\lim_{n\to\infty} k\langle v_1^2 \cdot \mathbf{1}_{\{|v_1|>\varepsilon\}}\rangle = 0. \tag{7.45}$$

[5]Here $\mathbf{1}_B$ stands for the characteristic function (indicator) of an event B.

Due to our previous constructions, we only need to verify Lindeberg's condition (7.45). Observe that

$$|v_1| \le \frac{\sup_{x\in\mathcal{M}} |S_p|}{\sqrt{k \operatorname{Var} S_p}} \le \frac{\operatorname{const} p}{\sqrt{n}}.$$

Choosing $a < 1/2$ ensures that $p/\sqrt{n} \to 0$, so that the set $\{x : |v_1| > \varepsilon\}$ will be empty (!) for all large enough n. This proves (7.45) and we obtain the CLT. □

We conclude this section with a vector CLT. Let $\mathbf{f} = (f_1, \dots, f_r)^T$ be a vector function on $\mathcal{M}$ whose components f_i are dynamically Hölder continuous. Consider the sum

$$\mathbf{S}_n = \{S_{1,n}, \dots, S_{r,n}\} \colon = \mathbf{f} + \mathbf{f} \circ \mathcal{F} + \dots + \mathbf{f} \circ \mathcal{F}^{n-1}$$

and an $r \times r$ matrix $\mathbf{V} = \{v_{ij}\}$ with components

$$v_{ij} = \lim_{n\to\infty} \frac{1}{n} \operatorname{Cov}(S_{i,n}, S_{j,n}) = \sum_{n=-\infty}^{\infty} \mathbf{C}_{f_i f_j}(n). \tag{7.46}$$

Note that $v_{ii} = \sigma_{f_i}^2$.

EXERCISE 7.53. Prove the second identity in (7.46). Verify that the matrix $\mathbf{V}$ is symmetric and positive semidefinite.

THEOREM 7.54 (Vector CLT). *If the matrix $\mathbf{V}$ is nonsingular, i.e. $\det \mathbf{V} \neq 0$, the vector $(\mathbf{S}_n - n\langle \mathbf{f}\rangle)/\sqrt{n}$ converges in distribution to the r-dimensional normal law with zero mean and covariance matrix $\mathbf{V}$.*

PROOF. The following lemma provides a useful criterion for the degeneracy of the matrix $\mathbf{V}$:

LEMMA 7.55. *The matrix $\mathbf{V}$ is singular, i.e. $\det \mathbf{V} = 0$, iff there is a linear combination $h = a_1 f_1 + \dots + a_r f_r$, with some nonzero vector $\mathbf{a} = (a_1, \dots, a_r)^T \neq 0$, which is cohomologous to a constant; i.e. $h = K + g - g \circ \mathcal{F}$ a.e. for some $K \in \mathbb{R}$ and $g \in L^2_\mu(\mathcal{M})$.*

PROOF. We prove the lemma first. Observe that $\det \mathbf{V} = 0$ is equivalent to $\sigma_h^2 = 0$ for some $\mathbf{a} \neq 0$. Now Remark 7.5 implies the lemma. □

Theorem 7.54 can be proven in two ways. First, we can just repeat the proof of Theorem 7.43 with obvious changes. Alternatively, we can reduce Theorem 7.54 to Theorem 7.43 by a trick. First we may assume $\langle \mathbf{f} \rangle = 0$ (otherwise we replace $\mathbf{f}$ with $\mathbf{f} - \langle \mathbf{f}\rangle$). Then for any linear combination $h = \mathbf{a}^T\mathbf{f} = a_1 f_1 + \dots + a_r f_r$ we have $\sigma_h \neq 0$ (due to Lemma 7.55), and so by Theorem 7.43 the function

$$R_n \colon = \mathbf{a}^T \mathbf{S}_n/\sqrt{n} = (h + \dots + h \circ \mathcal{F}^{n-1})/\sqrt{n}$$

converges in distribution to a normal law with zero mean and variance σ_h^2. By the Lévy continuity theorem (see Appendix B) its characteristic function $\varphi_{R_n}(t)$ converges pointwise, as $n \to \infty$, to

$$\exp(-\sigma_h^2 t^2/2) = \exp(-\mathbf{a}^T \mathbf{V} \mathbf{a} t^2/2)$$

for any $t \in \mathbb{R}$. Observe that

$$\varphi_{R_n}(t) = \langle \exp(\mathbf{i} t \mathbf{a}^T \mathbf{S}_n/\sqrt{n})\rangle = \varphi_{\mathbf{S}_n/\sqrt{n}}(a_1 t, \dots, a_r t),$$

where

$$\varphi_{\mathbf{S}_n/\sqrt{n}}(t_1, \dots, t_r) = \langle \exp(\mathbf{i}\mathbf{t}^T \mathbf{S}_n/\sqrt{n})\rangle$$

denotes the characteristic function of the vector variable $\mathbf{S}_n/\sqrt{n}$ and $\mathbf{t} = (t_1, \dots, t_r)^T$ its vector argument. Thus

$$\lim_{n\to\infty} \varphi_{\mathbf{S}_n/\sqrt{n}}(t_1, \dots, t_r) = \exp(-\mathbf{t}^T\mathbf{V}\mathbf{t}/2),$$

which is exactly the characteristic function of the normal distribution with zero mean and covariance matrix $\mathbf{V}$, so that Theorem 7.54 again follows by the Lévy continuity theorem. □

7.9. Other limit theorems

The Central Limit Theorem (CLT) is just the first step in the investigation of statistical properties of the sequence $\mathcal{X}_n = f \circ F^n$ introduced by (7.1) and resulting from a chaotic map $F\colon M \to M$ and an appropriate observable $f\colon M \to \mathbb{R}$.

The next step, after the CLT, is the invariance principle. We use the notation of Section 7.2. For $N \geq 1$ and $x \in M$ consider the continuous function $W_N(s;x)$ of $s \in [0,1]$ defined by

$$W_N\Big(\frac{n}{N};x\Big) = \frac{S_n(x) - n\langle f\rangle}{\sigma_f\sqrt{N}}$$

at rational points $s = n/N$ and by linear interpolation in between (the graph of $W_N(s;x)$ is then a polygonal line). The invariant measure μ and the family $\{W_N(s;x), x \in M\}$ induce a probability measure on the space of continuous functions on $[0,1]$.

DEFINITION 7.56. We say that the sequence (7.1) satisfies the *Weak Invariance Principle* (WIP) if the above measure weakly converges, as $N \to \infty$, to the Wiener measure. We say that (7.1) satisfies the *Almost Sure Invariance Principle* (ASIP) if there is a standard Wiener process (a Brownian motion) $B(s;x)$ on M with respect to the measure μ so that for some $\lambda > 0$

$$|W_N(s;x) - B(s;x)| = \mathcal{O}(N^{-\lambda})$$

for μ-almost all $x \in M$.

The invariance principle asserts that the accumulation function S_n, after a proper rescaling of space and time, converges to the Wiener process (the classical Brownian motion). To prove the invariance principle we use a general theorem established by Philipp and Stout [**PhS75**], which we present here in a form adapted to our needs.

Let ξ_0 be a finite or countable partition of $\mathcal{M}$. Denote

$$\xi_n = \mathcal{F}^n(\xi_0), \qquad \xi_m^n = \xi_m \vee \cdots \vee \xi_n$$

for all $m < n$ (we allow $m = -\infty$ and $n = \infty$). Each partition ξ_m^n corresponds to a σ-algebra $\mathfrak{F}_m^n$ consisting of measurable ξ_m^n-sets (mod 0); see Appendix A. We assume that $\xi_{-\infty}^{\infty} = \varepsilon$ is the partition into individual points; i.e. $\mathfrak{F}_{-\infty}^{\infty}$ coincides with the σ-algebra of all measurable sets in $\mathcal{M}$.

Let $f\colon \mathcal{M} \to \mathbb{R}$ be a measurable function. For $m \geq 1$, denote by $\bar{f}_m = \mathbb{E}(f|\xi_{-m}^m)$ the conditional expectation of f with respect to the partition ξ_{-m}^m; i.e.

$$\bar{f}_m(x) = \frac{1}{\mu(B)}\int_B f\,d\mu, \qquad \forall x \in B \quad \forall B \in \xi_{-m}^m.$$

The following theorem is proved in [**PhS75**]; see Theorem 7.1 and the Remark on p. 81 in that book.

THEOREM 7.57 ([**PhS75**]). *Suppose that there exist constants* $0 < \delta < 2$ *and* $C > 0$ *such that*

$$\langle |f(x)|^{2+\delta} \rangle < \infty \tag{7.47}$$

and for all $m \geq 1$

$$\langle |f(x) - \bar{f}_m|^{2+\delta} \rangle \leq C m^{-(2+7/\delta)}. \tag{7.48}$$

Moreover, suppose that $\sigma_f^2 > 0$ *and*

$$\operatorname{Var} S_n = n\sigma_f^2 + \mathcal{O}\big(n^{1-\delta/30}\big) \tag{7.49}$$

as $n \to \infty$. *Finally, suppose that for all* $n \geq 1$

$$|\mu(A \cap B) - \mu(A)\mu(B)| \leq C n^{-168(1+2/\delta)} \tag{7.50}$$

for all $A \in \mathfrak{F}_{-\infty}^0$ *and* $B \in \mathfrak{F}_n^\infty$. *Then the sequence* $\mathcal{X}_n = f \circ \mathcal{F}^n$ *satisfies the ASIP and the WIP.*

The ASIP actually implies the WIP, as well as many other limit theorems. We mention only some implications of the ASIP and refer the reader to [**PhS75**, Section 1] for more.

COROLLARY 7.58 (Integral tests). *Let* $\phi(t)$ *be a positive nondecreasing real-valued function. Then*

$$\mu\Big(\frac{S_n - n\langle f \rangle}{\sqrt{n}} > \phi(n) \ \text{ infinitely often}\Big) = 0 \text{ or } 1$$

according to whether

$$\int_1^\infty \frac{\phi(t)}{t}\, e^{-\phi^2(t)/2}\, dt$$

converges or diverges. **Furthermore, put** $M_n = \max_{1 \leq i \leq n} |S_i - i\langle f \rangle|$. *Then*

$$\mu\big(M_n/\sqrt{n} < \phi^{-1}(n) \ \text{ infinitely often}\big) = 0 \text{ or } 1$$

according to whether

$$\int_1^\infty \frac{\phi^2(t)}{t}\, e^{-8\pi^{-2}\phi^2(t)}\, dt$$

converges or diverges.

COROLLARY 7.59 (Law of Iterated Logarithm). *For* μ*-almost every point* $x \in \mathcal{M}$

$$\limsup_{n \to \infty} \frac{S_n - n\langle f \rangle}{\sqrt{2n\sigma_f^2 \log\log n}} = 1.$$

It remains to prove the following:

LEMMA 7.60. *The collision map* $\mathcal{F} \colon \mathcal{M} \to \mathcal{M}$ *for dispersing billiards of category A and dynamically Hölder continuous observables* $f \in \mathcal{H}$ *such that* $\sigma_f^2 > 0$ *satisfy all the conditions of Theorem* 7.57.

PROOF. Observe that (7.47) holds for any $\delta \in (0, 2)$ because $\|f\|_\infty < \infty$, and (7.49) follows from (7.37). To prove (7.48) and (7.50) we need to construct the partition ξ_0.

Let ξ_0 be the partition of $\mathcal{M}$ into the connected components of the set $\mathcal{M} \setminus \mathcal{S}^{\mathbb{H}}$. That is, the elements of ξ_0 are domains on which the map $\mathcal{F}_{\mathbb{H}} \colon \mathcal{M}_{\mathbb{H}} \to \mathcal{M}_{\mathbb{H}}$ is smooth. Recall that every element of ξ_0 is a curvilinear polygon bounded by stable curves, and the interior angles of these polygons do not exceed π (see Exercises 4.49 and 5.16).

Then $\xi_1 = \mathcal{F}(\xi_0)$ is the partition of $\mathcal{M}$ into the connected components of the set $\mathcal{M} \setminus \mathcal{S}_{-1}^{\mathbb{H}}$, so its elements are similar curvilinear polygons bounded by unstable curves. It is easy to see that for any $A \in \xi_0$ and $B \in \xi_1$ the intersection $A \cap B$ is a curvilinear polygon; i.e. the elements of $\xi_0^1 = \xi_0 \vee \xi_1$ are the connected components of the set $\mathcal{M} \setminus (\mathcal{S}^{\mathbb{H}} \cup \mathcal{S}_{-1}^{\mathbb{H}})$.

EXERCISE 7.61. Verify that for any $p, q > 0$ the elements of the partition ξ_{-p}^q are the connected components of the set $\mathcal{M} \setminus (\mathcal{S}_{p+1}^{\mathbb{H}} \cup \mathcal{S}_{-q}^{\mathbb{H}})$. Hint: Note that $\mathcal{F}^{-q}(\xi_{-p}^q) = \xi_{-p-q}^0$; then use induction on $p+q$.

Therefore, the elements of the partition ξ_{-m}^m are the connected components of the set $\mathcal{M} \setminus (\mathcal{S}_{m+1}^{\mathbb{H}} \cup \mathcal{S}_{-m}^{\mathbb{H}})$. Also, they are curvilinear polygons bounded by stable and unstable curves whose interior angles do not exceed π.

EXERCISE 7.62. Let $A \in \xi_{-m}^m$. Show that for any two points $x, y \in A$ there is a point $z \in A$ such that x and z belong to one unstable curve and y and z belong to one stable curve. Hint: Show that any two points in the same element of the partition ξ_{-m}^0 lie on one stable curve.

Now, in the notation of the above exercise, $\mathbf{s}_+(x,z) > m$ and $\mathbf{s}_-(y,z) > m$; thus $|f(x) - f(y)| \le 2K_f \vartheta_f^m$. It follows that

$$\langle |f(x) - \bar{f}_m|^{2+\delta} \rangle \le \text{const}\, \vartheta_f^{(2+\delta)m}.$$

This implies (7.48).

Lastly we prove the main hypothesis (7.50). Observe that ξ_1^∞ coincides with the partition $\xi_{\mathbb{H}}^u$ of $\mathcal{M}$ into (maximal) unstable H-manifolds introduced in Section 5.4. Similarly, $\xi_{-\infty}^0$ coincides with the partition $\xi_{\mathbb{H}}^s$ of $\mathcal{M}$ into (maximal) stable H-manifolds. Hence any set $A \in \mathfrak{F}_{-\infty}^0$ is a union of some stable H-manifolds. Similarly, for any set $B \in \mathfrak{F}_n^\infty$ its preimage $\mathcal{F}^{-n+1}(B) \in \mathfrak{F}_1^\infty$ will be a union of some unstable H-manifolds. Now Corollary 7.39 easily implies that

$$|\mu(A \cap B) - \mu(A)\mu(B)| \le C_0 \theta_0^{n-1}.$$

This proves (7.50). □

7.10. Statistics of collisions and diffusion

Here we derive several facts from the CLT that have clear physical interpretation. As before, our analysis is restricted to dispersing billiards of category A, but here this restriction is not just a matter of convenience: the results of this section do not extend to other categories of dispersing billiards.

Recall that the return function $\tau(x)$ is dynamically Hölder continuous; see Exercise 7.29. Therefore it satisfies the CLT; i.e. there exists $\sigma_f^2 \ge 0$ such that

$$\lim_{n\to\infty} \mu\left\{ \frac{t_n - n\langle\tau\rangle}{\sqrt{n}} \le z \right\} = \frac{1}{\sqrt{2\pi}\sigma_f} \int_{-\infty}^{z} e^{-\frac{s^2}{2\sigma_f^2}}\, ds \tag{7.51}$$

for all $-\infty < z < \infty$. Here

$$t_n = \tau(x) + \cdots + \tau(\mathcal{F}^{n-1}x)$$

is the time of the nth collision.

REMARK 7.63. Actually, $\sigma_\tau^2 > 0$. Indeed, if $\sigma_\tau^2 = 0$, then $\tau = K + g - g \circ \mathcal{F}$ due to Remark 7.5. It is a standard fact that if the ceiling function of a suspension flow Φ^t is cohomologous to a constant, then Φ^t is isomorphic to a suspension flow under a constant ceiling function; see [**Gur65**]. Such a flow cannot be mixing (cf. Exercise 6.29), but the billiard flow Φ^t is mixing; see Section 6.11. This proves $\sigma_\tau^2 > 0$.

Now consider the trajectory $\Phi^t(x)$ of a point $x \in \mathcal{M}$ during a time interval $(0, T)$ and denote by $n_x(T)$ the number of collisions it experiences. Recall that for a.e. $x \in \mathcal{M}$

$$\lim_{T\to\infty} T/n_x(T) = \bar{\tau} = \langle \tau \rangle \tag{7.52}$$

(see Exercises 2.37 and 2.38), and $\bar{\tau}$ is given by (2.32).

THEOREM 7.64. *We have*

$$\lim_{T\to\infty} \mu\left\{\frac{n_x(T) - T/\bar{\tau}}{\sqrt{T}} \le z\right\} = \frac{1}{\sqrt{2\pi}\sigma} \int_{-\infty}^{z} e^{-\frac{s^2}{2\sigma^2}}\, ds \tag{7.53}$$

for all $-\infty < z < \infty$. *Here* $\sigma^2 = \sigma_f^2/\bar{\tau}^3$.

PROOF. Observe that for every $Z > 0$ we have

$$\mu(n_x \le Z) = \mu(t_{[Z]+1} > T),$$

where $[Z]$ denotes the integral part of Z. Now (7.53) easily follows from (7.51) and (7.52); we leave the verification to the reader as an exercise. □

Next we recall the vector function $\mathbf{L}\colon \mathcal{M} \to \mathbb{R}^2$ introduced in Section 2.14. It can also be defined by

$$\mathbf{L}(x) = \tau(x)\mathbf{v}(x)$$

where $\mathbf{v}(x)$ is the unit velocity vector of the billiard particle as it moves from x to $\mathcal{F}(x)$. Observe that $\|\mathbf{L}(x)\| = \tau(x)$.

The sum

$$\mathbf{\Delta}_n = \mathbf{L} + \mathbf{L} \circ \mathcal{F} + \cdots + \mathbf{L} \circ \mathcal{F}^{n-1}$$

has the following physical meaning. As we explained in Section 4.1, every billiard trajectory on a billiard table $\mathcal{D} \subset \mathrm{Tor}^2$ of category A can be lifted from the torus to its universal cover $\mathbb{R}^2$; see an example in Fig. 4.2. Then we get a trajectory moving on an unbounded table with infinitely many periodically located scatterers. This unbounded billiard system is known as a planar periodic Lorentz gas; see the discussion in Section 4.1. It models the dynamics of a gas of electrons bouncing between heavy molecules in metals (the scatterers represent a crystalline molecular structure). Now $\mathbf{\Delta}_n$ is the vector displacement, on the plane $\mathbb{R}^2$, of the moving electron (billiard particle) in the course of n successive collisions with the scatterers. Recall that $\langle \mathbf{L} \rangle = 0$ (see Exercise 2.44 in Section 2.14), hence $\langle \mathbf{\Delta}_n \rangle = 0$.

Denote the components of the function $\mathbf{L}$ by $\mathbf{L} = (L_1, L_2)$. The argument used earlier in Exercise 7.29 now easily yields the dynamical Hölder continuity for L_1 and L_2. Then the vector CLT (Theorem 7.54) applies and gives the following:

THEOREM 7.65. *The vector* $\mathbf{\Delta}_n/\sqrt{n}$ *converges in distribution to a 2D normal law with zero mean and covariance matrix* $\mathbf{V} = \{v_{ij}\}$ *whose entries are*

$$v_{ij} = \sum_{n=-\infty}^{\infty} \mathbf{C}_{L_i L_j}(n).$$

That is, for any rectangle $B \subset \mathbb{R}^2$

$$\lim_{n\to\infty} \mu(\boldsymbol{\Delta}_n/\sqrt{n} \in B) = \frac{1}{2\pi\sqrt{\det \mathbf{V}}} \iint_B e^{-\frac{1}{2}\mathbf{x}^T\mathbf{V}^{-1}\mathbf{x}}\, dx\, dy$$

where $\mathbf{x} = (x, y)^T$.

REMARK 7.66. The covariance matrix $\mathbf{V}$ is nonsingular for any billiard table of category A, but the proof of this fact is quite involved. We refer the reader to [**BSC91**, Subsection 7.2] and [**BS96**, Section 5].

REMARK 7.67. On the contrary, for any bounded billiard table $\mathcal{D} \subset \mathbb{R}^2$ the covariance matrix $\mathbf{V}$ is singular. This follows from the obvious fact that the functions $\boldsymbol{\Delta}_n(x)$ are uniformly bounded.

Now suppose the initial point $x \in \mathcal{M}$ is selected randomly and lifted to the unit square $[0,1] \times [0,1]$ on the universal cover $\mathbb{R}^2$. Denote by $\mathbf{q}_0 = \pi_q(x)$ its initial position in that square. Then the position of its lifted trajectory at the nth collision, call it $\mathbf{q}_n$, is a 2D vector that satisfies $\mathbf{q}_n = \mathbf{q}_0 + \boldsymbol{\Delta}_n$. By the triangle inequality

$$\|\mathbf{q}_n - \boldsymbol{\Delta}_n\| \le \|\mathbf{q}_0\| \le \text{const} = \sqrt{2};$$

hence the asymptotic behavior of $\mathbf{q}_n$ is identical to that of $\boldsymbol{\Delta}_n$ described by Theorem 7.65. Precisely, $\mathbf{q}_n/\sqrt{n}$ converges to the 2D normal law with zero mean and covariance matrix $\mathbf{V}$.

Next we describe a typical experiment of statistical physics. Suppose N particles (electrons) $x_1, \dots, x_N$ are selected randomly on the boundaries of the scatterers in the unit square $[0,1] \times [0,1] \subset \mathbb{R}^2$; see Fig. 7.7. We allow each particle to move along its trajectory up to the nth collision and record their final locations $\mathbf{q}_1(n), \dots, \mathbf{q}_N(n)$. Then the distribution of the resulting vectors will be approximately normal with zero mean and covariance matrix $n\mathbf{V}$.

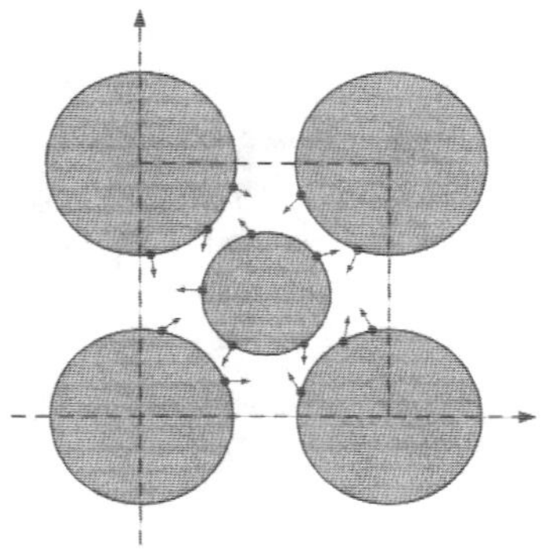

FIGURE 7.7. Initial selection of random particles in an experiment.

This is a classical diffusion process, in which moving particles jitter fast between the immovable obstacles but overall slowly diffuse (spread out) in all directions randomly. The covariance matrix $\mathbf{V}$ reflects the asymptotic behavior of the diffusion process and is called the *diffusion matrix*; it characterizes the shape of the distribution of the particles. The eigenvalues of $\mathbf{V}$ specify the speed of the diffusion, and their ratio determines the eccentricity.

A critical reader should notice that our hypothetical experiment is not quite realistic: the moving particles are not synchronized, as each may come to its nth collision at a different time. It is more natural to take a snapshot of the whole process at a large time $T > 0$, i.e. examine the positions of the moving particles at time T (rather than at their nth collisions).

Motivated by this task, we denote the location of the moving billiard particle in the planar Lorentz gas at time T by $\mathbf{q}(T)$ and consider its asymptotic distribution as $T \to \infty$.

THEOREM 7.68. *The vector* $\mathbf{q}(T)/\sqrt{T}$ *converges in distribution to a 2D normal law with zero mean and covariance matrix* $\bar{\tau}^{-1}\mathbf{V}$.

PROOF. Our argument follows that of [**BS81**, Section 3]. Given $T > 0$, we fix $m = [T/\bar{\tau}]$. Again we denote by $n(T) = n_x(T)$ the number of collisions the trajectory of the point $x \in \mathcal{M}$ experiences during time T. Then $\mathbf{q}(T) = \mathbf{q}_{n(T)} + \mathcal{O}(1)$, since the horizon is bounded for billiards of category A. Therefore $T = m\bar{\tau} + \mathcal{O}(1)$; hence

$$\frac{\mathbf{q}(T)}{\sqrt{T}} = \left(\frac{\mathbf{q}_m}{\sqrt{m\bar{\tau}}} + \frac{\mathbf{q}_{n(T)} - \mathbf{q}_m}{\sqrt{m\bar{\tau}}} + \mathcal{O}\Big(\frac{1}{\sqrt{m}}\Big)\right) \times \left(1 + \mathcal{O}\Big(\frac{1}{\sqrt{m}}\Big)\right).$$

Clearly, the terms $\mathcal{O}(1/\sqrt{m})$ cannot affect the limit distribution, so we drop them. Due to Theorem 7.65, the first term $\mathbf{q}_m/\sqrt{m\bar{\tau}}$ converges in distribution to the 2D normal law with zero mean and covariance matrix $\bar{\tau}^{-1}\mathbf{V}$.

It remains to show that the middle term $(\mathbf{q}_{n(T)} - \mathbf{q}_m)/\sqrt{m\bar{\tau}}$ converges to zero in probability, and we then apply Lemma 7.47. First, due to Theorem 7.64, for any $\varepsilon > 0$ there is $A = A_\varepsilon > 0$ such that

$$\mu\big(|n_x(T) - m| \geq A\sqrt{m}\big) \leq \varepsilon.$$

Now the desired result would follow if both expressions

$$\max_{1\leq j\leq A\sqrt{m}} \frac{1}{\sqrt{m}} \left|\sum_{i=m}^{m+j} \mathbf{L}(\mathcal{F}^i x)\right| \quad \text{and} \quad \max_{1\leq j\leq A\sqrt{m}} \frac{1}{\sqrt{m}} \left|\sum_{i=m-j}^{m} \mathbf{L}(\mathcal{F}^i x)\right|$$

converged to zero in probability. Because $\mathcal{F}$ preserves the measure μ, it suffices to consider

$$\max_{1\leq j\leq A\sqrt{m}} \frac{1}{\sqrt{m}} \left|\sum_{i=0}^{j} \mathbf{L}(\mathcal{F}^i x)\right| \quad \text{and} \quad \max_{1\leq j\leq A\sqrt{m}} \frac{1}{\sqrt{m}} \left|\sum_{i=-j}^{0} \mathbf{L}(\mathcal{F}^i x)\right|.$$

EXERCISE 7.69. Use the Birkhoff Ergodic Theorem to show that the above two expressions converge to zero in probability.

This proves Theorem 7.68. □

We now return to our imaginative experiments. Suppose again N initial particles (electrons) $x_1, \dots, x_N$ are selected randomly in the unit square $[0,1] \times [0,1] \subset \mathbb{R}^2$. We allow each electron to move along its trajectory up to time T and record their final locations as $\mathbf{q}_1(T), \dots, \mathbf{q}_N(T)$. Then the distribution of the resulting vectors will be approximately normal with zero mean and covariance matrix $n\bar{\tau}^{-1}\mathbf{V}$. Thus one may call $\bar{\tau}^{-1}\mathbf{V}$ the *real time diffusion matrix.* Note that it is proportional to the *discrete time diffusion matrix* $\mathbf{V}$.

REMARK 7.70. Theorems 7.65 and 7.68 do not apply to dispersing billiards of type B. Without horizon, the return function $\tau(x)$ and the displacement function $\mathbf{L}(x)$ are unbounded and have finite first moments but infinite second moments (Exercise 2.41). In that case the billiard particle may jump far between successive collisions; one calls this *ballistic motion.* Thus, the particle tends to travel faster than it does in the classical diffusion process (this behavior now may be related to the natural phenomena known as 'superdiffusion' and 'superconductivity'). Heuristic arguments [**Bl92**] and numerical experiments [**ZE97**] long suggested that the right scaling for the displacement function $\mathbf{q}_n$ is $\sqrt{n \log n}$ rather than $\sqrt{n}$; i.e. $\mathbf{q}_n/\sqrt{n \log n}$ should converge to a limit distribution. A proof of this conjecture was recently announced in [**SV04**].

7.11. Solid rectangles and Cantor rectangles

Here we begin our rather long proof of Coupling Lemma 7.24. It is based on various constructions, and we start with the most basic one, that of rectangles.

In Section 6.5 we already used 'rectangular-shaped' domains bounded by two unstable H-manifolds and two stable H-manifolds (recall Exercise 6.11).

DEFINITION 7.71. Any closed region $\mathfrak{Q} \subset \mathcal{M}$ bounded by two unstable H-manifolds and two stable H-manifolds is called a *solid rectangle.* Its boundary consists of four smooth curves, which we naturally call the *u-sides* and *s-sides*[6] of $\mathfrak{Q}$; see Fig. 7.8.

A solid rectangle $\mathfrak{Q}$ is the closure of its interior. For any interior point $x \in \mathfrak{Q}$ the unstable H-manifold $W^u_{\mathbb{H}}(x)$ either crosses both s-sides of $\mathfrak{Q}$ (like in Fig. 7.8) or terminates somewhere in $\mathfrak{Q}$ (maybe on its s-side). In the first case we say that $W^u_{\mathbb{H}}(x)$ *fully crosses* the solid rectangle $\mathfrak{Q}$. Similar notions apply to stable H-manifolds.

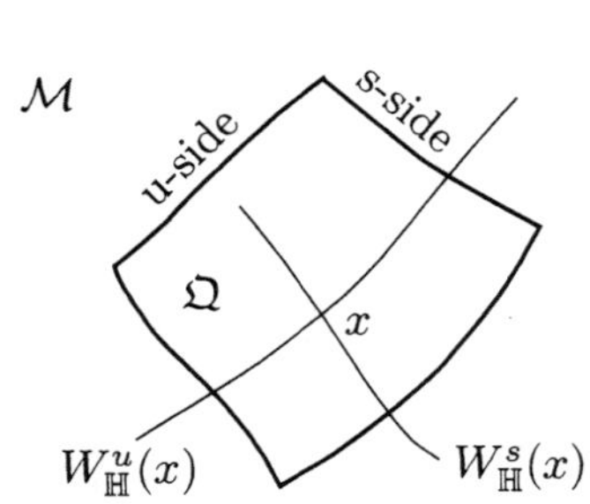

FIGURE 7.8. A solid rectangle $\mathfrak{Q}$. The H-manifold $W^u_{\mathbb{H}}(x)$ fully crosses $\mathfrak{Q}$, but $W^s_{\mathbb{H}}(x)$ terminates inside $\mathfrak{Q}$.

Given a solid rectangle $\mathfrak{Q} \subset \mathcal{M}$, denote by $\mathfrak{R} = \mathfrak{R}(\mathfrak{Q})$ the set of points $x \in \mathfrak{Q}$ such that *both* H-manifolds $W^u_{\mathbb{H}}(x)$ and $W^s_{\mathbb{H}}(x)$ fully cross $\mathfrak{Q}$. The set $\mathfrak{R}$ has a natural direct-product structure described next.

Recall that stable manifolds are decreasing curves in $\mathcal{M}$ and unstable manifolds are increasing curves; hence any stable manifold W^s can intersect any unstable

[6]For technical reasons, we assume that the four corner points of a solid rectangle are *not* the endpoints of the H-manifolds making its sides; i.e. those H-manifolds can be extended beyond the corner points.

manifold W^u in only one point; i.e. $W^s \cap W^u$ is either a singleton or an empty set. For any $x, y \in \mathcal{M}$ we put

$$[x,y]\colon = W^s_{\mathbb{H}}(x) \cap W^u_{\mathbb{H}}(y).$$

EXERCISE 7.72. Show that if $x, y \in \mathfrak{R}(\mathfrak{Q})$, then both points $[x, y]$ and $[y, x]$ exist and belong to $\mathfrak{R}(\mathfrak{Q})$.

Let $\mathfrak{Q}$ be a solid rectangle. Pick a point $z \in \mathfrak{R} = \mathfrak{R}(\mathfrak{Q})$ and put $C = W^u_{\mathbb{H}}(z) \cap \mathfrak{R}$ and $D = W^s_{\mathbb{H}}(z) \cap \mathfrak{R}$. Then

$$\mathfrak{R} = [C, D]\colon = \{[x, y]\colon x \in C,\ y \in D\}.$$

Also, for every $w \in \mathfrak{R}$ there is a unique representation $w = [x, y]$, where $x \in C$ and $y \in D$, so the sets C and D play the role of coordinate axes in $\mathfrak{R}$. Thus, $\mathfrak{R}$ has a direct-product structure, but it is quite different from $\mathfrak{Q}$; see the next exercise.

EXERCISE 7.73. Show that $\mathfrak{R}(\mathfrak{Q})$ is closed but nowhere dense; i.e. it is a Cantor-like set. Hint: For closedness, see Exercise 5.24. Then recall that arbitrarily short stable and unstable manifolds are dense in $\mathcal{M}$ (Section 4.13).

This motivates the following notion:

DEFINITION 7.74. A closed subset $\mathfrak{R} \subset \mathcal{M}$ is called a *Cantor rectangle* if for any $x, y \in \mathfrak{R}$ we have $\emptyset \neq [x, y] \in \mathfrak{R}$.

Any rectangle $\mathfrak{R}$ has a direct-product structure $\mathfrak{R} = [C, D]$, where $C = W^u_{\mathbb{H}}(z) \cap \mathfrak{R}$ and $D = W^s_{\mathbb{H}}(z) \cap \mathfrak{R}$ for some $z \in \mathfrak{R}$, as before. Rectangles are nowhere dense.

REMARK 7.75. The term ‘rectangle’ has an interesting origin. When Markov partitions were first constructed for linear hyperbolic toral automorphisms [**AW67**], their elements were true rectangles or parallelograms. So elements of Markov partitions were referred to by these names. In later constructions of Markov partitions (for Axiom A maps [**Bow70**] and billiards [**BS80, BSC90**]), their elements were Cantor-like sets with a direct-product structure, just like our Cantor rectangles, but they were still called ‘rectangles’ or ‘parallelograms’ by tradition.

Most of the time we will deal with Cantor rectangles $\mathfrak{R} \subset \mathcal{M}$ and only occasionally with solid rectangles $\mathfrak{Q} \subset \mathcal{M}$. For brevity, we will say *rectangle* meaning a Cantor rectangle.

Let $\mathfrak{R}$ be a rectangle and $z \in \mathfrak{R}$. The set $C = W^u_{\mathbb{H}}(z) \cap \mathfrak{R}$ is closed and lies on an increasing curve; thus it has two extreme points, x_1 and x_2 (see Fig. 7.9). Similarly, let y_1 and y_2 denote the extreme points of the set $D = W^s_{\mathbb{H}}(z) \cap \mathfrak{R}$.

EXERCISE 7.76. Verify that the two stable H-manifolds $W^s_{\mathbb{H}}(x_i)$, $i = 1, 2$, and two unstable H-manifolds $W^u_{\mathbb{H}}(y_i)$, $i = 1, 2$, enclose a solid rectangle $\mathfrak{Q}$ containing $\mathfrak{R}$. It is a minimal such rectangle; i.e. if $\mathfrak{Q}' \supset \mathfrak{R}$ is another solid rectangle, then $\mathfrak{Q} \subset \mathfrak{Q}'$.

We denote the above solid rectangle by $\mathfrak{Q}(\mathfrak{R})$ and call it the *hull* of the rectangle $\mathfrak{R}$.

DEFINITION 7.77. Let $\mathfrak{R}$ be a rectangle. A rectangle $\mathfrak{R}_1 \subset \mathfrak{R}$ is called a *u-subrectangle* if $W^u_{\mathbb{H}}(x) \cap \mathfrak{R} = W^u_{\mathbb{H}}(x) \cap \mathfrak{R}_1$ for any point $x \in \mathfrak{R}_1$. Similarly, $\mathfrak{R}_2 \subset \mathfrak{R}$ is called an *s-subrectangle* if $W^s_{\mathbb{H}}(x) \cap \mathfrak{R} = W^s_{\mathbb{H}}(x) \cap \mathfrak{R}_2$ for any point $x \in \mathfrak{R}_2$.

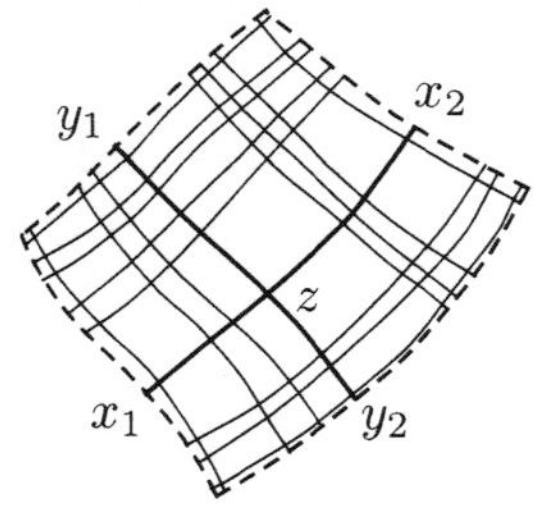

FIGURE 7.9. The solid rectangle $\mathfrak{Q}(\mathfrak{R}) \supset \mathfrak{R}$ is bounded by four dashed lines.

EXERCISE 7.78. Prove that the image $\mathcal{F}^n(\mathfrak{R})$ of a rectangle $\mathfrak{R}$ for any $n \in \mathbb{Z}$ is a finite or countable union of rectangles $\{\mathfrak{R}_i\}$. For $n > 0$, their preimages $\mathcal{F}^{-n}(\mathfrak{R}_i)$ are s-subrectangles in $\mathfrak{R}$. For $n < 0$, they are u-subrectangles in $\mathfrak{R}$.

EXERCISE 7.79. Verify that $\mu(\mathfrak{R}) > 0$ if and only if for any (and hence, for every) point $z \in \mathfrak{R}$ we have[7] $|W^u_{\mathbb{H}}(z) \cap \mathfrak{R}| > 0$ and $|W^s_{\mathbb{H}}(z) \cap \mathfrak{R}| > 0$. Hint: Use the absolute continuity (Sections 5.8 and 6.3).

In the following sections, we will deal only with rectangles of positive measure. We call

$$\rho^u(\mathfrak{R}) = \inf_{x \in \mathfrak{R}} \frac{|W^u_{\mathbb{H}}(x) \cap \mathfrak{R}|}{|W^u_{\mathbb{H}}(x) \cap \mathfrak{Q}(\mathfrak{R})|} \tag{7.54}$$

the (minimal) *u-density* of $\mathfrak{R}$. Similarly the (minimal) *s-density* $\rho^s(\mathfrak{R})$ is defined, and we call

$$\rho(\mathfrak{R}) = \min\{\rho^u(\mathfrak{R}), \rho^s(\mathfrak{R})\}$$

the (minimal) *density* of the rectangle $\mathfrak{R}$.

EXERCISE 7.80. Show that $\mu(\mathfrak{R}) > 0$ if and only if $\rho(\mathfrak{R}) > 0$. Hint: Note that the fraction in (7.54) is a continuous function of x defined on a closed set $\mathfrak{R}$; hence it attains its minimum. Then use absolute continuity.

Obviously, $0 \le \rho^{u,s}(\mathfrak{R}) < 1$. If $\rho(\mathfrak{R})$ is close to one, then the rectangle $\mathfrak{R}$ is "very dense"; i.e. it occupies nearly the entire available area of its hull $\mathfrak{Q}(\mathfrak{R})$.

Sinai's Fundamental Theorem 5.70 ensures that plenty of very dense rectangles exist everywhere in $\mathcal{M}$ (we note that this is true for billiards of categories A, B and C, but our arguments, as usual, are restricted to category A).

PROPOSITION 7.81. *For any point $x \in \mathcal{M}$ that has nonvanishing H-manifolds $W^u_{\mathbb{H}}(x)$ and $W^s_{\mathbb{H}}(x)$, there is a closed rectangle $\mathfrak{R} \ni x$ of positive measure. Moreover, for any $\delta > 0$, we can find a rectangle $\mathfrak{R} \ni x$ with density $\rho(\mathfrak{R}) > 1 - \delta$ and such that the point x divides the curves $W^u_{\mathbb{H}}(x) \cap \mathfrak{Q}(\mathfrak{R})$ and $W^s_{\mathbb{H}}(x) \cap \mathfrak{Q}(\mathfrak{R})$ in the ratio between $0.5 - \delta$ and $0.5 + \delta$; i.e. x is almost a geometric center of $\mathfrak{Q}(\mathfrak{R})$.*

PROOF. Since $W^u_{\mathbb{H}}(x)$ and $W^s_{\mathbb{H}}(x)$ exist, we have $x \in \mathcal{M} \setminus (\mathcal{S}^{\mathbb{H}}_{\infty} \cup \mathcal{S}^{\mathbb{H}}_{-\infty})$. Now the claim easily follows from Fundamental Theorem 5.70. □

[7] We denote by $|W|$ the length of a curve W. Thus for any subset $B \subset W$ the expression $|B|$ means the one-dimensional Lebesgue measure of B.

7.12. A 'magnet' rectangle

Here we continue the proof of Coupling Lemma 7.24 by constructing a special rectangle $\mathfrak{R}_*$ whose stable manifolds will be used to 'connect' (or 'couple') points of $\mathcal{F}^n(\cup_\alpha W_\alpha)$ with those of $\mathcal{F}^n(\cup_\beta W_\beta)$. The rectangle $\mathfrak{R}_*$, like a magnet, will 'attract' the H-components of the images of our proper standard families.

The rectangle $\mathfrak{R}_*$ will be constructed similarly to those in Proposition 7.81, but with some extra care to ensure better control over its stable manifolds.

First, let $W^u_{\mathbb{H}} \subset \mathcal{M}$ be an unstable H-manifold. As in Section 5.9, for every point $x \in W^u_{\mathbb{H}}$ and $n \geq 0$ we denote by $r_n(x)$ the distance from the point $\mathcal{F}^n(x)$ to the nearest endpoint of the H-component of $\mathcal{F}^n(W^u_{\mathbb{H}})$ that contains the point $\mathcal{F}^n(x)$. Also, recall that $r^s_{\mathbb{H}}(x)$ denotes the distance, measured along the (maximal) stable H-manifold $W^s_{\mathbb{H}}(x)$, from x to the nearest endpoint of $W^s_{\mathbb{H}}(x)$; see Section 5.12. In that section we also proved that

$$r^s_{\mathbb{H}}(x) \geq \min_{n\geq 0} \tilde{C}^{-1}\Lambda^n\, r_n(x), \tag{7.55}$$

where $\Lambda > 1$ was the minimal expansion factor (4.19) and $\tilde{C} = \tilde{C}(\mathcal{D}) > 0$ was a constant.

Now let $\tilde{W} \subset W^u_{\mathbb{H}}$ be a subcurve (to be chosen later). Given $\kappa > 0$, we now put

$$\tilde{W}_\kappa\colon = \tilde{W} \setminus \cup_{n\geq 0}\{x \in \tilde{W}\colon r_n(x) < \tilde{C}\Lambda^{-n}\kappa\}. \tag{7.56}$$

Note that the subset $\tilde{W}_\kappa \subset \tilde{W}$ is closed. Due to (7.55), we have $r^s_{\mathbb{H}}(x) \geq \kappa$ for every $x \in \tilde{W}_\kappa$. We denote by

$$\mathfrak{S}_\kappa(\tilde{W}) = \{W^s_{\mathbb{H}}(x)\colon x \in \tilde{W}_\kappa\}$$

the set of the corresponding stable H-manifolds (all of them extend the distance $\geq \kappa$ from $\tilde{W}$ on both sides).

Since $\tilde{W}_\kappa$ is a closed set, it has two extreme points, x_1 and x_2, on the curve $\tilde{W}$ which correspond to two extreme H-manifolds $W^s_{\mathbb{H}}(x_1)$ and $W^s_{\mathbb{H}}(x_2)$ in the family $\mathfrak{S}_\kappa(\tilde{W})$. We say that an unstable curve W *fully crosses* our family $\mathfrak{S}_\kappa(\tilde{W})$ if it crosses *all* the H-manifolds $W^s_{\mathbb{H}} \in \mathfrak{S}_\kappa(\tilde{W})$.

Denote by $\mathcal{G}^u_\kappa(\tilde{W})$ the family of all unstable H-manifolds $W^u_{\mathbb{H}} \subset \mathcal{M}$ that fully cross the family $\mathfrak{S}_\kappa(\tilde{W})$. Lastly let

$$\mathfrak{R}_\kappa(\tilde{W}) = \cup_{W^s_{\mathbb{H}} \in \mathfrak{S}_\kappa(\tilde{W})} \cup_{W^u_{\mathbb{H}} \in \mathcal{G}^u_\kappa(\tilde{W})} W^s_{\mathbb{H}} \cap W^u_{\mathbb{H}}$$

denote the rectangle made by our two families of stable and unstable H-manifolds. Note that it is a closed rectangle.

PROPOSITION 7.82. *For any $\delta > 0$ there are a subcurve $\tilde{W} \subset W^u_{\mathbb{H}}$ and a $\kappa > 0$ such that the rectangle $\mathfrak{R}_* = \mathfrak{R}_\kappa(\tilde{W})$ has density $\rho(\mathfrak{R}_*) > 1 - \delta$.*

PROOF. First let $\tilde{W} = W^s_{\mathbb{H}}$. It follows from the results of Section 5.12 that $|\tilde{W} \setminus \cup_{\kappa>0}\tilde{W}_\kappa| = 0$. Choose $\kappa > 0$ such that $|\tilde{W}_\kappa| > 0$ and pick a Lebesgue density point $z \in \tilde{W}_\kappa$ on the curve $\tilde{W}$. Now reduce $\tilde{W}$ so that it becomes a small neighborhood of z. The rest of the proof is the same as in Proposition 7.81. □

From now on we choose a small $\delta > 0$ and fix the corresponding $\tilde{W} \subset W^u_{\mathbb{H}}$ and $\kappa > 0$ and the special rectangle $\mathfrak{R}_* = \mathfrak{R}_\kappa(\tilde{W})$ constructed in the above proposition. We denote $\mathfrak{S} = \mathfrak{S}_\kappa(\tilde{W})$ for brevity, and slightly abusing notation we also denote by $\mathfrak{S}$ the union of all the H-manifolds $W^s_{\mathbb{H}} \in \mathfrak{S}$.

Note that $\tilde{W}_\kappa = \tilde{W} \cap \mathfrak{S}$. For any unstable curve W that fully crosses $\mathfrak{S}$ we set $W_\kappa\colon = W \cap \mathfrak{S}$.

Next, for any standard pair (W, ν) and any $n \geq 0$ denote by $W_{n,i}$ all the H-components of $\mathcal{F}^n(W)$ that fully cross $\mathfrak{S}$ and put

$$W_{n,*} = \cup_i \mathcal{F}^{-n}(W_{n,i} \cap \mathfrak{S}). \tag{7.57}$$

PROPOSITION 7.83. *There are constants $n_1 \geq 1$ and $d_1 > 0$ such that for any proper standard pair and any $n \geq n_1$ we have*

$$\nu(W_{n,*}) \geq d_1. \tag{7.58}$$

This easily extends to proper standard families, with obvious notation:

COROLLARY 7.84. *There are constants $n_0 \geq 1$ and $d_0 > 0$ such that for any proper standard family $\mathcal{G} = \{(W_\alpha, \nu_\alpha)\}$ and any $n \geq n_0$ we have*

$$\mu_{\mathcal{G}}\left(\cup_\alpha W_{\alpha,n,*}\right) \geq d_0. \tag{7.59}$$

The proof of Proposition 7.83 follows the argument of [**BSC91**, Theorem 3.13]. It is based on the mixing property of $\mathcal{F}$ established in Section 6.7. Actually, the mixing of $\mathcal{F}$ would imply (7.58) relatively easily if we had included *all* the H-components of $\mathcal{F}^n(W)$ in (7.57). But since we are interested only in those that fully cross $\mathfrak{S}$ (partial crossings do not count here), we need an additional construction to ensure full crossing.

DEFINITION 7.85. We say that an unstable curve W^u *properly crosses* a rectangle $\mathfrak{R}$ if

(a) W^u crosses both s-sides of $\mathfrak{Q}(\mathfrak{R})$.
(b) For every $x \in \mathfrak{R}$ we have $W^u \cap W^u_{\mathbb{H}}(x) \cap \mathfrak{Q}(\mathfrak{R}) = \emptyset$; i.e. W^u does not cross any unstable manifolds of $\mathfrak{R}$.
(c) For every $x \in \mathfrak{R}$ the point $W^u \cap W^s_{\mathbb{H}}(x)$ divides the curve $W^s_{\mathbb{H}}(x) \cap \mathfrak{Q}(\mathfrak{R})$ in the ratio between 0.1 and 0.9; i.e. W^u does not come too close to either u-side of $\mathfrak{Q}(\mathfrak{R})$.
(d) We have $\rho^s(\mathfrak{R}) > 0.99$.

EXERCISE 7.86. Show that any unstable curve W^u properly crosses a rectangle $\mathfrak{R}$. Hint: If the curve W^u does not coincide with any singularity curve $S \subset \mathcal{S}^{\mathbb{H}}_{-\infty}$, use Proposition 7.81. Otherwise a more careful argument is needed, which goes along the line of the proof of Theorem 6.14; we leave it to the reader.

LEMMA 7.87. *For any $q > 0$ there is a finite collection of rectangles $\mathfrak{R}_1, \dots, \mathfrak{R}_m$ such that any unstable curve W^u of length $\geq q$ properly crosses at least one of them.*

PROOF. The Hausdorff distance between unstable curves $W_1, W_2 \subset \mathcal{M}$ can be defined by

$$\operatorname{dist}_{\mathrm{H}}(W_1, W_2) = \max_{i=1,2} \sup_{x \in W_i} \operatorname{dist}(x, W_{3-i}).$$

If a sequence of unstable curves W_i converges in the Hausdorff metric to a curve $W \subset \mathcal{M}$, then W is also an unstable curve. Hence the set of closed unstable curves of length $\geq q$ with Hausdorff metric is compact. Also, it is easy to see that if an unstable curve W properly crosses a rectangle $\mathfrak{R}$, then all unstable curves sufficiently close to W in the Hausdorff metric also properly cross $\mathfrak{R}$. Thus each rectangle $\mathfrak{R}$ is properly crossed by an open (though maybe empty) set of unstable curves. Now the lemma follows by the standard compactness argument. $\square$

PROOF OF PROPOSITION 7.83. Let us choose and fix a finite collection of rectangles $\mathfrak{R}_1, \dots, \mathfrak{R}_m$ such that the unstable curve of every proper standard pair properly crosses one of them.

Let (W, ν) be a proper standard pair; it properly crosses a rectangle $\mathfrak{R}_k$ for some $1 \le k \le m$. For any $n \ge 1$ the set $\mathcal{F}^n(\mathfrak{R}_k)$ is a finite or countable union of rectangles; we denote them by $\mathfrak{R}_{k,n,i}$, $i \ge 1$. Recall that $\mathfrak{R}^-_{k,n,i} := \mathcal{F}^{-n}(\mathfrak{R}_{k,n,i})$ is an s-subrectangle in $\mathfrak{R}_k$ for every i; see Exercise 7.78.

EXERCISE 7.88. Show that the map $\mathcal{F}^n_{\mathbb{H}}$ is smooth on every solid rectangle $\mathfrak{Q}(\mathfrak{R}^-_{k,n,i})$. On the other hand, any two solid rectangles $\mathfrak{Q}(\mathfrak{R}^-_{k,n,i})$ and $\mathfrak{Q}(\mathfrak{R}^-_{k,n,j})$ with $i \ne j$ are separated by at least one singularity curve of the set $\mathcal{S}^{\mathbb{H}}_n$ running from one u-side of $\mathfrak{Q}(\mathfrak{R}_k)$ to the other (and thus crossing W); see Fig. 7.10. Hint: Use the continuation property of singularities (Section 4.9 and Exercise 5.16).

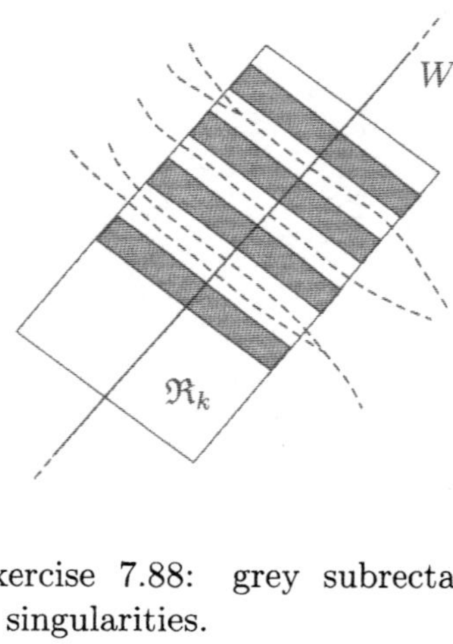

FIGURE 7.10. Exercise 7.88: grey subrectangles are $\mathfrak{R}^-_{k,n,j}$'s; dashed curves are singularities.

Now denote by $W_{n,j}$ all the H-components of $\mathcal{F}^n(W)$. Due to the previous exercise, each solid rectangle $\mathfrak{Q}(\mathfrak{R}_{k,n,i})$ is crossed by exactly one H-component $W_{n,j}$, and every H-component $W_{n,j}$ crosses at most one solid rectangle $\mathfrak{Q}(\mathfrak{R}_{k,n,i})$. We may renumber them so that each solid rectangle $\mathfrak{Q}(\mathfrak{R}_{k,n,i})$ is crossed by $W_{n,i}$.

We now return to our special rectangle $\mathfrak{R}_*$ (the magnet). It follows from the absolute continuity that μ-almost every point $y \in \mathfrak{R}_*$ is a Lebesgue density point of the set $W^s_{\mathbb{H}}(y) \cap \mathfrak{R}_*$ on the curve $W^s_{\mathbb{H}}(y)$. We need to work on a subset $\mathfrak{P}_* \subset \mathfrak{R}_*$ where the density condition is 'uniform'.

COROLLARY 7.89. *For any $\varepsilon > 0$ there is a $\delta_* > 0$ and a subset $\mathfrak{P}_* \subset \mathfrak{R}_*$ such that $\mu(\mathfrak{P}_*) > 0$ and for any stable H-manifold $W^s_{\mathbb{H}}$ such that $W^s_{\mathbb{H}} \cap \mathfrak{P}_* \ne \emptyset$ and $|W^s_{\mathbb{H}}| < \delta_*$ we have $|W^s_{\mathbb{H}} \cap \mathfrak{R}_*|/|W^s_{\mathbb{H}}| \ge 1 - \varepsilon$.*

We fix a small $\varepsilon > 0$ and the respective $\delta_* > 0$ and $\mathfrak{P}_*$.

The mixing of $\mathcal{F}$ implies that for all $n \ge n'_1$ we have

$$\mu(\mathfrak{P}_* \cap \mathcal{F}^n(\mathfrak{R}_k)) = \sum_i \mu(\mathfrak{P}_* \cap \mathfrak{R}_{k,n,i}) \ge d' \tag{7.60}$$

for some constants $n'_1 \ge 1$ and $d' > 0$, which we may assume are independent of k, i.e. of the standard pair (W, ν), due to the finiteness of the collection $\{\mathfrak{R}_k\}$.

LEMMA 7.90. *If* $\mathfrak{P}_* \cap \mathfrak{R}_{k,n,i} \neq \emptyset$, *then the corresponding H-component* $W_{n,i}$ *fully crosses the rectangle* $\mathfrak{R}_*$ *(here we assume that* n *is large enough,* $n \geq n_1''$ *for some constant* $n_1'' \geq 1$*).*

PROOF. Let $\mathfrak{P}_* \cap \mathfrak{R}_{k,n,i} \neq \emptyset$. Pick a point $y \in \mathfrak{P}_* \cap \mathfrak{R}_{k,n,i}$ and consider the stable H-manifold $\hat{W}^s_{\mathbb{H}} := W^s_{\mathbb{H}}(\mathcal{F}^{-n}y) \cap \mathfrak{Q}(\mathfrak{R}_k)$. If n is large enough, then its image $W^s_{\mathbb{H}} := \mathcal{F}^n(\hat{W}^s_{\mathbb{H}})$ is shorter than δ_*; hence $|W^s_{\mathbb{H}} \cap \mathfrak{R}_*|/|W^s_{\mathbb{H}}| \geq 1 - \varepsilon$ due to Corollary 7.89. If ε is small enough, then the properties (c) and (d) of proper crossing (Definition 7.85) guarantee that the curve $W^s_{\mathbb{H}}$ contains plenty of points belonging to the set $\mathfrak{R}_* \cap \mathfrak{R}_{k,n,i}$. We actually need only two points, one on each side of the curve $W_{n,i}$; we call them y_1 and y_2. Now the intersection $W^s_{\mathbb{H}} \cap W_{n,i}$ lies between two points

$$y_1, y_2 \in W^s_{\mathbb{H}} \cap \mathfrak{R}_* \cap \mathfrak{R}_{k,n,i}.$$

Let $\mathfrak{Q}^\diamond$ denote the solid rectangle bounded by the two s-sides of $\mathfrak{Q}(\mathfrak{R}_*)$ and the two unstable H-manifolds $W^u_{\mathbb{H}}(y_i)$, $i = 1, 2$ (see Fig. 7.11).

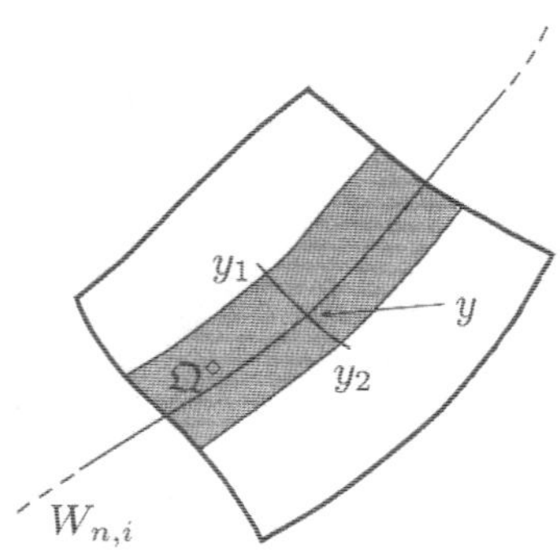

FIGURE 7.11. Proof of Lemma 7.90.

EXERCISE 7.91. Show that the map $\mathcal{F}_{\mathbb{H}}^{-n}$ is smooth on $\mathfrak{Q}^\diamond$. Hint: Use the continuation property of singularities (Section 4.9 and Exercise 5.16).

The smoothness of $\mathcal{F}_{\mathbb{H}}^{-n}$ within $\mathfrak{Q}^\diamond$ guarantees that $W_{n,i}$ cannot terminate there. It cannot cross the u-sides of $\mathfrak{Q}^\diamond$ either, due to property (b) of proper crossing (Definition 7.85), so it has to fully cross $\mathfrak{R}_*$. Lemma 7.90 is proved. □

To complete the proof of Proposition 7.83 we write

$$\begin{aligned}
\nu(W_{n,*}) &= \sum_i \nu\big(\mathcal{F}^{-n}(W_{n,i} \cap \mathfrak{R}_*)\big) \\
&\geq \sum_i \nu\big(\mathcal{F}^{-n}(W_{n,i} \cap \mathfrak{R}_* \cap \mathfrak{R}_{k,n,i})\big) \\
&\geq c_k \sum_i \mu\big(\mathcal{F}^{-n}(\mathfrak{R}_* \cap \mathfrak{R}_{k,n,i})\big) \\
&\geq c_{\min} d' > 0.
\end{aligned}$$

Here we used distortion bounds, absolute continuity, and (7.60). The constant $c_k > 0$ (coming from distortion bounds and absolute continuity) depends on k, but

$c_{\min}: = \min_{1\le k\le m} c_k > 0$ is independent of k, i.e. of the standard pair (W, ν). This holds for all $n \ge \max\{n_1', n_2''\}$. □

REMARK 7.92. In Proposition 7.83 we assumed that the standard pair (W, ν) is proper. If it is not, then we have to wait until its image $\mathcal{F}^m(W)$ becomes a proper standard family for some $m \ge 1$ and then apply Corollary 7.84, with n_0 replaced by $n_0 + m$. Recall that the waiting time here is $m = \varkappa\big|\ln|W|\big|$; see Remark 7.35.

7.13. Gaps, recovery, and stopping

We now outline the general strategy of our proof of Coupling Lemma 7.24. Assume, for simplicity, that $n_0 = 100$ and $d_0 = 0.01$ in Corollary 7.84. Given a proper standard family $\mathcal{G} = \{(W_\alpha, \nu_\alpha)\}$, after 100 iterations of $\mathcal{F}$ at least 1% of its image will be "on the magnet". The same applies to the other proper standard family $\mathcal{E} = \{(W_\beta, \nu_\beta)\}$, so at this time 1% of their images can be connected, or 'coupled', by the stable H-manifolds $W^s_{\mathbb{H}} \in \mathfrak{S}$. What do we do with the remaining 99% of their images?

If those remaining pieces made proper standard families, we would run another 100 iterations of $\mathcal{F}$ and again couple 1% of their images. In this way after $100n$ iterations of $\mathcal{F}$ we would couple all but 0.99^n of the images of the initial standard families $\mathcal{G}$ and $\mathcal{E}$. That would give us the exponential tail bound (7.15) and prove the Coupling Lemma.

There is a complication, however, in this plan: after each coupling the remaining 99% of the images will make a standard family that is *not proper*. Indeed, for any H-component W that fully crosses $\mathfrak{S}$, the 'coupled' points $x \in W \cap \mathfrak{R}_*$ will make a closed nowhere dense (Cantor-like) set on the curve W, so its complement will be a countable union of very short curves, among which there will be plenty of arbitrarily short ones. Each curve will need a certain time to grow (recover in size to become a proper standard family), and smaller pieces will grow longer before full recovery. To account for the necessary growth before recovery, we need additional constructions described next.

Let $\mathcal{P} = (W, \nu)$ be a standard pair such that W fully crosses the family $\mathfrak{S}$ (the magnet) constructed in the previous section. Then $W_\kappa = W \cap \mathfrak{S}$ is a closed nowhere dense set on the curve W, and its complement $W \setminus W_\kappa$ consists of infinitely many intervals; we call them *gaps in* W_κ. These gaps naturally correspond to the connected components of the set $\tilde{W} \setminus \tilde{W}_\kappa$ (gaps in $\tilde{W}_\kappa$) (see Fig. 7.12), which are created by the points $x \in \tilde{W}$ satisfying $r_n(x) < \tilde{C}\Lambda^{-n}\kappa$, in accordance with (7.56).

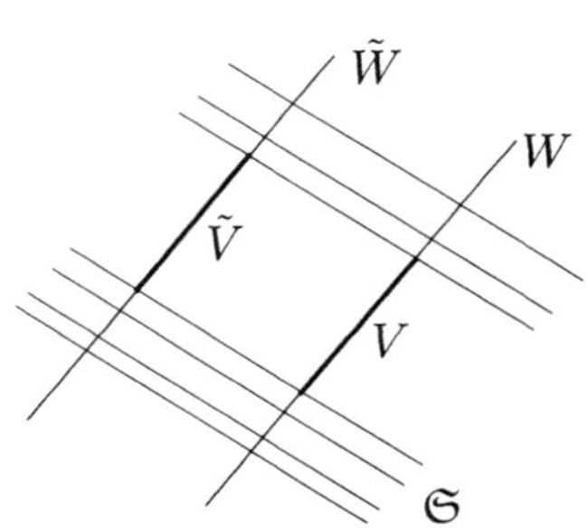

FIGURE 7.12. A gap V in $W \setminus W_\kappa$ corresponds to a gap $\tilde{V}$ in $\tilde{W} \setminus \tilde{W}_\kappa$.

Let $\tilde{V} \subset \tilde{W} \setminus \tilde{W}_\kappa$ be an interval. We call

$$n = \min\{i \geq 1 \colon r_i(x) < \tilde{C}\Lambda^{-i}\kappa \text{ for some } x \in \tilde{V}\}$$

the *rank* of $\tilde{V}$. Clearly every gap in $\tilde{W}_k$ has a rank.

EXERCISE 7.93. Show that if $\operatorname{rank} \tilde{V} = n$, then $\mathcal{F}^{n-1}(\tilde{V})$ is a curve of length $\geq C\Lambda^{-n}$ for some constant $C = C(\mathcal{D}) > 0$. Hint: Consider the H-component of $\mathcal{F}^{n-1}(\tilde{W})$ containing $\mathcal{F}^{n-1}(\tilde{V})$, show that it intersects the $(C\Lambda^{-n})$-neighborhood of the singularity set $\mathcal{S}^{\mathbb{H}}$ for some constant $C > 0$, then argue that it has to cross this neighborhood completely (for if it terminates somewhere inside that neighborhood, then it must have been torn by the singularities sometime earlier).

Now every gap $V \subset W \setminus W_\kappa$ corresponds to a gap $\tilde{V} \subset \tilde{W} \setminus \tilde{W}_\kappa$ that has some rank $n \geq 1$; in this case we also say that V itself has rank n. The endpoints of V are linked to those of $\tilde{V}$ by stable H-manifolds $W^s_{\mathbb{H}} \in \mathfrak{S}$ (see Fig. 7.12); hence their images $\mathcal{F}^{n-1}(V)$ and $\mathcal{F}^{n-1}(\tilde{V})$ must be exponentially close to each other (the distance between them is $\ll \Lambda^{-n}$). Therefore $\left|\mathcal{F}^{n-1}(V)\right| \geq \frac{1}{2}C\Lambda^{-n}$. It follows that the set $\mathcal{F}^{n(2+\varkappa \ln \Lambda)}(V)$ will be a proper standard family in accordance with Remarks 7.35 and 7.92.

Accordingly, we define *recovery time* function $\mathbf{r}_{\mathcal{P}}(x)$ on $W \setminus W_\kappa$ by setting $\mathbf{r}_{\mathcal{P}}(x) = [n(2 + \varkappa \ln \Lambda)]$, where n is the rank of the gap $V \subset W \setminus W_\kappa$ containing the point x (note that the function $\mathbf{r}_{\mathcal{P}}(x)$ is now constant on every gap).

EXERCISE 7.94. Show that for all $n \geq 1$

$$\nu\bigl(x \in W \setminus W_\kappa \colon \mathbf{r}_{\mathcal{P}}(x) > n\bigr) \leq \operatorname{const} \Lambda^{-n}\, \nu(W \setminus W_\kappa). \tag{7.61}$$

Hint: Use the second Growth Lemma (Theorem 5.53).

Next, let $\mathbf{s}_{\mathcal{P}}(x)$ be another function on $W \setminus W_\kappa$ (with values in $\mathbb{N}$) such that

$$\mathbf{s}_{\mathcal{P}}(x) \equiv \text{const} \qquad \text{on every gap } V \subset W \setminus W_\kappa \tag{7.62}$$

and

$$\mathbf{s}_{\mathcal{P}}(x) \geq \mathbf{r}_{\mathcal{P}}(x) + n_0. \tag{7.63}$$

Since both functions $\mathbf{r}_{\mathcal{P}}$ and $\mathbf{s}_{\mathcal{P}}$ are constant on each gap V, we will occasionally denote their values by $\mathbf{r}_{\mathcal{P}}(V)$ and $\mathbf{s}_{\mathcal{P}}(V)$, respectively.

Now applying Corollary 7.84 to every gap $V \subset W \setminus W_\kappa$ (more precisely, to its image $\mathcal{F}^{\mathbf{r}_{\mathcal{P}}(V)}(V)$; cf. Remarks 7.35 and 7.92) gives

$$\nu\bigl(V_{\mathbf{s}_{\mathcal{P}}(V),*}\bigr) \geq d_0\, \nu(V), \tag{7.64}$$

in the notation of (7.57). In other words at time $n = \mathbf{s}_{\mathcal{P}}(V)$ the d_0-fraction of the image $\mathcal{F}^n(V)$ will be "on the magnet". At this time the image $\mathcal{F}^n(V)$ may be 'stopped' and part of it (which intersects the magnet $\mathfrak{S}$) may be coupled with the corresponding image of another proper standard family (this will be done in the next section). We call $\mathbf{s}_{\mathcal{P}}$ the *stopping time* function.

Observe that the stopping time $\mathbf{s}_{\mathcal{P}}$ is not yet completely specified by (7.62) and (7.63); that is the purpose of the next proposition.

PROPOSITION 7.95. *We can define the stopping time function $\mathbf{s}_{\mathcal{P}}(x)$ on $W \setminus W_\kappa$ (see the remark below) so that for all $n \in \mathbb{N}$*

$$\frac{\nu\bigl(x \in W \setminus W_\kappa \colon \mathbf{s}_{\mathcal{P}}(x) = n\bigr)}{\nu(W \setminus W_\kappa)} = q_n, \tag{7.65}$$

where $\{q_n\}$ *is a sequence satisfying*

$$\sum q_n = 1 \quad \text{and} \quad q_n < \text{const}\,\theta^n \tag{7.66}$$

for some $\theta \in (\Lambda^{-1}, 1)$. *Furthermore, the sequence* $\{q_n\}$ *is independent of* $\mathcal{P}$*; i.e. it is the same for all standard pairs* $\mathcal{P} = (W, \nu)$ *such that* W *fully crosses the family* $\mathfrak{S}$.

We first describe the idea of the proof. Due to (7.61), it is easy to define $\mathbf{s}_{\mathcal{P}}$ so that for all $n > 0$

$$\frac{\nu\big(x \in W \setminus W_\kappa \colon \mathbf{s}_{\mathcal{P}}(x) = n\big)}{\nu(W \setminus W_\kappa)} \leq \text{const}\,\theta^n. \tag{7.67}$$

We still have considerable flexibility in defining the function $\mathbf{s}_{\mathcal{P}}$, and we want to adjust it so that it will satisfy (7.65) with a sequence $\{q_n\}$ being independent of $\mathcal{P}$. Our idea is illustrated by the following exercise:

EXERCISE 7.96. Let $g_\alpha \colon [0,1] \to \mathbb{N}$ be a (countable or uncountable) family of functions whose values are natural numbers such that for all $n \geq 1$

$$\mathbf{m}\big(x \in [0,1] \colon g_\alpha(x) \geq n\big) \leq e^{-n}$$

where $\mathbf{m}$ stands for the Lebesgue measure on $[0,1]$. Show that there is a sequence $\{q_n\}$ such that $q_n \leq e^{-n}$ and a family of functions $f_\alpha \colon [0,1] \to \mathbb{N}$ such that

$$f_\alpha(x) \geq g_\alpha(x) \qquad \forall \alpha \quad \forall x$$

and

$$\mathbf{m}\big(x \in [0,1] \colon f_\alpha(x) = n\big) = q_n \qquad \forall \alpha \quad \forall n \in \mathbb{N}$$

(note that the sequence $\{q_n\}$ must be independent of α). Hint: Consider the sequence $m_{\alpha,n} = \mathbf{m}\big(x \colon g_\alpha(x) = n\big)$.

The above exercise is relatively simple, compared to Proposition 7.95, as the functions f_α there need be defined on a continuous interval $[0,1]$. In our case the stopping time functions $\mathbf{s}_{\mathcal{P}}$ must be constant on every gap by (7.62), so they are effectively defined on a discrete (countable) set of gaps. Thus, their distribution, in the sense of (7.65) is harder to adjust.

PROOF OF PROPOSITION 7.95. We make the domain of our stopping time functions effectively continuous by splitting every gap V into an uncountable family of 'thinner' curves with the help of rectangles described in Section 7.5. Precisely, we replace each gap V with a rectangle $V \times [0,1]$. Then we can divide the latter into subrectangles $V \times I_j$, where $I_j \subset [0,1]$ are some subintervals, and define $\mathbf{s}_{\mathcal{P}}$ so that it takes a different value on each subrectangle I_j. The sizes of the subintervals $I_j \subset [0,1]$ must be selected to ensure (7.65), as well as (7.66), and this can be done just like in Exercise 7.96 because we can choose the endpoints of I_j's from the continuous set $[0,1]$.

Since our $\mathbf{s}_{\mathcal{P}}$ is now constant on every subrectangle $V \times I_j$, the latter can be then collapsed to a curve V_j, which geometrically coincides with the gap V, but carries a measure different from ν: it carries the measure ν multiplied by $|I_j|$. In the end we will have a countable family of curves $\{V_j\}$, and the function $\mathbf{s}_{\mathcal{P}}$ will be constant on each of them, as desired. $\square$

REMARK 7.97. Observe that in the proof of Proposition 7.95 we had to split some of the components of $W \subset W_\kappa$ into finitely or countably many (geometrically identical) curves, each having a different weight, and define the stopping time function $\mathbf{s}_{\mathcal{P}}$ separately for each curve. This small correction is needed to make the proposition precise.

7.14. Construction of coupling map

We now turn to the construction of the coupling map $\Theta\colon \cup_\alpha \hat{W}_\alpha \to \cup_\beta \hat{W}_\beta$ involved in Lemma 7.24. This will be done recursively, in an algorithm-like manner. The first two steps of our construction will be described in detail, and then it will be clear how it proceeds.

Recall that we are given two proper standard families $\mathcal{G} = (W_\alpha, \nu_\alpha)$, $\alpha \in \mathfrak{A}$, and $\mathcal{E} = (W_\beta, \nu_\beta)$, $\beta \in \mathfrak{B}$, with the corresponding measures $\mu_{\mathcal{G}}$ and $\mu_{\mathcal{E}}$. We denote by $\hat{\mathcal{G}} = (\hat{W}_\alpha, \hat{\nu}_\alpha)$ and $\hat{\mathcal{E}} = (\hat{W}_\beta, \hat{\nu}_\beta)$ the respective families of rectangles, and we have two probability measures $\hat{\mu}_{\mathcal{G}}$ and $\hat{\mu}_{\mathcal{E}}$ on the unions $\cup_\alpha \hat{W}_\alpha$ and $\cup_\beta \hat{W}_\beta$, respectively.

We define the first stopping time function $\mathbf{s}_0$ on the unions $\cup_\alpha \hat{W}_\alpha$ and $\cup_\beta \hat{W}_\beta$ to be constant $\mathbf{s}_0(x,t) \equiv n_0$. At time $\mathbf{s}_0 = n_0$ some of the H-components of the images $\mathcal{F}^{\mathbf{s}_0}(\cup_\alpha W_\alpha)$ and $\mathcal{F}^{\mathbf{s}_0}(\cup_\beta W_\beta)$ will fully cross the magnet $\mathfrak{S}$, and the total measure of the respective intersections with $\mathfrak{S}$ will be $\geq d_0$ due to Corollary 7.84.

For every H-component $W_{\alpha,\mathbf{s}_0,i}$ of $\mathcal{F}^{\mathbf{s}_0}(W_\alpha)$ that fully crosses $\mathfrak{S}$ we consider the corresponding rectangle $\hat{W}_{\alpha,\mathbf{s}_0,i} = W_{\alpha,\mathbf{s}_0,i} \times [0,1]$. We now split off a subrectangle $W_{\alpha,\mathbf{s}_0,i} \times [0, \boldsymbol{\tau}_{\alpha,i}]$ with some $0 < \boldsymbol{\tau}_{\alpha,i} \leq 0.5$ so that

$$\hat{\mu}_{\mathcal{G}}(\cup_\alpha \tilde{W}_{\alpha,1}) = d\colon = d_0/2, \tag{7.68}$$

where

$$\tilde{W}_{\alpha,1} = \{(x,t) \in \hat{W}_\alpha \colon \mathcal{F}^{\mathbf{s}_0}(x) \in W_{\alpha,\mathbf{s}_0,i} \cap \mathfrak{S} \quad \text{and} \quad t \in [0, \boldsymbol{\tau}_{\alpha,i}] \text{ for some } i\}.$$

This can be done easily due to Corollary 7.84: if $\mu_{\mathcal{G}}(\cup_\alpha W_{\alpha,\mathbf{s}_0,*}) = d_0$ in (7.59), we simply set $\boldsymbol{\tau}_{\alpha,i} = 0.5$ for all α and i. If the inequality in (7.59) is strict, we have "too much of a good thing". Then we lower some of $\boldsymbol{\tau}_{\alpha,i}$'s to make (7.68) exact.

We will put tildes over W's that denote subsets of $\cup_\alpha \hat{W}_\alpha$ on which we are currently defining the coupling map Θ. At the first step of our construction, Θ must take points $(x,t) \in \cup_\alpha \tilde{W}_{\alpha,1}$ to points $(y,s) \in \cup_\beta \hat{W}_\beta$ such that $\mathcal{F}^{\mathbf{s}_0}(x)$ and $\mathcal{F}^{\mathbf{s}_0}(y)$ lie on the same stable H-manifold $W^s_{\mathbb{H}} \in \mathfrak{S}$. It also must preserve measure (i.e., take $\hat{\mu}_{\mathcal{G}}$ to $\hat{\mu}_{\mathcal{E}}$). To correctly define Θ on the set $\cup_\alpha \tilde{W}_{\alpha,1}$ we will first describe its image, which we will denote by $\cup_\beta \tilde{W}_{\beta,1}$ (here the index 1 refers, of course, to the *first* step of our construction).

One may be tempted to define $\tilde{W}_{\beta,1}$'s in the same way as we defined $\tilde{W}_{\alpha,1}$'s above. In that case the sets $\cup_\beta \tilde{W}_{\beta,1}$ and $\cup_\alpha \tilde{W}_{\alpha,1}$ would have the same overall measure ($= d$), and their $\mathcal{F}^{\mathbf{s}_0}$-images would lie on the same stable H-manifolds $W^s_{\mathbb{H}} \in \mathfrak{S}$, but this may not suffice, as such a map may not preserve measure. Indeed, for some $W^s_{\mathbb{H}} \in \mathfrak{S}$ the intersections $W^s_{\mathbb{H}} \cap \mathcal{F}^{\mathbf{s}_0}(\cup_\alpha \tilde{W}_{\alpha,1})$ and $W^s_{\mathbb{H}} \cap \mathcal{F}^{\mathbf{s}_0}(\cup_\beta \tilde{W}_{\beta,1})$ may carry a different 'amount' of measures $\hat{\mu}_{\mathcal{G}}$ and $\hat{\mu}_{\mathcal{E}}$, respectively. The reason for this possible 'mismatch' is explained next.

EXERCISE 7.98. Suppose that the density of the measure $\mathcal{F}^{\mathbf{s}_0}(\mu_{\mathcal{G}})$ is constant (uniform) on every H-component of $\mathcal{F}^{s_0}(W_\alpha)$ that fully crosses the magnet $\mathfrak{S}$. Make a similar assumption on the density of the other measure $\mathcal{F}^{\mathbf{s}_0}(\mu_{\mathcal{E}})$. In addition, suppose that the Jacobian of the holonomy maps between the H-components of $\mathcal{F}^{\mathbf{s}_0}(W_\alpha)$ and those of $\mathcal{F}^{\mathbf{s}_0}(W_\beta)$ that fully cross the magnet $\mathcal{G}$ is constant (say, equal to one). Show that uniform densities on the H-components of $\mathcal{F}^{\mathbf{s}_0}(W_\alpha)$ will be taken to uniform densities on the H-components of $\mathcal{F}^{\mathbf{s}_0}(W_\beta)$. Conclude that there will be no mismatch; i.e. the map Θ would preserve measure.

The above exercise describes an 'ideal' (impossible) situation. In reality, the densities of our measures may vary along our H-components, and the Jacobian of the holonomy map may also vary and differ from one. To deal with these phenomena, we first assume, without loss of generality, that the diameter of the 'special rectangle' $\mathfrak{R}_*$ constructed in Section 7.12 is very small, so that the corresponding oscillations of the densities are small (say, the ratio of the densities at different points on the same H-component is between 0.99 and 1.01) and the Jacobian of the holonomy map takes values in a narrow interval around one, say, in $[0.99, 1.01]$.

We now define the set $\cup_\beta \tilde{W}_{\beta,1}$ as follows. For every H-component $W_{\beta,\mathbf{s}_0,j} \subset \mathcal{F}^{\mathbf{s}_0}(W_\beta)$ that fully crosses the magnet $\mathfrak{S}$ we will construct a function $\boldsymbol{\tau}_{\beta,j}(y)$ on $W_{\beta,\mathbf{s}_0,j} \cap \mathfrak{S}$, with values in the interval $[0, 0.6]$, and then put

$$\tilde{W}_{\beta,1} = \big\{(y,t) \in \hat{W}_\beta \colon \mathcal{F}^{\mathbf{s}_0}(y) \in W_{\beta,\mathbf{s}_0,j} \cap \mathfrak{S} \\ \text{and } \ t \in [0, \boldsymbol{\tau}_{\beta,j}(\mathcal{F}^{\mathbf{s}_0} y)] \ \text{ for some } j\big\}.$$

The functions $\boldsymbol{\tau}_{\beta,j}$ can be constructed so that for every stable H-manifold $W^s_{\mathbb{H}} \in \mathfrak{S}$ the intersections $W^s_{\mathbb{H}} \cap \mathcal{F}^{\mathbf{s}_0}(\cup_\alpha \tilde{W}_{\alpha,1})$ and $W^s_{\mathbb{H}} \cap \mathcal{F}^{\mathbf{s}_0}(\cup_\beta \tilde{W}_{\beta,1})$ carry the same 'amount' of measures $\hat{\mu}_{\mathcal{G}}$ and $\hat{\mu}_{\mathcal{E}}$, respectively. This is the reason why we need to give some room to the functions $\boldsymbol{\tau}_{\beta,j}$ so that their values can be adjusted accordingly, and this is why we allow them to go up to 0.6 (as compared to $\boldsymbol{\tau}_{\alpha,i} \leq 0.5$).

Also, since the densities of measures for standard pairs are dynamically Hölder continuous (7.11), and so is the Jacobian of the holonomy map (Proposition 5.48), it follows that the functions $\boldsymbol{\tau}_{\beta,j}(y)$ will be dynamically Hölder continuous as well; i.e. they will satisfy

$$|\ln \boldsymbol{\tau}_{\beta,j}(y) - \ln \boldsymbol{\tau}_{\beta,j}(z)| \leq C_0 \boldsymbol{\theta}^{s_+(y,z)} \tag{7.69}$$

for some constant $C_0 > 0$.

We now naturally define the coupling map $\Theta \colon \cup_\alpha \tilde{W}_{\alpha,1} \to \cup_\beta \tilde{W}_{\beta,1}$ that preserves measure and couples points whose $\mathcal{F}^{\mathbf{s}_0}$-images lie on the same stable manifold of the $\mathfrak{S}$ family. We note that

$$\hat{\mu}_{\mathcal{G}}\big(\cup_\alpha \tilde{W}_{\alpha,1}\big) = \hat{\mu}_{\mathcal{E}}\big(\cup_\beta \tilde{W}_{\beta,1}\big) = d, \tag{7.70}$$

where the constant $d = d_0/2$ was introduced in (7.68). Lastly we set the coupling time function $\Upsilon(x,t) = \mathbf{s}_0$ on $\cup_\alpha \tilde{W}_{\alpha,1}$. This concludes the first step of our construction of the coupling map Θ.

Before we move on to the second step, we need to 'take an inventory' of the remaining (uncoupled) parts of the families $\mathcal{G}$ and $\mathcal{E}$ and represent their images at time $\mathbf{s}_0$ by unions of some rectangles. To this end we first define a constant function $\boldsymbol{\tau}_{\alpha,i}(x)$ on every H-component $W_{\alpha,\mathbf{s}_0,i}$ of $\mathcal{F}^{\mathbf{s}_0}(W_\alpha)$ that fully crosses $\mathfrak{S}$ so that $\boldsymbol{\tau}_{\alpha,i}(x) \equiv \boldsymbol{\tau}_{\alpha,i}$, where $\boldsymbol{\tau}_{\alpha,i}$ is the constant chosen earlier (before equation (7.68)).

On the contrary, the function $\boldsymbol{\tau}_{\beta,j}(y)$ defined earlier on the Cantor-like subset $W_{\beta,\mathbf{s}_0,j} \cap \mathfrak{S} \subset W_{\beta,\mathbf{s}_0,j}$ is *not* constant. We now extend it to the entire curve $W_{\beta,\mathbf{s}_0,j}$

by linear interpolation (we make it linear on every gap $V \subset W_{\beta,\mathbf{s}_0,j} \setminus \mathfrak{S}$ and overall continuous). The graph of $\boldsymbol{\tau}_{\beta,j}$ divides the rectangle $W_{\beta,\mathbf{s}_0,j} \times [0,1]$ into two parts ('subrectangles', each with one irregular side; see Fig. 7.13).

EXERCISE 7.99. Verify that the function $\boldsymbol{\tau}_{\beta,j}(y)$, after its extension to $W_{\beta,\mathbf{s}_0,j}$, is still dynamically Hölder continuous in the sense of (7.69).

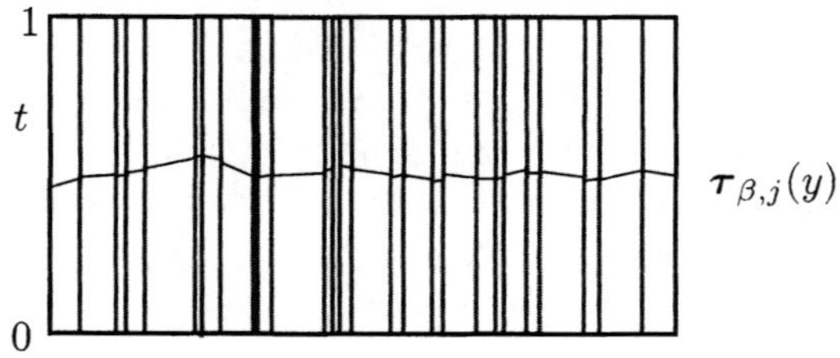

FIGURE 7.13. The partition of a rectangle over an H-component $W_{\beta,\mathbf{s}_0,j}$: the irregular line in the middle is the graph of the function $\boldsymbol{\tau}_{\beta,j}(y)$; it separates the 'upper subrectangle' (of the second type) from lower trapezoids (of the third type).

Now the 'uncoupled' sets $\cup_\alpha \mathcal{F}^{\mathbf{s}_0}\big(\hat{W}_\alpha \setminus \tilde{W}_{\alpha,1}\big)$ and $\cup_\beta \mathcal{F}^{\mathbf{s}_0}\big(\hat{W}_\beta \setminus \tilde{W}_{\beta,1}\big)$ consist of connected components of three types.

First, there are rectangles corresponding to the H-components of $\mathcal{F}^{\mathbf{s}_0}(W_\alpha)$ and $\mathcal{F}^{\mathbf{s}_0}(W_\beta)$ that do not fully cross the magnet $\mathfrak{S}$; we did not modify them in any way.

Second, there are the 'upper subrectangles'

$$\{(x,t)\colon x \in W_{\alpha,\mathbf{s}_0,i} \ \&\ t \in [\boldsymbol{\tau}_{\alpha,i}(x), 1]\} \tag{7.71}$$

and similar regions

$$\{(y,t)\colon y \in W_{\beta,\mathbf{s}_0,j} \ \&\ t \in [\boldsymbol{\tau}_{\beta,j}(x), 1]\} \tag{7.72}$$

(the latter are not genuine subrectangles; they have one 'jagged' side as shown in Fig. 7.13). All of the regions (7.71)–(7.72) have sufficiently long bases (longer than the size of the special rectangle $\mathfrak{R}_*$ in its unstable direction).

Third, there are the 'lower subrectangles'

$$\{(x,t)\colon x \in V \ \&\ t \in [0, \boldsymbol{\tau}_{\alpha,i}(x)]\}$$

constructed over gaps $V \subset W_{\alpha,\mathbf{s}_0,i} \setminus \mathfrak{S}$ and similar regions

$$\{(y,t)\colon y \in V' \ \&\ t \in [0, \boldsymbol{\tau}_{\beta,j}(y)]\}$$

constructed over gaps $V' \subset W_{\beta,\mathbf{s}_0,j} \setminus \mathfrak{S}$ (the latter are trapezoids as shown in Fig. 7.13).

Next we 'rectify' the irregular sides of the rectangles of the second and third type by a simple algorithm: it consists of stretching each vertical (i.e. parallel to the t axis) fiber inside every rectangle accordingly. More precisely, given a 'rectangle' $\hat{W}_1 = \{(x,t)\colon x \in W \ \&\ t \in [0, \boldsymbol{\tau}(x)]\}$, where W is an unstable curve and $\boldsymbol{\tau}(x)\colon W \to [0,1]$ is a continuous function equipped with a probability measure $d\hat{\nu}(x,t) = \rho(x)\,dx\,dt$, we transform the interval $[0, \boldsymbol{\tau}(x)]$ onto the unit interval $[0,1]$

linearly at every point $x \in W$ and thus obtain a full-height rectangle $\hat{W} = W \times [0, 1]$ with measure

$$d\hat{\nu}_1(x,t) = \rho_1(x)\,dx\,dt, \qquad \rho_1(x) = \boldsymbol{\tau}(x)\,\rho(x). \tag{7.73}$$

Recall that the 'ceiling function' $\boldsymbol{\tau}(x)$ is dynamically Hölder continuous for the rectangles of the second and third type (cf. (7.69) and Exercise 7.99), and so is every regular density $\rho(x)$ according to (7.11). Hence the new density $\rho_1(x)$ defined by (7.73) will also be dynamically Hölder continuous, precisely

$$|\ln \rho_1(x) - \ln \rho_1(y)| \le (C_0 + C_{\mathrm{r}})\,\boldsymbol{\theta}^{\mathbf{s}_+(x,y)}. \tag{7.74}$$

Of course the constant $C = C_0 + C_{\mathrm{r}}$ is larger than C_{r} in (7.11), so the density $\rho_1(x)$ is not necessarily regular (yet). But its images will smooth out and become regular soon enough.

EXERCISE 7.100. Show that there is a constant $m_0 \ge 1$ such that the curve W with the density $\rho_1(x)$ on it, as defined by (7.73), will be transformed by $\mathcal{F}^{m_0}$ into H-components on which the induced densities will be regular. Hint: Use the distortion bounds of Section 5.6 and recall that C_{r} in (7.11) is assumed to be large enough. In fact, by making C_{r} larger, if necessary, we can even ensure that $m_0 = 1$; i.e. the densities regularize right away.

Thus the remaining (uncoupled) sets

$$\cup_\alpha \mathcal{F}^{\mathbf{s}_0}(\hat{W}_\alpha \setminus \tilde{W}_{\alpha,1}) \quad \text{and} \quad \cup_\beta\, \mathcal{F}^{\mathbf{s}_0}(\hat{W}_\beta \setminus \tilde{W}_{\beta,1}) \tag{7.75}$$

are unions of rectangles of the full (unit) height. We denote the families of those rectangles by $\hat{\mathcal{G}}_1 = \{\hat{W}_{\alpha,1}\}$ and $\hat{\mathcal{E}}_1 = \{\hat{W}_{\beta,1}\}$, respectively. (Note that the $\mathcal{F}^{\mathbf{s}_0}$-image of an 'old rectangle' $\hat{W}_\alpha$ may contain countably many 'new' rectangles $\hat{W}_{\alpha,1}$. Thus we have to reindex our families, so the new indices α and β may not correspond to the old ones used for the original families $\mathcal{G}$ and $\mathcal{E}$, but this will cause no harm.)

So we get two new families $\hat{\mathcal{G}}_1$ and $\hat{\mathcal{E}}_1$, each carrying a measure induced by the $\mathcal{F}^{\mathbf{s}_0}$-image of the original measure ($\hat{\mu}_{\mathcal{G}}$ or $\hat{\mu}_{\mathcal{E}}$). Conditioning the induced measures on the new families $\hat{\mathcal{G}}_1$ and $\hat{\mathcal{E}}_1$ gives two probability measures on them, which we call $\hat{\mu}_{\hat{\mathcal{G}}_1}$ and $\hat{\mu}_{\hat{\mathcal{E}}_1}$, respectively. The densities of the new measures $\hat{\mu}_{\hat{\mathcal{G}}_1}$ and $\hat{\mu}_{\hat{\mathcal{E}}_1}$ may not be regular, but their images become regular in just $\le m_0$ iterations on $\mathcal{F}$ (or even in just one step; see Exercise 7.100), so we disregard this slight inconvenience.

The main complication, as we explained in Section 7.13, is that the new families $\hat{\mathcal{G}}_1$ and $\hat{\mathcal{E}}_1$ may not be proper, because they contain tons of arbitrarily small rectangles of the third type created in the gaps of the magnet $\mathfrak{S}$. Of course, if we condition the measures $\hat{\mu}_{\hat{\mathcal{G}}_1}$ and $\hat{\mu}_{\hat{\mathcal{E}}_1}$ onto the union of rectangles of the first and second type, then the so reduced families (albeit not necessarily proper either) will obviously recover and become proper standard families in just a few iterations of $\mathcal{F}$. We leave the verification of this simple fact to the reader. So we may assume that the rectangles of the first and second type make a proper standard family already.

However, on the rectangles of the third type, the recovery time may vary greatly, according to Section 7.13. We define the stopping time function $\mathbf{s}_1(x,t)$ on the rectangles of the third type as described in Proposition 7.95. The function $\mathbf{s}_1$ takes values in $\mathbb{N}$, is constant on every rectangle, and corresponds to the time when the image of the rectangle becomes a proper standard family plus an extra n_0 iterations of $\mathcal{F}$.

We need also to define the stopping time function $\mathbf{s}_1$ on the rectangles of the first and second types, so that it takes values in $\mathbb{N}$, is constant on every rectangle, and its overall distribution on *all* rectangles matches the one described in Proposition 7.95; i.e.

$$\hat{\mu}_{\hat{\mathcal{G}}_1}\big(\cup_\alpha \hat{W}_{\alpha,1}\colon \mathbf{s}_1 = n\big) = q_n \qquad \forall n \in \mathbb{N} \tag{7.76}$$

and

$$\hat{\mu}_{\hat{\mathcal{E}}_1}\big(\cup_\beta \hat{W}_{\beta,1}\colon \mathbf{s}_1 = n\big) = q_n \qquad \forall n \in \mathbb{N} \tag{7.77}$$

with the same sequence $\{q_n\}$ as in (7.65)–(7.66).

In order to define such a function $\mathbf{s}_1$ and ensure (7.76) and (7.77) we may need to split some rectangles $\hat{W}_{\alpha,1}$ and $\hat{W}_{\beta,1}$ of the first and second type into 'thinner' subrectangles, as we did in the proof of Proposition 7.95, and define $\mathbf{s}_1$ separately on every subrectangle. After going through the proof of Proposition 7.95, this should be an easy task, and we leave the details to the reader.

Now we are in a position very similar to the one we were in at the beginning of this section. For every rectangle $\hat{W}_{\alpha,1}$, the set $\mathcal{F}^{\mathbf{s}_1}(\hat{W}_{\alpha,1})$ with the induced measure will be a proper family, it will contain H-components fully crossing the magnet $\mathfrak{S}$, and their intersections with $\mathfrak{S}$ will have a relative measure $\geq d_0$ due to Corollary 7.84:

$$\frac{\hat{\mu}_{\hat{\mathcal{G}}_1}(\hat{W}_{\alpha,1,\mathbf{s}_1,*})}{\hat{\mu}_{\hat{\mathcal{G}}_1}(\hat{W}_{\alpha,1})} \geq d_0 \quad \text{and} \quad \frac{\hat{\mu}_{\hat{\mathcal{E}}_1}(\hat{W}_{\beta,1,\mathbf{s}_1,*})}{\hat{\mu}_{\hat{\mathcal{E}}_1}(\hat{W}_{\beta,1})} \geq d_0 \tag{7.78}$$

for every α and β, in the notation of (7.57)–(7.58). We note that $\mathbf{s}_1$ is constant on every rectangle $\hat{W}_{\alpha,1}$ and $\hat{W}_{\beta,1}$, so these notation make sense.

Of course, the value of $\mathbf{s}_1$ in (7.78) depends on α (or β). In what follows, we will group rectangles $\hat{W}_{\alpha,1}$ and $\hat{W}_{\beta,1}$ on which the function $\mathbf{s}_1$ takes the same value. In particular, we have

$$\hat{\mu}_{\hat{\mathcal{G}}_1}\big(\cup_\alpha \hat{W}_{\alpha,1,\mathbf{s}_1,*}\colon \mathbf{s}_1 = n\big) \geq d_0\, \hat{\mu}_{\hat{\mathcal{G}}_1}\big(\cup_\alpha \hat{W}_{\alpha,1}\colon \mathbf{s}_1 = n\big) = d_0 q_n \tag{7.79}$$

and, similarly,

$$\hat{\mu}_{\hat{\mathcal{E}}_1}\big(\cup_\beta \hat{W}_{\beta,1,\mathbf{s}_1,*}\colon \mathbf{s}_1 = n\big) \geq d_0\, \hat{\mu}_{\hat{\mathcal{E}}_1}\big(\cup_\beta \hat{W}_{\beta,1}\colon \mathbf{s}_1 = n\big) = d_0 q_n \tag{7.80}$$

due to (7.76)–(7.78).

Next, for every $n \geq 1$ we again consider all the rectangles $\{\hat{W}_{\alpha,1}\}$ and $\{\hat{W}_{\beta,1}\}$ on which the function $\mathbf{s}_1$ takes value n. Their images at time n will contain H-components that fully cross the magnet $\mathfrak{S}$, and the relative measure of their intersections with $\mathfrak{S}$ will be $\geq d_0$ due to (7.79)–(7.80). At that time we apply our coupling procedure described in the first step and then link ('couple') their subsets of relative measure $d = d_0/2$, according to (7.70), so that

$$\hat{\mu}_{\hat{\mathcal{G}}_1}\big(\hat{\mathcal{G}}_2^{(n)}\big) = \hat{\mu}_{\hat{\mathcal{E}}_1}\big(\hat{\mathcal{E}}_2^{(n)}\big) = d q_n, \tag{7.81}$$

where

$$\hat{\mathcal{G}}_2^{(n)}\colon = \big\{(x,t) \in \cup_\alpha \hat{W}_{\alpha,1,\mathbf{s}_1,*}\colon \mathbf{s}_1 = n \;\&\; \mathcal{F}^n(x,t) \text{ is coupled}\big\}$$

and

$$\hat{\mathcal{E}}_2^{(n)}\colon = \big\{(y,t) \in \cup_\beta \hat{W}_{\beta,1,\mathbf{s}_1,*}\colon \mathbf{s}_1 = n \;\&\; \mathcal{F}^n(y,t) \text{ is coupled}\big\}.$$

Doing this for all $n \geq 1$ constitutes the second step of our construction. We denote by

$$\hat{\mathcal{G}}_2 = \cup_n \mathcal{F}^{-\mathbf{s}_0}\big(\hat{\mathcal{G}}_2^{(n)}\big) \quad \text{and} \quad \hat{\mathcal{E}}_2 = \cup_n \mathcal{F}^{-\mathbf{s}_0}\big(\hat{\mathcal{E}}_2^{(n)}\big)$$

the preimages of all the subsets 'coupled' during the second step of the construction. Note that $\hat{\mathcal{G}}_2 \subset \hat{\mathcal{G}}$ and $\hat{\mathcal{E}}_2 \subset \hat{\mathcal{E}}$. The coupling map Θ is thus extended to $\Theta\colon \hat{\mathcal{G}}_2 \to \hat{\mathcal{E}}_2$. We also define the coupling time function Υ on the set $\hat{\mathcal{G}}_2$ by

$$\Upsilon(x,t) = \mathbf{s}_0(x,t) + \mathbf{s}_1\big(\mathcal{F}^{\mathbf{s}_0}(x,t)\big).$$

It should be clear now how the construction of the coupling map proceeds as the above steps are repeated recursively.

7.15. Exponential tail bound

Finally, we prove clause B of Coupling Lemma 7.24. This will also imply that the coupling map Θ constructed in Section 7.14 is defined almost everywhere on the standard family $\hat{\mathcal{G}}$.

First we summarize the results of the constructions made in the previous section. For each $k \geq 1$, at the kth step we define the stopping time function $\mathbf{s}_{k-1}$ on the sets $\cup_\alpha \hat{W}_{\alpha,k-1}$ and $\cup_\beta \hat{W}_{\beta,k-1}$ of yet uncoupled points. Then we 'couple' some points of their images $\cup_\alpha \mathcal{F}^{\mathbf{s}_{k-1}}\big(\hat{W}_{\alpha,k-1}\big)$ and $\cup_\beta \mathcal{F}^{\mathbf{s}_{k-1}}\big(\hat{W}_{\beta,k-1}\big)$. Then we denote by $\hat{\mathcal{G}}_k$ and $\hat{\mathcal{E}}_k$ the preimages of just 'coupled' subsets, on the original families $\mathcal{G}$ and $\mathcal{E}$. Lastly we extend the coupling time function Υ to the set $\hat{\mathcal{E}}_k$ by

$$\Upsilon(z) = \mathbf{s}_0(z) + \mathbf{s}_1(\mathcal{F}^{\mathbf{s}_0} z) + \cdots + \mathbf{s}_{k-1}(\mathcal{F}^{\mathbf{s}_0 + \cdots + \mathbf{s}_{k-2}} z),$$

where $z = (x,t) \in \hat{\mathcal{G}}_k$. Observe that the point $\mathcal{F}^{\Upsilon(z)}(z)$ and its partner $\mathcal{F}^{\Upsilon(z)}\big(\Theta(z)\big)$ lie on the same stable H-manifold, which proves claim A of Lemma 7.24 (assuming that Θ is indeed defined almost everywhere).

Next we rewrite the 'measure' relations (7.76)–(7.77) and (7.81) for the kth step. For brevity, we identify the set $\hat{\mathcal{G}}_k$ and $\hat{\mathcal{E}}_k$ with their images; i.e. we consider all our stopping time functions as defined on the original families $\hat{\mathcal{G}}$ and $\hat{\mathcal{E}}$. Then (7.76)–(7.77) generalize to the following 'conditional probability' formula

$$\hat{\mu}_{\mathcal{Y}}(\mathbf{s}_k = n / \mathbf{s}_{k-1} = n_{k-1}, \dots, \mathbf{s}_1 = n_1, \mathbf{s}_0 = n_0) = q_n \tag{7.82}$$

where $\mathcal{Y} = \mathcal{G}$ or $\mathcal{E}$, and (7.81) generalize to another 'conditional probability' formula

$$\hat{\mu}_{\mathcal{Y}}\big(\hat{\mathcal{Y}}_k / \mathbf{s}_{k-1} = n_{k-1}, \dots, \mathbf{s}_1 = n_1, \mathbf{s}_0 = n_0\big) = d, \tag{7.83}$$

where again $\mathcal{Y} = \mathcal{G}$ or $\mathcal{E}$ (note that we need to divide by q_n).

The following argument is standard in the studies of random walks (but we do not assume here that the reader is familiar with it). Let

$$\bar{p}_n \colon = \hat{\mu}_{\mathcal{G}}\big((x,t) \in \cup_\alpha \hat{W}_\alpha \colon \Upsilon(x,t) = n\big) \tag{7.84}$$

denote the fraction of points coupled exactly at time n (i.e., at the nth iteration of $\mathcal{F}$ rather than at the nth step of our construction). For example, $\bar{p}_i = 0$ for $i < n_0$ and $\bar{p}_{n_0} = d$. Then, due to (7.83)

$$p_n \colon = \bar{p}_n / d$$

will be the fraction of points *stopped* at time n, i.e.

$$p_n = \hat{\mu}_{\mathcal{G}}\big((x,t) \in \cup_\alpha \hat{W}_\alpha \colon \mathbf{s}_0 + \mathbf{s}_1 + \cdots + \mathbf{s}_k = n \text{ for some } k\big).$$

Due to (7.82) and (7.83) we have the following 'convolution law':

$$p_{n+n_0} = (1-d)\Big(q_n + (1-d)\sum_{i=1}^{n-1} q_{n-i}\, p_{n_0+i}\Big) \qquad \forall n \geq 1. \tag{7.85}$$

Its verification is rather straightforward, and we leave it to the reader as an enjoyable exercise.

Now consider two complex analytic functions:

$$P(z) = \sum_{n=1}^{\infty} p_{n_0+n} z^n \quad \text{and} \quad Q(z) = \sum_{n=1}^{\infty} q_n z^n.$$

Since $|p_n| \leq 1$ and $|q_n| \leq 1$, these functions are defined at least on the open unit disk $\{z\colon |z| < 1\}$ in the complex plane $\mathbb{C}$. Moreover, (7.66) easily implies that $|Q(z)| \leq 1$ for all $|z| \leq 1$ and that $Q(z)$ is analytic in a slightly larger complex disk $\{z\colon |z| < \theta^{-1}\}$; here $\theta^{-1} > 1$.

EXERCISE 7.101. Verify that (7.85) implies

$$P(z) = (1-d)\, Q(z) + (1-d)^2\, P(z)\, Q(z).$$

Due to the previous exercise, we have

$$P(z) = \frac{(1-d)\, Q(z)}{1-(1-d)^2\, Q(z)}. \tag{7.86}$$

EXERCISE 7.102. Prove that $P(z)$ is analytic in some complex disk of radius greater than one; i.e. in $\{z\colon |z| < 1+\delta\}$ for some $\delta > 0$. Note: You need to show that the denominator of (7.86) does not vanish in a vicinity of the unit disk.

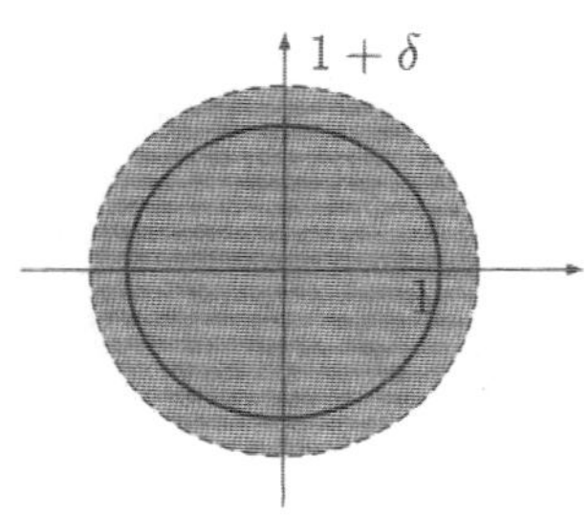

FIGURE 7.14. The domain of analyticity of $P(z)$.

Since $P(z)$ is analytic in the disk $\{z\colon |z| < 1+\delta\} \subset \mathbb{C}$, we obtain an exponential tail bound for p_n:

$$|p_n| \leq \text{const}(1+\delta')^{-n}$$

for any $\delta' < \delta$. A similar bound then follows for $\bar{p}_n = d\, p_n$ introduced in (7.84). Coupling Lemma 7.24 is now proved. $\square$

CHAPTER 8

Bunimovich billiards

8.1. Introduction

Recall that the boundary of generic billiard tables consists of curves of three types: dispersing (concave), focusing (convex), and flat; see Section 2.1. In the previous chapters we have seen that billiards with all-dispersing boundary are strongly chaotic:[1] hyperbolic, ergodic, mixing, and Bernoulli. They also have positive entropy and good statistical properties.

Billiards with all-flat boundary, i.e. polygons, are never hyperbolic: all their Lyapunov exponents are zero (Exercise 3.21); hence their entropy vanishes (as follows from our general results in Section 3.12). Interestingly, though, billiards in generic polygons are ergodic [**KMS86**], but this appears to be a peculiarity. Even being ergodic, they do not exhibit a genuinely chaotic behavior. It remains unknown whether billiards in polygons can ever be mixing, and in our opinion this is very unlikely. In any case, polygonal billiards make a very special class, and they are beyond the scope of this book.

What about billiards with all-focusing (convex) boundary? In Chapter 1 we learned that billiards in the two simplest convex domains, circles and ellipses, are completely regular. More precisely, almost every billiard trajectory in a circle and in an ellipse has a caustic, so it belongs to a one-dimensional invariant curve in the collision space $\mathcal{M}$.

Moreover, in 1973 Lazutkin [**La73**] proved that billiards in generic strictly convex domains with smooth boundary had caustics (precisely, the trajectories with caustics made a set of positive measure in the collision space); hence they could not be ergodic. It seemed therefore that chaos could not be produced by focusing (concave) boundaries – or at least, in order to produce chaos, focusing boundaries must be very special. And since Sinai's discovery of dispersing billiards in 1970, they remained, for several years, the only known class of chaotic billiards.

It came then as a big surprise when in the middle 1970's Sinai's student L. Bunimovich [**Bu74a, Bu74b, Bu79**] constructed chaotic billiards whose boundaries were all-focusing and consisted of circular arcs! (What can be simpler than a circular arc?) Bunimovich billiards are the subject of this chapter.

8.2. Defocusing mechanism

First we describe the idea behind Bunimovich's constructions. Recall that dispersing (divergent) wave fronts grow (expand), while focusing (convergent) wave fronts shrink (contract); cf. Section 3.7. In dispersing billiards, dispersing wave fronts remain dispersing in the future (Lemma 4.11); hence they monotonically

[1] We have proved this for categories A and B only, but the most important chaotic properties (ergodicity and Bernoulli) are valid for other categories, too.

grow in time (Lemma 4.14). This essentially guarantees the existence of a positive Lyapunov exponent and ultimately the hyperbolicity of the dynamics.

In billiards with focusing boundaries, dispersing wave fronts usually become focusing sooner or later (recall, for instance, Exercise 3.32); thus there is no class of wave fronts that would grow monotonically all the time. However, we are interested in the hyperbolicity of the collision map $\mathcal{F}$, so we care only about the size of the wave front *at collisions* (what happens between collisions is irrelevant). Thus we need to find wave fronts that grow 'from collision to collision'.

Using the notation of Sections 3.6–3.7, let $X \in \Omega$ be a point and $(0, d\xi, d\omega) \in \mathcal{T}_X^{\perp}\Omega$ a tangent vector. Such a vector specifies an infinitesimal wave front passing through the point X. Denote its image at time t by $(0, d\xi_t, d\omega_t) = D\Phi^t(0, d\xi, d\omega)$. Recall that $|d\xi_t|$ depends on t continuously, while $|d\omega_t|$ remains constant between collisions and changes discontinuously at every collision; see (3.25) and (3.27).

Suppose the trajectory of the point X experiences two consecutive collisions with the boundary $\partial\mathcal{D}$ at times t and $t + \tau$. Then (3.35) implies

$$\frac{|d\xi_{t+\tau}|}{|d\xi_t|} = |1 + \tau\mathcal{B}^+|,$$

where $\mathcal{B}^+ = d\omega_{t+0}/d\xi_t$ is the curvature of the wave front right after the first collision (at time t).

The wave front expands 'from collision to collision' if $|1+\tau\mathcal{B}^+| > 1$, which may occur in two different ways:

$$\tau\mathcal{B}^+ > 0 \qquad \text{or} \qquad \tau\mathcal{B}^+ < -2. \tag{8.1}$$

The first condition corresponds to dispersing wave fronts, while the second to focusing wave fronts. Thus, a focusing wave front expands 'from collision to collision' provided

$$\mathcal{B}^+ < -2/\tau. \tag{8.2}$$

Recall that a focusing wave front with curvature $\mathcal{B}_t < 0$ passes through a focusing point and defocuses at time $t^* = t - 1/\mathcal{B}_t$; see Section 3.8.

EXERCISE 8.1. Verify that (8.2) is equivalent to $t^* < t + \tau/2$; i.e. the front must defocus *before* it reaches the midpoint between collisions. Also use (3.31) to check that $\mathcal{B}_s > 0$ for $s \in (t + t^*, t + \tau)$; i.e. after passing the focusing point the front becomes and remains dispersing.

Geometrically, a focusing wave front contracts between the first collision at time t and the focusing point at time t^*; then it begins expansion and keeps expanding till the next collision at time $t + \tau$. Since the expansion lasts longer than the contraction, the front not only recovers its original size but gains some extra before it arrives at the next collision. This is called the *defocusing mechanism* illustrated by Fig. 8.1.

It remains to ensure that the defocusing mechanism will continue working at all collisions for a certain class of wave fronts so that the billiard map $\mathcal{F}$ will be hyperbolic. This will be done in the next section.

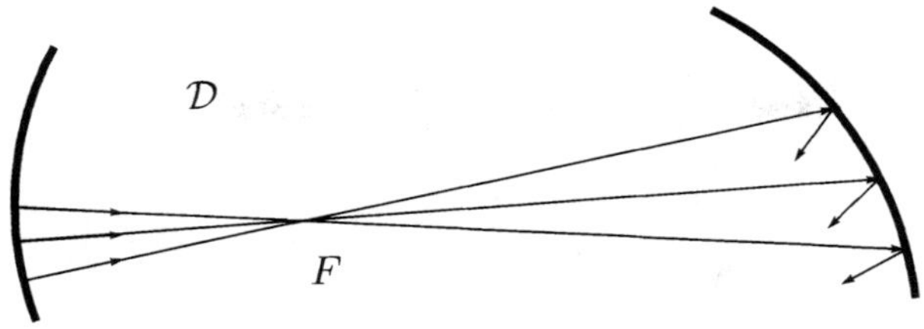

FIGURE 8.1. A focusing front defocuses at the point F and grows larger by the next collision.

8.3. Bunimovich tables

From now on we assume that every focusing boundary component of our billiard table $\mathcal{D}$ is a circular arc but not a full circle (we also allow $\partial\mathcal{D}$ to have dispersing and/or flat components).

Let $\Gamma_i \subset \partial\mathcal{D}$ be a focusing arc of radius $\mathbf{r} > 0$ and let a wave front make several consecutive collisions with Γ_i at times $t_n = t + n\tau$ for $n = 0, 1, \dots, m$. Observe that the collision angle φ and the intercollision time $\tau = 2\mathbf{r}\cos\varphi$ remain constant during this series of collisions; see Fig. 8.2.

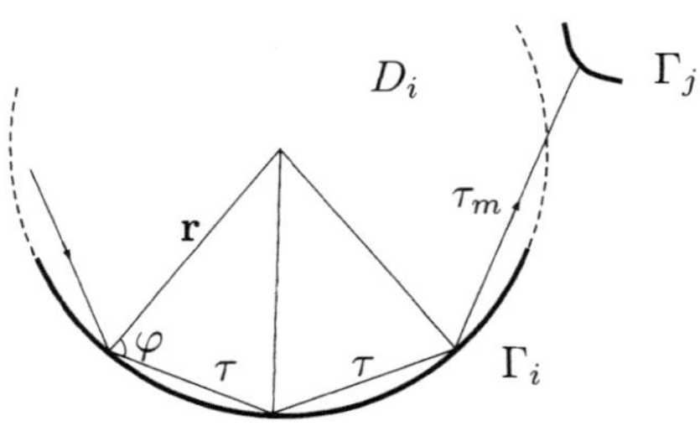

FIGURE 8.2. A series of reflections at a focusing circular arc Γ_i (here $m = 2$).

Let $\mathcal{B}_n^+$ denote the curvature of our wave front immediately after the nth collision. According to (3.33),

$$\mathcal{B}_{n+1}^+ = \mathcal{R} + \frac{1}{\tau + 1/\mathcal{B}_n^+} = -\frac{2}{\mathbf{r}\cos\varphi} + \frac{1}{2\mathbf{r}\cos\varphi + 1/\mathcal{B}_n^+}$$

(note that the curvature of the arc $\mathcal{K} = -1/\mathbf{r}$ is negative according to our convention in Section 2.1).

EXERCISE 8.2. Assume that the expanding condition (8.2) holds at the initial collision at time t_0, i.e. $\mathcal{B}_0^+ < -2/\tau$. Verify that $\mathcal{B}_1^+ < -2/\tau$. Hence, by induction, $\mathcal{B}_n^+ < -2/\tau$ for all $n = 0, 1, \dots, m$.

Thus, the expanding condition (8.2) automatically propagates in the course of successive collisions with a focusing circular arc.

Now suppose our wave front leaves the arc Γ_i at time t_m and moves on to another boundary component Γ_j. Denote by $t_{m+1} > t_m$ the time of the first collision with Γ_j. Then the expanding condition (8.2) at the last collision with Γ_i will read $\mathcal{B}_m^+ < -2/\tau_m$, where $\tau_m = t_{m+1} - t_m$ is the (new) intercollision time; see Fig. 8.2. Recall that $\mathcal{B}_m^+ < -2/\tau$ (Exercise 8.2). To ensure that $\mathcal{B}_m^+ < -2/\tau_m$, we need to assume that $\tau_m \geq \tau$; i.e. the collision point with Γ_j lies beyond the open disk D_i whose boundary contains the arc Γ_i (see Fig. 8.2).

If we want to construct a billiard table with hyperbolic behavior, we need the above property for every billiard trajectory coming off Γ_i. Thus we have to assume that the open disk D_i does not intersect any other components of the boundary $\partial\mathcal{D}$, i.e. $D_i \subset \mathcal{D}$. In other words, the arc Γ_j can be completed to a disk within the domain $\mathcal{D}$. This will be the main characteristic property of Bunimovich billiards, which we call the *B-condition.*

Next, we need to verify the expanding condition (8.1) for subsequent collisions (with Γ_j). If Γ_j is dispersing or flat, then, since the wave front falling upon it is dispersing, it will remain dispersing after the collision, so the first condition of (8.1) will hold.

Suppose Γ_j is another focusing boundary component; then it is a circular arc of radius $\mathbf{r}' > 0$. At the first collision with Γ_j the curvature of the wave front is transformed by

$$\mathcal{B}_{m+1}^+ = -\frac{2}{\mathbf{r}'\cos\varphi'} + \frac{1}{\tau_m + 1/\mathcal{B}_m^+}$$

where φ' is the corresponding collision angle. The expanding condition (8.2) now reads $\mathcal{B}_{m+1}^+ < -2/\tau_{m+1}$, where $\tau_{m+1} = t_{m+2} - t_{m+1}$ is the subsequent intercollision time.

EXERCISE 8.3. Recall that (8.2) holds for the last collision with Γ_i, i.e. $\mathcal{B}_m^+ < -2/\tau_m$. Show that if $\tau_m \geq 2\mathbf{r}'\cos\varphi'$ and $\tau_{m+1} \geq 2\mathbf{r}'\cos\varphi'$, then $\mathcal{B}_{m+1}^+ < -2/\tau_{m+1}$.

This exercise confirms that the expanding condition (8.2) will propagate from collisions with Γ_i to collisions with another focusing arc Γ_j if the latter also satisfies the B-condition.

DEFINITION 8.4. We say that $\mathcal{D}$ is a *Bunimovich billiard table* if every focusing boundary component $\Gamma_i \subset \partial\mathcal{D}$ is a circular arc (but not a full circle) which satisfies the B-condition.

REMARK 8.5. It is possible that two or more focusing arcs $\Gamma_i \subset \partial\mathcal{D}$ lie on the same circle, but we do not allow them to cover the entire circle.

REMARK 8.6. If a corner point of a Bunimovich table is made by two adjacent boundary components and one of them (or both) is a circular arc, then the interior angle is $\geq \pi$.

Examples of Bunimovich billiard tables are shown in Fig. 8.3.

8.4. Hyperbolicity

In the previous section, trying to ensure hyperbolicity for the collision map $\mathcal{F}$, we defined a class of tables with focusing boundary components for which certain wave fronts keep expanding from collision to collision. Here we prove that the resulting billiard map $\mathcal{F}$ is indeed hyperbolic.

It is convenient to replace (8.1) with a slightly stronger condition. Suppose an infinitesimal wave front collides with $\partial\mathcal{D}$ at a point $x = (r, \varphi)$. Denote its

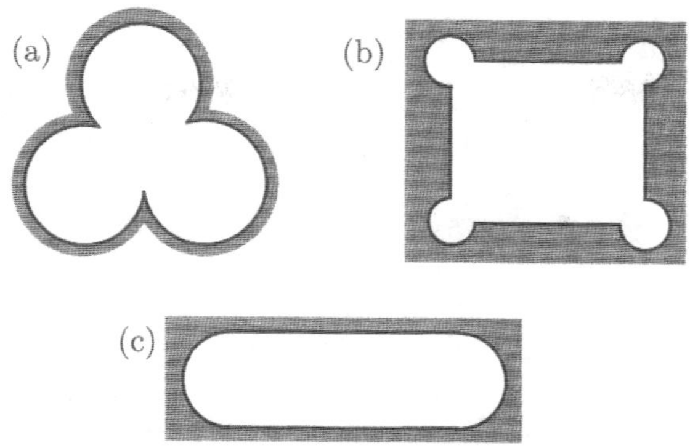

FIGURE 8.3. Bunimovich billiard tables: (a) 'flower', (b) 'pool table with pockets', and (c) 'stadium'.

precollisional and postcollisional curvature by $\mathcal{B}^-$ and $\mathcal{B}^+$, respectively. The front induces a tangent line $L \subset \mathcal{T}_x\mathcal{M}$ (see Section 3.10) whose slope $\mathcal{V} = d\varphi/dr$ is given by (3.39):

$$\mathcal{V} = \mathcal{B}^- \cos\varphi + \mathcal{K} = \mathcal{B}^+ \cos\varphi - \mathcal{K}, \tag{8.3}$$

where $\mathcal{K}$ denotes the curvature of the boundary $\partial\mathcal{D}$ at the point x.

DEFINITION 8.7. We say that the colliding wave front and the corresponding tangent vector are *unstable* if $\mathcal{B}^- > 0$ whenever $\mathcal{K} > 0$ and

$$\mathcal{B}^- \in \left(0, \frac{1}{\mathbf{r}\cos\varphi}\right) \tag{8.4}$$

whenever $\mathcal{K} = -1/\mathbf{r} < 0$.

Denote by $\mathcal{C}_x^u \subset \mathcal{T}_x\mathcal{M}$ the unstable cone at $x \in \mathcal{M}$, i.e. the closure of the set of all unstable tangent vectors:

$$\mathcal{C}_x^u = \{(dr, d\varphi) \in \mathcal{T}_x\mathcal{M} \colon \ \mathcal{K} \le d\varphi/dr \le \infty\}$$

for dispersing and flat boundary components and

$$\mathcal{C}_x^u = \{(dr, d\varphi) \in \mathcal{T}_x\mathcal{M} \colon \ -1/\mathbf{r} \le d\varphi/dr \le 0\}$$

for focusing arcs.

EXERCISE 8.8. Verify that for Bunimovich billiard tables unstable wave fronts satisfy the expanding condition (8.1). That is, for dispersing and flat boundary components $\mathcal{B}^- > 0$ implies $\mathcal{B}^+ > 0$, and for focusing arcs (8.4) implies (8.2).

THEOREM 8.9. *Let $\mathcal{D}$ be a Bunimovich billiard table. Then the family $\mathcal{C}_x^u$ of unstable cones is $D\mathcal{F}$-invariant. Equivalently, an unstable wave front remains unstable at all subsequent collisions in the future.*

PROOF. To prove that $D\mathcal{F}(\mathcal{C}_x^u) \subset \mathcal{C}_{\mathcal{F}x}^u$ we need to consider four cases, as each of the two points, $x = (r, \varphi)$ and $x_1 = (r_1, \varphi_1) = \mathcal{F}(x)$, may belong to either a focusing arc or a dispersing/flat boundary component. We consider only the least trivial case, where both points lie on focusing arcs. Denote their radii by $\mathbf{r}$ and

$\mathbf{r}_1$ (if they lie on the same arc, then $\mathbf{r}_1 = \mathbf{r}$). The B-condition implies that the intercollision time τ satisfies

$$\tau \geq 2\max\{\mathbf{r}\cos\varphi, \mathbf{r}_1\cos\varphi_1\}. \tag{8.5}$$

Then the curvature of our front before the collision at x_1 is

$$\mathcal{B}_1^- = \frac{1}{\tau + 1/\mathcal{B}^+} < \frac{1}{\tau - \tau/2} \leq \frac{1}{\mathbf{r}_1 \cos\varphi_1}$$

since $1/\mathcal{B}^+ \in (-\tau/2, 0)$ according to Exercise 8.8. We leave three other cases to the reader. □

Thus we have a family of $D\mathcal{F}$-invariant cones. However, this does not imply hyperbolicity yet. In fact, the above theorem holds also for circular billiards, which are not at all hyperbolic. So, we may have an invariant cone family but no hyperbolicity; see Section 3.13. Hyperbolicity requires that the cone field is strictly invariant, at least eventually; cf. Remark 3.53.

In the Bunimovich billiards the strict invariance of unstable cones, i.e. the property

$$D\mathcal{F}(\mathcal{C}_x^u) \subset \operatorname{int} \mathcal{C}_{\mathcal{F}(x)}^u \tag{8.6}$$

holds in two cases:

Case 1. The point x belongs to a dispersing boundary component.

Case 2. The point x belongs to a focusing component Γ_i, and its image $\mathcal{F}(x)$ belongs to either a dispersing or a flat component or to a focusing arc Γ_j that does not lie on the same circle as Γ_i (cf. Remark 8.5).

EXERCISE 8.10. Verify the strict invariance in the previous two cases. Hint: Use direct inspection going slightly beyond the proof of Theorem 8.9.

Accordingly we introduce the notion of *essential collisions*:

DEFINITION 8.11. The collision at a point $x \in \mathcal{M}$ is said to be *essential* if it occurs either on a dispersing boundary component or on a focusing arc, but in that case either the previous collision, $\mathcal{F}^{-1}(x)$, or the following collision, $\mathcal{F}(x)$, occurs on any other boundary component that is not an arc of the same circle.

It must be checked now that almost every trajectory on a Bunimovich billiard table does contain essential collisions. If it does, we have hyperbolicity.

EXERCISE 8.12. Show that if almost every trajectory contains an essential collision, then for almost every trajectory essential collisions occur with a positive frequency. That is, if m_n denotes the number of essential collisions during the first n collisions, then $\liminf_{n\to\infty} m_n/n > 0$. Hint: Use the 'frequencies of visits' function; see Appendix C.

Simple geometric analysis shows that there might be two types of ('nonhyperbolic') trajectories that fail to contain essential collisions. We describe them below:

Type A. These are trajectories which reflect only at focusing components lying on the same circle.

Type B. These are trajectories which collide only with flat boundary components.

EXERCISE 8.13. Show that every trajectory of type A is periodic and such trajectories make a set of zero measure. Hint: Use the results of Section 1.1.

Concerning trajectories of type B, it remains an open (and challenging) problem to prove that they make a null set. Of course, this problem is essentially related to polygonal billiard tables. A partial solution is obtained in [**GKT95**]:

THEOREM 8.14. *If $\partial\mathcal{D}$ is a convex simply connected polygon, then for every point $x \in \mathcal{M}$ either $\mathcal{F}^n(x) = x$ (it is periodic) or the closure of the set $\{\mathcal{F}^n(x)\}_{n=0}^{\infty}$ contains at least one vertex of the polygon.*

While we do not know, in general, if type B trajectories make a null set, this condition might be easy to check for particular tables. For example, this is the case if the billiard table has just one (or no) flat boundary components; see Fig. 8.3 (a). Also, this is the case if there are exactly two parallel flat boundary components (see Fig. 8.3 (c)) or if there are two mutually perpendicular pairs of parallel flat boundary components (see Fig. 8.3 (b)).

We summarize our discussion:

THEOREM 8.15. *Let $\mathcal{D}$ be a Bunimovich billiard table. Assume that the type B trajectories make a set of zero measure. Then the billiard map $\mathcal{F}$ is hyperbolic.*

REMARK 8.16. Unlike dispersing billiards, billiards with at least one focusing arc are never uniformly hyperbolic. During a long series of successive nonessential collisions, unstable tangent vectors grow slowly (more slowly than exponentially), thus preventing any uniform expansion.

In Bunimovich billiards, there are three types of long series of nonessential collisions (see illustration in Fig. 8.4):

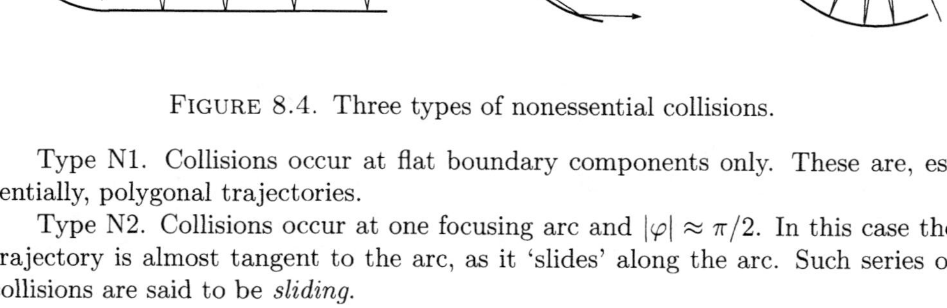

FIGURE 8.4. Three types of nonessential collisions.

Type N1. Collisions occur at flat boundary components only. These are, essentially, polygonal trajectories.

Type N2. Collisions occur at one focusing arc and $|\varphi| \approx \pi/2$. In this case the trajectory is almost tangent to the arc, as it 'slides' along the arc. Such series of collisions are said to be *sliding*.

Type N3. Collisions occur at one focusing arc (or several arcs lying on the same circle) and $|\varphi|$ is not near $\pi/2$. In this case the trajectory is close to a periodic orbit in a circle.

EXERCISE 8.17. Show that trajectories of type N3 can occur only on a focusing arc larger than half-circle or on several focusing arcs lying on the same circle.

8.5. Unstable wave fronts and continued fractions

Let $X \in \Omega$ be a hyperbolic point (see Definition 3.2) and $dX^u = (0, d\xi^u, d\omega^u) \in \mathcal{T}_X^\perp\Omega$ denote an unstable tangent vector at X. Then dX^u specifies an infinitesimal wave front (a 'germ' of the unstable manifold at X). We will show here that the curvature of that front, $\mathcal{B}^u(X) = d\omega^u/d\xi^u$, can be expressed by a continued fraction. The latter has the same structure as that in Section 4.7; we recall it below.

Denote by $x = \mathbf{P}^+(X)$ the first collision point on the trajectory of X (cf. (3.17)) and $x_n = (r_n, \varphi_n) = \mathcal{F}^n(x)$ for $n \in \mathbb{Z}$ all the other collision points. Let $\mathcal{K}_n = \mathcal{K}(r_n)$ denote the curvature of $\partial\mathcal{D}$ at the nth collision point, $\tau_n = \tau(x_n)$ the free path, and $\mathcal{R}_n = \mathcal{R}(x_n) = 2\mathcal{K}_n/\cos\varphi_n$ the collision parameter. We also denote by $\mathcal{B}_t$ the curvature of our wave front at time t and by $\mathcal{B}_n^\pm$ its curvature before ($-$) and after ($+$) the nth collision.

The continued fraction representing $\mathcal{B}^u(X)$ has the form

$$\frac{1}{|t^-(X)| + \cfrac{1}{\mathcal{R}_{-1} + \cfrac{1}{\tau_{-2} + \cfrac{1}{\mathcal{R}_{-2} + \cfrac{1}{\tau_{-3} + \cfrac{1}{\ddots}}}}}} \tag{8.7}$$

where $t^-(X)$ is the time of the most recent collision in the past; cf. (3.16). Note that we may have $\mathcal{B}^u(X) = \infty$, which corresponds to a focusing point on the unstable manifold.

The fraction (8.7) is identical to that appearing in (4.42), but its analysis will be more complicated, as now some $\mathcal{R}_{-n}$ may be negative.

PROPOSITION 8.18. *If $X \in \Omega$ is a hyperbolic point, then the continued fraction (8.7) converges and equals $\mathcal{B}^u(X)$.*

PROOF. We start as in the proof of Proposition 4.35. Due to (3.31), it is enough to prove it for $X \in \mathcal{M}$, in which case $x = X$ and $|t^-(X)| = \tau_{-1}$, and then (8.7) will represent the precollisional curvature $\mathcal{B}^{u-}(X)$.

The convergence of an infinite continued fraction means that the sequence of its convergents has a limit. The even-numbered convergents

$$\mathcal{B}^{\mathrm{e}}_{-n} = \cfrac{1}{\tau_{-1} + \cfrac{1}{\mathcal{R}_{-1} + \cfrac{1}{\ddots + \cfrac{1}{\mathcal{R}_{-n}}}}},$$

correspond to wave fronts whose preimages before the nth collision in the past are flat (i.e., $\mathcal{B}^-_{-n} = 0$); recall (3.34). The odd-numbered convergents

$$\mathcal{B}^{\mathrm{o}}_{-n} = \cfrac{1}{\tau_{-1} + \cfrac{1}{\mathcal{R}_{-1} + \cfrac{1}{\ddots + \cfrac{1}{\tau_{-n}}}}}.$$

correspond to wave fronts whose preimages before the nth collision in the past focus (i.e. $\mathcal{B}^{\pm}_{-n} = \infty$).

If the nth collision in the past occurs at a dispersing or flat boundary component ($\mathcal{K}_{-n} \geq 0$), then both $\mathcal{B}^{\mathrm{e}}_{-n}$ and $\mathcal{B}^{\mathrm{o}}_{-n}$ correspond to the images of unstable wave fronts constructed in the remote past, so they must approach $\mathcal{B}^u(X)$, as $n \to \infty$, due to (3.51).

If the nth collision occurs at a focusing arc, then the wave front with $\mathcal{B}^-_{-n} = \infty$ (which yields the odd-numbered convergent) is not unstable, as it fails to satisfy (8.4), so we consider its image.

Due to (3.31) and (3.33) we have

$$\mathcal{B}^-_{-n+1} = \mathbb{F}_n(\mathcal{B}^-_{-n}) := \cfrac{1}{\tau_{-n} + \cfrac{1}{\mathcal{R}_{-n} + \mathcal{B}^-_{-n}}}. \tag{8.8}$$

In this formula, we treat τ_{-n} and $\mathcal{R}_{-n}$ as 'fixed' parameters of the given collision, and $\mathcal{B}^-_{-n}$ as a 'variable' curvature of arbitrary wave front. The formula (8.8) describes the transformation of the curvature of the front.

Suppose the nth reflection occurs at a focusing arc of radius $\mathbf{r}_{-n}$; then $\mathcal{R}_{-n} = -\frac{2}{\mathbf{r}_{-n}\cos\varphi_{-n}}$ and $\tau_{-n} \geq 2\,\mathbf{r}_{-n}\cos\varphi_{-n}$ due to (8.5). The graph of the function $\mathbb{F}_n(x)$ is shown in Fig. 8.5.

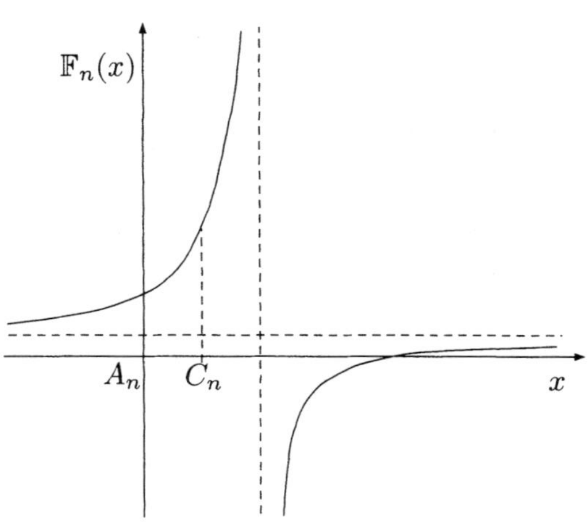

FIGURE 8.5. The function $y = \mathbb{F}_n(x)$ describing the transformation of the curvature of a wave front.

Now the even-numbered convergent $\mathcal{B}^{\mathrm{e}}_{-n}$ is generated by the wave front with $\mathcal{B}^{-}_{-n} = 0$ (so we mark the point $A_n = 0$ on the horizontal axis); the odd-numbered convergent $\mathcal{B}^{\mathrm{o}}_{-n}$ is generated by the wave front with $\mathcal{B}^{-}_{-n} = \infty$ (so we put $B_n = \infty$); and all the unstable wave fronts satisfy $0 < \mathcal{B}^{-}_{-n} < \frac{1}{\mathbf{r}_{-n}\cos\varphi_{-n}}$, due to (8.4), so we mark the point $C_n = \frac{1}{\mathbf{r}_{-n}\cos\varphi_{-n}}$. Note that our function is monotonically increasing on the interval (A_n, C_n) containing all unstable fronts.

The points A_n, B_n, and C_n are mapped by $\mathbb{F}_n$ to

$$A_{n,1}\colon = \mathbb{F}_n(A_n) = \frac{1}{\tau_{-n} - \mathbf{r}_{-n}\cos\varphi_{-n}/2},$$

$$B_{n,1}\colon = \mathbb{F}_n(B_n) = \frac{1}{\tau_{-n}},$$

$$C_{n,1}\colon = \mathbb{F}_n(C_n) = \frac{1}{\tau_{-n} - \mathbf{r}_{-n}\cos\varphi_{-n}}.$$

Note that $0 = A_{n-1} < B_{n,1} < A_{n,1} < C_{n,1}$. Therefore the image of the wave front with $\mathcal{B}^{-}_{-n} = \infty$ at the very next collision will be unstable, so its further image at time $t = 0$ will converge to $\mathcal{B}^u(X)$, as $n \to \infty$. This completes the proof of Proposition 8.18. □

8.6. Some more continued fractions

Here, as in Section 4.7, we will show that the continued fraction (8.7) actually converges at every point $X \in \Omega$ whose past semitrajectory is well defined.

We continue our analysis of the transformation $\mathbb{F}_n\colon \mathcal{B}^{-}_{-n} \mapsto \mathcal{B}^{-}_{-n+1}$ plotted in Fig. 8.5. If the $(n-1)$st reflection also occurs at a focusing arc (of radius $\mathbf{r}_{-n+1}$), then $C_{n,1} \le C_{n-1} = \frac{1}{\mathbf{r}_{-n+1}\cos\varphi_{-n+1}}$ due to the invariance of the unstable cones (note that if the nth and $(n-1)$st collisions occur at the same circle, then $C_{n,1} = C_{n-1}$). Next, at the $(n-2)$nd reflection, the images of our points will satisfy

$$0 = A_{n-2} < B_{n-1,1} < A_{n-1,1} < B_{n,2} < A_{n,2} < C_{n,2} \le C_{n-1,1} \le C_{n-2},$$

where $B_{n,2} = \mathbb{F}_{n-1}(B_{n,1})$, etc.; see diagram in Fig. 8.6.

Suppose for a moment that all the subsequent reflections occur at focusing arcs. Then at the point X we will have

$$0 < B_{1,1} < A_{1,1} < B_{2,2} < A_{2,2} < \cdots < C_{2,2} < C_{1,1}, \tag{8.9}$$

where $A_{k,k} = \mathcal{B}^{\mathrm{e}}_{-k}$ and $B_{k,k} = \mathcal{B}^{\mathrm{o}}_{-k}$, and $C_{k,k}$ represents the image of the wave front satisfying $\mathcal{B}^{-}_{-k} = \frac{1}{\mathbf{r}_{-k}\cos\varphi_{-k}}$.

We see that the even-numbered and odd-numbered convergents alternate and make a monotonically increasing sequence, which necessarily converges. We will demonstrate below that the sequence $\{C_{k,k}\}$ converges to the same limit.

Now if the past semitrajectory of X contains collisions at both focusing and dispersing/neutral boundary components, then the resulting picture will look like

$$0 < B_{1,1} < A_{1,1} < B_{2,2} < A_{2,2} < \cdots < D_{2,2} < D_{1,1}, \tag{8.10}$$

where each convergent $B_{k,k}$ is included only if the kth collision occurs at a focusing arc, and

$$D_{k,k} = \begin{cases} C_{k,k} & \text{for collisions at focusing arcs} \\ B_{k,k} & \text{for collisions at dispersing/flat boundary.} \end{cases}$$

We leave it as an exercise for the reader to derive (8.10).

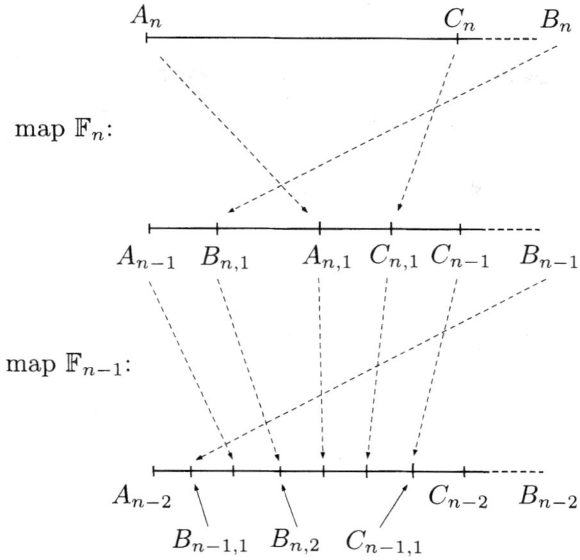

FIGURE 8.6. Diagram of the points A_n, B_n, C_n and their images.

Observe that this picture is quite different from the one we had for dispersing billiards; cf. (4.44). Back there, the even-numbered convergent $\mathcal{B}^{\mathrm{e}}_{-n}$ always represented the *least-expanding* unstable front (which started out as flat, $\mathcal{B}^{-}_{-n} = 0$), and the odd-numbered convergent $\mathcal{B}^{\mathrm{o}}_{-n}$ always represented the *most-expanding* unstable front (which started as an 'explosion', $\mathcal{B}^{-}_{-n} = \infty$), so all the other expanding fronts were in between (according to Exercise 4.13).

Here, when the nth collision occurs at a focusing arc, the least-expanding unstable front will correspond to $C_{n,n}$, rather than $A_{n,n}$ (or $B_{n,n}$), so the interval $[B_{n,n}, A_{n,n}] = [\mathcal{B}^{\mathrm{o}}_{-n}, \mathcal{B}^{\mathrm{e}}_{-n}]$ does not cover the entire range of unstable fronts originating at the nth collision, hence the need to include $C_{n,n}$ in our analysis.

To prove the convergence of the fraction (8.7) it is enough to show that

$$\lim_{k\to\infty} A_{k,k} = \lim_{k\to\infty} D_{k,k}.$$

Our proof is rather 'dynamical' and differs from Bunimovich's original argument [**Bu74a**]. We establish an analogue of Lemma 4.36:

LEMMA 8.19. *Suppose the past semitrajectory of X is well defined (i.e. $t_{-n} \to -\infty$ as $n \to \infty$). Then for $n \geq 2$*

$$|A_{n,n} - D_{n,n}| \leq 1/|t_{-n}|. \tag{8.11}$$

PROOF. The quantities $A_{n,n}$ and $D_{n,n}$ are represented by two finite continued fractions that differ only in the last term. As in the proof of Lemma 4.36

$$D_{n,n} - A_{n,n} = U_1 V_1 U_2 V_2 \cdots U_{n-1} V_{n-1} \left[\frac{1}{\tau_{-n} + 1/\mathcal{B}^{+}_{-n}} - \frac{1}{\tau_{-n} + 1/\mathcal{R}_{-n}} \right],$$

where U_i comes from the decomposition of $D_{n,n}$:

$$U_i = \cfrac{1}{1 + \tau_{-i}\left(\mathcal{R}_{-i} + \cfrac{1}{\tau_{-i-1} + \cfrac{1}{\ddots + \cfrac{1}{\tau_{-n} + 1/\mathcal{B}^{+}_{-n}}}}\right)}$$

and V_i comes from the decomposition of $A_{n,n}$:

$$V_i = \cfrac{1}{1 + \tau_{-i}\left(\mathcal{R}_{-i} + \cfrac{1}{\tau_{-i-1} + \cfrac{1}{\ddots + \cfrac{1}{\tau_{-n} + 1/\mathcal{R}_{-n}}}}\right)}.$$

Here

$$\mathcal{B}^{+}_{-n} = \begin{cases} \infty & \text{if } D_{n,n} = B_{n,n} \\ -\frac{1}{\mathbf{r}_{-n}\cos\varphi_{-n}} & \text{if } D_{n,n} = C_{n,n}. \end{cases}$$

Note that U_i and V_i are (up to the sign) the reciprocals of expansion factors of some unstable wave fronts (in the p-metric); hence $|U_i| \le 1$ and $|V_i| \le 1$.

Next we establish a better estimate for $|U_i|$. We have seen in the proof of Lemma 4.36 that if all the collisions occur at dispersing boundary components, then $1/\mathcal{B}^{+}_{-n} = 0$ and

$$0 \le U_i \le \frac{\tau_{-i-1} + \cdots + \tau_{-n}}{\tau_{-i} + \cdots + \tau_{-n}}.$$

This means that the wave front starting as an explosion ($\mathcal{B}^{-}_{-n} = \infty$) at the nth collision will expand between the $(i+1)$st and ith collisions at least as much as a wave front moving from the explosion at time t_{-n} without any collisions. The latter (fictional) wave front will have curvature at time t_{-i} equal to $1/(\tau_{-i-1} + \cdots + \tau_{-n})$.

A similar property holds for Bunimovich billiards, but its derivation is more involved. We introduce an "averaged" curvature at the kth collisions by

$$\bar{\mathcal{B}}_{-k} = \tfrac{1}{2}|\mathcal{B}^{+}_{-k} + \mathcal{B}^{-}_{-k}|.$$

EXERCISE 8.20. Show that for unstable wave fronts $|1 + \tau_{-k}\mathcal{B}^{+}_{-k}| \ge 1 + \tau_{-k}\bar{\mathcal{B}}_{-k}$; i.e. the unstable wave front expands (in the p-metric) between the kth and $(k+1)$st collisions no less than a 'fictional' wave front whose curvature after the kth collision is $\bar{\mathcal{B}}_{-k}$.

LEMMA 8.21. *For unstable wave fronts we have* $\bar{\mathcal{B}}^{-1}_{-k+1} \le \bar{\mathcal{B}}^{-1}_{-k} + \tau_{-k}$.

In other words, the "averaged" curvature $\bar{\mathcal{B}}_{-k}$ of an unstable front decreases between collisions no more than does the regular curvature of a wave front. As for the latter we have $[\mathcal{B}^{-}_{-k+1}]^{-1} = [\mathcal{B}^{+}_{-k}]^{-1} + \tau_{-k}$ due to (3.31). Despite the transparent geometric meaning of this lemma, its proof is not simple.

PROOF. We need to consider four cases, as each of the two collisions involved can occur either at a dispersing/flat boundary component or at a focusing arc. We treat only the most difficult case here, where both collisions occur at focusing arcs (of radii $\mathbf{r}_{-k}$ and $\mathbf{r}_{-k+1}$, respectively), and leave three other cases to the reader as an exercise.

For brevity, denote $\mathbf{r}_{-k}\cos\varphi_{-k} = s_{-k}$ and $-1/\mathcal{B}^+_{-k} = u_{-k}$ (note that $0 < u_{-k} < s_{-k}$). Then the inequality claimed in the lemma takes the form

$$\frac{1}{\frac{1}{s_{-k+1}} - \frac{1}{\tau_{-k}-u_{-k}}} \le \frac{1}{\frac{1}{u_{-k}} - \frac{1}{s_{-k}}} + \tau_{-k}.$$

Clearing the denominators and using (3.31) and (3.33) give

$$\tau_{-k}(s_{-k} - u_{-k})(\tau_{-k} - 2s_{-k+1}) + (\tau_{-k} - s_{-k} - s_{-k+1})u_{-k}^2 \ge 0.$$

This is true since all the factors are nonnegative; cf. (8.5). Lemma 8.21 is proved. □

We return to the proof of Lemma 8.19. First assume that the nth reflection occurs at a dispersing or flat boundary component; then $\bar{\mathcal{B}}_{-n} = \mathcal{B}^+_{-n} = \infty$ and

$$\left|\frac{1}{\tau_{-n} + 1/\mathcal{B}^+_{-n}} - \frac{1}{\tau_{-n} + 1/\mathcal{R}_{-n}}\right| \le \frac{1}{\tau_{-n}},$$

since $\mathcal{R}_{-n} \ge 0$. Combining the results of the above exercise and lemma gives $\bar{\mathcal{B}}^{-1}_{-i} \le \tau_{-n} + \cdots + \tau_{-i-1}$ and

$$\begin{aligned}
\left|U_1 \cdots U_{n-1}[\tau_{-n}]^{-1}\right| &\le \frac{1}{\tau_{-n}} \times \prod_{i=1}^{n-1} \frac{1}{1 + \tau_{-i}\bar{\mathcal{B}}_{-i}} \\
&\le \frac{1}{\tau_{-n}} \times \prod_{i=1}^{n-1} \frac{\tau_{-i-1} + \cdots + \tau_{-n}}{\tau_{-i} + \cdots + \tau_{-n}} \\
&= \frac{1}{\tau_{-1} + \cdots + \tau_{-n}} = \frac{1}{|t_{-n}|}.
\end{aligned}$$

Second, suppose the nth reflection occurs at a focusing arc (of radius $\mathbf{r}_{-n}$); then $1/\mathcal{B}^+_{-n} = 2/\mathcal{R}_{-n} = -\mathbf{r}_{-n}\cos\varphi_{-n}$.

EXERCISE 8.22. Show that

$$\frac{1}{\tau_{-n} - \mathbf{r}_{-n}\cos\varphi_{-n}} - \frac{1}{\tau_{-n} - \frac{1}{2}\mathbf{r}_{-n}\cos\varphi_{-n}} \le \frac{1}{\tau_{-n} + \mathbf{r}_{-n}\cos\varphi_{-n}}.$$

Hint: Remember that $\tau_{-n} \ge 2\mathbf{r}_{-n}\cos\varphi_{-n}$.

Now we will estimate V_i instead of U_i. Since V_i corresponds to an unstable wave front that starts out flat ($\mathcal{B}^-_{-n} = 0$), $\bar{\mathcal{B}}^{-1}_{-n} = \mathbf{r}_{-n}\cos\varphi_{-n}$; hence

$$\bar{\mathcal{B}}^{-1}_i \le \mathbf{r}_{-n}\cos\varphi_{-n} + \tau_{-n} + \cdots + \tau_{-i-1}$$

and so

$$\begin{aligned}
\left|\frac{V_1 \cdots V_{n-1}}{\tau_{-n} + \mathbf{r}_{-n}\cos\varphi_{-n}}\right| &\le \frac{1}{\tau_{-n} + \mathbf{r}_{-n}\cos\varphi_{-n}} \times \prod_{i=1}^{n-1} \frac{1}{1 + \tau_{-i}\bar{\mathcal{B}}_{-i}} \\
&\le \frac{1}{\tau_{-1} + \cdots + \tau_{-n} + \mathbf{r}_{-n}\cos\varphi_{-n}} \le \frac{1}{|t_{-n}|}.
\end{aligned}$$

This proves Lemma 8.19. □

REMARK 8.23. The estimate (8.11) is sharp. For example, if all the n collisions occur at flat boundary components (which is possible in many Bunimovich billiards, including pool tables and stadia; see Fig. 8.3), then $A_{n,n} = 0$ and $D_{n,n} = 1/|t_{-n}|$. However, if the nth collision occurs at a focusing arc, the estimate (8.11) can be improved.

The following is a direct analogue of Corollary 4.37:

COROLLARY 8.24. *Let $x \in \mathcal{M}$. Suppose an unstable front at x has curvature $\mathcal{B}$ and its preimage at the time t_{-n} of the nth collision in the past is still unstable. Then $|\mathcal{B} - \mathcal{B}^u(x)| \le 1/|t_{-n}|$.*

Hence, just as in Exercise 4.38, the function $\mathcal{B}^u(X)$ is continuous on the set of points $X \in \Omega$ where the past semitrajectory of X exists and where $\mathcal{B}^u(X) \neq \infty$ (the latter happens at focusing points).

Due to the time reversibility, the slope $\mathcal{B}^s(X)$ of the stable subspace E^s_X, $X \in \Omega$, is expressed by a similar continued fraction (4.45), which is defined on the set of points $X \in \Omega$ that have future semitrajectories, and is continuous whenever $\mathcal{B}^s(X) \neq \infty$.

The unstable and stable subspaces $E^{u,s}_x \subset \mathcal{T}_x\mathcal{M}$ have slopes $\mathcal{V}^{u,s}(x)$, which are given by the expressions (4.46) and (4.48). The functions $\mathcal{V}^u(x)$ and $\mathcal{V}^s(x)$ are defined and continuous on the sets of points $x \in \mathcal{M}$ where all the past, respectively future, iterations of $\mathcal{F}$ are smooth. We note that stable and unstable manifolds never focus at the times of collisions; hence $|\mathcal{V}^{u,s}(x)| < \infty$.

We conclude this section by answering one question raised in Section 3.12:

PROPOSITION 8.25. *The function $\log^+ |\mathcal{B}^{u+}(x)|$ is integrable on $\mathcal{M}$.*

PROOF. Points $x \in \mathcal{M}$ on dispersing and flat boundary components can be handled exactly as in the proof of Proposition 4.39. For points on focusing arcs we have

$$\log^+ |\mathcal{B}^{u+}(x)| \le \log^+[2\mathcal{K}(x)/\cos\varphi(x)];$$

thus the proposition follows from the integrability of the function $|\log\cos\varphi|$ on $\mathcal{M}$; see (3.8). □

8.7. Reduction of nonessential collisions

The study of Bunimovich billiards largely parallels that of Sinai billiards. However, there is a new phenomenon in billiards with focusing arcs that we have not seen in Sinai billiards of categories A and B: it is nonuniform hyperbolicity, described at the end of Section 8.4. During a long series of nonessential collisions (of types N1, N2, and N3, as defined in Section 8.4), expansion and contraction occur very slowly (if at all); thus such collisions are "wasted" in terms of the hyperbolicity of the dynamics.

Fortunately, the map $\mathcal{F}$ has a very simple structure (it is effectively linear) during series of nonessential collisions.

EXERCISE 8.26. Let $x = (r, \varphi)$ and $\mathcal{F}(x) = (r_1, \varphi_1)$ be successive collision points lying on two flat components of the boundary $\partial\mathcal{D}$. Find explicit formulas for the map $\mathcal{F}$ in the vicinity of x; in particular, verify that φ_1 is a linear function of φ and r_1 is a linear function of r.

EXERCISE 8.27. Let $x = (r, \varphi)$ and $\mathcal{F}(x) = (r_1, \varphi_1)$ be successive collision points lying on the same focusing arc (or two focusing arcs on the same circle) of radius $\mathbf{r}$. Verify that $\varphi_1 = \varphi$ and

$$r_1 = r + \mathbf{r}(\pi - 2\varphi) \quad \text{if} \quad \varphi > 0$$

and

$$r_1 = r - \mathbf{r}(\pi + 2\varphi) \quad \text{if} \quad \varphi < 0;$$

in particular, r_1 is a linear function of both r and φ.

The simplicity of the map $\mathcal{F}$ during the series of nonessential collisions allows us to reduce those series as follows.

If the kth collision on the trajectory of $X \in \Omega$ occurs at a flat boundary component, then $\mathcal{R}_k = 0$, so the corresponding term in the continued fraction $\mathcal{B}^u(X)$, if $k < 0$, or $\mathcal{B}^s(X)$, if $k > 0$, vanishes. The simple trick

$$\tau_{k-1} + \cfrac{1}{0 + \cfrac{1}{\tau_k + \cdots}} = \tau_{k-1} + \tau_k + \cdots$$

allows us to eliminate all zeros from the continued fractions $\mathcal{B}^u(X)$ and $\mathcal{B}^s(X)$ without changing their values. The so reduced continued fractions correspond to a 'fictional' billiard trajectory that experiences the same collisions at the same times, as the real trajectory $\Phi^t(X)$, with nonflat boundary components but no collisions with flat boundaries (those are 'skipped').

Geometrically, such a fictional trajectory can be constructed with the help of the trick used in Sections 1.1 and 1.2. Whenever the real trajectory collides with a flat side Γ_j of the table $\mathcal{D}$, we reflect the entire table across Γ_j and let our trajectory move straight into the mirror image of $\mathcal{D}$; see Fig. 8.7 (we 'unfold' the trajectory). The new (imaginary) trajectory will then collide with the boundary of the new (imaginary) table, and every time it hits a flat side we repeat the table flipping trick, thus obtaining more and more mirror images of $\mathcal{D}$ where the imaginary trajectory will keep on moving.

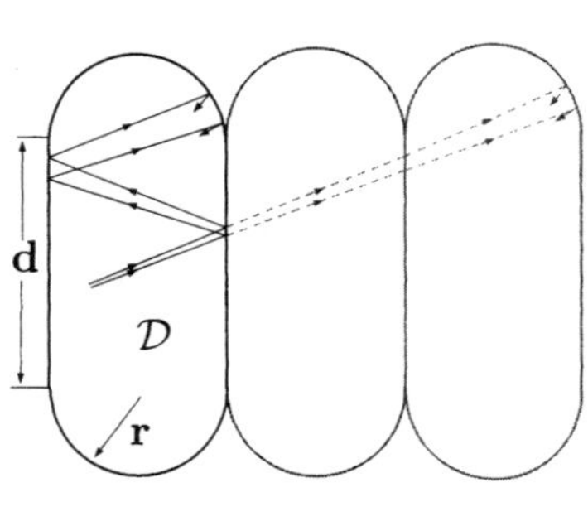

FIGURE 8.7. Imaginary trajectories (grey) moving without collisions at flat boundary components in the mirror images of $\mathcal{D}$.

In addition, let a wave front move along the trajectory of X. Then the table flipping trick gives us a new, imaginary wave front moving along the 'imaginary version' of the trajectory $\Phi^t(X)$; see Fig. 8.7. Every trajectory of that front moves straight through the flat boundary components into the mirror images of $\mathcal{D}$. Obviously, geometric characteristics of the imaginary front (its curvature and length) at any time will be identical to those of the real wave front (compare this to Exercise 3.33).

Thus, all collisions with flat boundary components can be effectively eliminated from the analysis, and the series of nonessential collisions of type N1 will no longer appear. The only 'side effect' will be increased intervals between collisions.

Next, suppose the trajectory of a point $X \in \Omega$ experiences a series of successive collisions with the same focusing arc (or several focusing arcs on the same circle) of radius $\mathbf{r}$. Denote the collision points by $x_n, x_{n+1}, \dots, x_{n+k}$. Then, according to Exercise 8.27

$$\varphi_n = \cdots = \varphi_{n+k} =: \varphi,$$

$$\tau_n = \tau_{n+1} = \cdots = \tau_{n+k-1} = 2\mathbf{r}\cos\varphi =: \tau,$$

and

$$\mathcal{R}_n = \mathcal{R}_{n+1} = \cdots = \mathcal{R}_{n+k} = \frac{2}{\mathbf{r}\cos\varphi} = \frac{4}{\tau}.$$

The corresponding stretch of the continued fraction can be 'telescoped' as follows:

$$-\frac{4}{\tau} + \cfrac{1}{\tau + \cfrac{1}{-\frac{4}{\tau} + \cfrac{1}{\tau + \ddots \cfrac{1}{-\frac{4}{\tau} + \ddots}}}} = -\frac{2}{\tau} + \cfrac{1}{-k\tau + \cfrac{1}{-\frac{2}{\tau} + \ddots}}. \tag{8.12}$$

EXERCISE 8.28. Verify (8.12). Hint: Use the induction on k; see also (1.3).

The reduced fraction on the right-hand side of (8.12) corresponds to an imaginary billiard trajectory evolving as follows: at time t_n, as the real trajectory makes its first collision of our series, the imaginary trajectory reflects off a circle of a double radius $2\mathbf{r}$; then it moves *backward* without collisions $k\tau$ units of time (this is exactly the time spent by the real trajectory between its first and last collisions in the series); and at time $t_{n+k} = t_n + k\tau$, as the real trajectory makes its last collision, the imaginary trajectory again reflects off a circles of radius $2\mathbf{r}$; see Fig. 8.8.

FIGURE 8.8. Imaginary trajectories ('fat' grey) moving 'backwards' after colliding with a circle of the double radius.

Now let an infinitesimal wave front move along the trajectory of X and experience $k+1$ successive collisions with our focusing arc. Denote by $\mathcal{B}_n^-$ its curvature before the very first collision. Then the expansion of the front in the p-metric, in the course of the whole series, is

$$\mathcal{J} = \prod_{i=0}^{k-1} \left|1 + \tau(-4/\tau + \mathcal{B}_{n+i}^-)\right|.$$

Every trajectory of our front has an imaginary version constructed as above, so we obtain an imaginary wave front evolving between two collisions with circles of radius $2\mathbf{r}$. Its expansion between those two collisions, in the p-metric, is

$$\mathcal{J}_{\mathrm{im}} = \left|1 - k\tau(-2/\tau + \mathcal{B}_n^-)\right|$$

according to (8.12).

EXERCISE 8.29. Prove that $\mathcal{J} = \mathcal{J}_{\mathrm{im}}$, so that the imaginary front expands just as much as the real wave front. Hint: Note that $-4/\tau + \tau\mathcal{B}_{n+i}^-$ has the form shown on the left-hand side of (8.12); thus is can be replaced by the reduced fraction similar to the one on the right-hand side of (8.12).

Thus, geometric characteristics of the imaginary front (its curvature and length) will be identical to those of the real wave front at the times of the very first and very last collisions of our series. The entire series of collisions with a focusing arc can be replaced by two collisions with imaginary arcs of the double radius. This entails two 'side effects': the interval between two collisions with (real or imaginary) arcs can no longer be arbitrarily small (see the exercise below) and the curvature of the front after the first collision with the arc is different from that of the real wave front.

We remark that if the wave front is unstable, i.e. $\mathcal{B}^-$ satisfies (8.4), then $-2/\tau + \mathcal{B}^- < 0$; hence the imaginary wave front becomes focusing after the first collision. But since it moves backwards, it will monotonically expand without passing through focusing points; see Fig. 8.8. That makes its dynamics similar to what we have seen in dispersing billiards.

EXERCISE 8.30. Show that for any focusing arc Γ_i there is a positive constant $c_i > 0$ such that in any series of $k+1 \geq 2$ successive collisions with Γ_i the interval $(= k\tau)$ between the very first and very last collision always exceeds c_i.

8.8. Stadia

The reduction of series of nonessential collisions makes it possible to establish a close analogy between Bunimovich billiards and dispersing billiards of categories A and B. Many facts that we presented in Chapters 4 and 5 extend to Bunimovich billiards as well: the smoothness of stable and unstable curves, the properties of singularity curves, the construction of stable and unstable manifolds, the formulas for u-SRB densities, distortion bounds, and absolute continuity.

On dispersing (and flat) boundary components, these basic properties usually take the same form as in the case of Sinai billiards, but on focusing arcs they take a different form and require special treatment. Recall that in Definition 8.7 we had to deal with dispersing components and focusing arcs distinctly; accordingly, we had two separate formulas for unstable cones. Whenever two successive collisions are involved, we have to consider four different combinations of collision types (recall our proofs of Theorem 8.9 and Lemma 8.21).

As a result, the proofs of many basic facts for Bunimovich billiards, even though similar in spirit to those for Sinai tables, are necessarily much longer and technically more involved. (For example, compare our treatment of the continued fractions in Section 4.7 for Sinai billiards and in Sections 8.5 and 8.6 for Bunimovich tables.) For this reason we will restrict our further analysis to a certain limited class of Bunimovich billiards – stadia.

A stadium is shown in Fig. 8.3(c); it is a table bounded by two equal parallel line segments and two equal semicircles (its boundary is C^1, but not C^2, smooth). It was introduced by Bunimovich in 1974 [**Bu74b**], but became popular in 1979 after the publication of his other article [**Bu79**]. It is this model that caught the eye of many physicists in the 1980s who were amazed to see such a simple example of a totally chaotic billiard. It remains chaotic no matter how short the parallel segments are, but if they vanish, the stadium turns into a disk where the billiard is completely integrable (Section 1.1). Thus one gets a continuous family of billiard tables where a transition from a completely regular regime to total chaos occurs instantly. Physicists did dozens of computer experiments investigating this miraculous transition and greatly publicized the stadia.

In what follows we restrict our analysis to stadia. Thus we will no longer consider dispersing boundary components, and flat sides will be effectively eliminated by the flipping table trick; see below. This allows us to avoid unpleasant complications resulting from several different types of collisions and at the same time capture all essential features of generic Bunimovich billiards with focusing arcs. Similar analysis could be done for other models, like flowers shown in Fig. 8.3(a) and pool tables with pockets shown in Fig. 8.3(b); we leave this as a project for the motivated reader.

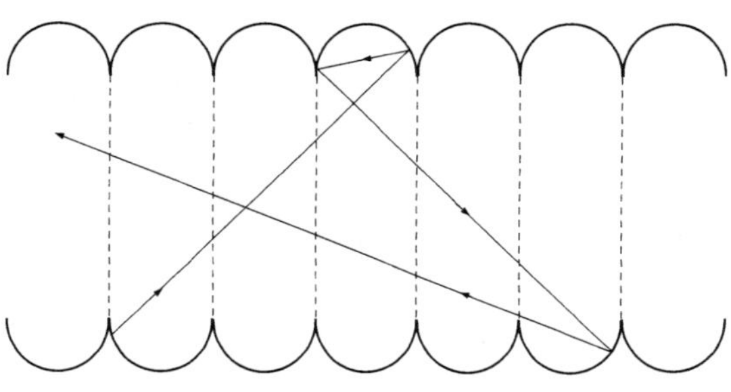

FIGURE 8.9. An infinite table made by identical stadia with common flat sides.

The flipping table trick was described in the previous section and illustrated in Fig. 8.7. Observe that due to a natural symmetry, the mirror image of a stadium $\mathcal{D}$ across its flat side is identical to $\mathcal{D}$ itself. Further images make an infinite row of identical stadia with common flat sides; see Fig. 8.9. Thus we obtain an unbounded billiard table $\mathcal{D}_\infty$ whose border consists of two rows of identical adjacent semicircles. Equivalently, we can identify the opposite flat sides of the original stadium and get a table $\mathcal{D}_0$ on a 2D torus; this will be a fundamental domain for the periodic table

$\mathcal{D}_\infty$. Observe that neither $\mathcal{D}_\infty$ nor $\mathcal{D}_0$ have flat boundary components. All collisions occur at semicircles only.

Furthermore, the reduction of nonessential collisions in the previous section suggests that we redefine the map $\mathcal{F}$ so that in every series of collisions with a focusing arc only the very first and the very last ones will be kept. Accordingly, we define a subset $\mathcal{M}_\diamond \subset \mathcal{M}$ consisting of points $x = (q, v)$ such that q belongs to a focusing arc and either its image $\mathcal{F}(x)$ or its preimage $\mathcal{F}^{-1}(x)$ does not lie on that same arc. Also, we consider the return map $\mathcal{F}_\diamond \colon \mathcal{M}_\diamond \to \mathcal{M}_\diamond$; i.e. $\mathcal{F}_\diamond(x) = \mathcal{F}^{n(x)}(x)$ where $n(x) = \min\{n > 0 \colon \mathcal{F}^n(x) \in \mathcal{M}_\diamond\}$. The map $\mathcal{F}_\diamond$ preserves the measure μ conditioned on $\mathcal{M}_\diamond$, which we denote by $\mu_\diamond = [\mu(\mathcal{M}_\diamond)]^{-1}\mu$.

The set $\mathcal{M}_\diamond$ consists of two identical hexagons (corresponding to two identical semicircles); one is shown in Fig. 8.10. This white nonconvex hexagon $AHBFGE$ is the union of two parallelograms: $CBDE$ contains all the points of *first* collisions with the semicircle and $ACFD$ contains all the points of *last* collisions with the semicircle (their intersection, the rhombus $CHGD$, consists of points of *only* collisions with the semicircle). The two dark triangles contain points of intermediate collisions, so they are not included in $\mathcal{M}_\diamond$.

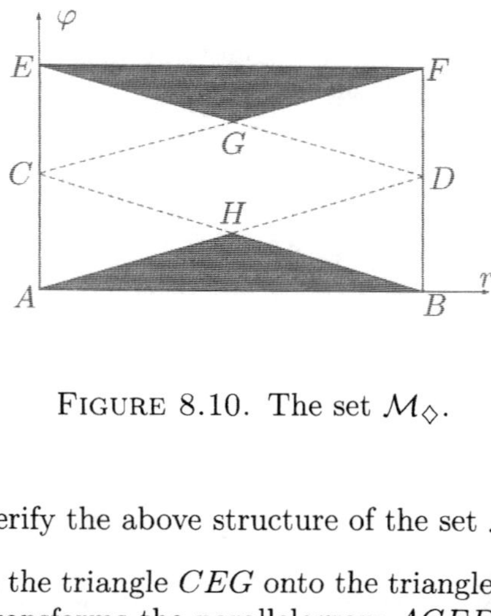

FIGURE 8.10. The set $\mathcal{M}_\diamond$.

EXERCISE 8.31. Verify the above structure of the set $\mathcal{M}_\diamond$.

The map $\mathcal{F}_\diamond$ takes the triangle CEG onto the triangle DFG; likewise, it takes BHD into ACH. It transforms the parallelogram $ACFD$ onto the parallelogram identical to $CBDE$ but lying within the other hexagon (corresponding to the opposite semicircle).

Next we describe the action of $\mathcal{F}_\diamond$ on the upper left triangle CEG. On the bigger triangle CEF, where $\varphi > 0$, the original map $\mathcal{F}$ acts by the rule

$$(r, \varphi) \mapsto (r + \mathbf{r}(\pi - 2\varphi), \varphi); \tag{8.13}$$

see Exercise 8.27. If a billiard trajectory makes k successive collisions with the semicircle (along which the φ coordinate remains unchanged), then

$$\varphi = \frac{\pi}{2} - \frac{\pi}{2k} + \mathcal{O}\Big(\frac{1}{k^2}\Big), \quad \text{so} \quad \cos\varphi = \frac{\pi}{2k} + \mathcal{O}\Big(\frac{1}{k^2}\Big). \tag{8.14}$$

Now the new map $\mathcal{F}_\diamond$ acts by the rule

$$r \mapsto r + k\mathbf{r}(\pi - 2\varphi). \tag{8.15}$$

It is linear on each region where k is constant, but there are countably many such regions in $CBDE$ corresponding to different k's; they are separated by lines consisting of trajectories hitting the endpoint of our semicircle on their last collision with it. So the new map $\mathcal{F}_\diamond$ is piecewise linear and has countably many discontinuity lines.

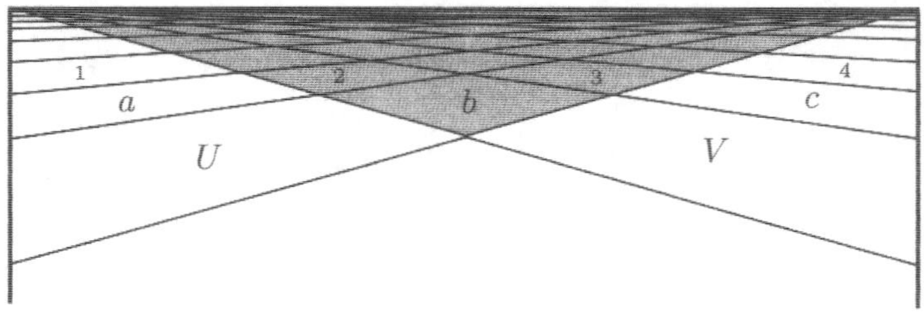

FIGURE 8.11. The action of $\mathcal{F}$.

The action of $\mathcal{F}$ and $\mathcal{F}_\diamond$ in the upper rectangle $EFDC$ (where $\varphi > 0$) is shown in Fig. 8.11. It is divided by two families of straight lines, each family converging to one top corner of the hexagon, into countably many quadrilaterals that make a checkerboard pattern. The original billiard map $\mathcal{F}$ transforms every quadrilateral (linearly) into the next one on the right in the same horizontal row. The new map $\mathcal{F}_\diamond$ sends the leftmost quadrilateral onto the rightmost one in the same row. This is illustrated by the following diagram (see Fig. 8.11):

$\mathcal{F}$	$\mathcal{F}_\diamond$
$U \mapsto V$	$U \mapsto V$
$a \mapsto b \mapsto c$	$a \mapsto c$
$1 \mapsto 2 \mapsto 3 \mapsto 4$	$1 \mapsto 4$

Observe that the lines in the triangle CEG are singularities for the map $\mathcal{F}_\diamond$, while those in DFG are singularities for its inverse $\mathcal{F}_\diamond^{-1}$. The number of quadrilaterals in each row equals k (the number of successive collisions with the semicircle).

The lower rectangle $ACDB$ (where $\varphi < 0$) is divided into quadrilaterals in a similar manner, but the maps $\mathcal{F}$ and $\mathcal{F}_\diamond$ move them from left to right.

EXERCISE 8.32. Verify the above description of the maps $\mathcal{F}$ and $\mathcal{F}_\diamond$.

REMARK 8.33. The above description of the map $\mathcal{F}$ (and the 'reduced' map $\mathcal{F}_\diamond$) applies to any focusing arc of any Bunimovich billiard table as long as the arc does not exceed semicircle. If the arc is larger than the semicircle, then in addition nonessential collisions of type N3 come into play. We leave this case to the reader as an exercise.

The map $\mathcal{F}_\diamond$ has other singularities which lie in the parallelogram $ACFD$; they correspond to trajectories running into an endpoint of the opposite semicircle.

EXERCISE 8.34. Show that the above singularities consist of two countable (symmetric) families of increasing curves that accumulate near the vertices C and D; see Fig. 8.12.

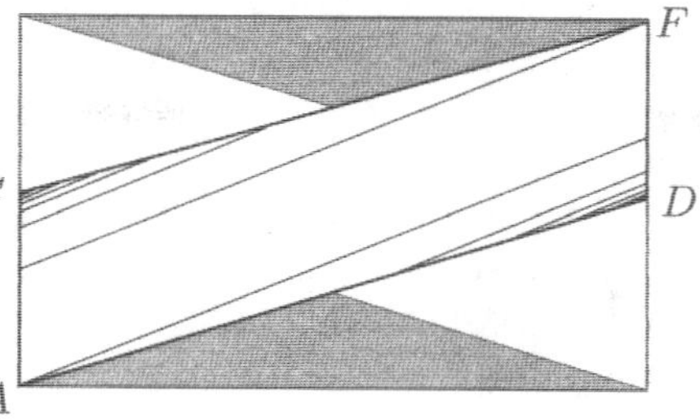

FIGURE 8.12. The singularities of $\mathcal{F}_\diamond$ (black lines) in the parallelogram $ACFD$.

8.9. Uniform hyperbolicity

In this section we prove that the return map $\mathcal{F}_\diamond\colon \mathcal{M}_\diamond \to \mathcal{M}_\diamond$ for any stadium is uniformly hyperbolic. We use our standard notation (adapted to the map $\mathcal{F}_\diamond$): given a point $x = (r, \varphi) \in \mathcal{M}_\diamond$, we denote by $x_n = (r_n, \varphi_n) = \mathcal{F}_\diamond^n(x)$ its images and by $\mathcal{R}_n$, τ_n, etc., the corresponding parameters at x_n. Note that n here is the counter of essential collisions only.

Given a tangent vector $dx = (dr, d\varphi) \in \mathcal{T}_x\mathcal{M}_\diamond$, we denote by $dx_n = (dr_n, d\varphi_n) = D_x\mathcal{F}_\diamond^n(dx)$ its image, by $\mathcal{V}_n = d\varphi_n/dr_n$ its slope, and by $\mathcal{B}_n^\pm$ the curvature of the corresponding wave front before and after the collision; recall the relation (8.3). Also denote by $\|dx_n\| = \left[(dr_n)^2 + (d\varphi_n)^2\right]^{1/2}$ the Euclidean norm and by $\|dx_n\|_p = |\cos\varphi_n\, dr_n|$ the p-norm of these tangent vectors.

REMARK 8.35. We apply our reduction scheme of Section 8.7 to every series of successive sliding collisions; thus the corresponding τ_n's become negative. It is important, though, that the expansion in the p-metric is always given by

$$\frac{\|dx_{n+1}\|_p}{\|dx_n\|_p} = |1 + \tau_n\mathcal{B}_n^+| = \frac{|\mathcal{B}_n^+|}{|\mathcal{B}_{n+1}^-|} \tag{8.16}$$

whether $\tau_n > 0$ or $\tau_n < 0$ (compare this to (8.12)).

The free path between (essential) collisions is uniformly bounded away from zero: $|\tau_n| \ge \tau_{\min} > 0$ (see Exercise 8.30). It is not bounded above, though. (Why?)

The slope $\mathcal{V} = d\varphi/dr$ of an unstable vector $(dr, d\varphi)$ is always negative; in fact, $-1/\mathbf{r} \le \mathcal{V} < 0$ (see (8.3) and (8.4)). In what follows, we also suppose that its preimage $(dr_{-1}, d\varphi_{-1})$ is an unstable vector, too.

Let x and x_{-1} belong to opposite semicircles. As both semicircles satisfy the B-condition, the focusing point of the wave front corresponding to our tangent vectors lies closer to x_{-1} than to x, and it divides the interval τ_{-1} between x_{-1} and x in the ratio $\rho \le 1 - c_0$, where $c_0 = c_0(\mathcal{D}) > 0$ is a constant. This implies

$$\frac{\|dx\|_p}{\|dx_{-1}\|_p} = \frac{|\mathcal{B}_{-1}^+|}{|\mathcal{B}^-|} = \rho^{-1} \ge (1 - c_0)^{-1} > 1.$$

Note also that $0 \le \mathcal{B}^- \le 2/\tau_{\min} = \text{const}$, and

$$\mathcal{B}^- \le \frac{1 - c_1}{\mathbf{r}\cos\varphi}, \tag{8.17}$$

where $c_1 = c_1(\mathcal{D}) > 0$ is a constant. Since the slope of our tangent vector is $\mathcal{V} = \mathcal{K} + \mathcal{B}^- \cos\varphi$ and $\mathcal{K} = -1/\mathbf{r} =$const, we have $-1/\mathbf{r} \le \mathcal{V} \le -c_1/\mathbf{r}$. This can be expressed as $\mathcal{V} \asymp -1$. (Recall that $F \asymp G$ means $0 < C_1 < F/G < C_2 < \infty$ for some positive constants C_1, C_2 depending only on $\mathcal{D}$; see Section 4.3.)

Now let x and x_{-1} belong to the same semicircle (they may be separated by other, nonessential, sliding collisions). In that case we need to assume that the second preimage of our tangent vector $(dr_{-2}, d\varphi_{-2})$ is unstable, too, so that we can apply (8.17) to x_{-1}. This implies

$$\mathcal{B}^+_{-1} = -\frac{1}{\mathbf{r}\cos\varphi_{-1}} + \mathcal{B}^-_{-1} \le -\frac{c_1}{\mathbf{r}\cos\varphi_{-1}}$$

according to (8.12). As $\tau_{-1} \le -\tau_{\min} < 0$, we have

$$\frac{\|dx\|_p}{\|dx_{-1}\|_p} = \left|1 + \tau_{-1}\mathcal{B}^+_{-1}\right| \ge 1 + \frac{c_1\tau_{\min}}{\mathbf{r}\cos\varphi_{-1}},$$

which is bounded away from 1. Also, in this case the time τ_{-1} has the lower bound $\tau_{-1} \ge -\tau_{\max} > -\infty$, and it follows that

$$-\frac{1}{\tau_{\min}} \le \mathcal{B}^- \le -\frac{1}{\tau_{\max} + c_1/\mathbf{r}},$$

i.e. $\mathcal{B}^- \asymp -1$. If the number k of sliding collisions between the points x_{-1} and x is large, then $\cos\varphi \asymp 1/k$ due to (8.14) and

$$\mathcal{B}^+ = -\frac{1}{\mathbf{r}\cos\varphi} + \mathcal{B}^- \asymp -k.$$

For the slope $\mathcal{V}$ we have

$$\mathcal{V} = -\mathcal{K} + \mathcal{B}^+ \cos\varphi = \mathcal{B}^- \cos\varphi \asymp -1/k$$

(this can also be derived directly from (8.14)–(8.15)).

We summarize the above discussion. There is always a lower bound for the slope of unstable vectors

$$-\infty < \text{const} \le \mathcal{V} < 0. \tag{8.18}$$

The expansion of unstable vectors in the p-metric is always uniform:

$$\frac{\|dx_{n+1}\|_p}{\|dx_n\|_p} = |1 + \tau_n\mathcal{B}^+_n| \ge \Lambda > 1, \tag{8.19}$$

where $\Lambda = \Lambda(\mathcal{D}) > 1$ is a constant. Moreover, when $\cos\varphi_n$ is small, a better estimate is available:

$$\frac{\|dx_{n+1}\|_p}{\|dx_n\|_p} = |1 + \tau_n\mathcal{B}^+_n| \asymp \frac{1}{\cos\varphi_n}. \tag{8.20}$$

For the expansion in the Euclidean metric we have, as in (4.18),

$$\frac{\|dx_n\|}{\|dx_0\|} = \frac{\|dx_n\|_p}{\|dx_0\|_p}\,\frac{\cos\varphi_0}{\cos\varphi_n}\,\frac{\sqrt{1+\mathcal{V}_n^2}}{\sqrt{1+\mathcal{V}_0^2}}. \tag{8.21}$$

The last fraction is uniformly bounded away from zero and infinity due to (8.18). To deal with the middle fraction, we use (8.20) with $n = 0$ and obtain

$$\frac{\|dx_n\|}{\|dx_0\|} \ge \text{const} \times \frac{\|dx_n\|_p}{\|dx_1\|_p} \ge \hat{c}\,\Lambda^n, \tag{8.22}$$

collision type:	in a series of k sliding collisions: first one	last one	before a series of m reflections at flat sides
τ	-1	1	m
$\cos\varphi$	$1/k$	$1/k$	1
$\mathcal{B}^-$	1	-1	$1/m$
$\mathcal{R}$	$-k$	$-k$	-1
$\mathcal{B}^+$	$-k$	$-k$	-1
$\mathcal{V} = d\varphi/dr$	-1	$-1/k$	-1
$\lvert 1+\tau\mathcal{B}^+\rvert$	k	k	m

TABLE 8.1. Asymptotics of various quantities at three types of 'problematic' collisions.

where $\hat{c} = \hat{c}(\mathcal{D}) > 0$ is a constant. Thus the hyperbolicity is uniform in the Euclidean metric, too. Note, however, that the expansion of unstable vectors is always monotone in the p-metric, but it may not be so in the Euclidean metric.

We also note that

$$\frac{\|dx_{n+1}\|}{\|dx_n\|} = \frac{\|dx_{n+1}\|_p}{\|dx_n\|_p}\,\frac{\cos\varphi_n}{\cos\varphi_{n+1}}\,\frac{\sqrt{1+\mathcal{V}_{n+1}^2}}{\sqrt{1+\mathcal{V}_m^2}} \asymp \frac{1}{\cos\varphi_{n+1}} \tag{8.23}$$

which is useful when $\cos\varphi_{n+1} \approx 0$.

There are three 'problematic' regions in the reduced collision space $\mathcal{M}_\diamond$ where the quantities we deal with approach zero or infinity. These are the first and the last collisions in a long series of $k \gg 1$ sliding collisions with a semicircle. Also, these are collisions preceding a long series of $m \gg 1$ reflections off the flat sides of the original stadium. Such points are located in the vicinities of the vertices E, F and C,D, respectively, in the hexagon in Fig. 8.10.

Table 8.1 presents asymptotic behavior (in the sense of $\asymp$) of various quantities at these three types of 'problematic' collisions (for example, at the first sliding collision $\tau \asymp -1$, $\cos\varphi \asymp 1/k$, etc.). All the formulas in the table follow immediately from our previous analysis or from each other.

The last line in Table 8.1 describes the expansion of unstable curves in the p-metric, and now we describe it in the Euclidean metric. If $x \in \mathcal{M}_\diamond$ is the first sliding collision, then $\cos\varphi = \cos\varphi_1$ ($\asymp 1/k$); hence $\|dx_1\|/\|dx\| \asymp k \asymp \|dx_1\|_p/\|dx\|_p$.

Let $x \in \mathcal{M}_\diamond$ be the last collision in a long series of $k \gg 1$ sliding collisions with a semicircle. Clearly, $\mathcal{F}_\diamond(x)$ is the first collision in another long series of $k_1 \gg 1$ of sliding collisions with the opposite semicircle.

EXERCISE 8.36. Show that $c\sqrt{k} \le k_1 \le Ck^2$ for some constants $c, C > 0$. (The situation here is similar to the type B dispersing billiard tables; cf. Section 4.10.)

Since $\cos\varphi_1 \asymp 1/k_1$, we obtain due to (8.23)

$$c\sqrt{k} \le \|dx_1\|/\|dx\| \le Ck^2 \tag{8.24}$$

for some constants $c, C > 0$. Lastly, let $x \in \mathcal{M}_\diamond$ precede a long series of $m \gg 1$ reflections off the flat sides of the original stadium. Then $\mathcal{F}_\diamond(x)$ also precedes another long series of $m_1 \gg 1$ reflections off the flat sides.

EXERCISE 8.37. Show that $cm \le m_1 \le Cm$ for some constants $c, C > 0$. In fact, more precise bounds hold: $m/3 + o(m) \le m_1 \le 3m + o(m)$. We will verify this by a different approach in Section 8.14.

As a result, $\|dx_1\|/\|dx\| \asymp m \asymp \|dx_1\|_p/\|dx\|_p$.

8.10. Stable and unstable curves

Here we consider stable and unstable curves in the reduced collision space $\mathcal{M}_\diamond$ of a stadium. Stable curves have positive slope and unstable curves have negative slope (which is opposite to the case of dispersing billiards!). These curves cannot self-intersect. Their lengths (in both metrics, $|\cdot|$ and $|\cdot|_p$) are uniformly bounded. If W^u is an unstable curve and W^s a stable curve, then their intersection $W^u \cap W^s$ is transversal and consists of at most one point.

We note that the singularity lines of the map $\mathcal{F}_\diamond$ are increasing (stable) curves (cf. Section 8.8), so each unstable curve intersects every singularity line in at most one point. Thus the image $\mathcal{F}_\diamond(W)$ of any unstable curve $W \subset \mathcal{M}_\diamond$ is a finite or countable union of unstable curves, which we call the *components* of $\mathcal{F}_\diamond(W)$.

Now we establish the C^2 smoothness of these curves, with uniformly bounded second derivatives, in the same sense as we did in Section 4.5 for dispersing billiards (here we do not go beyond the second derivative to avoid unnecessary complications). As in Section 4.5, our curves contain endpoints and the smoothness always extends to one-sided derivatives at those.

PROPOSITION 8.38. *For every C^2 smooth unstable curve $W \subset \mathcal{M}_\diamond$ there is an $n_W \ge 1$ such that for all $n > n_W$ every smooth component $W' \subset \mathcal{F}_\diamond^n(W)$ has its second derivative uniformly bounded:*

$$|d^2\varphi_n/dr_n^2| \le C \tag{8.25}$$

on W', where $C = C(\mathcal{D}) > 0$ is a constant.

PROOF. We will suppress the index n for brevity. Let $(r, \varphi(r))$ denote a point on W'. Differentiating (8.3) (in which $\mathcal{K} \equiv \text{const}$) gives

$$d^2\varphi/dr^2 = -\mathcal{B}^-\mathcal{V}\sin\varphi + \cos\varphi\, d\mathcal{B}^-/dr. \tag{8.26}$$

The first term is uniformly bounded; hence the second derivative would be bounded if we prove

LEMMA 8.39. *For every C^2 smooth unstable curve $W \subset \mathcal{M}_\diamond$ there is $n_W \ge 1$ such that for all $n > n_W$ every smooth component $W' \subset \mathcal{F}_\diamond^n(W)$ satisfies*

$$|d\mathcal{B}^-/dr| \le C'$$

on W', where $C' = C'(\mathcal{D}) > 0$ is a constant.

PROOF. Since W is C^2 smooth, $\sup_W |d^2\varphi/dr^2| < \infty$; hence

$$\sup_W |\cos\varphi\; d\mathcal{B}^-/dr| < \infty \tag{8.27}$$

due to (8.26). Using our standard notation, we have for $n \geq 1$

$$\mathcal{B}_n^- = \cfrac{1}{\tau_{n-1} + \cfrac{1}{\mathcal{R}_{n-1} + \mathcal{B}_{n-1}^-}};$$

hence using (3.31) and (3.33) gives

$$\frac{d\mathcal{B}_n^-}{dr_n} = \left[\mathcal{B}_n^-\right]^2 \frac{d\tau_{n-1}}{dr_n} + \frac{1}{[1+\tau_{n-1}\mathcal{B}_{n-1}^+]^2}\left(\frac{d\mathcal{R}_{n-1}}{dr_n} + \frac{d\mathcal{B}_{n-1}^-}{dr_n}\right). \tag{8.28}$$

First we check that

$$|d\tau_{n-1}/dr_n| \asymp 1, \tag{8.29}$$

which is straightforward for collision-free flights (recall the proof of Lemma 4.44). A different case appears when x_{n-1} is the first collision in a series of k sliding collisions with a semicircle. Then $\tau_{n-1} = 2(k-1)\mathbf{r}\cos\varphi_n$; hence $|d\tau_{n-1}/d\varphi_n| \asymp k$, which is unbounded. But $|dr_n/d\varphi_n| = 1/|\mathcal{V}_n| \asymp k$, so these two k's cancel each other out, proving (8.29). Thus the first term in (8.28) is uniformly bounded.

Next

$$\frac{d\mathcal{R}_{n-1}}{dr_n} = \frac{d\mathcal{R}_{n-1}}{dr_{n-1}} \times \frac{dr_{n-1}}{dr_n} = -\frac{2\,\mathcal{V}_{n-1}\sin\varphi_{n-1}}{\mathbf{r}\cos^2\varphi_{n-1}} \times \frac{dr_{n-1}}{dr_n}.$$

Note that $|dr_n| = \|dx_n\|/\sqrt{1+\mathcal{V}_n^2} \asymp \|dx_n\|$; hence $|dr_{n-1}/dr_n| \asymp \|dx_{n-1}\|/\|dx_n\|$, which is uniformly bounded due to uniform hyperbolicity (8.22). The small factor $\cos^2\varphi_{n-1}$ in the denominator is absorbed by $[1+\tau_{n-1}\mathcal{B}_{n-1}^+]^2$ due to (8.20); thus

$$\frac{1}{[1+\tau_{n-1}\mathcal{B}_{n-1}^+]^2}\left|\frac{d\mathcal{R}_{n-1}}{dr_n}\right| \leq \text{const.}$$

Next we represent $|dr_{n-1}/dr_n|$ as

$$\left|\frac{dr_{n-1}}{dr_n}\right| = \frac{\|dx_{n-1}\|_p}{\|dx_n\|_p} \times \frac{\cos\varphi_n}{\cos\varphi_{n-1}} = \frac{\theta_n w_n}{w_{n-1}}$$

where

$$\theta_n = |1+\tau_n\mathcal{B}_n^+|^{-1} = \|dx_n\|_p/\|dx_{n+1}\|_p$$

and

$$w_n = \theta_n^{-1}\cos\varphi_n = |1+\tau_n\mathcal{B}_n^+|\cos\varphi_n.$$

Observe that $\theta_n \leq \Lambda^{-1} < 1$ and w_n is bounded above and below due to (8.20), i.e.

$$0 < w_{\min} \leq w_n \leq w_{\max} < \infty.$$

Combining these estimates with (8.28) gives

$$\left|\frac{d\mathcal{B}_n^-}{dr_n}\right| \leq C'' + \theta_{n-1}^2\theta_n\frac{w_n}{w_{n-1}}\left|\frac{d\mathcal{B}_{n-1}^-}{dr_{n-1}}\right| \tag{8.30}$$

where $C'' = C''(\mathcal{D}) > 0$ is a constant. A simple recursive application of (8.30) gives

$$\left|\frac{d\mathcal{B}_n^-}{dr_n}\right| \leq \frac{w_{\max}}{w_{\min}}\left(\frac{C''}{1-\Lambda^{-3}} + \Lambda^{-3n+1}\theta_0\left|\frac{d\mathcal{B}^-}{dr}\right|\right).$$

Lastly, note that due to (8.27) and (8.20) we have $\sup_W \theta_0 |d\mathcal{B}^-/dr| < \infty$, which completes the proof of Lemma 8.39. □

Proposition 8.38 is now proved. □

REMARK 8.40. Note that the above argument is quite different from the proof of the corresponding Proposition 4.29 for the case of dispersing billiards: there we used the billiard flow, while here we work with the collision map only. The reason is that between collisions unstable wave fronts pass through focusing points where the curvature becomes infinite, which complicates the analysis. On the contrary, in Proposition 4.29 we chose to use the billiard flow because the analysis was simple enough and it provided somewhat sharper estimates.

Due to the time reversibility, stable curves all have similar (dual) properties.

Next, we describe singularities for the map $\mathcal{F}_\diamond$. Put $\hat{\mathcal{S}}_0 = \partial\mathcal{M}_\diamond$ and denote by $\hat{\mathcal{S}}_n$ the singularities of the map $\mathcal{F}_\diamond^n$. We have seen in Section 8.8 that singularity curves for the map $\mathcal{F}_\diamond$ are stable and those for $\mathcal{F}_\diamond^{-1}$ are unstable. It is easy to check directly that these curves have uniformly bounded curvature (some of them are actually straight lines!). Thus all the curves in $\hat{\mathcal{S}}_n \setminus \hat{\mathcal{S}}_0$ for $n \geq 1$ are stable and have uniformly bounded curvature, and all the curves in $\hat{\mathcal{S}}_{-n} \setminus \hat{\mathcal{S}}_0$ for $n \geq 1$ are unstable and also have uniformly bounded curvature.

It is easy to see that singularities of the map $\mathcal{F}_\diamond$ have the same general properties as they do in dispersing billiards. In particular, the alignment of singularity lines takes place, as in Section 4.8, because it follows from Corollary 8.24 (our analogue of Corollary 4.37). Next, the continuation of singularity lines holds; see Section 4.9. The singularity curves of $\hat{\mathcal{S}}_{-n} \cup \hat{\mathcal{S}}_m$ for any $m, n > 0$ divide $\mathcal{M}_\diamond$ into curvilinear polygons whose internal angles do not exceed π (the "convexity"); see again Section 4.9 (Exercise 4.49).

Note that the sets $\hat{\mathcal{S}}_\infty$ and $\hat{\mathcal{S}}_{-\infty}$ (cf. (4.53)) are dense in $\mathcal{M}_\diamond$ due to the hyperbolicity of the map $\mathcal{F}_\diamond$ (cf. Section 4.11).

8.11. Construction of stable and unstable manifolds

The properties of singularity curves described in the previous section allow us to construct stable and unstable manifolds for the stadia in exactly the same manner as in Section 4.11.

Let $x \in \mathcal{M}_\diamond \setminus \hat{\mathcal{S}}_{-\infty}$ and for any $n \geq 1$ denote by $\mathcal{Q}_{-n}(x)$ the connected component of the open set $\mathcal{M}_\diamond \setminus \hat{\mathcal{S}}_{-n}$ that contains x. The intersection of their closures

$$\tilde{W}^u(x) \colon = \cap_{n=1}^\infty \overline{\mathcal{Q}_{-n}}(x)$$

is a closed continuous monotonically increasing curve. The unstable manifold $W^u(x)$ is the curve $\tilde{W}^u(x)$ without its endpoints. Its slope at every point $y \in W^u(x)$ equals $\mathcal{V}^u(y) = \mathcal{K} + \mathcal{B}^{u-}(y)\cos\varphi$, where $\mathcal{B}^{u-}(y)$ is given by (8.7). The map $\mathcal{F}_\diamond^{-n}$ is smooth on $W^u(x)$ for all $n \geq 1$, and the preimages of $W^u(x)$ contract under $\mathcal{F}_\diamond^{-1}$ uniformly and exponentially in time. Proposition 8.38 ensures that unstable manifolds are C^2 smooth and have uniformly bounded derivatives.

Stable manifolds $W^s(x)$ for $x \in \mathcal{M}_\diamond$ are constructed likewise and have similar (dual) properties.

REMARK 8.41. We restricted our analysis to the C^2 smoothness, as this is sufficient for the proofs of basic ergodic and statistical properties of the stadia. However, as the boundary of the stadia is piecewise C^∞, we expect stable and unstable manifolds, as well as singularity curves, to be infinitely differentiable with the kth derivative being uniformly bounded by a constant $C_k(\mathcal{D}) > 0$ for every $k \geq 1$ (cf. Propositions 4.29 and 4.43). This can be proved by extending Proposition 8.38 to higher order derivatives, and we leave this as a (challenging) project for the reader.

Next we discuss the size of unstable manifolds. A point $x \in \mathcal{M}_\diamond$ divides $W^u(x)$ into two segments, and we denote by $r^u(x)$ the length (in the Euclidean metric) of the shorter one. If $W^u(x)$ is empty, then we put $r^u(x) = 0$.

THEOREM 8.42. *$W^u(x)$ exists (i.e. $r^u(x) > 0$) for almost every $x \in \mathcal{M}_\diamond$. Furthermore,*

$$\mu_\diamond\{x \in \mathcal{M}_\diamond : r^u(x) < \varepsilon\} \leq C\varepsilon \tag{8.31}$$

for some constant $C = C(\mathcal{D}) > 0$ and all $\varepsilon > 0$.

PROOF. The argument is very similar to our proofs of Theorem 4.66 and the estimate (4.59) in Theorem 4.75; we describe only two steps that are different here.

First, we define a piecewise constant function $E_u(x)$ on $\mathcal{M}_\diamond$ as follows: at a first sliding collision in a series of length k we set $E_u(x) = k$ and at a last collision in that series we set $E_u(x) = \sqrt{k}$; at a collision before a long series of m reflections at the flat sides $E_u(x) = m$. Then our estimates at the end of Section 8.9 imply that there is a small constant $\tilde{c} = \tilde{c}(\mathcal{D}) > 0$ such that $\tilde{c}E_u(x)$ is the lower bound on the factor of expansion of unstable vectors $dx \in \mathcal{T}_x\mathcal{M}_\diamond$ for all $x \in \mathcal{M}_\diamond$, i.e. $\|D_x\mathcal{F}_\diamond(dx)\| \geq \tilde{c}E_u(x)\,\|dx\|$.

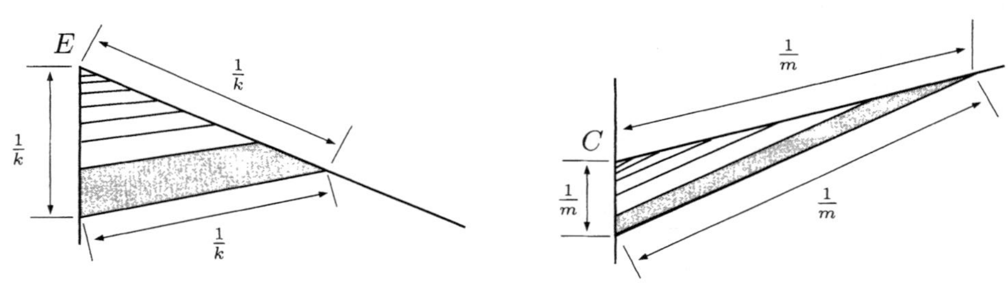

FIGURE 8.13. Singularity curves.

Second, we need to prove an analogue of (4.62); i.e.

$$\mu_\diamond\big(\tilde{\mathcal{U}}^u_\varepsilon(\hat{\mathcal{S}}_1)\big) < C\varepsilon \tag{8.32}$$

where

$$\tilde{\mathcal{U}}^u_\varepsilon(\hat{\mathcal{S}}_1) = \{x \in \mathcal{M}_\diamond : E_u(x)\, d^u(x, \hat{\mathcal{S}}_1) < \varepsilon\}$$

and $d^u(x, \hat{\mathcal{S}}_1)$ denotes the length of the shortest unstable curve that connects x with the set $\hat{\mathcal{S}}_1$.

This requires a detailed analysis of singularities in the vicinities of the points where the singularity lines accumulate, i.e. the points E, B, C, D in the hexagon shown in Fig. 8.10. A close-up of the vicinities of points E and C is shown in Fig. 8.13; the picture near points B and D is symmetric.

Points of first collisions in long series of k sliding collisions fill a grey quadrilateral on the left (near E); points before series of m reflections at the flat sides fill a grey quadrilateral on the right (near C).

EXERCISE 8.43. Show that the dimensions of these figures are, in the sense of $\asymp$, as indicated in Fig. 8.13. Thus the left quadrilateral has dimensions $k^{-1} \times k^{-2}$, and the right quadrilateral has dimensions $m^{-1} \times m^{-2}$.

The intersection of $\tilde{\mathcal{U}}_\varepsilon^u$ with the left quadrilateral lies within the $(ck^{-1/2}\varepsilon)$-neighborhood of its boundary, and the intersection of $\tilde{\mathcal{U}}_\varepsilon^u$ with the right quadrilateral lies within the $(cm^{-1}\varepsilon)$-neighborhood of its boundary, where $c > 0$ is a constant. Now

$$\mu_\diamond\big(\tilde{\mathcal{U}}_\varepsilon^u(\hat{\mathcal{S}}_1)\big) < c\sum_k k^{-5/2}\varepsilon + c\sum_m m^{-2}\varepsilon < C\varepsilon \tag{8.33}$$

for some constants $c, C > 0$ (note that the extra factor k^{-1} is included here due to the density $\cos\varphi \asymp 1/k$ of the measure $\mu_\diamond$). □

EXERCISE 8.44. Prove a linear tail bound similar to (8.31) for the p-metric:

$$\mu_\diamond\{x \in \mathcal{M}_\diamond : p^u(x) < \varepsilon\} \leq C\varepsilon$$

where $p^u(x)$ is the distance in the p-metric from x to the nearer endpoint of $W^u(x)$. Hint: Recall the proof of (4.60) in Section 4.14; the function $E_u(x)$ can be defined as above. Then put

$$\tilde{\mathcal{U}}_{p,\varepsilon}^u(\hat{\mathcal{S}}_1) = \{x \in \mathcal{M}_\diamond : E_u(x)\, d_p^u(x, \hat{\mathcal{S}}_1) < \varepsilon\}$$

where $d_p^u(x, \hat{\mathcal{S}}_1)$ denotes the length (in the p-metric) of the shortest unstable curve that connects x with the set $\hat{\mathcal{S}}_1$. Lastly show that $\mu_\diamond\big(\tilde{\mathcal{U}}_\varepsilon^u(\hat{\mathcal{S}}_1)\big) < C\varepsilon$. Note that (8.33) needs to be modified only slightly:

$$\mu_\diamond\big(\tilde{\mathcal{U}}_\varepsilon^u(\hat{\mathcal{S}}_1)\big) < c\sum_k k^{-3/2}\varepsilon + c\sum_m m^{-2}\varepsilon < C\varepsilon.$$

The following is an analogue of Exercise 4.50:

LEMMA 8.45. *Let $W \subset \mathcal{M}_\diamond$ be an unstable curve on which $\mathcal{F}_\diamond$ is smooth. Then $|\mathcal{F}_\diamond(W)| \leq C\sqrt{|W|}$, where $C = C(\mathcal{D}) > 0$ is a constant.*

PROOF. We denote by $x = (r, \varphi)$ points on the curve W and by $x_1 = (r_1, \varphi_1)$ points on its image $\mathcal{F}_\diamond(W)$. The claim is trivial unless $\cos\varphi_1 \approx 0$ (cf. (8.23)), so there are two cases to consider. First, if W consists of first collisions in a long series of k sliding collisions, then $|W| \leq \text{const} \cdot k^{-2}$ due to Exercise 8.43; hence

$$|\mathcal{F}_\diamond(W)| \asymp k|W| \leq \text{const} \cdot |W|^{1/2}.$$

The more difficult case is where W consists of last collisions in a long series of k sliding collisions. Then $\mathcal{F}_\diamond(W)$ consists of *first* collisions of the next sliding series (on the opposite semicircle); thus $\|dx_1\| \asymp dr_1 \asymp d\varphi_1$ on $\mathcal{F}_\diamond(W)$. We introduce a new variable $s = \cos\varphi_1$ on $\mathcal{F}_\diamond(W)$ and denote by $s_{\min}$ and $s_{\max}$ its extreme values on that curve. Then $\|dx_1\| \asymp ds$ on $\mathcal{F}_\diamond(W)$ and $|\mathcal{F}_\diamond(W)| \asymp s_{\max} - s_{\min}$. Denote by $\|dx_1\|/\|dx\| \asymp s$ the (local) factor of contraction of the curve $\mathcal{F}_\diamond(W)$ under the map $\mathcal{F}_\diamond^{-1}$; cf. (8.23). Now

$$|W| = \int_W \|dx\| \asymp \int_{\mathcal{F}_\diamond(W)} s\,\|dx_1\| \asymp \int_{s_{\min}}^{s_{\max}} s\,ds \asymp s_{\max}^2 - s_{\min}^2.$$

Now the lemma follows from the obvious: $(s_{\max} - s_{\min})^2 \le s_{\max}^2 - s_{\min}^2$. □

EXERCISE 8.46. Show that if a sequence of unstable manifolds $\{W_i^u\}$ converges, as $i \to \infty$, to a curve W (in the C^0 metric), then W (taken without its endpoints) is an unstable manifold itself. (Recall Exercise 4.71.)

EXERCISE 8.47. Show that for $\mu_\diamond$-almost every $x \in \mathcal{M}_\diamond$ both endpoints of $W^u(x)$ belong to $\hat{\mathcal{S}}_{-\infty}$. (Recall Exercise 4.73.)

We have constructed and described stable and unstable manifolds for the return map $\mathcal{F}_\diamond\colon \mathcal{M}_\diamond \to \mathcal{M}_\diamond$. Of course, they coincide with stable and unstable manifolds for the original map $\mathcal{F}$ on $\mathcal{M}_\diamond$ as well. By iterating them under the map $\mathcal{F}$ into the area $\mathcal{M} \setminus \mathcal{M}_\diamond$ we readily obtain stable and unstable manifolds for the map $\mathcal{F}$ in the entire collision space $\mathcal{M}$.

8.12. u-SRB densities and distortion bounds

Here we extend the basic results of Sections 5.1–5.6 to stadia. We begin with two general facts.

First, two measurable partitions ξ^u and ξ^s of the reduced space $\mathcal{M}_\diamond$ (as well as the entire space $\mathcal{M}$) can be constructed exactly as in Section 5.1 (word for word). Second, conditional measures on unstable manifolds $W = W^u(x)$ are absolutely continuous and their densities ρ_W (u-SRB densities) satisfy a fundamental formula analogous to (5.4):

$$\frac{\rho_W(y)}{\rho_W(z)} = \lim_{n\to\infty} \frac{\mathcal{J}_W\mathcal{F}_\diamond^{-n}(y)}{\mathcal{J}_W\mathcal{F}_\diamond^{-n}(z)} \tag{8.34}$$

for all $y, z \in W$. Here $\mathcal{J}_W\mathcal{F}_\diamond^{-n}(x) = \|D_x\mathcal{F}_\diamond^{-n}(dx)\|/\|dx\|$, where dx is a nonzero tangent vector to W at the point x.

The existence of the above limit for typical unstable manifolds W can be proved almost as in Section 5.2, but with a few important changes. First, taking the logarithm and using the chain rule reduce this problem to the convergence of the series

$$\sum_{n=0}^{\infty}\Big(\ln \mathcal{J}_{W_n}\mathcal{F}_\diamond^{-1}(y_n) - \ln \mathcal{J}_{W_n}\mathcal{F}_\diamond^{-1}(z_n)\Big), \tag{8.35}$$

where $W_n = \mathcal{F}_\diamond^{-n}(W)$ and $y_n = \mathcal{F}_\diamond^{-n}(y)$ (the same for z_n). Just as in Section 5.2, we pick $x \in W$ and denote $x_n = \mathcal{F}_\diamond^{-n}(x)$. Then

$$\begin{aligned}\mathcal{J}_{W_n}\mathcal{F}_\diamond^{-1}(x_n) &= \frac{1}{|1+\tau_{n+1}\mathcal{B}_{n+1}^+|}\,\frac{\cos\varphi_n}{\cos\varphi_{n+1}}\,\frac{\sqrt{1+\mathcal{V}_{n+1}^2}}{\sqrt{1+\mathcal{V}_n^2}}\\ &= \frac{\cos\varphi_n}{|2\mathcal{K}_{n+1}\tau_{n+1} + \cos\varphi_{n+1}\big(1+\tau_{n+1}\mathcal{B}_{n+1}^-\big)|}\,\frac{\sqrt{1+\mathcal{V}_{n+1}^2}}{\sqrt{1+\mathcal{V}_n^2}},\end{aligned}$$

which differs from (5.6) by only two absolute value signs. Therefore

$$\begin{aligned}\ln \mathcal{J}_{W_n}\mathcal{F}_{\diamond}^{-1}(x_n) &= \ln\cos\varphi_n + \tfrac{1}{2}\ln(1+\mathcal{V}_{n+1}^2) - \tfrac{1}{2}\ln(1+\mathcal{V}_n^2)\\ &\quad - \ln\bigl|2\mathcal{K}_{n+1}\tau_{n+1} + \cos\varphi_{n+1}\bigl(1+\tau_{n+1}\mathcal{B}_{n+1}^-\bigr)\bigr|,\end{aligned} \tag{8.36}$$

which differs from (5.7) by another absolute value sign. And again, due to the uniform hyperbolicity of $\mathcal{F}_\diamond$ we obtain

$$\bigl|\ln \mathcal{J}_{W_n}\mathcal{F}_{\diamond}^{-1}(y_n) - \ln \mathcal{J}_{W_n}\mathcal{F}_{\diamond}^{-1}(z_n)\bigr| \leq \hat{C}\Lambda^{-n}\max\Bigl|\frac{d}{dx_n}\ln \mathcal{J}_{W_n}\mathcal{F}_{\diamond}^{-1}(x_n)\Bigr|$$

where the maximum is taken over all $x_n \in W_n$ and d/dx_n denotes the derivative with respect to the Euclidean length on W_n. Next, the same considerations as in Section 5.2 ensure that the derivatives of all the expressions on the right-hand side of (8.36), except the first one, are uniformly bounded. The derivative of the first term is

$$\Bigl|\frac{d}{dx_n}\ln\cos\varphi_n\Bigr| = \Bigl|-\frac{\sin\varphi_n}{\cos\varphi_n}\frac{d\varphi_n}{dr_n}\frac{dr_n}{dx_n}\Bigr| \leq \frac{\text{const}\cdot|\mathcal{V}_n|}{\cos\varphi_n}.$$

For dispersing billiards we had $\mathcal{V}_n \asymp 1$ and thus arrived at (5.8), but here we can do a little better: namely, we get

$$\frac{d}{dx_n}\ln \mathcal{J}_{W_n}\mathcal{F}_{\diamond}^{-1}(x_n) = P_n(x_n) + \frac{|\mathcal{V}_n|}{\cos\varphi_n}Q_n(x_n) \tag{8.37}$$

where P_n and Q_n are uniformly bounded functions of x_n (unlike Section 5.2, we do not need their derivatives). This improvement of (5.8) will be important for distortion bounds below. The rest of the proof of the convergence of the series (8.35) is identical to that in Section 5.2, so we will not repeat it.

Next we derive distortion bounds. As in Section 5.6, let W be an unstable curve such that $W_n = \mathcal{F}_{\diamond}^{-n}(W)$ is also an unstable curve for all $0 \leq n \leq N-1$. In addition, we assume that $W \cap \hat{\mathcal{S}}_1 = \emptyset$. For every point $x \in W$ we put $x_n = \mathcal{F}_{\diamond}^{-n}(x)$ and denote by $\mathcal{J}_W\mathcal{F}_{\diamond}^{-n}(x)$ the factor of contraction of W under the map $\mathcal{F}_{\diamond}^{-n}$ at x. The following is an analogue of Lemma 5.27:

LEMMA 8.48. *For every $y, z \in W$ and every $1 \leq n \leq N$ we have*

$$C_{\mathrm{d}}^{-1} \leq e^{-C|W|^{1/2}} \leq \frac{\mathcal{J}_W\mathcal{F}_{\diamond}^{-n}(y)}{\mathcal{J}_W\mathcal{F}_{\diamond}^{-n}(z)} \leq e^{C|W|^{1/2}} \leq C_{\mathrm{d}},$$

where $C = C(\mathcal{D}) > 0$ and $C_{\mathrm{d}} = C_{\mathrm{d}}(\mathcal{D}) > 0$ are constants.

REMARK 8.49. We do not need homogeneity strips constructed in Section 5.3 for dispersing billiards – distortion bounds hold for all unstable curves, provided they do not intersect $\hat{\mathcal{S}}_1$. The reason is that the singularity lines of $\hat{\mathcal{S}}_1$ already divide the 'hazardous' area where $\cos\varphi \approx 0$ into sufficiently narrow strips (see Fig. 8.13), which ensures the homogeneity of unstable curves in each strip.

REMARK 8.50. The bounds in Lemma 8.48 are apparently stronger than those in Lemma 5.27, because $|W|^{1/2} < |W|^{1/3}$. This does not mean, however, that billiards in stadia are 'more regular' than dispersing billiards: the power $1/3$ was an artifact of our rather arbitrary construction of the homogeneity strips; in fact we can easily modify it to change the power from $1/3$ to $1/2$, but there is no need for that.

PROOF. Taking the logarithm and using the chain rule give

$$\begin{aligned}\left|\ln \mathcal{J}_W\mathcal{F}_\diamond^{-n}(y) - \ln \mathcal{J}_W\mathcal{F}_\diamond^{-n}(z)\right| &\le \sum_{i=0}^{n-1}\left|\ln \mathcal{J}_{W_i}\mathcal{F}_\diamond^{-1}(y_i) - \ln \mathcal{J}_{W_i}\mathcal{F}_\diamond^{-1}(z_i)\right| \\ &\le \sum_{i=0}^{n-1} |W_i| \max\left|\frac{d}{dx_i}\ln \mathcal{J}_{W_i}\mathcal{F}_\diamond^{-1}(x_i)\right| \\ &\le \text{const} \sum_{i=0}^{n-1} |W_i|(1+|\mathcal{V}_i|/\cos\varphi_i)\end{aligned}$$

where at the last step we used (8.37). Now we claim that

$$|W_i|\,|\mathcal{V}_i|/\cos\varphi_i \le \text{const}\,|W_i|^{1/2}. \tag{8.38}$$

This is trivial unless $\cos\varphi_i$ is small, which happens only at first and last collisions in long series of sliding collisions. At a first sliding collision $|W_i| \asymp k^{-2}$ (see Exercise 8.43), and $\cos\varphi_i \asymp k^{-1}$ and $|\mathcal{V}_i| \asymp 1$ (here k is the number of collisions in the series), which implies (8.38). At a last sliding collision $|W_i| \asymp k^{-1}$ and $\cos\varphi_i \asymp k^{-1}$, but we also have $|\mathcal{V}_i| \asymp k^{-1}$ (see Table 8.1), which completes the proof of (8.38). Hence

$$\begin{aligned}\left|\ln \mathcal{J}_W\mathcal{F}_\diamond^{-n}(y) - \ln \mathcal{J}_W\mathcal{F}_\diamond^{-n}(z)\right| &\le \text{const} \sum_{i=0}^{n-1} |W_i|^{1/2} \\ &\le \text{const}\,|W|^{1/2},\end{aligned}$$

where at the last step we used the uniform hyperbolicity of $\mathcal{F}_\diamond$. Now Lemma 8.48 is proved. □

COROLLARY 8.51. *For every $x \in W$ and every $1 \le n \le N$*

$$C_{\mathrm{d}}^{-1} \le \frac{\mathcal{J}_W\mathcal{F}_\diamond^{-n}(x)}{|W_n|/|W|} \le C_{\mathrm{d}}.$$

The following is an analogue of Theorem 5.29:

THEOREM 8.52 (Distortion bounds). *For every unstable manifold $W \subset \mathcal{M}_\diamond$ that does not intersect $\hat{\mathcal{S}}_1$*

$$\left|\frac{d}{dx}\ln \rho_W(x)\right| \le \frac{C}{|W|^{1/2}},$$

where $C = C(\mathcal{D}) > 0$ is a constant.

Again note that the bounds in Theorem 8.52 are stronger than those in Theorem 5.29, because $|W|^{1/2} < |W|^{1/3}$.

PROOF. As in the proof of Theorem 5.29, we fix a point $\bar{x} \in W$. Then due to (8.34) and the chain rule

$$\ln \rho_W(x) = \ln \rho_W(\bar{x}) + \sum_{n=0}^{\infty}\bigl(\ln \mathcal{J}_{W_n}\mathcal{F}_\diamond^{-1}(x_n) - \ln \mathcal{J}_{W_n}\mathcal{F}_\diamond^{-1}(\bar{x}_n)\bigr);$$

hence

$$\begin{aligned}\frac{d}{dx}\ln\rho_W(x) &= \sum_{n=0}^{\infty}\frac{d}{dx}\ln\mathcal{J}_{W_n}\mathcal{F}_{\diamond}^{-1}(x_n)\\ &= \sum_{n=0}^{\infty}\frac{dx_n}{dx}\frac{d}{dx_n}\ln\mathcal{J}_{W_n}\mathcal{F}_{\diamond}^{-1}(x_n)\\ &= \sum_{n=0}^{\infty}\mathcal{J}_W\mathcal{F}_{\diamond}^{-n}(x)\left[P_n(x_n)+\frac{|\mathcal{V}_n|\,Q_n(x_n)}{\cos\varphi_n}\right],\end{aligned}$$

where P_n and Q_n are uniformly bounded functions of x_n from (8.37). Therefore,

$$\begin{aligned}\left|\frac{d}{dx}\ln\rho_W(x)\right| &\leq \text{const}\sum_{n=0}^{\infty}\mathcal{J}_W\mathcal{F}_{\diamond}^{-n}(x)\left[1+\mathcal{V}_n/\cos\varphi_n\right]\\ &\leq \text{const}\sum_{n=0}^{\infty}\frac{|W_n|/|W|}{|W_n|^{1/2}}\\ &\leq \text{const}\sum_{n=0}^{\infty}\frac{\Lambda^{-n/2}}{|W|^{1/2}}\end{aligned}$$

where we used (8.38), Corollary 8.51, and the uniform hyperbolicity of $\mathcal{F}_{\diamond}$; cf. (8.22). Theorem 8.52 is proved. □

COROLLARY 8.53. *For every $x, y \in W$*

$$C_{\mathrm{d}}^{-1} \leq e^{-C|W|^{1/2}} \leq \frac{\rho_W(x)}{\rho_W(y)} \leq e^{C|W|^{1/2}} \leq C_{\mathrm{d}},$$

where $C = C(\mathcal{D}) > 0$ is a constant.

Thus we derived all the basic distortion bounds for stadia, along the lines of Section 5.6. We will not estimate higher order derivatives here to avoid unnecessary complications.

8.13. Absolute continuity

Here we estimate the Jacobian of the holonomy map (for general definitions and discussions, we refer the reader to Section 5.7). Let $W^1, W^2 \subset \mathcal{M}_{\diamond}$ be two unstable curves. Denote

$$W_*^i = \{x \in W^i \colon W^s(x) \cap W^{3-i} \neq \emptyset\}$$

for $i = 1, 2$. The holonomy map $\mathbf{h}\colon W_*^1 \to W_*^2$ takes points $x \in W_*^1$ to $\bar{x} = W^s(x) \cap W^2$. The Jacobian of $\mathbf{h}$, with respect to the Lebesgue measures on W^1 and W^2, satisfies a basic formula analogous to (5.21):

$$J\mathbf{h}(x) = \lim_{n\to\infty}\frac{\mathcal{J}_{W^1}\mathcal{F}_{\diamond}^n(x)}{\mathcal{J}_{W^2}\mathcal{F}_{\diamond}^n(\mathbf{h}(x))}. \tag{8.39}$$

While the general theory of hyperbolic maps [**KS86**] guarantees that the Jacobian is positive and finite for almost every point $x \in W_*^1$, it is important to obtain explicit bounds on it.

We assume that W^1 and W^2 are C^2 smooth curves with uniformly bounded first and second derivatives (then their future images will have uniformly bounded derivatives as well, due to Proposition 8.38).

Let $x \in W^1_*$ and $\bar{x} = \mathbf{h}(x) \in W^2_*$. We assume that the segment of $W^s(x)$ between the points x and $\bar{x}$ does not cross $\hat{\mathcal{S}}_{-1}$ (this will replace the assumption of homogeneity of $W^s(x)$ that we had to make for dispersing billiards in Section 5.8; this is in accordance with Remark 8.49). We put $\delta = \operatorname{dist}(x, \bar{x})$ and denote by $\gamma = |\mathcal{V}(x) - \mathcal{V}(\bar{x})|$ the 'angle' between the tangent vectors to the curves W^1 and W^2 at the points x and $\bar{x}$, respectively.

The following is an analogue of Theorem 5.42:

THEOREM 8.54. *The Jacobian of the holonomy map* $\mathbf{h}$ *is uniformly bounded*

$$C^{-1} \leq J\mathbf{h}(x) \leq C,$$

for all $x \in W^1_*$*; here* $C = C(\mathcal{D}) > 1$ *is a constant. Moreover,*

$$A^{-\gamma-\delta^{1/2}} \leq J\mathbf{h}(x) \leq A^{\gamma+\delta^{1/2}}, \tag{8.40}$$

where $A = A(\mathcal{D}) > 1$ *is a constant.*

PROOF. Our argument follows the lines of the proof of Theorem 5.42. Denote $W^i_n = \mathcal{F}^n_\diamond(W^i)$ for $i = 1, 2$ and $n \geq 1$. Also denote $x_n = \mathcal{F}^n_\diamond(x)$ and $\bar{x}_n = \mathcal{F}^n_\diamond(\bar{x})$ and put $\delta_n = \operatorname{dist}(x_n, \bar{x}_n)$. Due to the uniform hyperbolicity of $\mathcal{F}_\diamond$ (cf. (8.22)), $\delta_n \leq \hat{C}\delta\Lambda^{-n}$, where $\hat{C} = 1/\hat{c}$.

Taking the logarithm of (8.39) and using the chain rule give

$$\ln J\mathbf{h}(x) = \sum_{n=0}^{\infty} \Big(\ln \mathcal{J}_{W^1_n}\mathcal{F}_\diamond(x_n) - \ln \mathcal{J}_{W^2_n}\mathcal{F}_\diamond(\bar{x}_n) \Big). \tag{8.41}$$

Due to (8.21), (8.16), and (8.3)

$$\begin{aligned} \mathcal{J}_{W^1_n}\mathcal{F}_\diamond(x_n) &= |1 + \tau_n \mathcal{B}^+_n| \, \frac{\cos\varphi_n}{\cos\varphi_{n+1}} \, \frac{\sqrt{1+\mathcal{V}^2_{n+1}}}{\sqrt{1+\mathcal{V}^2_n}} \\ &= \frac{\big|\cos\varphi_n + \tau_n(\mathcal{K}_n + \mathcal{V}_n)\big|}{\cos\varphi_{n+1}} \, \frac{\sqrt{1+\mathcal{V}^2_{n+1}}}{\sqrt{1+\mathcal{V}^2_n}}, \end{aligned}$$

where, as usual, $\tau_n = \tau(x_n)$, and $\mathcal{B}_n$, $\mathcal{K}_n$, etc., are taken at the point x_n. Therefore

$$\begin{aligned} \ln \mathcal{J}_{W^1_n}\mathcal{F}_\diamond(x_n) &= -\ln\cos\varphi_{n+1} + \tfrac{1}{2}\ln(1+\mathcal{V}^2_{n+1}) - \tfrac{1}{2}\ln(1+\mathcal{V}^2_n) \\ &\quad + \ln\big|\cos\varphi_n + \tau_n(\mathcal{K}_n + \mathcal{V}_n)\big|. \end{aligned} \tag{8.42}$$

Using similar notation at $\bar{x}_n$ we get

$$\begin{aligned} \ln \mathcal{J}_{W^2_n}\mathcal{F}_\diamond(\bar{x}_n) &= -\ln\cos\bar{\varphi}_{n+1} + \tfrac{1}{2}\ln(1+\bar{\mathcal{V}}^2_{n+1}) - \tfrac{1}{2}\ln(1+\bar{\mathcal{V}}^2_n) \\ &\quad + \ln\big|\cos\bar{\varphi}_n + \bar{\tau}_n(\mathcal{K}_n + \bar{\mathcal{V}}_n)\big|. \end{aligned} \tag{8.43}$$

(Note that $\mathcal{K}_n$ is the same at both points x_n and $\bar{x}_n$.) Comparing the first terms of the above expressions gives

$$\begin{aligned} \big|\ln\cos\varphi_{n+1} - \ln\cos\bar{\varphi}_{n+1}\big| &\leq \frac{\operatorname{const}|\varphi_{n+1} - \bar{\varphi}_{n+1}|}{\cos\varphi_{n+1}} \\ &\leq \operatorname{const}\delta^{1/2}_{n+1} \end{aligned}$$

where we used (8.38), more precisely, its 'time reversal' counterpart. All the other terms in (8.42) and (8.43) are uniformly bounded and differentiable; thus it easily follows that

$$\big|\ln \mathcal{J}_{W_n^1}\mathcal{F}_\diamond(x_n) - \ln \mathcal{J}_{W_n^2}\mathcal{F}_\diamond(\bar{x}_n)\big| \le C\big(\delta_{n+1}^{1/2} + \gamma_n + \gamma_{n+1} + \delta_n\big) \tag{8.44}$$

where $C > 0$ is a constant and $\gamma_n = |\Delta\mathcal{V}_n| = |\mathcal{V}_n - \bar{\mathcal{V}}_n|$ stands for the 'angle' between the tangent vectors to the curves W_n^1 and W_n^2 at the points x_n and $\bar{x}_n$, respectively (note that $\gamma_0 = \gamma$).

Due to Lemma 8.45 (more precisely, due to its 'time reversal' counterpart), we have $\delta_n \le C\delta_{n+1}^{1/2}$; thus the term δ_n in (8.44) may be dropped. It remains to estimate γ_n for all $n \ge 0$. For brevity, we denote $\Delta\mathcal{B}_n^- = \mathcal{B}_n^- - \bar{\mathcal{B}}_n^-$, $\Delta\tau_n = \tau_n - \bar{\tau}_n$, etc. First we estimate $\Delta\mathcal{B}_{n+1}^-$ by using (3.31):

$$\begin{aligned}
\Delta\mathcal{B}_{n+1}^- &= \frac{1}{\tau_n + 1/\mathcal{B}_n^+} - \frac{1}{\bar{\tau}_n + 1/\bar{\mathcal{B}}_n^+} \\
&= -\frac{1}{\tau_n + 1/\mathcal{B}_n^+}\,\frac{1}{\bar{\tau}_n + 1/\bar{\mathcal{B}}_n^+}\left(\Delta\tau_n + \frac{1}{\mathcal{B}_n^+} - \frac{1}{\bar{\mathcal{B}}_n^+}\right) \\
&= -\frac{\Delta\tau_n}{\big(\tau_n + 1/\mathcal{B}_n^+\big)\big(\bar{\tau}_n + 1/\bar{\mathcal{B}}_n^+\big)} + \frac{\Delta\mathcal{R}_n + \Delta\mathcal{B}_n^-}{\big(1 + \tau_n\mathcal{B}_n^+\big)\big(1 + \bar{\tau}_n\bar{\mathcal{B}}_n^+\big)}
\end{aligned} \tag{8.45}$$

(recall that $\mathcal{B}_n^+ = \mathcal{R}_n + \mathcal{B}_n^-$).

EXERCISE 8.55. Show that $|\tau_n + 1/\mathcal{B}_n^+| \ge \tau_{\min}/2$. Hint: If the points x_n and x_{n+1} belong to the opposite semicircles, use (8.2). For collisions at the same semicircle, the situation is similar to dispersing billiards due to our reduction scheme in Section 8.7.

Now the first term in (8.45) is bounded by

$$\text{const}\,\Delta\tau_n \le \text{const}(\delta_n + \delta_{n+1}) \le \text{const}\,\delta_n.$$

Next,

$$\frac{\Delta\mathcal{R}_n}{\big(1 + \tau_n\mathcal{B}_n^+\big)\big(1 + \bar{\tau}_n\bar{\mathcal{B}}_n^+\big)} \le \frac{2\mathcal{K}_n\cos\bar{\varphi}_n - 2\mathcal{K}_n\cos\varphi_n}{\big(\cos\varphi_n + \tau_n(\mathcal{K}_n + \mathcal{V}_n)\big)\big(\cos\bar{\varphi}_n + \bar{\tau}_n(\mathcal{K}_n + \bar{\mathcal{V}}_n)\big)}.$$

Since the denominator is bounded away from zero, due to (8.20), the absolute value of the fraction is $\mathcal{O}(\delta_n)$; thus (8.45) reduces to

$$\Delta\mathcal{B}_{n+1}^- = \tilde{Q}_n^{(1)} + \frac{\Delta\mathcal{B}_n^-}{\big(1 + \tau_n\mathcal{B}_n^+\big)\big(1 + \bar{\tau}_n\bar{\mathcal{B}}_n^+\big)},$$

where $|\tilde{Q}_n^{(1)}| \le \text{const}\,\delta_n$. Next, due to (8.3)

$$\begin{aligned}
\Delta\mathcal{V}_{n+1} &= \mathcal{B}_{n+1}^-\cos\varphi_{n+1} - \bar{\mathcal{B}}_{n+1}^-\cos\bar{\varphi}_{n+1} \\
&= \tilde{Q}_{n+1}^{(2)} + \cos\varphi_{n+1}\,\Delta\mathcal{B}_{n+1}^-,
\end{aligned}$$

where $|\tilde{Q}_{n+1}^{(2)}| \le \text{const}\,\delta_{n+1}$. Combining the last two estimates gives

$$\Delta\mathcal{V}_{n+1} = \tilde{Q}_n^{(3)} + \frac{\cos\varphi_{n+1}}{\cos\varphi_n}\,\frac{\Delta\mathcal{V}_n}{\big(1 + \tau_n\mathcal{B}_n^+\big)\big(1 + \bar{\tau}_n\bar{\mathcal{B}}_n^+\big)},$$

where $|\tilde{Q}_n^{(3)}| \le \text{const}\,\delta_n$. Now consider the fraction

$$u_n := \frac{\cos\varphi_{n+1}}{\cos\varphi_n\big(1 + \tau_n\mathcal{B}_n^+\big)\big(1 + \bar{\tau}_n\bar{\mathcal{B}}_n^+\big)}.$$

It is easy to verify, just as in Exercises 5.43 and 5.44, that for all $0 \le k \le n$

$$|u_n u_{n-1} \cdots u_k| \le C/\Lambda^{n-k}$$

where $C = C(\mathcal{D}) > 0$ is a constant, and

$$\begin{aligned} \gamma_n = |\Delta \mathcal{V}_n| &\le \text{const}\Big(\sum_{k=0}^{n} \delta_k/\Lambda^{n-k} + \gamma/\Lambda^n \Big) \\ &\le \text{const}\Big(\delta n/\Lambda^n + \gamma/\Lambda^n \Big). \end{aligned} \tag{8.46}$$

Combining (8.46) with (8.44) yields

$$\Big| \ln \mathcal{J}_{W_n^1} \mathcal{F}_\diamond(x_n) - \ln \mathcal{J}_{W_n^2} \mathcal{F}_\diamond(\bar{x}_n) \Big| \le C \Big(\frac{\delta^{1/2}}{\Lambda^{n/3}} + \frac{\delta n}{\Lambda^n} + \frac{\gamma}{\Lambda^n} \Big), \tag{8.47}$$

where $C = C(\mathcal{D}) > 0$ is a constant. Summing up over n gives

$$\big| \ln J\mathbf{h}(x) \big| \le \text{const}\big(\gamma + \delta^{1/2}\big),$$

which completes the proof of the theorem. □

As in Section 5.8, the bound (8.40) can be improved to

$$A^{-\delta^{1/2}} \le J\mathbf{h}(x) \le A^{\delta^{1/2}}$$

if the curves W^1 and W^2 do not intersect and extend beyond the points x and $\bar{x}$ by at least the distance $D\delta^{1/2}$, where $D = D(\mathcal{D}) > 0$ is a large constant.

Lastly we show, as in Section 5.8, that the Jacobian $J\mathbf{h}(x)$, as a function of $x \in W_*^1$ is 'dynamically Hölder continuous'. Let $\mathcal{Q}_n(x)$ again denote the open connected component of the set $\mathcal{M}_\diamond \setminus \hat{\mathcal{S}}_n$ containing the point x. For two points $x, y \in \mathcal{M}_\diamond$ denote by

$$\mathbf{s}_+(x, y) = \min\{n \ge 0 \colon y \notin \mathcal{Q}_n(x)\} \tag{8.48}$$

the 'separation time'.

PROPOSITION 8.56. *There are constants $C > 0$ and $\boldsymbol{\theta} \in (0, 1)$ such that*

$$|\ln J\mathbf{h}(x) - \ln J\mathbf{h}(y)| \le C\boldsymbol{\theta}^{\mathbf{s}_+(x,y)}.$$

PROOF. Denote $\bar{x} = \mathbf{h}(x)$ and $\bar{y} = \mathbf{h}(y)$. Observe that $\mathbf{s}_+(\bar{x}, \bar{y}) = \mathbf{s}_+(x, y)$; this follows from the continuation property of singularity lines (Section 8.10). Using the notation of (8.41) and the triangle inequality gives

$$\begin{aligned} \Delta \colon &= |\ln J\mathbf{h}(x) - \ln J\mathbf{h}(y)| \\ &\le \sum_{n=0}^{\infty} |\ln \mathcal{J}_{W_n^1} \mathcal{F}_\diamond(x_n) - \ln \mathcal{J}_{W_n^2} \mathcal{F}_\diamond(\bar{x}_n) - \ln \mathcal{J}_{W_n^1} \mathcal{F}_\diamond(y_n) + \ln \mathcal{J}_{W_n^2} \mathcal{F}_\diamond(\bar{x}_n)|. \end{aligned}$$

Let $m = \mathbf{s}_+(x, y)/2$. Then we apply the estimate (8.47) to all $n > m$ and Lemma 8.48 (on distortion bounds) to all $n \le m$. This proves our claim with $\boldsymbol{\theta} = \Lambda^{-1/4}$. □

8.14. Growth lemmas

In the previous sections we extended to Bunimovich stadia many basic constructions done earlier (in Chapters 4 and 5) for dispersing billiards, including estimates for the size of stable and unstable manifolds, their uniform C^2 smoothness, distortion bounds, absolute continuity, and the regularity of the holonomy map.

As we said, all these basic results could be further extended to other classes of Bunimovich billiard tables, but we refrained from doing this here, in order to keep our presentation reasonably short. The reader is encouraged to work on such proofs and even publish them, as they have yet to appear in print.

At this point we are at a critical juncture, though. In order to continue our program of study developed for dispersing billiards in Chapters 4–7, we need an analogue of the first growth lemma (Theorem 5.52), which is a crucial element in all our arguments at the end of Chapter 5, as well as in Chapters 6 and 7.

For example, the second and third growth lemmas (Theorems 5.53 and 5.61) follow from the first one in a general manner (see our proofs in Chapter 5). Theorems 5.66 and 5.67, as well as Fundamental Theorem 5.70, can then be easily extended to Bunimovich billiards, as their proofs are based on growth lemmas and other basic facts established earlier (though certain technical details need be adjusted to the new class of billiards). Then we will be in an excellent position to attack ergodic and statistical properties by adapting our arguments of Chapters 6 and 7 to Bunimovich tables.

However, the first growth lemma presents a serious challenge. Unlike other basic facts proved earlier, this lemma is very delicate and sensitive to the details of the dynamics. Before we turn to its proof, we need to discuss various aspects of this lemma in the context of Bunimovich billiards.

First of all, the growth lemma is designed for uniformly hyperbolic maps, according to Section 5.9. It simply fails for nonuniformly hyperbolic systems, even without singularities:

EXERCISE 8.57. Let $F\colon M \to M$ be a smooth hyperbolic map on a 2D manifold. Suppose that there is a constant $C > 0$ such that for any $\delta > 0$ and $n \geq 1$ there is an unstable curve W of length $|W| < \delta$ such that $|F^n(W)| < C|W|$. Also assume distortion bounds in the sense of Lemma 8.48. Show that an analogue of Theorem 5.52 for such a map would fail. Hint: Consider $\varepsilon \ll |W|$ and choose n so that $c_1\vartheta_1^n < 1/C$. Also note that due to the distortion bounds, the map F^n becomes almost linear on W, as $\delta \to 0$.

Thus the first growth lemma cannot hold for the original billiard map $\mathcal{F}$ in Bunimovich tables, so we again need to use the reduced map $\mathcal{F}_\diamond$. It is uniformly hyperbolic, but there are other problems in the way of the growth lemma.

Consider a small unstable curve W in the 'problematic' area of $\mathcal{M}_\diamond$ where the singularity lines accumulate. For example, let W terminate at the vertex E of the hexagon $AHBFGE$ (the corresponding corner of $\mathcal{M}_\diamond$ is shown in Fig. 8.13 and in the left of Fig. 8.14). Then W is divided into pieces of lengths $\asymp 1/k^2$ for $k_1 < k < \infty$, where $k_1 \asymp 1/|W|$; each piece is expanded by $\mathcal{F}_\diamond$ with a factor of $\asymp k$.

EXERCISE 8.58. Verify that, in the notation of Theorem 5.52,

$$\mathbf{m}_W\big(r_1(x) < \varepsilon\big) \asymp \varepsilon|\ln\varepsilon|$$

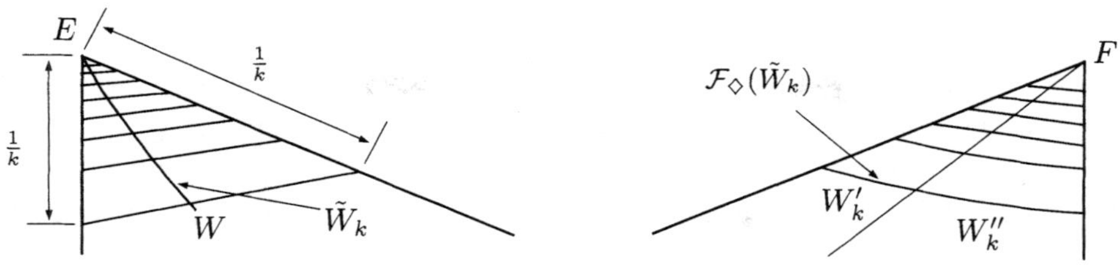

FIGURE 8.14. An unstable curve W (left) and its images $\mathcal{F}_\diamond(W)$ (right).

for all $\varepsilon \ll |W|$; hence the estimate claimed in that theorem must fail.

The reason for the failure of the first growth lemma for the map $\mathcal{F}_\diamond$ becomes clear as we recall its proof (for dispersing billiards) in Section 5.10: the key element of that proof is Lemma 5.56, particularly the inequality (5.38). In terms of $\mathcal{F}_\diamond$, it means the following: given an unstable curve $W \subset \mathcal{M}_\diamond$, denote by W_k all the smooth components of $\mathcal{F}_\diamond(W)$ and put

$$\lambda(W_k) = \max_{y \in W_k} \mathcal{J}_{W_k} \mathcal{F}_\diamond^{-1}(y).$$

Then an analogue of (5.38) would be

$$\liminf_{\delta \to 0} \sup_{W:\, |W| < \delta} \sum_k \lambda(W_k) < 1. \tag{8.49}$$

This property fails for stadia in a major way. Indeed, for unstable curves W terminating at the vertex E of the hexagon $AHBFGE$ (as explained above) the image $\mathcal{F}_\diamond(W)$ has countably many components $\{W_k\}$, $k \geq \mathcal{O}(1/|W|)$, for which $\lambda(W_k) \sim \text{const}/k$. Thus the series in (8.49) actually diverges!

In other words, even though the map $\mathcal{F}_\diamond$ is uniformly hyperbolic, its hyperbolicity is not 'strong enough' to overcome fragmentation caused by singularities in the way required by the first growth lemma. To remedy the situation, we will further reduce the space $\mathcal{M}_\diamond$ so as to strengthen the hyperbolicity of the return map.

Let $\mathcal{M}_\clubsuit \subset \mathcal{M}_\diamond$ consist of only *first* collisions at focusing arcs. This reduces hexagons to parallelograms: the space $\mathcal{M}_\clubsuit$ will be the union of the parallelogram $ECBD$ (Fig. 8.10) and an identical parallelogram corresponding to the opposite semicircle. We record two simple facts about the corresponding return map $\mathcal{F}_\clubsuit : \mathcal{M}_\clubsuit \to \mathcal{M}_\clubsuit$:

(1) Unstable curves and unstable manifolds for the map $\mathcal{F}_\clubsuit$ are, automatically, unstable curves/manifolds for the map $\mathcal{F}_\diamond$ as well.
(2) For all $x \in \mathcal{M}_\clubsuit$ we have $\mathcal{F}_\clubsuit^n(x) = \mathcal{F}_\diamond^{n'}(x)$ for some $n \leq n' \leq 2n$.

Based on these facts, it is not hard to extend all our previous results from $\mathcal{F}_\diamond$ to $\mathcal{F}_\clubsuit$. We proceed to verify the key inequality (8.49) for the new map $\mathcal{F}_\clubsuit$.

Again let a small unstable curve $W \subset \mathcal{M}_\clubsuit$ terminate at the vertex E (of the parallelogram $ECBD$). As we noted, W will be cut by the singularities of the map $\mathcal{F}_\diamond$ into countably many pieces; denote them by $\{\tilde{W}_k\}$, where $k \geq k_W \sim 1/|W|$. The image $\mathcal{F}_\diamond(\tilde{W}_k)$ of each piece will lie near the vertex F; see Fig. 8.14. There it will be cut by $\mathcal{F}_\diamond$ into two subcurves along a unique singularity line of $\mathcal{F}_\diamond$ stretching down and left from the vertex F (that line already appeared in Fig. 8.12), and we denote those subcurves by W_k' and W_k''.

We note that $\mathcal{F}_\clubsuit = \mathcal{F}_\diamond^2$ on W. The first iteration of $\mathcal{F}_\diamond$ expands each piece $\tilde{W}_k$ of W by a factor $\asymp k$, and the second iteration of $\mathcal{F}_\diamond$ expands the subcurves W_k' and W_k'' by a factor $\geq \text{const}\sqrt{k}$, according to (8.24). Hence if we denote by $W_k^{(1)} = \mathcal{F}_\diamond(W_k')$ and $W_k^{(2)} = \mathcal{F}_\diamond(W_k'')$ for $k \geq k_W$ (where $k_W \sim 1/|W|$) all the components of $\mathcal{F}_\clubsuit(W)$, then $\lambda(W_k^{(i)}) \leq \text{const}\, k^{-3/2}$ for $i = 1, 2$. Therefore the series in (8.49) will converge and its sum will be $\mathcal{O}\big(k_W^{-1/2}\big) = \mathcal{O}\big(|W|^{1/2}\big)$. Now the smallness of W guarantees that this sum will be less than one.

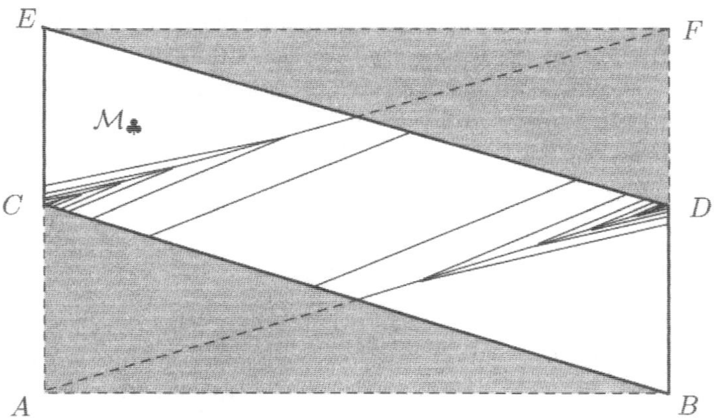

FIGURE 8.15. The parallelogram $ECBD$ of the set $\mathcal{M}_\clubsuit$ and its singularity lines near the vertices C and D.

There is another problematic area where singularity lines accumulate: the vicinity of the point C; cf. Fig. 8.10. Singularity lines of the map $\mathcal{F}_\diamond$ near C are shown in Fig. 8.12. The new map $\mathcal{F}_\clubsuit$ has more singularity lines near C; they are shown in Fig. 8.15 (compare it to Fig. 8.13). There are two types of singularity lines here: the old ones below the line CF and the new ones above the line CF. The reason for the second set of singularity lines is the following: points below the line CF are mapped by the original billiard map $\mathcal{F}$ directly to a flat side of the stadium, while the trajectories of points above CF experience one more collision with the same semicircle followed by a long series of reflections off the flat sides of the stadium (an example of such a trajectory is shown in Fig. 8.16).

The singularity lines of the map $\mathcal{F}_\clubsuit$ divide the vicinity of the point C in the space $\mathcal{M}_\clubsuit$ into cells of two types: $D_{1,m}$ (lying below CD) and $D_{2,m}$ (lying above CF), where $m \geq 1$ is the number of consecutive reflections at the flat sides of the stadium along the trajectories of points $x \in D_{i,m}$, $i = 1, 2$, before they reach the opposite semicircle. These cells are shown in Fig. 8.17; the coordinates of their vertices are also indicated (to the leading order, so that $\mathbf{d}/m$ means $\mathbf{d}/m + o(1/m)$, etc.). Here $\mathbf{d}$ denotes the length of each flat side of the stadium; see Fig. 8.7.

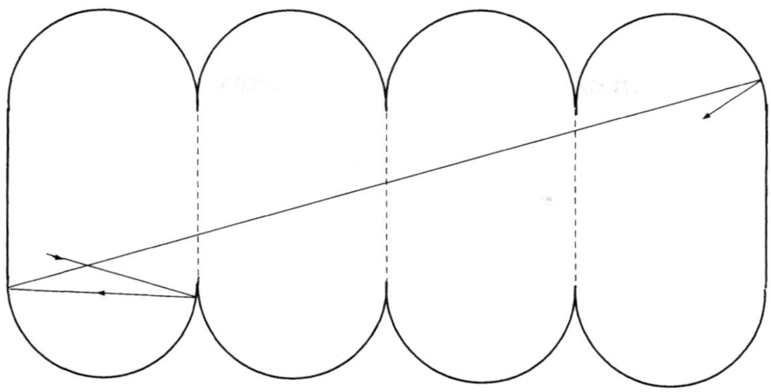

FIGURE 8.16. A trajectory bouncing off a semicircle before a long series of 'flat collisions'.

EXERCISE 8.59. Verify the above statements. Also verify the values of the coordinates of the vertices of the cells. Hint: All this should be done by direct inspection involving nothing but elementary geometry.

Next we assume that W is a 'second-generation' unstable curve; i.e. its preimage $W_{-1} = \mathcal{F}_{\clubsuit}^{-1}(W)$ is an unstable curve, too. Then W_{-1} itself lies in a cell $D_{i,m'}$ for some $i = 1, 2$ and $m' \geq 1$ (of course, W_{-1} lies on the opposite semicircle of $\partial\mathcal{D}$, but the corresponding cell structure is identical).

Fig. 8.17 shows the image $\mathcal{F}_{\clubsuit}(D_{i,m'})$ of the cell $D_{i,m'}$ superimposed on the singularity curves of the map $\mathcal{F}_{\clubsuit}$. The domain $\mathcal{F}_{\clubsuit}(D_{i,m'})$ consists of points landed on a semicircle after m' successive reflections off the flat sides of the stadium. By the involution (Section 2.14), points $(r, \varphi) \in \mathcal{F}_{\clubsuit}(D_{i,m'})$ correspond to $(r, -\varphi) \in D_{1,m'} \cup \mathcal{F}_{\diamondsuit}(D_{2,m'})$; see Fig. 8.17 (note that $\mathcal{F}_{\diamondsuit}(D_{2,m'})$ is the continuation of $D_{1,m'}$ below the line CB). Thus the domain $\mathcal{F}_{\clubsuit}(D_{i,m'})$ is symmetric to $D_{1,m'} \cup \mathcal{F}_{\diamondsuit}(D_{2,m'})$ about the horizontal line $\varphi = 0$. We also note that the images $\mathcal{F}_{\clubsuit}(D_{i,m'})$ for $i = 1$ and $i = 2$ look similar; in fact they are adjacent to each other.

It is not hard to see that the domain $\mathcal{F}_{\clubsuit}(D_{i,m'})$ (and thus the curve W in it) can only intersect cells $D_{i,m}$ with

$$m'/3 + o(m') \leq m \leq 3m' + o(m') \tag{8.50}$$

(this is another solution to Exercise 8.37). The coefficients 1/3 and 3 here are important for our further estimates.

LEMMA 8.60. *Let $W_{-1} \subset D_{i,m'}$ be an unstable curve for some large $m' \gg 1$ and $i = 1, 2$. Put $W = \mathcal{F}_{\clubsuit}(W_{-1}) \subset \mathcal{F}_{\clubsuit}(D_{i,m'})$. Then the subcurve $W \cap D_{1,m}$ is expanded under the map $\mathcal{F}_{\clubsuit}$ by a factor $4m + o(m)$, and the subcurve $W \cap D_{2,m}$ is expanded under the map $\mathcal{F}_{\clubsuit}$ by a factor $8m + o(m)$.*

PROOF. The curve W has slope $d\varphi/dr = -1/\mathbf{r} + o(1)$, as is clear from Fig. 8.17. There is an alternative argument: recall that $d\varphi/dr = \mathcal{V} = \mathcal{K} + \mathcal{B}^- \cos\varphi$ due to (8.3), where $\mathcal{K} = -1/\mathbf{r}$ and $\mathcal{B} = \mathcal{O}(1/m')$. Now the geometric parameters of cells shown in Fig. 8.17 suggest that the intersection $W \cap D_{1,m}$ has dimensions

$$|\Delta r| = \frac{\mathbf{d}}{4m^2} + o(1/m^2) \qquad \text{and} \qquad |\Delta\varphi| = \frac{\mathbf{d}}{4\mathbf{r}m^2} + o(1/m^2).$$

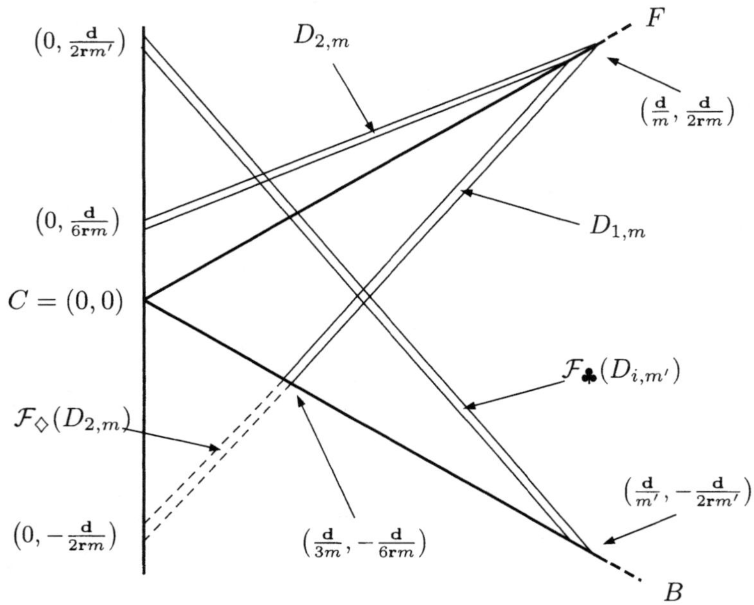

FIGURE 8.17. The cells $D_{1,m}$ and $D_{2,m}$ and the image $\mathcal{F}_\clubsuit(D_{i,m'})$ near the vertex C. The vertex C has coordinates $r = 0$ and $\varphi = 0$; other coordinates are shown, to the leading order.

The map $\mathcal{F}_\clubsuit$ stretches this curve all the way across the domain $\mathcal{F}_\clubsuit(D_{i,m})$, from top to bottom; hence its image has dimensions

$$|\Delta r| = \frac{\mathbf{d}}{m} + o(1/m) \qquad \text{and} \qquad |\Delta\varphi| = \frac{\mathbf{d}}{\mathbf{r}m} + o(1/m). \tag{8.51}$$

Thus the factor of expansion of the curve $W \cap D_{1,m}$ by the map $\mathcal{F}_\clubsuit$ will be $4m + o(m)$.

Similarly, the intersection $W \cap D_{2,m}$ has dimensions

$$|\Delta r| = \frac{\mathbf{d}}{8m^2} + o(1/m^2) \qquad \text{and} \qquad |\Delta\varphi| = \frac{\mathbf{d}}{8\mathbf{r}m^2} + o(1/m^2).$$

The map $\mathcal{F}_\clubsuit$ stretches this curve all the way across the domain $\mathcal{F}_\clubsuit(D_{2,m})$, from top to bottom; hence its image has dimensions given by (8.51). Thus the factor of expansion of the curve $W \cap D_{2,m}$ by the map $\mathcal{F}_\clubsuit$ will be $8m + o(m)$. Our argument here is somewhat sketchy and makes reference to Fig. 8.17 at crucial steps, but the reader is invited to complete the details. □

Summarizing the above estimates we conclude that any sufficiently small unstable curve $W \subset \mathcal{M}_\clubsuit$ in the vicinity of the vertex C is transformed by $\mathcal{F}_\clubsuit$ into pieces $W_{i,m} = \mathcal{F}_\clubsuit(W \cap D_{i,m})$ with $i = 1, 2$ and $m'/3 + o(m') \le m \le 3m' + o(m')$ for some $m' \ge 1$. If we ignore second order terms in our estimates, the series in (8.49) takes form

$$\sum_{m=m'/3}^{3m'} \left[\lambda(W_{1,m}) + \lambda(W_{2,m})\right] = \sum_{m=m'/3}^{3m'} \left(\frac{1}{4m} + \frac{1}{8m}\right) \le \frac{3}{8}\ln 9 + o(1) < 1. \tag{8.52}$$

EXERCISE 8.61. Verify that the second order terms (both in the upper and lower indices of the summation and in the expansion factors) will not affect the final conclusion: the sum remains $< 1-\varepsilon$ for some $\varepsilon > 0$ and all large enough m'.

It is amazing how delicate our analysis is; a slight alteration in the numerical values of the coefficients $(3, 1/3, 4, 8)$ may jeopardize the final conclusion (i.e., the sum might readily exceed one). This demonstrates how sensitive the first growth lemma is to the details of the map.

A related remark: we could define the space $\mathcal{M}_{\clubsuit}$ so that it consisted of *last* (rather than *first*) collisions with focusing arcs (it would then be represented by the parallelogram $ACFD$ in Fig. 8.15). This definition was actually adopted in [**BSC91, Mar04**]. In that case the first growth lemma would technically fail (!), but fortunately it could be modified and salvaged with some extra work [**Mar04**].

REMARK 8.62. However, the first growth lemma truly fails for certain types of Bunimovich billiards, as is described in [**CZ05**, Section 10]. This happens in generic Bunimovich tables whose boundaries contain an arc larger than a semicircle. Also, this failure occurs in a modified stadia bounded by two equal parallel lines and two identical arcs, each *smaller* than a semicircle. In both cases the corresponding sum in (8.49) diverges for any power of the map $\mathcal{F}_{\clubsuit}$. It is not clear whether (or how) the growth lemmas could be modified and salvaged in these cases.

Back to the stadium; the proof of the key estimate (8.49) is now almost complete. We verified it for unstable curves in all 'problematic' regions where singularity curves accumulate. However, other parts of $\mathcal{M}_{\clubsuit}$ may, rather surprisingly, cause trouble, too. Suppose an unstable curve W crosses just one singularity line S, and the corresponding two pieces of W are expanded by factors 1.8 and 1.9 (which is a pretty strong expansion!). Then (8.49) clearly fails, as $(1.8)^{-1} + (1.9)^{-1} > 1$.

The above situation is not at all unusual, and it does occur in many Bunimovich billiards, including stadia. This again shows how delicate the estimate (8.49) is and how easily it breaks down.

To deal with this new complication, let us recall our general discussion in Section 5.9. It demonstrates that a single singularity line S (dividing small unstable curves into two pieces) may wreck havoc, indeed, if the images of W consistently cross S at every iteration of the map. However, if the images of W move without fragmentation for some time, they may recover in size, and the accumulated expansion factor may became large enough so that (8.49) holds true.

This observation suggests considering a higher iterate of the map $\mathcal{F}_{\clubsuit}$. The expansion factor of $\mathcal{F}_{\clubsuit}^n$ grows exponentially with n due to the uniform hyperbolicity. Of course, the number of singularity lines grows exponentially, too, and their structure becomes increasingly more complex. But we are interested only in the *local complexity* of singularity lines, as defined below, since we need to deal with small unstable curves only; remember that $\delta \to 0$ in (8.49).

Let us define the local complexity of singularities. Given a small $\varepsilon > 0$, denote by $\mathcal{M}_{\clubsuit}^{\varepsilon}$ the part of the space $\mathcal{M}_{\clubsuit}$ without ε-neighborhoods of the vertices E, C, B, D, where infinitely many singularity lines accumulate and where we already verified the key estimate (8.49). Given $n \geq 1$ and an unstable curve W, denote by $K(W, n)$ the number of pieces W is divided into by the singularity lines of the map $\mathcal{F}_{\clubsuit}^n$. Then the local complexity of singularities is defined by

$$K_n(\mathcal{M}_{\clubsuit}^{\varepsilon}) = \liminf_{\delta\to 0} \sup_{W\subset \mathcal{M}_{\clubsuit}^{\varepsilon}:\, |W|<\delta} K(W,n).$$

The local complexity is related to the multiplicity of intersections of singularity lines. Suppose that no more than $K_n'(\mathcal{M}_{\clubsuit}^{\varepsilon})$ singularity lines of the map $\mathcal{F}_{\clubsuit}^n$ intersect (meet) at any one point $x \in \mathcal{M}_{\clubsuit}^{\varepsilon}$.

EXERCISE 8.63. Verify that $K_n(\mathcal{M}_{\clubsuit}^{\varepsilon}) \le K_n'(\mathcal{M}_{\clubsuit}^{\varepsilon}) + 1$. Hint: Every singularity line can intersect any unstable curve W only once. Also use the compactness of singularity lines.

It is now clear that the local complexity does not have to grow fast with n. For example, if there are no multiple intersections (say, no more than two or three singularity lines intersect at any point $x \in \mathcal{M}_{\clubsuit}^{\varepsilon}$), then K_n remains constant!

For many billiards the local complexity of singularities grows at most linearly; see [**BSC90**, Section 8] or [**CY00**, pages 110–112] or [**Ch01**, Section 5]. It appears that the linear grows takes place whenever the billiard flow Φ^t is continuous (though not necessarily differentiable); the stadium is one of those cases. However, the proof of that linear bound is rather complicated, so we omit it. We state, without proof, the following fact:

PROPOSITION 8.64 ([**CZ05**]). *For any stadium there are constants $a, b > 0$ such that $K_n(\mathcal{M}_{\clubsuit}^{\varepsilon}) \le a + bn$ for every $\varepsilon > 0$.*

We now complete the proof of (8.49) for the map $\mathcal{F}_{\clubsuit}^n$ for some n. Combining the above proposition with the uniform hyperbolicity (8.21) shows that the corresponding sum in (8.49) is less than $K_n(\mathcal{M}_{\clubsuit}^{\varepsilon})\,\hat{c}^{-1}\,\Lambda^{-n} \le (a+bn)\,\hat{c}^{-1}\,\Lambda^{-n}$. If n is large enough, this product is less than one.

Thus we proved (8.49) for the map $\mathcal{F}_{\clubsuit}^n$ with a sufficiently large n. We fix that value of n. Now the first growth lemma (for the map $\mathcal{F}_{\clubsuit}^n$) can be derived in the same way as in Section 5.10.

This section again demonstrates how complicated Bunimovich billiards are if one follows all the proofs in a rigorous manner. With considerable effort we managed to derive the first growth lemma for the stadia (omitting some parts, such as the proof of Proposition 8.64) and only after an additional reduction of the collision space and replacing the return map with its high enough iterate. We also remarked that the first growth lemma failed for some other types of Bunimovich billiards.

8.15. Ergodicity and statistical properties

What next? Further analysis of stadia (not to mention other Bunimovich tables) may require one or two more chapters of the book, but then it will grow out of proportion. So from now on we will survey only the existing results and just indicate methods of proofs.

Speaking of stadia, the first growth lemma is proved. As we mentioned already, the second and third growth lemmas and Theorems 5.66, 5.67, and 5.70 (all for the map $\mathcal{F}_{\clubsuit}^n$) then follow.

The analysis of Chapter 6 can be applied to the map $\mathcal{F}_{\clubsuit}^n$, and we can establish its ergodicity. Hence $\mathcal{F}_{\clubsuit}$ is ergodic as well. Since $\mathcal{F}_{\clubsuit}$ is the return map obtained from the original billiard map $\mathcal{F}\colon \mathcal{M} \to \mathcal{M}$, the latter will be ergodic, too. The mixing and Bernoulli properties can be established with relatively little extra work, as in Chapter 6.

The proof of Bernoulli property for the billiard flow requires the use of 4-loops as in Chapter 6. The key element is the nonintegrability of stable and unstable manifolds, which can be verified geometrically, as the corresponding wave fronts have curvature of opposite sign (cf. Section 6.11). Note, however, that the last property holds only when the billiard particle is far enough from the focusing arcs; near those arcs both wave fronts have curvature of the same sign.

The statistical properties of the map $\mathcal{F}_{\clubsuit}\colon \mathcal{M}_{\clubsuit} \to \mathcal{M}_{\clubsuit}$ can be analyzed as in Chapter 7. Ultimately, we can establish exponential decay of correlations and the entire bouquet of limit theorems for the map $\mathcal{F}_{\clubsuit}$ (and Hölder continuous functions on $\mathcal{M}_{\clubsuit}$).

Unlike ergodicity, however, the statistical properties do not extend from the return map $\mathcal{F}_{\clubsuit}$ to the original collision map $\mathcal{F}$. Indeed, only a few iterations of $\mathcal{F}_{\clubsuit}$ may correspond to arbitrarily many iterations of $\mathcal{F}$; thus the mixing rates of $\mathcal{F}$ might be much slower than those of $\mathcal{F}_{\clubsuit}$.

It has long been predicted (based on heuristic arguments) that the correlation function $\mathbf{C}_{f,g}(n)$ (see (7.5)) for Hölder continuous observables f and g in stadia should decay as $\sim 1/n$. And because the series $\sum_n 1/n$ diverges, it was expected that the Central Limit Theorem (CLT) should fail in accordance with our remark after (7.9).

Only recently these conjectures have been (at least partially) proved. A slightly weaker than $1/n$ bound on correlations was established in [**Mar04, CZ05**]:

$$|\mathbf{C}_{f,g}(n)| \le \frac{\text{const} \cdot \ln^2 n}{n}.$$

It appears that the arguments of [**Mar04, CZ05**] can be improved so that a sharper bound (close to the expected const$/n$) would follow, and this work is currently under way.

The CLT, in its classical version stated in Section 7.2, was disproved in [**BG06**]. Instead, a rather surprising non-classical CLT for the partial sums S_n (see (7.2)) was obtained. Let the stadium be bounded by two semicircles of radius $\mathbf{r} = 1$ and two line segments Γ_1 and Γ_2 of length $\mathbf{d} > 0$ each. Given a Hölder continuous observable $f(r, \varphi)$ on the collision space $\mathcal{M}$, denote by

$$\mathbf{I}_f = \frac{1}{2\mathbf{d}} \int_{\Gamma_1 \cup \Gamma_2} f(r, 0)\, dr$$

its average value on the set of normal vectors attached to Γ_1 and Γ_2 (those normal vectors exactly generate nonhyperbolic trajectories that never hit semicircles).

THEOREM 8.65 ([**BG06**]). *If* $\mathbf{I}_f \neq 0$, *then* $S_n/\sqrt{n \ln n} \to \mathcal{N}(0, \sigma_f^2)$, *where*

$$\sigma_f^2 = \frac{4 + 3\ln 3}{4 - 3\ln 3} \times \frac{\mathbf{d}^2 \mathbf{I}_f^2}{4(\pi + \mathbf{d})}.$$

If $\mathbf{I}_f = 0$, *then* $S_n/\sqrt{n} \to \mathcal{N}(0, \sigma_f^2)$ *for some* $\sigma_f^2 \ge 0$.

Here we observe a nonclassical asymptotic $\sqrt{n \ln n}$ in the main case $\mathbf{I}_f \neq 0$; it is similar to the superdiffusion in Lorentz gases without horizon (cf. Remark 7.70).

Estimates on correlation functions have recently been obtained for a few more classes of Bunimovich billiards [**CZ05**], but the status of the CLT and other limit theorems is limited to the above remarkable result.

While statistical properties for some Bunimovich billiards were obtained only very recently, the ergodicity and mixing properties for *all* classes of hyperbolic Bunimovich billiards have long been accepted facts [**Bu74a, Bu74b, Bu79**]; see also [**Bu90**]. We have seen, however, that Sinai's original scheme for proving ergodicity via his 'fundamental theorem' (our Theorem 5.70) may run aground here, as it is based on the first growth lemma (or its analogues), which tends to fail in some cases.

Instead, the ergodicity (and mixing) can be obtained by an alternative scheme, which seems to work for all Bunimovich billiards. It was designed in the middle of the 1980s by Sinai and his school [**Sin79, SC87**] for hard ball gases (much more complicated than any planar billiards studied in this book). It was then modified and generalized in [**KSS90, Ch93, Mar93a, LW95**]. Due to its flexibility it was quickly applied to various classes of planar chaotic billiards with nonuniform hyperbolicity [**Sz92, Mar93b, DM01, DMM03**].

This new scheme is based on an elaborate construction of 'regular coverings' in the neighborhood of hyperbolic phase points. It is fairly complicated, and its full description, with proofs, might span a chapter longer than our Chapter 6. On the other hand, it requires minimal information about the underlying map. The latter needed only to be a hyperbolic map with singularities and to have a smooth invariant measure. While standard absolute continuity is supplied by the general Pesin and Katok-Strelcyn theory [**Pes77a, KS86**], no specific bounds on the Jacobian of the holonomy map, no specific distortion bounds, and no growth lemmas were necessary. The 'regular coverings' method needs fairly strong assumptions on the singularities of the map, but those happen to be readily verifiable for nearly all classes of planar chaotic billiards.

We present the 'regular coverings' method in Section 9.11. For good tutorials on this method we refer the reader to [**KSS90, LW95, Va92**].

CHAPTER 9

General focusing chaotic billiards

This chapter is mostly a survey of recent results on planar chaotic billiards with focusing boundary components. Complete proofs can be found in original papers [**Wo86, Mar88, Don91, Bu92**], which are clearly written and readily accessible; see also the previous book by the same authors [**CM03**].

After Bunimovich's discovery of the defocusing mechanism in billiards with focusing arcs [**Bu74a**] in 1974 (and especially after the wide publicity gained by his stadium [**Bu79**] after 1979), many mathematicians tried to construct other chaotic billiards with focusing boundary components. It was not easy to do, because the circle was too rigid a figure whose properties played an essential role in Bunimovich's calculations.

Only in the middle 1980s Wojtkowski [**Wo86**] and then Markarian [**Mar88**] constructed new classes of hyperbolic billiards with focusing boundaries other than circular arcs. Later Donnay [**Don91**] and Bunimovich [**Bu92**] summarized those constructions and presented a unified theory for a general class of focusing curves (called absolutely focusing arcs by Bunimovich). All these recent results were obtained by using either invariant cones or quadratic forms; see Section 3.13 and the first three sections here (both methods are essentially equivalent).

However, our knowledge of chaotic billiards with focusing boundaries is primarily limited to hyperbolicity. In some cases ergodicity and mixing have been established [**Sz92, Mar93b, DM01, DMM03**], but these results are far from general. (All the proofs of ergodicity used 'regular coverings' mentioned in the previous section.)

Further properties of billiards with focusing boundaries remain virtually unexplored. No technical estimates similar to those in our Chapter 5 (on distortion bounds, the Jacobian of the holonomy map, size of invariant manifolds, etc.) are available. Nothing is known about the statistical properties of these billiards (though it is quite clear that their mixing must be pretty slow).

This chapter consists of two large parts. The first one (Sections 9.1–9.7) is devoted to hyperbolicity, presenting all major results in this direction, their interrelation, and plenty of examples. The second part (Sections 9.8–9.11) is devoted to ergodicity. The first three sections (9.8–9.10) provide necessary technical background, and Section 9.11 outlines a ready-to-use program for proving ergodicity, along with all existing applications.

We adopt the following notation; cf. (2.5) and (2.17):

$$\mathcal{M}_+ = \bigcup_{\Gamma_i \subset \Gamma_+} \mathcal{M}_i, \quad \mathcal{M}_- = \bigcup_{\Gamma_i \subset \Gamma_-} \mathcal{M}_i, \quad \mathcal{M}_0 = \bigcup_{\Gamma_i \subset \Gamma_0} \mathcal{M}_i. \tag{9.1}$$

9.1. Hyperbolicity via cone techniques

These methods are very well explained in papers by Wojtkowski [**Wo86**] and Donnay [**Don91**]. They include detailed studies of the evolution of wave fronts bouncing off focusing boundaries of a billiard table. The key point is to ensure that such fronts come off the boundary as focusing fronts, then defocus and arrive at the next collision as dispersing fronts (cf. Section 8.2).

Wojtkowski proved the hyperbolicity of billiards with focusing boundaries satisfying a general condition (see (9.11) below) via an explicit geometric construction of invariant cones. Later more general classes of billiard tables were proved to be hyperbolic using more sophisticated cone fields (see Section 9.6).

In what follows, we use notation and results of Section 3.7.

9.1.1. Focusing times. We introduce focusing times for infinitesimal families of trajectories, which will be used as projective coordinates.

A *variation* $\eta(\alpha)$ is a one-parameter smooth family of lines in $\mathbb{R}^2$. Let $z \in \mathcal{M}$, $u \in \mathcal{T}_z\mathcal{M}$, and $\xi : (-\varepsilon, \varepsilon) \to \mathcal{M}$ be a curve such that $\xi(0) = z$ and $\xi'(0) = u$. We associate with u the variation

$$\eta_+(\alpha) = \{q(\alpha) + tv(\alpha),\ t \in \mathbb{R},\ |\alpha| < \varepsilon\};$$

here $\xi(\alpha) = (q(\alpha), v(\alpha))$. We say that $\eta_+(\alpha)$ *focuses at time* f^+ if there exists $f^+ = \lim_{\alpha \to 0} t(\alpha)$, where $t(\alpha)$ is determined by the intersection of the lines $\eta_+(\alpha)$ and $\eta_+(0) = q(0) + tv(0)$. Also we consider the variation

$$\eta_-(\alpha) = q(\alpha) + t\tilde{v}(\alpha),\ t \in \mathbb{R}$$

where the vector $\tilde{v}(a)$ is obtained by reflecting $v(\alpha)$ across the tangent line to $\partial\mathcal{D}$ at the point $q(\alpha) \in \partial\mathcal{D}$. We define the *focusing* time f^- of the variation $\eta_-(\alpha)$ in the same way as before.

Although for each vector $u \in \mathcal{T}_z\mathcal{M}$, we can construct infinitely many distinct variations $\eta_+(\alpha)$ ($\eta_-(\alpha)$), all of which will focus at the same time f^+ (resp., f^-). An alternative definition of the focusing times is the following: $\eta_+(\alpha)$ focuses if the vector $\partial\eta_+/\partial\alpha_{|\alpha=0}$ is parallel to u for some $t = f^+ \in \mathbb{R}$. We say that the vector u focuses forward (backward) if the variation $\eta_+(\alpha)$ (respectively, $\eta_-(\alpha)$) focuses.

Consider two vector fields $X_r = \partial/\partial r$ and $X_\varphi = \partial/\partial\varphi$ on $\mathcal{M}$. Given a point $z = (r, \varphi) \in \mathcal{M}$ and a vector $u = u_r\,\partial/\partial r + u_\varphi\,\partial/\partial_\varphi \in \mathcal{T}_z\mathcal{M}$ with $u_r, u_\varphi \in \mathbb{R}$, the *forward* and *backward focusing times* $f^+(u), f^-(u)$ of u are given by (see [**Wo86**])

$$f^\pm(u) = \begin{cases} \frac{-\cos\varphi}{\mathcal{K}(r) \pm m(u)} & \text{if } u_r \neq 0 \\ 0 & \text{if } u_r = 0 \end{cases} \tag{9.2}$$

where $m(u) = u_\varphi/u_r$ is the slope of u.

We say that u is *divergent* (*convergent*) if $f^+(u)$ is negative (positive) and u is *flat* if $f^+(u) = \infty$. Later on, when we consider cone fields, we will use the notation $f^\pm(z, u)$, rather than $f^\pm(u)$, to indicate the dependence on the point z.

9.1.2. Reflection Law. Let $z = (r, \varphi) \in \mathcal{M} \setminus \mathcal{S}_1$ so that the point $\mathcal{F}z = (r_1, \varphi_1)$ is well defined. For any $u \in \mathcal{T}_z\mathcal{M}$, let $f_0 = f^+(u)$ and $f_1 = f^+(D_z\mathcal{F}u)$. Then the mirror equation (3.33) can be rewritten as

$$\frac{1}{f_1} + \frac{1}{\tau(z) - f_0} = \frac{-2\mathcal{K}(r_1)}{\cos\varphi_1}. \tag{9.3}$$

The curvature $\mathcal{B} = -1/f^+$ changes its sign from $-$ to $+$ when the corresponding wave front focuses. If a convergent wave front arrives at a dispersing component, i.e. if $\mathcal{K}(r_1) > 0$, it will become, after the reflection, either "less convergent" (its focusing time will be shorter than it would have been without reflection) or divergent.

Under suitable conditions, the focusing time f has an important ordering property that will be crucial for the proof of hyperbolicity of billiard maps.

LEMMA 9.1 (Ordering property). *Let $z \in \mathcal{M} \setminus \mathcal{S}_1$ and $u, w \in \mathcal{T}_z\mathcal{M}$. Assume that $0 < f^+(z, w) < \tau(z)$ and $0 < f^+(\mathcal{F}z, D_z\mathcal{F}w)$. Then*

$$0 \leq f^+(z, u) \leq f^+(z, w) \Rightarrow 0 < f^+(\mathcal{F}z, D_z\mathcal{F}u) \leq f^+(\mathcal{F}z, D_z\mathcal{F}w).$$

The implication is also true if we replace the inequalities with strict inequalities.

PROOF. By formula (9.3), we have that $1/f^+(D_z\mathcal{F}w) < 1/f^+(D_z\mathcal{F}u)$. □

One can see immediately that f^+ is a projective coordinate in $\mathcal{T}_z\mathcal{M}$. A cone $\mathcal{C}(z) \subset \mathcal{T}_z\mathcal{M}$ can be identified with a closed interval of values of f^+:

$$\mathcal{C}(z) = \{u \in \mathcal{T}_z\mathcal{M} : f_1 \leq f^+(u) \leq f_2\}$$

for some real numbers f_1, f_2 such that $-\infty \leq f_1 < f_2 \leq +\infty$.

DEFINITION 9.2. Let $\mathcal{C}(z)$ be a cone field on $\mathcal{M}_-$ and $\mathcal{C}'(z)$ be the complementary cone field. For every $z \in \mathcal{M}_-$, define

$$f^+(z) = \sup_{u \in C(z)} f^+(z, u), \qquad f^-(z) = \sup_{u \in C'(z)} f^-(z, u).$$

The next result gives a simple criterion which helps to check whether a cone field on $\mathcal{M}_-$ is invariant. It was first proved in [**Don91**].

LEMMA 9.3 (Focusing Lemma [**Don91**]). *Let $z \in \mathcal{M}_- \setminus \mathcal{S}_1$ such that $\mathcal{F}z \in \mathcal{M}_-$. Suppose that $0 \leq f^+(z, u) \leq \tau(z)$ for every $u \in \mathcal{C}(z)$ and $0 < f^-(\mathcal{F}z) \leq \tau(z)$. Then*

$$0 < f^+(z) + f^-(\mathcal{F}z) \leq \tau(z) \Rightarrow D_z\mathcal{F}\mathcal{C}(z) \subseteq \mathcal{C}(\mathcal{F}z),$$

and if the inequality on the left-hand side is strict, then we get strict invariance: $D_z\mathcal{F}\mathcal{C}(z) \subset \operatorname{int}(\mathcal{C}(\mathcal{F}z))$.

PROOF. Let $w_1 \in C(\mathcal{F}z)$ such that $f^+(\mathcal{F}z, w_1) = f^+(\mathcal{F}z)$; in other words w_1 belongs to one of the edges of $C(\mathcal{F}z)$. If $w = D_{\mathcal{F}z}\mathcal{F}^{-1}w_1$, then $f^+(z, w) = \tau - f^-(\mathcal{F}z)$. Now if $u \in C(z)$, then $f^+(z, u) \leq f^+(z)$. By the hypothesis we have $f^+ \leq (<)\tau - f^-(\mathcal{F}z) = f^+(z, w)$ so that $f^+(z, u) \leq (<) f^+(z, w)$. To complete the proof one applies Lemma 9.1. □

We now state a version of the Focusing Lemma for trajectories that hit non-focusing boundary components.

Let $\tilde{\gamma}_1, \tilde{\gamma}_2$ be focusing arcs of $\partial\mathcal{D}$ (they are not necessarily different). Suppose that the trajectory of a point z leaves $\tilde{\gamma}_1$ and returns to the focusing part of $\partial\mathcal{D}$ at time $n \geq 1$ at a point $z' = (q', v')$, $q' \in \tilde{\gamma}_2$. That is, consider an orbit $\{z, \mathcal{F}z, \ldots, \mathcal{F}^n z\}, n \geq 1$, such that $\pi_q(z) \in \tilde{\gamma}_1$, $\pi_q(\mathcal{F}^k z) \notin \mathcal{M}_-$ for $k = 1, \cdots, n-1$ and $\pi_q(T^n z) = z' \in \tilde{\gamma}_2$ (the projection π_q was introduced in Section 2.5). We denote by $\lambda(z)$ the length of the trajectory segment between z and z', i.e., $\lambda(z) = \sum_{i=0}^{n-1} \tau(\mathcal{F}^i z)$.

LEMMA 9.4. *Under the above conditions, if* $u \in \mathcal{C}(z)$, $u' = D_z\mathcal{F}^n u$, *and* $w \in \mathcal{C}(z')'$, *then*

$$0 < f^+(z) + f^-(z') \le \lambda(z) \quad \Rightarrow \quad m(u') > m(w)$$

where $m(u)$ *again denotes the slope of the vector* u.

PROOF. It follows from the original proof of Lemma 1.6 in [**Don91**] and the remarks immediately after our formula (9.3). □

9.2. Hyperbolicity via quadratic forms

Here we state another set of sufficient conditions for hyperbolicity of smooth maps with singularities. While the cone techniques carry a strong geometric flavor, these conditions are more analytic. Still they are simple enough to verify, and they have been successfully used in the studies of many physically interesting models, including billiards (see Remark 3.54).

9.2.1. Smooth maps with singularities. As in Section 3.1, let M be a compact Riemannian manifold (perhaps, with boundary and corners), $N \subset M$ an open and dense subset and $F\colon N \to M$ a C^r (with $r \ge 2$) diffeomorphism of N onto $F(N)$. Note that all the iterations of F are defined on the set

$$\tilde{N} = \cap_{n=-\infty}^{\infty} F^n(N).$$

Assume that F preserves a probability measure μ on M and $\mu(\tilde{N}) = 1$.

Let us denote by $\mathcal{S}_1 = \partial N$ the singularity set for the map F and by $\mathcal{S}_{-1} = \partial F(N)$ the singularity set for the map F^{-1}. For $n \ge 1$ we put $\mathcal{S}_n = \mathcal{S}_1 \cup F^{-1}(\mathcal{S}_1) \cup \cdots \cup F^{-n+1}(\mathcal{S}_1)$, the singularity set for F^n, and similarly $\mathcal{S}_{-n} = \mathcal{S}_{-1} \cup F(\mathcal{S}_{-1}) \cup \cdots \cup F^{n-1}(\mathcal{S}_{-1})$, the singularity set for F^{-n}.

9.2.2. Quadratic forms. We recall that a quadratic form Q in $\mathbb{R}^d$ is a function $Q\colon \mathbb{R}^d \to \mathbb{R}$ such that $Q(u) = Q_2(u,u)$, where Q_2 is a bilinear symmetric function on $\mathbb{R}^d \times \mathbb{R}^d$. Equivalently, $Q\colon \mathbb{R}^d \to \mathbb{R}$ is a quadratic form if there is a symmetric matrix A such that $Q(u) = u^T A u$ for $u \in \mathbb{R}^d$ (here u^T means transposition of a column-vector u).

A *quadratic form* Q on M is a function $Q : \mathcal{T}M \to \mathbb{R}$ such that its restriction Q_x to $\mathcal{T}_x M$ at μ-almost every point $x \in M$ is a quadratic form in the usual sense.

We say that a quadratic form Q is *nondegenerate* at x if for every nonzero vector $u \in \mathcal{T}_x M$, there exists a $v \in \mathcal{T}_x M$ such that $Q_2(v,u) \ne 0$ (equivalently, $\det A \ne 0$ for the corresponding symmetric matrix A). We say that Q is *positive* (nonnegative) if at every point x the form Q_x is positive definite (positive semidefinite); i.e. $Q_x(u) > 0$ (respectively, $Q_x(u) \ge 0$) for all $0 \ne u \in \mathcal{T}_x M$. For a form Q and $x \in M$, let $K_Q(x)$ be the number of positive eigenvalues of the matrix that defines the form Q_x (i.e., the maximal dimension of a subspace of $\mathcal{T}_x M$ on which the form is positive).

We denote by $F^{\#}Q$ (the pullback of Q by F) the function defined by $(F^{\#}Q)_x u = Q_{F(x)}(D_x F u)$. One can easily verify that $F^{\#}Q$ is also a quadratic form, and that $F^{\#}Q$ is nondegenerate at x iff Q is nondegenerate at $F(x)$. We note that $P = F^{\#}Q - Q$ is a quadratic form, too.

For a quadratic form Q defined on the orbit of a point x, we put

$$S_x\colon = \{u \in \mathcal{T}_x M\colon Q(D_x F^n u) < 0, \ \forall n \ge 0\}$$

$$U_x\colon = \{u \in \mathcal{T}_x M\colon Q(D_x F^n u) > 0, \ \forall n \le 0\}.$$

9.2.3. Pesin Region. The set of all hyperbolic points in $\tilde{N}$ is often called the *Pesin region* of F:

$$\Sigma(F) = \{x \in \tilde{N} \colon \lambda_i(x) \neq 0, \text{ for every } i\},$$

where $\lambda_i(x)$ denotes all the Lyapunov exponents of the map F at x. We note that the Pesin region $\Sigma(F)$ is invariant under F.

THEOREM 9.5 ([**Mar94**]). *Let $Q : \mathcal{T}M \to \mathbb{R}$ be a nondegenerate quadratic form such that*
(i) *Q_x depends measurably on x.*
(ii) *For every $x \in \tilde{N}$ we have $0 < K_Q(x) < \dim M$ and $K_Q(F^n(x)) = K_Q(x)$ for all $n \in \mathbb{Z}$.*
(iii) *$P_x = (F^{\#}Q - Q)_x$ is positive for every $x \in \tilde{N}$.*
Then $\mu(\Sigma(F)) = 1$. Moreover, for every $x \in \Sigma(F)$ we have $S_x = E^s_x$ and $U_x = E^u_x$.

REMARK 9.6. Condition (iii) can be slightly relaxed: it is enough to require that $P \geq 0$ and P is positive eventually; i.e. for almost every $x \in M$ there exists $k = k(x) \in \mathbb{N}$ such that $Q(D_xF^{k+1}u) - Q(D_xF^k u) > 0$ for every nonzero vector $u \in \mathcal{T}_xM$.

We note that the requirement $K_Q(F^n(x)) = K_Q(x)$ in hypothesis (ii) is automatically satisfied in two important cases:
(a) M has dimension 2. In that case $K_Q(x) = 1$ for all $x \in \tilde{N}$.
(b) Q is continuous on M and M is connected. In this case $K_Q(x)$ is a continuous integral-valued function on M; hence it is constant.

If the quadratic form Q is continuous on M, the spaces E^u_x and E^s_x depend on x continuously. In this case the theorem was proved in [**Mar88**]. Note that the converse also holds:

THEOREM 9.7 ([**Mar94**]). *If $\mu(\Sigma(F)) = 1$, there exists a quadratic form $Q : \mathcal{T}M \to \mathbb{R}$ such that the conditions* (i) - (iii) *of Theorem* 9.5 *are satisfied.*

9.3. Quadratic forms in billiards

Here we develop a general machinery that applies to practically any chaotic billiard and allows us to verify hyperbolicity.

Let $x = (q, v) \in \mathcal{M}$ be a point with coordinates (r, φ) and $v = (dr, d\varphi) \in \mathcal{T}_x\mathcal{M}$ a tangent vector. We now introduce new coordinates (U, V) in the tangent space $\mathcal{T}_x\mathcal{M}$ with the help of the orthogonal projection of $(dr, d\varphi)$ on the subspace perpendicular to v:

$$U = \cos\varphi\, dr \quad \text{and} \quad V = \mathcal{K}\, dr + d\varphi.$$

U is the restriction to $\mathcal{M}$ of a transversal Jacobi field along a billiard trajectory, and V is its derivative. For this reason U is sometimes denoted by J and V by J'.

EXERCISE 9.8. Let $u = (U, V)$. Show that the (forward) focusing time is $f^+(u) = -U/V$. Hint: Use formula (9.2).

REMARK 9.9. Consider the Riemannian metric ρ given by $U^2 + V^2$ on the tangent bundle of $\mathcal{M}$. It can be proved that the symplectic form $\omega = \cos\varphi\, dr \wedge d\varphi$ becomes $U \wedge V$ in the UV coordinates, thus the volume form induced by the metric ρ coincides with the symplectic form ω.

The evolution of (U, V) along a segment of trajectory of length τ between two consecutive collisions (at points (r, φ) and (r_1, φ_1)), as well as at the collision at the point (r_1, φ_1), is given, respectively, by

$$\begin{pmatrix} 1 & \tau \\ 0 & 1 \end{pmatrix} \quad \text{and} \quad \begin{pmatrix} -1 & 0 \\ \frac{-2\mathcal{K}(r_1)}{\cos\varphi_1} & -1 \end{pmatrix}. \tag{9.4}$$

Then, the derivative of $\mathcal{F}$ at $x = (r, \varphi)$ in the UV coordinates is

$$D_x\mathcal{F} = -\begin{pmatrix} 1 & \tau \\ \frac{2\mathcal{K}(r_1)}{\cos\varphi_1} & \frac{2\tau\mathcal{K}(r_1)}{\cos\varphi_1} + 1 \end{pmatrix}. \tag{9.5}$$

EXERCISE 9.10. Prove formulas (9.4) and (9.5). Hint: Use formula (2.26).

Now, if $D_x\mathcal{F}(U, V) = (U_1, V_1) \in \mathcal{T}_{\mathcal{F}x}\mathcal{M}$ is the image of our vector (U, V), with $U_1 = \cos\varphi_1\, dr_1$ and $V_1 = \mathcal{K}_1\, dr_1 + d\varphi_1$, then a simple computation gives

$$U = -(\tau\mathcal{K}_1 + \cos\varphi_1)\, dr_1 + \tau d\varphi_1 \quad \text{and} \quad V = \mathcal{K}_1\, dr_1 - d\varphi_1.$$

We now define a measurable nondegenerate quadratic form

$$Q_x(U, V) = aU^2 + 2b\, UV + cV^2 \tag{9.6}$$

where a, b, c are measurable functions of x (we will make them piecewise continuous below) and $ac - b^2 \neq 0$. The coefficients a, b, c should satisfy condition (iii) of Theorem 9.5 (see also Remark 9.6). We compute

$$\begin{aligned} P_x(U, V) &= (\mathcal{F}^{\#}Q - Q)_x(U, V) \\ &= a_1U_1^2 + 2b_1U_1V_1 + c_1V_1^1 - (aU^2 + 2b\, UV + cV^2). \end{aligned}$$

Denote $\varphi_1' = d\varphi_1/dr_1$. Then

$$\begin{aligned} P_x(U, V) &= (dr_1)^2[(a_1 - a)\cos^2\varphi_1 + 2(b_1 - b)(\mathcal{K}_1 + \varphi_1')(\cos\varphi_1) \\ &\quad + (c_1 - c)(\mathcal{K}_1 + \varphi_1')^2 + a(2\tau\cos\varphi_1(\mathcal{K}_1 + \varphi_1') - \tau^2(\mathcal{K}_1 - \varphi_1')^2) \\ &\quad + 2b(-2\mathcal{K}_1\cos\varphi_1 + \tau(\mathcal{K}_1 - \varphi_1')^2) + 4c\mathcal{K}_1\varphi_1']. \end{aligned}$$

Then the sign of P depends on the following quadratic polynomial (in φ_1'):

$$\begin{aligned} &(c_1 - c - a\tau^2 + 2b\tau)(\varphi_1')^2 \\ &+ 2[(b_1 - b)\cos\varphi_1 + \mathcal{K}_1(c_1 - c) + a\tau\cos\varphi + a\tau^2\mathcal{K}_1 - 2b\tau\mathcal{K}_1 + 2c\mathcal{K}_1]\varphi_1' \\ &+ (a_1 - a)\cos^2\varphi_1 + 2(b_1 - b)\mathcal{K}_1\cos\varphi_1 + (c_1 - c)\mathcal{K}_1^2 \\ &-2a\tau\mathcal{K}_1\cos\varphi_1 - a\tau^2\mathcal{K}_1^2 + 4b\mathcal{K}_1\cos\varphi_1 + 2b\tau\mathcal{K}_1^2. \end{aligned} \tag{9.7}$$

We set $b = b_1 = 1$ in order to have easy control of the nondegeneracy of Q.

EXERCISE 9.11. Let $\mathcal{K}_1 \neq 0$ and $E = -2\mathcal{K}_1^{-1}\cos\varphi$. Show that $P > 0$ for every φ_1' iff two conditions are satisfied:

$$c_1 - c - a\tau^2 + 2\tau > 0 \tag{9.8}$$

and

$$\begin{aligned} &(c - 2\tau + a\tau^2)[4c_1 - 2E + (a_1 - a)E^2/4] \\ &+ c_1E\,[2 - 2a\tau - (a_1 - a)E/4] + a^2\tau^2E^2/4 < 0. \end{aligned} \tag{9.9}$$

CONVENTION. For every focusing boundary component Γ and $q \in \Gamma$ we will denote by $R = 1/|\mathcal{K}(q)|$ the radius of curvature of Γ at q. The circle of radius R tangent to Γ at q 'from inside' $\mathcal{D}$ is the circle of curvature, or osculating circle. A similar circle of radius $R/2$ will be called the *circle of semi-curvature.* We will denote by $L = 2R\cos\varphi$ the length of the segment of the billiard trajectory coming out of q with angle φ within the circle of curvature.

In particular, if the boundary component containing the next collision point q_1 also is focusing (i.e. if $\mathcal{K}_1 < 0$), then $E = L_1 = 2R_1\cos\varphi_1$ is the time the trajectory arriving at (or departing from) q_1 spends in the circle of curvature to $\partial\mathcal{D}$ at q_1.

EXERCISE 9.12. Show that in the case $\mathcal{K}_1 = 0$ the condition (9.9) takes the form

$$(a_1 - a)(c - c_1 - 2\tau) + aa_1\tau^2 < 0. \tag{9.10}$$

REMARK 9.13. Our formalism gives another proof of the hyperbolicity for dispersing billiards, as well as billiards whose boundary components are dispersing and flat (provided the set of points whose trajectories never hit dispersing components has measure zero). Indeed, if we take $a \equiv c \equiv 0$, then (9.9) holds for dispersing boundaries (as $\mathcal{K} > 0$). On neutral components we consider the same quadratic form Q; then $P_x \geq 0$ (it is actually positive on $\mathcal{T}_x\mathcal{M}$, with the exception of one direction). Also, P becomes strictly positive whenever the trajectory hits a dispersing component.

9.4. Construction of hyperbolic billiards

We start with an interesting question posed in [**Wo86, Mar88, Don91**]: given a convex curve $\Gamma \subset \mathbb{R}^2$, can it be a focusing boundary component for a chaotic (hyperbolic) billiard? In other words, can one 'build' a chaotic billiard table using the given focusing arc Γ? We deduce geometric conditions on the curve Γ under which the answer is positive.

First we examine local conditions a focusing arc Γ must meet in order to be suitable for a construction of a hyperbolic billiard. The following heuristic observations will be important. We look at the behavior of expressions (9.8) and (9.9) during long series of successive (nearly sliding) reflections at Γ, with $\varphi \approx \pm\pi/2$. Then clearly $L \approx \tau \approx 0$, and a, c are continuous in a neighborhood of the points $(q_1, \pm\pi/2)$. Now if $c \gg \tau$, then the discriminant Δ of (9.7) of the quadratic polynomial satisfies $\Delta \approx 4cc_1 \approx 4c_1^2$. As we need $\Delta < 0$, we deduce that $\lim_{\varphi\to\pm\pi/2} c = 0$. If $c \ll \tau$, we have $\Delta \approx 4\tau E \approx 4\tau^2$. In conclusion, we must have $c \approx \tau$; this justifies our choice of $c = \tau$ (or L).

If $a \equiv 0$, $c = L$, and $c_1 = L_1$, then (9.9) immediately gives

$$L + L_1 < 2\tau. \tag{9.11}$$

If the curve Γ is C^4 smooth, the condition (9.11) is equivalent to

$$\frac{d^2R}{dr^2} < 0, \tag{9.12}$$

where r is our usual arc length coordinate on Γ and $R = -1/|\mathcal{K}|$ is the radius of its curvature.

EXERCISE 9.14. Prove that the conditions (9.11) and (9.12) on a focusing C^4 arc Γ are indeed equivalent. Hint (see [**Wo86**]): Let q_0, q_1 be distinct points of Γ, choose cartesian coordinates (x, y) on the plane in such a way that the x axis passes through q_0 and q_1 with q_0 being the origin and q_1 having a positive x-coordinate, and let the arc q_0q_1 of Γ lie below the x axis (so that $y \leq 0$). If $\Gamma(r) = (x(r), y(r))$ is parameterized by the arc length, denote by $\theta(r)$ the angle which the tangent line to Γ at point r makes with the x-axis. Note that the radius of curvature of Γ is $R = dr/d\theta$. Let $\tau = |q_0q_1|$. Then

$$L = \int_a^b \frac{dx}{dr}\,dr = R(b)\sin\theta(b) - R(a)\sin\theta(a) - \int_a^b \sin\theta\,\frac{dR}{dr}\,dr$$

and

$$2\tau - L - L_1 = \int_a^b y(r)\,\frac{d^2R}{dr^2}\,dr.$$

This concludes our hint to Exercise 9.14.

Next, if $a \equiv 0$, $c = \tau$, and $c_1 = \tau_1$, then we obtain the condition

$$L_1(\tau + \tau_1) < 2\tau\tau_1. \tag{9.13}$$

The above condition is 'locally equivalent' [**Mar88**] to

$$\frac{d^2R^{1/3}}{dr^2} > 0. \tag{9.14}$$

By local equivalence we mean that both conditions are equivalent for sufficiently short τ and τ_1. (This equivalence is weaker than the one between (9.11) and (9.12).)

EXERCISE 9.15. Prove that conditions (9.13) and (9.14) on a focusing C^4 arc Γ are 'locally' equivalent. Hint (see [**Mar88**]): Let $\psi = \pi/2 - \varphi$ and $\Gamma(\psi) = \rho(\psi)\,\Gamma'(0)\,e^{i\psi}$, with $\rho(0) = 0$ the equation defining the curve Γ in a neighborhood of a collision point $\Gamma(0)$. Use the Taylor expansion of τ_1, τ for small values of ψ. Recall the expression of the curvature in polar coordinates: $\mathcal{K}_1(\psi) = \frac{\rho^2+2\rho'^2-\rho\rho''}{(\rho^2+\rho'^2)^{3/2}}$. Then prove that (9.13) is equivalent to $3\rho''(0)^2\mathcal{K}_1 < 4[\rho'(0) + \rho'''(0)]$, which, in turn, is equivalent to $3\mathcal{K}_1\mathcal{K}_1'' < 2\mathcal{K}_1'^2$, which is equivalent to (9.14).

EXERCISE 9.16. Prove that the epicycloid, the hypocycloid, and the cycloid satisfy condition (9.12). In particular, the latter holds for the cardioid (a closed curve whose equation in polar coordinates is $r(t) = 1 + \cos t$, $-\pi \leq t \leq \pi$). The arc of ellipse $x = a\cos t$, $y = b\sin t$, $b^2 > a^2$, with $-\pi/4 < t < \pi/4$ also satisfies (9.12); see [**Wo86**].

EXERCISE 9.17. Consider the ellipse $x = a\cos t$, $y = b\sin t$, $b^2 > a^2$, $0 \leq t < 2\pi$. Prove that (a) the arc $-\pi/4 < t < \pi/4$ satisfies condition (9.12) and (b) the arc defined by $\sin^2 t > b^2/(b^2 + a^2)$ satisfies (9.13). Curiously, the arcs in parts (a) and (b) are disjoint! (See [**Mar88**].)

9.4.1. Short focusing curves. The following surprising result was obtained by Donnay [**Don91**] for C^6 arcs and independently by Markarian [**Mar94**] for C^4 arcs:

THEOREM 9.18 ([**Mar94**]). *Let $\Gamma \subset \mathbb{R}^2$ be a C^4 convex curve and $q \in \Gamma$. Then there exists a neighborhood $U(q)$ of q such that the shorter curve $\Gamma \cap U(q)$ can be a boundary component of a hyperbolic billiard.*

We call the above pieces $\Gamma \cap U(q)$ *short focusing arcs.* This theorem gives yet another way, in addition to (9.11) and (9.13), to construct focusing arcs suitable for inclusion into hyperbolic tables.

9.4.2. Construction of hyperbolic billiard tables. We have described classes of focusing arcs that *could be* used in the construction of hyperbolic billiards. This does not mean, of course, that any billiard table made of these arcs is hyperbolic. Here we describe how to assemble boundary pieces of various kinds in order to make a hyperbolic billiard table. Certain simple rules ("building code") must be followed. We need to verify conditions on interior angles made by adjacent components, distances between different components, etc. The following theorem shows how to construct plane billiards with hyperbolic behavior whose boundary components are of any type with the only requirement that focusing components satisfy either condition (9.11) or (9.13).

THEOREM 9.19 ([**Wo86, Mar88, CM03**]). *Hyperbolic billiards with C^3 boundary components can be constructed according to either of the following two strategies:*
(i) *Focusing boundary components satisfy (*9.11*). The circles of semi-curvature at each point of every focusing component must not contain points of other components or circles of semi-curvature of other focusing components. Adjacent focusing components must make interior angles greater than π. Focusing and dispersing adjacent components must make interior angles not less than π. Focusing and neutral adjacent components must make interior angles greater than $\pi/2$.*
(ii) *Focusing boundary components satisfy (*9.13*). The circles of curvature of focusing components must not contain points of other boundary components. Conditions on adjacent arcs are the same as in* (i)*.*
In addition, the set of trajectories that hit only flat boundary components must have measure zero.

Conditions (9.11) and (9.13) involve strict inequalities. For circular arcs, these inequalities become identities; thus circular arcs are not permitted by the above theorem. It is somewhat embarrassing to exclude Bunimovich billiards from our general theorem, so we add a little extension to it that includes Bunimovich billiards:

PROPOSITION 9.20. *In the above theorem, we can allow equalities in conditions (*9.11*) and (*9.13*). In this case, however, we need to require that almost every trajectory bounces off at least two distinct boundary components.*

The last requirement is included to rule out some trivial examples like a circular billiard (Section 1.1).

REMARK 9.21. We observe that the conditions in Theorem 9.19 are open in the C^3 metric. Hence, the nonflat components of the boundaries of such billiards can be perturbed in the C^3-topology and hyperbolicity will be preserved. If equalities are allowed, as in Proposition 9.20, then perturbations may destroy hyperbolic behavior of the new billiard table. This is the case when arcs of circles make part of the boundary. If circular arcs are perturbed, conditions (9.11) and (9.13) may break down, and the hyperbolicity may be lost.

However, it is proved in [**Mar94**] that circular arcs smaller than a half circle can be perturbed in the C^4 topology maintaining hyperbolic behavior. A similar result for C^6 perturbations was independently obtained by Donnay [**Don91**].

Theorem 9.19 remains valid if we also allow short focusing arcs described in Theorem 9.18; see [**CM03**].

EXERCISE 9.22. Consider an ellipse and cut it along the major axis. Move both halves apart along the minor axis. Join the endpoints of the two halves by two lines parallel to the minor axis. Call the resulting figure $\mathcal{D}$. Insert two internal 'walls' in $\mathcal{D}$ which are linear segments parallel to the minor axis and join the images of each original focus in both halves; see Fig. 9.1. Prove that if the two halves are moved sufficiently far apart, this billiard is hyperbolic (see precise conditions in [**Wo86**, Appendix C]). Then use the properties of elliptical billiards (Section 1.4) to verify that this billiard has at least three distinct ergodic components (the construction of two of them is indicated in Fig. 9.1; the third one is obtained by reversing the trajectory shown by the solid line). This remarkable example is described by Wojtkowski in [**Wo86**, Appendix C].

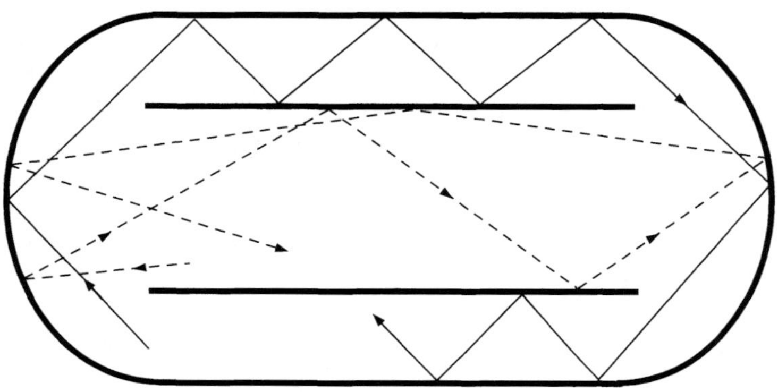

FIGURE 9.1. Wojtkowski's example of a hyperbolic but not ergodic billiard table constructed from an ellipse.

9.5. Absolutely focusing arcs

In this section we describe the largest known class of focusing arcs suitable for hyperbolic billiards. Recall that Bunimovich's construction was based on the defocusing mechanism (Section 8.2). Bunimovich required that unstable wave fronts expand (grow) monotonically in the p-metric between successive collisions, and this restricted his analysis to circular arcs only. Wojtkowski and Markarian kept the requirement of monotonic expansion of unstable fronts, in a sense, but made it more flexible by the use of different metrics in tangent spaces. Absolutely focusing arcs relax the monotonicity requirement altogether, as long as the defocusing mechanism per se works; i.e. the unstable wave fronts still pass through a focusing point between successive collisions at focusing arcs. Surprisingly, this suffices to prove asymptotic expansion of vectors in unstable cones and thus hyperbolicity.

Our presentation follows [**DMM06**]. Let $\Gamma \subset \mathbb{R}^2$ be a C^r focusing boundary component of a billiard table $\mathcal{D}$. Denote by p_1 and p_2 the endpoints of Γ and by $\alpha(\Gamma) = \int_0^{|\Gamma|} |\mathcal{K}(r)|\, dr$ the total angle by which the tangent line to Γ turns as one travels from p_1 to p_2 along Γ.

A C^3 focusing boundary component Γ is called an *absolutely focusing arc* if

AF1: $\alpha(\Gamma) \leq \pi$.

AF2: Any infinitesimal parallel wave front falling upon Γ from $\mathcal{D}$ will focus between every two consecutive collisions at Γ and leave Γ after the last collision with it as a focusing front.

Requirement **(AF1)** (along with Halpern's Theorem 2.13) guarantees that no billiard trajectory can have an infinite series of consecutive reflections at Γ. The (main) requirement **(AF2)** will guarantee hyperbolicity; see below.

EXERCISE 9.23. Prove that if Γ is a focusing component of the boundary that satisfies condition (9.11), then it also satisfies requirement **(AF2)**. Hint (see [**Wo86**]): Consider two wave fronts falling upon Γ from $\partial\mathcal{D}$: one is initially parallel and the other is initially characterized by a constant angle of incidence. The second front focuses at distance $L/2$ along the trajectory departing from the first reflection point. Compare the relative positions of their focusing points on the outgoing trajectory.

EXERCISE 9.24. Prove that if a focusing component $\Gamma \subset \partial\mathcal{D}$ satisfies condition (9.13), then it also satisfies the requirement **(AF2)**.

The above exercises show that the class of absolutely focusing arcs is larger than those satisfying (9.11) or (9.13). The following example illustrates this conclusion.

In Exercise 9.17 we described two disjoint arcs of the ellipse $x^2/a^2 + y^2/b^2 = 1$ with $a > b > 0$: one satisfies Wojtkowski's (but not Markarian's) condition (it is defined by $|x| \leq a/\sqrt{2}$ and $y > 0$), and the other satisfies Markarian's (but not Wojtkowski's) condition (it is defined by $x \geq b^4/(a^2 + b^2)$). On the other hand, Donnay proved (see Theorem 5 and Section 7 in [**Don91**] and also [**MOP96**]) that if $a/b < \sqrt{2}$, then the entire half-ellipse $x \geq 0$ will be an absolutely focusing arc. This arc contains half of Wojtkowski's arc and the entire Markarian's arc; see Fig. 9.2.

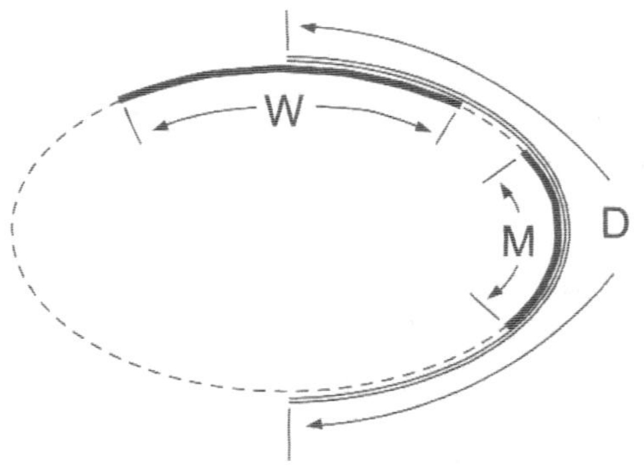

FIGURE 9.2. Wojtkowski's arc (top), Markarian's arc (right) and Donnay's half ellipse.

EXERCISE 9.25. Verify that Wojtkowski's arc and Markarian's arc of the ellipse described above are disjoint.

There are two other important properties of C^6 absolutely focusing curves:

(a) sufficiently short pieces of any convex curve (see Theorem 9.18) are absolutely focusing arcs [**Don91**, Theorem 1];

(b) in the space of focusing curves of class C^6 having the same length $|\Gamma| = L$ and satisfying the condition $\int_0^L |\mathcal{K}(r)|\,dr \leq \pi$, the subset of absolutely focusing curves is open [**Don91**, Theorem 4].

Donnay proved [**Don91**, Theorem 2] that absolutely focusing arcs can be part of hyperbolic billiards if they are joined by sufficiently long segments making a convex region. His proof is based on a subtle study of the properties of focusing wave fronts in integrable billiards. In the rest of this section we summarize Donnay's results. Following Donnay, we consider only absolutely focusing arcs of class C^6. We should however note that our results hold as well for Wojtkowski's and Markarian's curves, which are of class C^4 (the latter admit continuous or piecewise continuous families of invariant cones with finite focusing times [**Wo86, Mar88**], which is more than enough to be included in the next theorem).

The following theorem summarizes Donnay's results [**Don91**] and our recent extensions [**DMM06**]:

THEOREM 9.26. *Any billiard table with C^6 boundary components satisfying conditions* **(B1)**–**(B3)** *below is hyperbolic.*

Recall that $\mathcal{S}_{\pm\infty} = \cup_{n\geq 1}\mathcal{S}_{\pm n}$. Denote by N_∞^+ and N_∞^- the sets of points whose positive (resp., negative) semi-trajectory is defined at all times but which hits only flat components eventually, i.e. $\mathcal{F}^n(x) \in \mathcal{M}_0$ for all $n \geq n_x$ (resp., $n \leq -n_x$).

- **B1:** Each focusing boundary component is an absolutely focusing arc of class C^6 (C^4 Wojtkowski's and Markarian's focusing curves are allowed as well). Furthermore $\mathcal{D}$ is not a polygon, i.e, $\Gamma_+ \cup \Gamma_- \neq \emptyset$.
- **B2:** There is a certain positive lower bound on the length of segments of trajectories that start at a focusing component of $\partial\mathcal{D}$ and end at another nonflat component of $\partial\mathcal{D}$.
- **B3:** $\mu(N_\infty^+) = 0$ and $\mathbf{m}(N_\infty^+ \cap \mathcal{S}_{-\infty}) = 0$, here $\mathbf{m}$ denotes the one-dimensional Lebesgue measure on $\mathcal{S}_{-\infty}$.

Note that $N_\infty^- := \mathcal{I}(N_\infty^+)$, where $\mathcal{I}$ is the involution (Section 2.14), and the time-reversal of **(B3)** (see Remark 4.16) is $\mu(N_\infty^-) = \mathbf{m}(N_\infty^- \cap \mathcal{S}_\infty) = 0$. Condition **(B3)** holds for generic billiard tables because polygonal billiards are generically ergodic [**KMS86**].

Condition **(B2)** is somewhat vague here. It imposes restrictions on distances and angles between components of $\partial\mathcal{D}$, but the precise lower bound is not easy to give. When we deal with Wojtkowski's and Markarian's curves, **(B2)** has a simple geometrical formulation in terms of the relative position of the circles of semi-curvature of distinct focusing curves [**Bu79, Wo86**] or in terms of the distance of the circles of curvature of focusing arcs from other components of $\partial\mathcal{D}$; see [**Mar88**]. As we deal here with a larger class of focusing arcs, our condition **(B2)** has to be more general and, consequently, more involved.

In order to describe Donnay's unstable cones we need to introduce new notation.

Consider a focusing component $\Gamma_i \subset \partial\mathcal{D}$. Let $\mathcal{S}^*_{1,i} = \{z \in \mathcal{M}_i : \pi_q(\mathcal{F}z) \in \partial\Gamma_i\}$, and define inductively $\mathcal{S}^*_{n,i} = (\mathcal{F}^{-1}\mathcal{S}^*_{n-1,i} \cap \mathcal{M}_i) \cup \mathcal{S}^*_{1,i}$ for $n > 1$. The set $\mathcal{S}^*_{n,i}$ consists of elements of $\mathcal{M}_i$ having at most $n-1$ consecutive collisions with Γ_i

before hitting $\partial\Gamma_i$ (loosely speaking, it is the singular set for the map $\mathcal{F}^n$ restricted to trajectories moving only along Γ_i). Also let $\mathcal{S}^*_{\infty,i} = \cup_{n>0}\mathcal{S}^*_{n,i}$.

The sets $\mathcal{S}^*_{-n,i}$ and $\mathcal{S}^*_{-\infty,i}$ are defined similarly by considering backward iterations of $\mathcal{F}$. It follows from the time-reversal principle (Remark 4.16) that $\mathcal{S}^*_{-n,i} = \mathcal{I}(\mathcal{S}^*_{n,i})$ and $\mathcal{S}^*_{-\infty,i} = \mathcal{I}(\mathcal{S}^*_{\infty,i})$.

As usual, when we deal with singularity curves, it is important to establish their regularity, as we did in Sections 4.8 and 4.9 for dispersing billiards. Formal definitions and proofs will be given in Section 9.8; here we just mention those properties. The closures of singularity sets $\mathcal{S}^*_{\pm n,i}$ are unions of finitely many smooth compact curves. The components of $\mathcal{S}^*_{n,i}$ and $\mathcal{S}^*_{-n,i}$ are transversal to each other. For any $m, n \geq 1$ the set $\mathcal{S}^*_{m,i} \cup \mathcal{S}^*_{-n,i}$ partitions $\mathcal{M}_i$ into finitely many open subdomains with piecewise smooth boundaries.

Let $\mathcal{S}_0 = \emptyset$. The function $\varrho : \operatorname{int}\mathcal{M}_i \setminus \mathcal{S}^*_{\infty,i} \to \mathbb{N}$ is defined as follows:

$$\varrho(z) = \min\{k \geq 0 : z \notin \mathcal{S}_k \quad \text{and} \quad \pi_q(\mathcal{F}^k z) \notin \Gamma_i\}.$$

It gives the number of consecutive reflections of the trajectory of z along Γ_i before it leaves Γ_i. Thus $E\colon = \mathcal{I}[\varrho^{-1}(0)]$ is the subset of $\operatorname{int}\mathcal{M}_i$ consisting of vectors entering Γ_i (i.e., vectors leaving Γ_i immediately in the past). For every $k \geq 0$, the set $E_k\colon = E \cap \varrho^{-1}(k)$ consists of vectors entering $\operatorname{int}\mathcal{M}_i$ and leaving Γ_i after exactly k collisions. Clearly $E = \cup_{k\geq 0} E_k$, and by previous observation the closures of E and E_k are boxes.

9.6. Cone fields for absolutely focusing arcs

The construction of an unstable cone field $\mathcal{C}$ on $\mathcal{M}_-$ is due to Donnay [**Don91**] and is done separately for each focusing component Γ_i.

For every $z \in E$, let $X^-(z)$ and $X^+(z)$ be vectors of $\mathcal{T}_z\mathcal{M}_i$ such that their r-component is nonpositive and $f^-(z, X^-(z)) = f^+(\mathcal{F}^{\varrho(z)}z, D_z\mathcal{F}^{\varrho(z)}X^+(z)) = +\infty$. The last condition means that the evolution of $X^-(z)$ backward in time gives rise to a parallel variation and that the $\varrho(z)$ iterations of $X^+(z)$ forward in time (i.e., right as it leaves Γ_i) is a parallel variation. Given a tangent vector $u = (u_r, u_\varphi)$, again let $m(u) = u_\varphi/u_r$ denote its slope. Using the fact that Γ_i is absolutely focusing, one can easily check that

- $\mathcal{K}(z) = m(X^-(z)) < m(X^+(z)) < -\mathcal{K}(z)$ for any $z \in E$,
- X^- is continuous on E, and X^+ is continuous on every E_i.

Suppose now that there exists a unit vector field X_l on E such that

(1) $m(X^+(z)) > m(X_l(z)) > m(X^-(z))$ for any $z \in E$,
(2) $\sup_{z\in E_k} f^-(z, X_l(z)) < +\infty$ for every k,
(3) $\sup_{z\in E_k} f^+(\mathcal{F}^{\varrho(z)}z, D_z\mathcal{F}^{\varrho(z)}X_l(z)) < +\infty$ for every k.

These conditions say that the vector $X_l(z)$ leaves Γ_i after finitely many collisions, that it focuses between any two such collisions and after the last one, and that the focusing times before the first and the last collisions are uniformly bounded in $z \in E$.

Given two linearly independent vectors $X_1, X_2 \in \mathbb{R}^2$, we can define a cone in $\mathbb{R}^2$ by $\mathcal{C}(X_1, X_2) = \{aX_1 + bX_2 : ab \geq 0\}$. Using this notation we put

$$\mathcal{C}_E(z)\colon = \mathcal{C}\Big(X_l(z), \frac{\partial}{\partial\varphi}(z)\Big), \quad z \in E,$$

or, equivalently, $\mathcal{C}_E(z) = \{u \in \mathcal{T}_z E : m(u) \le m(X_l(z))\}$. Now Donnay's cone field on $\mathcal{M}_i$ is defined by

$$\mathcal{C} = \{D_z \mathcal{F}^k \mathcal{C}_E(z) \colon z \in E \quad \text{and} \quad 0 \le k \le \varrho(z)\}.$$

We see that $\mathcal{C}$ is defined on $\operatorname{int} \mathcal{M}_i \setminus (\mathcal{S}^*_{\infty,i} \cup \mathcal{S}^*_{-\infty,i})$. The ordering property (see Section 9.1) implies that every vector of $\mathcal{C}$ has forward and backward focusing times which are uniformly bounded.

To show the existence of a vector field X_l, we first observe that for every fixed $\bar{k} > 0$, any linear combination $aX^- + bX^+$ with $ab > 0$ will be a vector field satisfying Properties (1)–(3) on $\cup_{0 \le k \le \bar{k}} E_k$ (not on the entire E). We now introduce Lazutkin's coordinates (x, y) on $\mathcal{M}_i$: at every point $(r, \varphi) \in \mathcal{M}_i$

$$x = C_1 \int_0^r R^{-\frac{2}{3}}(t)\, dt, \qquad y = C_2 R^{\frac{1}{3}}(r) \sin \frac{\psi}{2}, \tag{9.15}$$

where $\psi = \pi/2 - \varphi$ (it is the angle with the oriented tangent to Γ_i), the constants C_1 and C_2 depend on Γ_i, and $R(r) = -\frac{1}{\mathcal{K}(r)}$ is the radius of curvature of Γ_i at the point r. We refer to [**Don91**] for more details.

Lemma 5.9 of [**Don91**] implies that the vector field $\partial/\partial x$ has Properties (1)–(2) on $\cup_{\bar{k} \le k} E_k$ for $\bar{k} > 0$ sufficiently large depending on Γ_i. Therefore the vector field X_l can be chosen as

$$X_l(z) = \begin{cases} aX^-(z) + bX^+(z) & \text{if } z \in \bigcup_{m=0}^{\bar{m}} E_m, \\ \frac{\partial}{\partial x}(z) & \text{otherwise,} \end{cases}$$

for any positive a, b and $\bar{m} > 0$ sufficiently large.

In this way, we obtain a cone field on $\mathcal{M}_i$ which depends on the parameters $a, b, \bar{m}$. To finish, we have to select specific values for them. As far as a and b are concerned, we simply take any two positive reals; whereas for $\bar{m}$, we take a positive integer which is, as already pointed out, as large as prescribed by Lemma 5.9 of [**Don91**] (there an explicit condition is formulated in terms of the closeness of the angle ψ to 0 or π; large m's correspond to ψ's close to 0 or π). This choice is technical; it will simplify the proof of some part of the Local Ergodic Theorem below.

We now introduce a quantity f_i which measures how far the focusing component Γ_i has to be placed relative to other nonflat components of Γ in order to obtain hyperbolicity. This quantity plays an important role in Condition **(B2)** of the previous section. Let

$$f_i^{\pm} = \sup_{z \in M_i} f^{\pm}(z) \qquad \text{and} \qquad f_i = \max\{f_i^+, f_i^-\}. \tag{9.16}$$

It follows from Property (2) that f_i is finite. Let $f = \max_i f_i$ where the max is taken over all focusing components of $\partial\mathcal{D}$.

REMARK 9.27. Observe that the choice of $a, b, \bar{m}$ affects the value of f_i. We do not know how to choose these parameters in order to obtain the optimal, i.e., the smallest f_i. For a discussion on this issue, see also Section 4 of [**Don91**].

To complete the construction of $\mathcal{C}$ on $\mathcal{M}$, it remains to define it on $\mathcal{M}_+ \cup \mathcal{M}_0$, which is easy:

$$\mathcal{C}(z) = \begin{cases} \{u \in \mathcal{T}_z M : -|d(z)| \le f^+(z, u) \le 0\} & \text{if } z \in \mathcal{M}_+, \\ \{u \in \mathcal{T}_z M : f^+(z, u) \le 0\} & \text{if } z \in \mathcal{M}_0. \end{cases}$$

In other words, for $z \in \mathcal{M}_+$, the cone $\mathcal{C}(z)$ consists of divergent wave fronts which focus inside the half-osculating disk of $\partial\mathcal{D}$ at the point $\pi_q(z)$, whereas for $z \in \mathcal{M}_0$ the cone $\mathcal{C}(z)$ consists of all divergent wave fronts. We could have defined $\mathcal{C}$ equivalently as follows:

$$\mathcal{C}(z) = \{(u_r, u_\varphi) \in \mathcal{T}_z\mathcal{M} : u_r u_\varphi \geq 0\}$$

for all $z \in \mathcal{M}_+ \cup \mathcal{M}_0$. This completes the construction of the cone field $\mathcal{C}$ for billiards satisfying **(B1)**.

9.7. Continued fractions

Here we generalize the results of Sections 8.5 and 8.6, where we proved the convergence of billiard-related continued fractions for Bunimovich billiards.

THEOREM 9.28. *Let $\mathcal{D}$ be a billiard table satisfying the conditions of Theorem 9.19. Then the continued fractions (4.42) and (4.45) converge at every point $X \in \Omega$ where the past and, respectively, future semitrajectory is defined at all times. These fractions give the slopes of the lines E^u_x and E^s_x, respectively (wherever the latter exist).*

PROOF. We recall a general theorem on the convergence of continued fractions proved in [**Bu92**, Theorems 10 and 11]. A fraction

$$A: = \cfrac{1}{a_1 + \cfrac{1}{a_2 + \cfrac{1}{a_3 + \cfrac{1}{a_4 + \ddots}}}} \tag{9.17}$$

converges if the following two conditions hold:

(a) all the odd elements a_{2n+1} are positive and their sum is infinite,

(b) for any two negative elements $a_{2n'}, a_{2n''}$, $n' < n''$, for which there are no other negative elements in between, we have

$$a_{2n'+1} \geq |a_{2n'}|^{-1}(2 + \delta_{n'}) + |a_{2n''}|^{-1}\Big(2 - \frac{\delta_{n''}}{1 + \delta_{n''}}\Big) \tag{9.18}$$

for some $\delta_l \geq -1$, $l = 1, 2, \ldots$. In fact we will need only the case where all the even elements are negative and $\delta_l = 0$. Then (9.18) takes a particularly simple form:

$$a_{2n+1} \geq 2|a_{2n}|^{-1} + 2|a_{2n+2}|^{-1}. \tag{9.19}$$

We prove the theorem in the stable case; the unstable case is obtained by the usual time reversibility (Remark 4.16).

For simplicity, we restrict our arguments to billiard tables without dispersing components; i.e. let $\mathcal{M}_+ = \emptyset$. Collisions with flat boundary components can be eliminated by the 'table flipping trick' described in Section 8.7, and the corresponding continued fraction will be reduced. After that, all the odd elements $a_{2n+1} = \tau_n$ (intercollision times) will be positive, and all the even elements $a_{2n} = \mathcal{R}_n = 2\mathcal{K}_n/\cos\varphi_n$ will be negative, as $\mathcal{K} < 0$ on the focusing boundary. Since the future semitrajectory of $X \in \Omega$ is defined at all times, we have $\sum_n a_{2n+1} = \infty$.

Now let the billiard table consist of Wojtkowski's focusing arcs. Then condition (9.11) can be rewritten as

$$\tau_n \geq 2|\mathcal{R}_n|^{-1} + 2|\mathcal{R}_{n+1}|^{-1}, \tag{9.20}$$

which, rather miraculously, coincides with (9.19). The convergence is proved.

Let the billiard table consist of Markarian's focusing arcs. Then condition (9.13) can be rewritten as

$$|\mathcal{R}_n| \geq 2\tau_{n-1}^{-1} + 2\tau_n^{-1}$$

or equivalently

$$|a_{2n}| \geq 2a_{2n-1}^{-1} + 2a_{2n+1}^{-1}. \tag{9.21}$$

This differs from (9.19), but we use a trick. Fraction (9.17) converges if and only if

$$B\colon = \cfrac{1}{b_1 + \cfrac{1}{b_2 + \cfrac{1}{b_3 + \cfrac{1}{b_4 + \ddots}}}}, \qquad b_n = -a_{n+1} \tag{9.22}$$

converges, because $A = 1/(a_1 - B)$. In the new fraction (9.22) the odd elements $b_{2n-1} = -\mathcal{R}_n$ are positive and the even elements $b_{2n} = -\tau_n$ are negative. Markarian's condition (9.21) takes the form

$$b_{2n-1} \geq 2|b_{2n-2}|^{-1} + 2|b_{2n}|^{-1},$$

which coincides with (9.19).

It remains to verify that the sum of all odd elements is infinite, i.e. $\sum b_{2n-1} = \sum |\mathcal{R}_n| = \infty$. If there are infinitely many collisions at focusing arcs, then this fact follows from the obvious lower bound

$$|\mathcal{R}_n| \geq \min_{q \in \Gamma_-} 2|\mathcal{K}(q)| = \text{const} > 0.$$

On the other hand, if there is only a finite number of collisions at Γ_-, then our fraction (9.17), after the reduction, will be simply finite. □

Lastly, we note that hyperbolic billiards with focusing boundaries have positive entropy, which can be computed by general formulas developed in Section 3.12; see [**CM92**] and [**Ch97**] for a more detailed account of this issue.

9.8. Singularities

Singularity sets $\mathcal{S}_{\pm n}$ of the collision map $\mathcal{F}\colon \mathcal{M} \to \mathcal{M}$ in a general billiard were defined in Chapter 2. For the reader's convenience, we recall the main formulas:

$$\mathcal{S}_0 = \partial \mathcal{M}$$

$$\mathcal{S}_1 = \mathcal{S}_0 \cup \{x \in \operatorname{int} \mathcal{M}\colon \mathcal{F}(x) \notin \operatorname{int} \mathcal{M}\}$$

$$\mathcal{S}_{-1} = \mathcal{S}_0 \cup \{x \in \operatorname{int} \mathcal{M}\colon \mathcal{F}^{-1}(x) \notin \operatorname{int} \mathcal{M}\}$$

$$\mathcal{S}_{n+1} = \mathcal{S}_n \cup \mathcal{F}^{-1}(\mathcal{S}_n) \quad \text{and} \quad \mathcal{S}_{-(n+1)} = \mathcal{S}_{-n} \cup \mathcal{F}(\mathcal{S}_{-n}).$$

The last two sets are the singularity sets for $\mathcal{F}^{n+1}$ and $\mathcal{F}^{-(n+1)}$, respectively.

The properties of singularities for dispersing billiards were studied in Sections 4.8 and 4.9. We proved that $\mathcal{S}_\pm$ consisted of smooth compact curves with uniformly bounded curvature that only intersect at their endpoints. In this section we prove similar regularity properties for the smooth components of $\mathcal{S}_{\pm 1}$.

As was observed in Exercise 4.40, we have $x = (r, \varphi) \in \mathcal{S}_n$ iff $(r, -\varphi) \in \mathcal{S}_{-n}$. Once again, this symmetry allows us to restrict our analysis to the case $n > 0$; i.e. we consider the singularities $\mathcal{S}_1$.

Observe that $\mathcal{S}_1 \setminus \mathcal{S}_0$ consists of points whose next collision is either grazing (which can occur only at a dispersing component of $\partial\mathcal{D}$) or at a corner point of $\partial\mathcal{D}$. Accordingly $\mathcal{S}_1 = \mathcal{S}_0 \cup \mathcal{S}_g \cup \mathcal{S}_c$ where $\mathcal{S}_g = \{(r, \varphi) : |\varphi_1| = \pi/2\}$ and $\mathcal{S}_c = \{(r, \varphi) : r_1$ is an endpoint of some $\Gamma_i \subset \partial\mathcal{D}\}$ (here g stands for 'grazing' and c for 'corner'). Note that $\mathcal{S}_0$, $\mathcal{S}_g$, $\mathcal{S}_c$ may have common points.

THEOREM 9.29. *For any planar billiard table $\partial\mathcal{D} \subset \mathbb{R}^2$ satisfying Assumptions A1–A4 of Chapter 2, the set $\mathcal{S}_1$ is a finite union of smooth compact curves that intersect only at their endpoints. Each set $\mathcal{S}_g$ and $\mathcal{S}_c$ separately is a union of finitely many smooth (but not necessarily compact) curves whose closures are of class $C^{\ell-1}$ and finitely many isolated points.*

EXERCISE 9.30. Find billiard tables for which the set $\mathcal{S}_c$ contains isolated points (try billiards with cusps). Find billiard tables for which the set $\mathcal{S}_g$ (or $\mathcal{S}_c$) contains curves whose endpoints belong to the other set $\mathcal{S}_c$ (respectively, $\mathcal{S}_g$).

PROOF. For any $z = (q, v) \in \mathcal{M}$ we denote by $q_1 = q + \tau(z)v$ and $z_1 = \mathcal{F}(z) = (r_1, \varphi_1)$ the next collision point.

We begin with S_c. In general, this set is not compact. This is the case if the boundary contains a focusing arc γ one of whose ends is $\bar{q}$ ($r = a_i$). Then $S_c(\bar{q}, \gamma) = \{(q, \varphi(q)) : q \in \gamma, q_1 = \bar{q}\}$ is not compact, but it is evident that as q approaches $\bar{q}$, $\varphi(q)$ approaches $\pm\pi/2$, say $\pi/2$. Then $S_c(\bar{q}, \gamma) \cup (\bar{q}, \pi/2) \subset S_c \cup \mathcal{S}_0$ is compact.

Let $\bar{q}$ be the end of an arc of the boundary. Except for the case of isolated points, there always exists at $\bar{q}$ some "sector of visibility" of the boundary Γ. As it is the union of a finite number of smooth closed arcs and of closed curves, it is sufficient to prove that if γ is a sufficiently small closed boundary piece and if it is visible from $\bar{q}$, then $S_c(\bar{q}, \gamma)$ is a closed arc of class $C^{\ell-1}$. "γ is visible from $\bar{q}$" means that for all points $q \in \gamma$ the interval going from q to $\bar{q}$ has no common point with $\Gamma \setminus \{q, \bar{q}\}$. If $\bar{q} \notin \gamma$ one can reduce the study to three cases depending on γ being focusing, dispersing or flat, but the more interesting one happens when $\bar{q} \in \gamma$, focusing. We will consider in detail the last case; the previous cases are similar.

Let γ be defined by $y = f(x), x \in [0, a], f(x) \geq 0$ in an orthogonal system of coordinates (x, y), $f(0) = 0, f''(x) > 0$, $(0,0) = \bar{q}$ and for $q \in \gamma$, $q = (t, f(t)), 0 \leq t \leq a$. Let r be the arc length parameter of γ beginning at $\bar{q}$. For each $r \neq 0$, let $\varphi(r)$ be the unique angle such that $q_1 = q_1(r, \varphi) = \bar{q}$, $\varphi < \pi/2$, and define $\bar{\varphi} = \pi/2 + \varphi(r)$.

Clearly to prove that $S_c(\bar{q}, \gamma) = \{(r, \varphi(r)), 0 \leq r|\gamma|\} \cup (0, -\pi/2)$ is an arc of class $C^{\ell-1}$ is equivalent to proving the same for $C_b = \{(t, \bar{\varphi}(t), 0 < t \leq a\} \cup (0,0)$. $\bar{\varphi}(t) = \alpha(t) - \beta(t)$ where $\alpha(t)$ and $\beta(t)$ are the angles with the x axis of the tangent line to γ at q and the line $q\bar{q}$. Consequently, for $0 < t \leq a$ one has

$$\bar{\varphi}(t) = \arctan f'(t) - \arctan \frac{f(t)}{t} .$$

Finally, if we define $\bar{\varphi}(0) = 0$, as $f' \in C^{\ell-1}([0,a])$, then $\bar{\varphi} \in C^{\ell-1}([0,a])$, and C_b is a closed arc of class $C^{\ell-1}$.

Let us now consider the set S_b. As in the previous case, S_b is in general noncompact and may also contain a finite number of isolated points. To prove our assertion it is enough to consider two boundary pieces γ, γ_1 fulfilling the following condition: for every $q \in \gamma$ there corresponds a unique $\varphi(q)$ such that the half straight line $L(q, \varphi)$ is tangent to γ_1 at the point $q_1 = q_1(q, \varphi)$.

Let $(x(r), y(r))$, $(x_1(r_1), y_1(r_1))$ be parameterizations by arc length of the boundary pieces γ, γ_1, with $a < r < b$, $c < r_1 < d$. If $q = (x(r^0), y(r^0))$, $q_1 = (x_1(r^1), y_1(r^1))$, then

$$\det \begin{pmatrix} x_1(r^1) - x(r^0) & y_1(r^1) - y(r^0) \\ x_1'(r^1) & y_1'(r^1) \end{pmatrix} = D(r^0, r^1) = 0.$$

From the Implicit Function Theorem, in some neighborhood U of r^0 there exists a function $r_1 = r_1(r)$ of class C^r such that $r_1(r^0) = r^1$ and $D(r, r_1(r)) = 0$ for all $r \in U$ if $\frac{\partial D}{\partial r_1}(r^0, r^1) \neq 0$. This condition is satisfied because (see Exercise 9.31)

$$\frac{\partial D}{\partial r_1}(r^0, r^1) = \det \begin{pmatrix} x_1(r^1) - x(r^0) & y_1(r^1) - y(r^0) \\ x_1''(r^1) & y_1''(r^1) \end{pmatrix}. \tag{9.23}$$

This immediately implies that if γ and γ_1 are two disjoint boundary curves, then the set S_b is a closed arc of class $C^{\ell-1}$.

To finish the proof it remains to consider the case of two sufficiently small intersecting curves having a unique common end w.

We begin considering an orthogonal system of coordinates (x, y). The boundary curves γ_1, γ are the graphs of C^3 functions $y = g(x) \leq 0, b < x \leq 0$ and $y = f(x) \geq 0, x \in I$, respectively (0 is one of the boundary points of the interval $I \subset \mathbb{R}$). They touch at the point $w = (0, 0)$, i.e., $f(0) = g(0) = 0$. γ_1 is a dispersing curve, tangent to the x-axis: $g'(0) = 0$ and $g''(x) < 0$. It is clear that we must consider separately the cases where the function $f(x)$ is defined for values of $x \leq 0$ or $x \geq 0$.

Consider the function $F(t, u) = f(t) - g(u) - g'(u)(t - u)$. It is of class $C^{\ell-1}$ and $F(0, 0) = 0$. If a billiard trajectory leaving γ at $(t, f(t))$ with angle $\varphi(t)$ is tangent to γ_1 at $(u, g(u))$, then $F(t, u) = 0$. The regularity of the arc $(t, \varphi(t)), t \in I$, of S_b defined by these trajectories depends on the regularity of the implicit function $u = u(t)$. In fact, for each u a $t \in I$ is defined, and as they are inverse functions of each other, $t(u), u(t)$ are in the same regularity class. The regularity of $\varphi(t)$ will be explained in detail after the study of the regularity of $u(t)$.

The C^1 function $u = u(t)$ satisfying $F(t, u(t)) = 0$ in a neighborhood of $(0, 0)$ exists, by the Implicit Function Theorem, if $\frac{\partial F}{\partial u} = F_u \neq 0$ for every $(t, u) \neq (0, 0)$ and there exists the limit (eventually ∞) of $u'(t) = -\frac{F_t}{F_u}(t, u(t))$ when $t \to 0$ and $F(t, u) = 0$. We have

$$\frac{F_t}{F_u} = \frac{f'(t) - g'(u)}{g''(u)(u - t)} := h(t).$$

Then F_u is different from zero if $u \neq t$ ($g''(u) < 0$ because our dispersing curves cannot have zero curvature). The limit of $h(t)$ obviously exists ($= \infty$) if $f'(0) \neq g'(0) = 0$. So, we must consider only the case $f'(0) = g'(0) = 0$.

The interior angle can be zero or π. If it is zero, Γ_0 can be flat or dispersing, since Assumption A4 in Section 2.4 does not allow cusps formed by focusing and dispersing arcs (in fact, as much as we know, this type of cusp has never been studied). In these cases, $y = f(x) \geq 0, -a < x \leq 0$.

If the interior angle is π and Γ_0 is flat or dispersing, γ, γ_1 do not see themselves. Finally, we must consider the case with interior angle equal to π and γ focusing. We will consider in detail this case; the other two merit similar analysis.

γ is the graph of a function $y = f(x) \geq 0, 0 \leq x < a$. Let $f(t) = A\frac{t^2}{2} + f'''\frac{t^3}{6}$, $g(u) = B\frac{u^2}{2} + g'''\frac{u^3}{6}$, $A > 0, B < 0$, where f''', g''' represent the third derivatives of f and g at some points $0 \leq t_1 \leq t,\ u \leq u_1 \leq 0$, respectively. We also consider the quadratic approximations of both functions $f_*(t) = At^2/2$, $g_*(u) = Bu^2/2$. Let $u_*(t)$ be the solution of $F_*(t,u) = f_*(t) - g_*(u) - g_*'(u)(t-u)$ and

$$\frac{f_*'(t) - g_*'(u_*)}{g_*''(u_*)(u_* - t)} = h_*(t) \quad (\text{if } f(x) = 0 \text{ and } h_*(t) = -2).$$

Once again we recall that $h_*(t) = -u_*'(t), t > 0$. Then $u(t)$ will be $C^{\ell-1}$ smooth at $t = 0$ if we prove that $h_*(t)$ does and $h(t) - h_*(t) \to 0$ as $t \to 0$. We have

$$h(t) - h_*(t) =$$
$$\frac{Bt(A-B)(u-u_*) + \frac{1}{2}g'''u[Bu(t-u_*) - (u-t)(At-Bu_*)] + \frac{1}{2}Bf'''t^2[u_*-t]}{B(u_*-t)(B+g'''u)(u-t)}.$$

First we concentrate our attention on the terms of the numerator not containing $u - u_*$. Each of them includes the product of three infinitesimal values. As $t - u$, $t - u_*$ (factors in the denominator) are greater than $t, -u, -u_*$ (all of them going to zero as t does), we have that this part of the quotient goes to zero. On the other hand, computing the difference $F(t,u) - F_*(t,u_*)$ we obtain

$$u - u_* = \frac{\frac{1}{2}f'''t^2(-2t/3+u) - \frac{1}{2}g'''u^2(-2u/3+t)}{B[t-(u+u_*)/2]}.$$

The same argument as before allows us to deduce also that the part of the quotient containing the factor $t(u - u_*)$ goes to zero. So, we have proved that $h(t) - h_*(t)$ goes to zero as $t \to 0$. Then, we need only compute

$$h_*(t) = \frac{At - Bu_*}{B(u_*-t)} = -1 + \frac{(A-B)t}{B(u_*-t)}.$$

Making some more tedious computations we obtain

$$t = \frac{Bu_* + \sqrt{B^2u_*^2 - ABu_*^2}}{A}, \qquad \frac{t}{u_*-t} = \frac{B - \sqrt{B^2-AB}}{A - B + \sqrt{B^2-AB}}.$$

(Observe that this value is different from zero and ∞ because $-B, A > 0$.) This finishes the proof of the regularity of $u(t)$.

Now, in the case we are presently studying, $\varphi(t) = -\pi/2 + \arctan f'(t) - \arctan g'(u)$. Then

$$\varphi'(t) = \frac{f''(t)}{1+[f'(t)]^2} - \frac{g''(u)u'(t)}{1+[g'(t)]^2}$$

whose limit, as $t \to 0$, exists and equals (in this configuration it is different from zero):

$$A + Bh_*(0) = (A-B)\left(1 - \frac{B+\sqrt{B^2-AB}}{A-B-\sqrt{B^2-AB}}\right).$$

This finishes the proof. ☐

EXERCISE 9.31. Verify formula (9.23) and deduce that the determinant is $\neq 0$. Hint: $(x_1''(r^1), y_1''(r^1))$ is orthogonal to $(x_1'(r^1), y_1'(r^1))$.

REMARK 9.32. In Theorem 6.1 of [**KS86**, Part V] it is proved that for a more general class of billiard tables, which includes focusing and dispersing boundary components whose curvature may vanish, the set $\mathcal{S}_1$ is the union of a finite number of arcs (not necessarily closed) and closed curves, all of class $C^{\ell-1}$ and of finite length, and of finite number of isolated points. Our theorem, though under some more restrictive conditions, establishes a little more: the $C^{\ell-1}$ smoothness of the closures of the smooth components of $\mathcal{S}_1$.

9.9. Application of Pesin and Katok-Strelcyn theory

Here we return to the general class of smooth maps with singularities introduced in Section 3.1 and used again in 9.2.

Recall (Section 3.1) that a point $p \in M$ is hyperbolic if its Lyapunov exponents exist and none of them equals zero; in this case $\mathcal{T}_pM = E_p^s \oplus E_p^u$, where

$$E_p^s = \oplus_{\lambda_p^{(i)}<0} E_i \quad \text{and} \quad E_p^u = \oplus_{\lambda_p^{(i)}>0} E_i. \tag{9.24}$$

If none of the Lyapunov exponents lie in $(-\lambda, \lambda)$, then we can deduce (see Exercise 3.5) that for any $\varepsilon > 0$ there is $K(p, \varepsilon) > 0$ such that for all $n \geq 1$

$$\|D_pF^n v\| \leq K(p,\varepsilon)\, e^{-n(\lambda-\varepsilon)} \|v\| \qquad \forall v \in E_p^s, \tag{9.25}$$

$$\|D_pF^{-n} v\| \leq K(p,\varepsilon)\, e^{-n(\lambda-\varepsilon)} \|v\| \qquad \forall v \in E_p^u. \tag{9.26}$$

For nonuniformly hyperbolic maps (see Section 3.1), the constant $K(p, \varepsilon)$ depends on the point p. We note that when $K(p, \varepsilon)$ is large, the effect of contraction of stable vectors and that of expansion of unstable vectors can be seen only for large enough n. So, the contraction and expansion are only asymptotic, and for arbitrarily long periods of time neither may occur. Moreover, stable vectors may temporarily expand under D_pF^n. This is all very characteristic of nonuniformly hyperbolic systems.

EXERCISE 9.33. Compare the previous analysis with the behavior of long trajectories sliding along the arcs of circles or in between parallel lines in a Bunimovich stadium. Hint: Use the results of Section 8.7.

A systematic study of nonuniformly hyperbolic diffeomorphisms with absolutely continuous invariant measures was done by Pesin in the mid-seventies [**Pes76, Pes77a, BP01**]. In the 1980s, Pesin's theory was extended to more general nonuniformly hyperbolic maps by Ledrappier, Katok, Ruelle, and Young, among others. We follow the exposition by Katok and Strelcyn [**KS86**] that covers a large class of nonuniformly hyperbolic maps with singularities.

Let F be a smooth map with singularities as defined in Section 9.2. We additionally require that the class of smoothness of F be at least $r \geq 2$ and that it satisfy the hypothesis of Oseledets's Theorem (Theorem 3.1).

Let $d(x, y)$ be the Riemannian distance between points on M, and $d(x, \mathcal{S})$ denote the distance from $x \in M$ to the singularity set $\mathcal{S}$.

For $x \in N$, we denote by $\exp_x\colon \mathcal{T}_xN \to N$ the exponential map (it is defined by $\exp_x(v) = \gamma(x, v, 1)$, where $\gamma(x, v, t)$ is the geodesic in N satisfying $\gamma(x, v, 0) = x$ and $\dot\gamma(x, v, 0) = v$). Let $R(x, N)$ be the radius of injectivity of the map $\exp_x\colon \mathcal{T}_xN \to N$; i.e. $R(x, N) = \sup\{r\colon \exp_x$ is one-to-one on the ball $B_r(0) \subset \mathcal{T}_xN\}$. The map $\exp_x$ is defined and injective on the ball $B_{r(x,N)}(0) \subset \mathcal{T}_xN$ where $r(x, N) = \min\{R(x, N), d(x, \mathcal{S})\}$.

We define $F_{ox} = \exp_{F(x)}^{-1} \circ F \circ \exp_x$. This is a proper way to represent F in a linear coordinate system near the point x. This map is well defined in a neighborhood of $0 \in \mathcal{T}_x N$.

We now make two additional assumptions, following Katok and Strelcyn [**KS86**]:

KS1: There are constants $C_1 > 0$ and $a > 0$ such that for all $\varepsilon > 0$ the μ-measure of the ε-neighborhood (in the Riemannian metric) of $\mathcal{S}$ satisfies

$$\mu(U_\varepsilon(\mathcal{S})) \leq C_1\, \varepsilon^a;$$

i.e. the measure μ does not build up too much near the singularity set S.

KS2: There are constants $C_2 > 0$ and $b > 0$ such that for every $x \in N$ and $v \in \mathcal{T}_x N$, $\|v\| \leq r(x, N)$, we have

$$\|F''_{ox}(v)\| \leq C_2\, d(\exp_x(v), \mathcal{S})^{-b};$$

i.e. the second derivative F''_{ox} does not grow too fast near the singularity set $\mathcal{S}$.

EXERCISE 9.34. Let $\tilde{N}_\kappa = \{x \in \tilde{N}\colon\ d(f^n(x), \mathcal{S}) > c(x)\,|n|^{-\kappa}$ for some $c(x) > 0$ and all $0 \neq n \in \mathbb{Z}\}$. Show that $\mu(\tilde{N}_\kappa) = 1$ for some $\kappa > 0$. Hint: Note that the set $\tilde{N}^*_{\kappa,n}\colon = \{d(f^n(x), \mathcal{S}) < |n|^{-\kappa}\}$ coincides with $F^{-n}(U_{|n|^{-\kappa}}(\mathcal{S}))$; hence $\mu(\tilde{N}^*_{\kappa,n}) = \mu(U_{|n|^{-\kappa}}(\mathcal{S}))$. Now put $\kappa = 2/a$ and use **(KS1)** to prove that $\sum_n \mu(\tilde{N}^*_{\kappa,n}) < +\infty$. Note that $\tilde{N}_\kappa$ consists of points that belong to only finitely many sets $\tilde{N}^*_{\kappa,n}$.

To estimate the entropy from above, Katok and Strelcyn used the following extra assumption:

KS3: There are constants $C_3 > 0$ and $d \geq 1$ such that for every $x \in N$

$$\|D_x F\| \leq C_3\, d(x, S)^{-d}.$$

Our main goal is to verify that the collision map $\mathcal{F}\colon \mathcal{M} \to \mathcal{M}$ satisfies all these three Katok-Strelcyn assumptions, **(KS1)–(KS3)**. Certain steps are left to the reader as exercises.

EXERCISE 9.35. Verify that the collision map $\mathcal{F}$ is a smooth map with singularities; in particular $N = \mathcal{M} \setminus \mathcal{S}_1 = \operatorname{int} \mathcal{M} \setminus \mathcal{S}_1$.

EXERCISE 9.36. Prove that $\mathcal{F}\colon \mathcal{M}\setminus\mathcal{S}_1 \to \mathcal{M}\setminus\mathcal{S}_{-1}$ satisfies assumption **(KS1)**. Hint: Use Theorem 9.29 and observe that the density $\cos\varphi$ of the measure μ with respect to the Lebesgue measure is bounded.

EXERCISE 9.37. Let $(x(r), y(r))$ and $(x_1(r_1), y_1(r_1))$ denote parameterization by the arc length of the boundary components γ and γ_1, with $a < r < b$, $c < r_1 < d$, as in the proof of Theorem 9.29 (regularity of S_g) and

$$\tau = \sqrt{[x_1(r_1) - x(r)]^2 + [y(r_1) - y(r)]^2}.$$

Prove that

$$\frac{\partial \tau}{\partial \varphi} = \frac{\partial \tau}{\partial r_1}(r, r_1(r, \varphi)) \times \frac{\partial r_1}{\partial \varphi} = \frac{[x_1(r_1) - x(r)]x_1'(r_1) + [y(r_1) - y(r)]y_1'(r_1)}{\cos\varphi_1}$$

and

$$\frac{\partial \tau}{\partial r} = \frac{[x_1(r_1) - x(r)][x_1'(r_1)\partial r_1/\partial r - x'(r)] + [y(r_1) - y(r)][y_1'(r_1)\partial r_1/\partial r - y'(r)]}{\tau}.$$

Hint: Use, repeatedly, the formula (2.26).

EXERCISE 9.38. Show that for every $x \in \mathcal{M} \setminus \mathcal{S}_1$ the following estimations hold:

$$\|D_x\mathcal{F}\| \le \frac{a_1}{\cos\varphi_1} \quad \text{and} \quad \|D_x^2\mathcal{F}\| \le \frac{a_2}{\cos^3\varphi_1},$$

where $\mathcal{F}(x) = (r_1, \varphi_1)$ and $a_1, a_2 > 0$ are constants. Hint: Use (2.26). For the second estimate take the derivative of its terms, use the results in the previous exercise, and use the fact that the boundary $\partial\mathcal{D}$ is of class C^3.

THEOREM 9.39. *The collision map* $\mathcal{F} : \mathcal{M} \setminus \mathcal{S}_1 \to \mathcal{M} \setminus \mathcal{S}_{-1}$ *satisfies the assumptions* **(KS1)**–**(KS3)**.

PROOF. The fact that $\mathcal{F}$ is a smooth map with singularities was proved in Exercise 9.35. The $C^{\ell-1}$ smoothness was established in Theorem 2.33. The hypotheses of Oseledets's theorem were verified in Theorem 3.6. Assumption **(KS1)** was proved in Exercise 9.36.

It remains to estimate the rate of growth of the first and second derivatives of $\mathcal{F}$. They were explicitly computed in Theorem 7.2 of [**KS86**, Part V]. Due to the results of Exercises 9.37 and 9.38, it is enough to prove that $\cos\varphi_1(x) \ge b\, d(x, \mathcal{S}_1)$, where $b > 0$ is a constant, for every $x \in \mathcal{M} \setminus \mathcal{S}_1$. We will not give all the details of the proof, but make the following structural observations.

(a) Since $\mathcal{F} : \mathcal{M} \setminus \mathcal{S}_1 \to \mathcal{M} \setminus \mathcal{S}_{-1}$ is an homeomorphic embedding, for any compact set $C \subset \mathcal{M} \setminus \mathcal{S}_1$ there exists a positive number b_C such that for every $x \in C$, $\cos\varphi_1(x) \ge b_C\; d(x, \mathcal{S}_1)$.

(b) Since $\mathcal{S}_1$ is compact (Theorem 9.29), it suffices to prove that for any $P = (q, \varphi) \in \mathcal{S}_1$ there exists a neighborhood $U(P)$ and a number b_P such that

$$\cos\varphi_1(x) \ge b_P\; d(x, \mathcal{S}_1) \tag{9.27}$$

for every $x \in U(P) \cap (\mathcal{M} \setminus \mathcal{S}_1)$. This inequality is not evident only in the case

$$\inf\{\cos\varphi_1(x) : x \in U(P) \cap (\mathcal{M} \setminus \mathcal{S}_1)\} = 0. \tag{9.28}$$

(c) It is sufficient to find all configurations for which (9.28) holds and prove (9.27) for them. The following three cases exhaust all possibilities such that (9.28) takes place (in the last two cases we consider only the points P such that the billiard trajectories going from $\gamma \setminus \{q\}$ to $\gamma_1 \setminus \{\bar{q}\}$ exist):

(i) q belongs to a focusing component and $\varphi = \pm\pi/2$;
(ii) q is a common point of two closed curves $\gamma, \gamma_1 \subset \partial\mathcal{D}$;
(iii) q is an endpoint a closed curve $\gamma \subset \partial\mathcal{D}$ and $\bar{q}$ is an endpoint of another closed curve $\gamma_1 \subset \partial\mathcal{D}$, which is disjoint from γ, and such that the line $q\bar{q}$ is tangent to γ_1 at $\bar{q}$.

(d) Taking into account the convexity and mutual disposition of γ and γ_1, 36 different cases can be distinguished and are listed in of [**KS86**, Part V, Section 8].

(e) We will not give the details of the completely geometrical (and elementary) proofs of (9.27) for each of these cases. They follow the lines of the proof of Theorem 9.29. We merely observe here that our billiard tables satisfy the hypotheses of Proposition 7.1 and Lemmas 7.3 and 7.4 of [**KS86**, Part V], which are used in this proof. □

9.10. Invariant manifolds and absolute continuity

Parts I and II of [**KS86**] are devoted to the proof of the existence of invariant manifolds and their absolute continuity for hyperbolic systems satisfying all the assumptions in the previous section, with the exception of **(KS3)**. Then one concludes that the same properties hold for our hyperbolic billiard systems.

In this section we continue working with smooth maps with singularities satisfying all the assumptions indicated in the previous section; we will also assume that the set of hyperbolic points has positive measure. At every point $x \in \Sigma(F)$ we have the usual subspaces E^s_x and E^u_x defined by (9.24), and we let

$$\lambda^+(x) = \min\{\lambda_i(x) > 0\} \quad \text{and} \quad \lambda^-(x) = \max\{\lambda_i(x) < 0\}$$

be the smallest (in absolute value) positive and negative exponents, respectively. Our collision map $\mathcal{F}$ satisfies all these assumptions, and furthermore $\lambda^-(x) = -\lambda^+(x)$ due to Lemma 3.9.

The following theorem is, in fact, a generalization of our early conditions (9.25) and (9.26); cf. also Exercise 3.5.

THEOREM 9.40. *Given $\varepsilon > 0$, there is a measurable function $C(x,\varepsilon)$ on $\Sigma(F)$ such that for all $x \in \Sigma(F)$, $n \geq 1$ and $m \in \mathbb{Z}$*

$$\|D_{F^m x} F^n v\| \leq C(x,\varepsilon)\, e^{\lambda^-(x)n + \varepsilon n + \varepsilon|m|} \|v\| \quad \forall \in E^s_x$$

$$\|D_{F^m x} F^n v\| \geq C^{-1}(x,\varepsilon)\, e^{\lambda^+(x)n - \varepsilon n - \varepsilon|m|} \|v\| \quad \forall v \in E^u_x$$

and the angle $\gamma(F^m x)$ between $E^s_{F^m x}$ and $E^u_{F^m x}$ satisfies

$$\gamma(F^m x) \geq C^{-1}(x,\varepsilon)\, e^{-\varepsilon|m|}.$$

The next theorem concerns stable and unstable manifolds.

THEOREM 9.41. *For μ-almost every $x \in \Sigma(F)$ there is a $\delta(x) \in (0, r(x,N))$ such that for any small $\varepsilon > 0$ the set*

$$W^s(x) = \{y \in \exp_x B_{\delta(x)}(0)\colon \limsup_{n\to\infty} \frac{1}{n} \log \operatorname{dist}(F^n(x), F^n(y)) \leq \lambda^-(x) + \varepsilon\}$$

is a C^r differentiable (stable) manifold. Similarly,

$$W^u(x) = \{y \in \exp_x B_{\delta(x)}(0)\colon \liminf_{n\to-\infty} \frac{1}{n} \log \operatorname{dist}(F^n(x), F^n(y)) \geq \lambda^+(x) - \varepsilon\}$$

is a C^r differentiable (unstable) manifold. We also have

$$\mathcal{T}_x W^s(x) = E^s_x \quad \text{and} \quad \mathcal{T}_x W^u(x) = E^u_x.$$

We note that the stable and unstable manifolds at x are transversal; i.e. they intersect in the point x alone, and the angle between them is positive. We also note that $W^s(x)$ and $W^u(x)$ exist almost everywhere, but not necessarily everywhere, on $\Sigma(F)$. Their existence is an important issue, so we provide some more details. These manifolds fail to exist at points $x \in \Sigma(F)$ whose trajectories approach the singularity set $\mathcal{S}$ too fast. More specifically, if

$$d(F^{n_k}x, \mathcal{S}) < \beta^{n_k} \quad \text{for some} \quad \beta < e^{-\lambda_{m(x)}(x)}$$

and for some infinite sequence $n_k \to \infty$, then $W^s(x)$ cannot exist. Indeed, if it did exist, it would have contracted too slowly under the iterates of F, and so it would hit the set $\mathcal{S}$ sooner or later; i.e. for some $n_k > 0$ we would have $F^{n_k}(W^s(x)) \cap \mathcal{S} \neq \emptyset$, which is impossible by the definition of $W^s(x)$ given in the above theorem.

Fortunately, our assumption **(KS1)** guarantees that for almost every point $x \in N$ there are $c(x) > 0$ and $\kappa > 0$ such that

$$d(F^n(x), S) > c(x)\,|n|^{-\kappa}$$

for all $0 \neq n \in \mathbb{Z}$ (see Exercise 9.34). This fact, in turn, can be used to derive the existence of $W^s(x)$ and $W^u(x)$ and even estimate their size, so we omit further details.

9.10.1. Absolute continuity. We define the holonomy map as in (H6) of Section 6.3 (see also Section 5.7). Part II of [**KS86**] is devoted to the proof of the following:

THEOREM 9.42. *For any measurable set $A \subset V^1_*$ such that $\mathbf{m}_1(A) = 0$ one has $m_2(\mathbf{h}(A)) = 0$, where $\mathbf{m}_i$, $i = 1, 2$, denotes the Lebesgue measure on V^i.*

Theorems 9.41 and 9.42 correspond to the results of Section 4.11 and 5.8 for dispersing billiards and those in Sections 8.10 and 8.13 for Bunimovich stadia. The general theorems presented here are less precise than our earlier results for more specific billiards, but they will be sufficient for the next section.

9.11. Ergodicity via 'regular coverings'

In Chapter 6 we proved ergodicity for dispersing billiards using the original Sinai's method (dating back to 1970). In Chapter 8 we applied it again to Bunimovich's billiards, but with limited success. The limitations of this approach were discussed in the introduction to Chapter 6 and in Section 8.15. Here we present a more recent and more powerful scheme for proving ergodicity (based on the construction of 'regular coverings') developed also by Sinai and his school in the mid-1980s. We have already described various aspects of both methods in Chapter 6 and in Section 8.15. Here we make a more formal presentation.

9.11.1. Assumptions. We return to general smooth maps with singularities defined in Section 3.1 and used in Section 9.2. Let $F\colon N \to M$ be such a map, and let F satisfy all Katok-Strelcyn conditions (see Section 9.9). Since we intend to apply this method to the collision map $\mathcal{F}$, we may assume that $\dim M = 2$; this assumption will allow us to simplify some technical conditions.

We endow M with a nondegenerate Riemannian metric $\|\cdot\|$ and assume that the invariant measure μ is absolutely continuous with respect to the Lebesgue measure on M and has a bounded density, which is positive on $\operatorname{int} M$. Lastly, let the map F be hyperbolic (but not necessarily uniformly hyperbolic), i.e. $\mu(\Sigma(F)) = 1$.

Our main purpose is to describe sufficient conditions under which F is ergodic. We will not prove ergodicity per se, as this part is very difficult; the reader will be referred to other surveys on this part. We note only that ergodicity does not follow by any means from hyperbolicity: the Pesin region $\Sigma(F)$ decomposes only into finitely or countably many ergodic components Σ_i of positive measure. The sets Σ_i may be rather complicated and very irregular [**KS86**, Section 13 in Part II].

A commonly used scheme for proving ergodicity consists of two major steps. First, one establishes 'local ergodicity'; i.e. one shows that for certain typical points $x \in M$ there is a neighborhood $\mathcal{U}_x \subset M$ that belongs (mod 0) to one ergodic component of F. Second, one needs to verify that those typical points cover an arcwise connected set of full measure (which implies 'global ergodicity' by an easy argument). This strategy was applied in Sections 6.5 and 6.6, respectively, where we proved ergodicity for dispersing billiards.

We start with local ergodicity, which is usually the more difficult of the two steps. We will actually present two sets of sufficient conditions for local ergodicity. One was developed by Sinai and Chernov [**SC87**], well explained in [**KSS90**], and further generalized by Chernov in [**Ch93**]. The other set is an adaption of the first one to symplectic maps (which of course include all billiards) by Liverani and Wojtkowski [**LW95**]. Each of these two versions of 'local ergodic theorem' has certain advantages, so we present both of them.

The proof of local ergodicity assuming either set of sufficient conditions is very complicated (this is where 'regular coverings' come into play), and it is beyond the scope of this book.

9.11.2. Local Ergodicity (Sinai-Chernov version). We state the local ergodic theorem in its current, most general form following [**Ch93**]. Since the map F is hyperbolic, there are one-dimensional stable and unstable spaces, E^s_x and E^u_x, respectively, at almost every point $x \in M$.

DEFINITION 9.43. A point $x \in N$ is said to be *u-essential* if for any $A > 1$ there exist an $n \geq 1$ and a neighborhood $\mathcal{U}_x$ of the point x such that for all $y \in \mathcal{U}_x$

$$\|D_y F^n(u)\| > A\|u\|, \quad u \in E^u_y$$

(whenever the space E^u_y exists). Similarly, *s-essential* points are defined by replacing F^n with F^{-n} and E^u_y with E^s_y.

DEFINITION 9.44. A point $x \in N$ is said to be *SC-sufficient* if there exist $B > 1$ and two integers $n < m$, such that $F^n(x)$ is defined, and there is a neighborhood $\mathcal{U}$ of the point $F^n(x)$ such that for all $y \in U$ we have

$$\|D_y F^{m-n}(u)\| > B\|u\|, \quad u \in E^u_y \quad \text{and} \quad \|D_y F^{m-n}(v)\| < B^{-1}\|v\|, \quad v \in E^s_y$$

(whenever the spaces E^u_y and E^s_y exist).

We now give a set of sufficient conditions for ergodicity:

- **SC1:** (Expansion) Unstable (stable) vectors expand (respectively, contract) monotonically:[1] $\|D_x F(u)\| \geq \|u\|$ for all $u \in E^u_x$ and $\|(D_x F^{-1}(u)\| \geq \|u\|$ for all $u \in E^s_x$.
- **SC2:** (Continuity) The families of stable and unstable subspaces E^s_x and E^u_x are continuous on their domains. Furthermore, the limit spaces $\lim_{y \to x} E^u_y$ and $\lim_{y \to x} E^s_y$ are always transversal to each other at every SC-sufficient point x, even if E^u_x or E^s_x does not exist.
- **SC3:** (Regularity) For any $n \neq 0$ the set $\mathcal{S}_n$ is a finite[2] union of smooth curves which may intersect only at their endpoints. The curvature of all the curves in $\mathcal{S}_n$ is bounded by a constant $c_n < \infty$.
- **SC4:** (Double singularities) For any $n \geq 1$ the intersection $\mathcal{S}_n \cap \mathcal{S}_{-n}$ is a finite[3] set.

[1]A metric for which this property holds is called *adapted* or *Lyapunov metric*. We constructed an adapted metric for dispersing billiards in Section 5.10.

[2]We may assume that $\mathcal{S}_n$ is a countable union of smooth curves, but they must accumulate at finitely many points in M.

[3]We may assume that the intersection $\mathcal{S}_n \cap \mathcal{S}_{-n}$ is countable but has finitely many limit points.

SC5: (Transversality) At almost every point $x \in \mathcal{S}_1$ the subspace E_x^u is defined and transversal to S_1, and at almost every point $x \in S_{-1}$ the subspace E_x^s is defined and transversal to S_{-1}.

SC6: (Thickness of neighborhoods of singularities) For any $\delta > 0$ let $\mathcal{U}_\delta$ be the δ-neighborhood of the set $\mathcal{S}_1 \cup \mathcal{S}_{-1}$; then $\mu(\mathcal{U}_\delta) \leq \text{const} \cdot \delta$.

SC7: (Ansatz) Almost every point of $\mathcal{S}_{-1}$ (with respect to the Lebesgue length on it) is u-essential, and almost every point of S_1 is s-essential.

THEOREM 9.45 (Local ergodic theorem [**Ch93**]). *Under the above conditions, every SC-sufficient point $x \in M$ has an open neighborhood $\mathcal{U}_x$ that belongs (mod 0) to one ergodic component.*

The first version of this theorem [**SC87, KSS90**] assumed a condition stronger than **(SC5)** (it is Proper Alignment; see **(LW4)** below). It was relaxed later by Chernov [**Ch93**].

We also note that the assumption on the curvature in **CS3** was overlooked in all these papers; its necessity was pointed out recently in [**BCST02**].

9.11.3. Local Ergodicity (Liverani-Wojtkowski version). Motivated by applications in Hamiltonian mechanics (which included but were not limited to billiards), Liverani and Wojtkowski adapted the local ergodic theorem to symplectic maps. In the 2D case presented here the symplecticity of F is essentially equivalent to the absolute continuity of μ, which we have already assumed.

The local ergodic theorem by Liverani and Wojtkowski requires more restrictive assumptions on the map F than those made in Section 3.1 (and again in Section 9.2). We need to assume that there are two finite partitions of M

$$M = B_1^+ \cup \cdots \cup B_r^+ = B_1^- \cup \cdots \cup B_r^-$$

where $B_i^\pm$ are compacts sets, int $B_i^\pm$ is connected and dense in $B_i^\pm$, and the boundaries $\partial B_i^\pm$ are made by finitely many compact smooth curves which may intersect only at their endpoints. The sets $B_i^\pm$ play the role of 'building blocks' (or 'boxes').

Now suppose a map F is defined separately on each domain B_i^+, $1 \leq i \leq r$, so that F is a C^r $(r \geq 2)$ diffeomorphism of the interior of B_i^+ onto the interior of B_i^- and a homeomorphism of B_i^+ onto B_i^-.

Observe that the sets $N = \cup_i \operatorname{int} B_i^+$ and $F(N) = \cup_i \operatorname{int} B_i^-$ are open and dense in M. We assume that on each connected component of N the map F can be extended by continuity to the boundary ∂N, and on each connected component of $F(N)$ the inverse map F^{-1} can be extended by continuity to the boundary $\partial F(N)$. Since the boundaries of B_i^+ may have common points, the function F may be multivalued on $\cup_i \partial B_i^+$.

The hyperbolicity must be ensured by a field of unstable cones $\{\mathcal{C}\}$ that is invariant and eventually strictly invariant (Section 3.13). We suppose cones $\{\mathcal{C}_z\}$ are defined in the interior of each 'box' B_i^+. Let X_z, Y_z be unit vectors making the edges of $\mathcal{C}_z$, i.e.

$$\mathcal{C}_z = \{u \in \mathcal{T}_z M \colon u = aX_z + bY_z, \quad ab \geq 0\}$$

(here a, b play the roles of coordinates of the vector u). For each cone $\mathcal{C}_z$, there is an associated quadratic form given by $Q_z(u) = A(X_z, Y_z)\, ab$, where $A(X_z, Y_z)$ is the area of the parallelogram spanned by X_z and Y_z. The quantity

$$\sigma(D_zF^n) = \inf_{u\in\mathrm{int}(\mathcal{C}_z)} \sqrt{\frac{Q_{F^nz}(D_zF^nu)}{Q_z(u)}}$$

measures the amount of expansion under D_zF^n inside the unstable cone. Similarly we can define σ for F^{-1} by replacing $\mathcal{C}_z$ with its complementary cone.

DEFINITION 9.46. A point $z \in M$ is said to be *LW-sufficient* if there exists an $n > 0$ such that $z \notin \mathcal{S}_n \cap \mathcal{S}_{-n}$ and either $\sigma(D_zF^n) > 3$ or $\sigma(D_zF^{-n}) > 3$.

Since the cone family $\{\mathcal{C}\}$ is eventually strictly invariant, the set of sufficient points of M has full μ-measure [**LW95**, §7F].

To apply the local ergodic theorem, we need to verify the following conditions.

LW1: (Noncontraction property) There exists a constant $\rho > 0$ such that for every $n \geq 1$, every $z \in M \setminus \mathcal{S}_n$ and $u \in \mathcal{C}_z$ we have $\|D_zF^nu\| \geq \rho\|u\|$.

LW2: (Continuity) The restriction of $\mathcal{C}$ to the interior of every box B_i^+ is continuous.[4]

LW3: (Regularity) For any $n \neq 0$ the set $\mathcal{S}_n$ is a finite union of smooth compact curves which may only intersect at their endpoints. The curvature of all curves in $\mathcal{S}_n$ is bounded by a constant $c_n < \infty$.

LW4: (Proper Alignment) The tangent space to $\mathcal{S}_{-1}$ at any point $z \in \mathcal{S}_{-1}$ is strictly contained in $\mathcal{C}(z)$, and the tangent space to $\mathcal{S}_1$ at any point $z \in \mathcal{S}_1$ is strictly contained in the complementary cone $\mathcal{C}'(z)$.

LW5: (Ansatz) For almost every point $z \in \mathcal{S}_1$ (with respect to the Lebesgue measure on $\mathcal{S}_1$) we have $\lim_{n\to\infty} \sigma(D_zF^{-n}) = \infty$. Similarly, for almost every point $z \in \mathcal{S}_{-1}$ we have $\lim_{n\to\infty} \sigma(D_zF^n) = \infty$.

THEOREM 9.47 (Local ergodic theorem [**LW95**]). *Under the above conditions, every LW-sufficient point x in a box B_i^+ has an open neighborhood $\mathcal{U}_x$ in B_i^+ that belongs (mod 0) to one ergodic component.*

9.11.4. Comparison. The Liverani-Wojtkowski (LW) theorem handles the hyperbolicity via invariant cones, while the Sinai-Chernov (SC) theorem deals directly with stable and unstable subspaces (this is, perhaps, an insignificant difference reflecting the taste of the authors).

More importantly, the SC theorem requires monotonic expansion and contraction **(SC5)**, while the LW strategy carries on with a less restrictive noncontraction property **(LW1)**.

On the other hand, the LW theorem imposes more stringent requirements on the singularities of the map F, which must be a *finite* union of very regular curves. The SC theorem allows less regular singularity sets with possibly countably many curves (we have seen such maps in Section 4.10 studying dispersing billiards of category B). This is why the SC theorem has to make extra assumptions **(SC4)**, **(SC5)**, and **(SC6)**, which in the case of the LW setting are unnecessary; they simply follow from other conditions.

Next, **(SC5)** is a weaker property than **(LW4)**; this improvement of the SC theorem was made in [**Ch93**]. And the definition of SC-sufficiency requires only

[4] It appears that this condition can be relaxed [**DMM06**]: it is enough to assume that each box B_i^+ can be decomposed into countably many sub-boxes such that the restriction of $\mathcal{C}$ to the interior of every sub-box is continuous.

some expansion along the trajectory of the point, while the LW-sufficiency needs the expansion factor > 3.

9.11.5. Global ergodicity. Suppose a local ergodic theorem (in either version) is proved. It tells us, roughly speaking, that the ergodic components Σ_i of F are (mod 0) unions of open balls around sufficient points. Now we make our last assumption:

G1: (Abundance of sufficient points) The set $G \subset M$ of sufficient points is arcwise connected and has full measure.

The version of sufficiency (SC or LW) must correspond to the version of the local ergodic theorem established for the given map F.

THEOREM 9.48 (Global ergodic theorem). *Under all the above conditions, the map F is ergodic.*

The proof is almost identical to that of Proposition 6.19, so we leave it to the reader as an easy exercise.

After proving ergodicity one may proceed to mixing and Bernoulli properties. According to a general ergodic decomposition theorem (see Section 6.7), each ergodic component Σ_i can be partitioned into finitely many disjoint measurable sets $\Sigma_{i,j}$, $1 \le j \le J_i$, so that $F(\Sigma_{i,j}) = \Sigma_{i,j+1}$ for $1 \le j < J_i$ and $F(\Sigma_{i,J_i}) = \Sigma_{i,1}$ and the map F^{J_i} restricted to $\Sigma_{i,j}$ is K-mixing for every $1 \le j \le J_i$.

So, it is enough to verify that F^n is ergodic for every $n \ge 1$ to deduce that F is K-mixing. Then the general results by Chernov and Haskell [**CH96**] and Ornstein and Weiss [**OW98**] should imply that F is Bernoulli.

9.11.6. Applications. The first attempt to implement the above program was made in [**Sz92**], where it was argued that all focusing billiards satisfying Wojtkowski's condition (9.11) were ergodic. Markarian [**Mar93a**] dealt with just one billiard table satisfying (9.11), the cardioid (see Exercise 9.16), but gave a detailed proof of ergodicity. These first works used the Sinai-Chernov version of the local ergodic theorem.

Del Magno [**DM01**] followed the Liverani-Wojtkowski approach to show that the billiard in a 'truncated' ellipse

$$x^2/a^2 + y^2 = 1, \quad |y| \le h < 1 \tag{9.29}$$

was ergodic provided $h < 1/\sqrt{1+a^2}$. Later he and Markarian [**DMM03**] extended this result to the so-called 'elliptic stadia', regions bounded by the two elliptical arcs satisfying (9.29) with $1 < a < \sqrt{4-2\sqrt{2}}$ and $h < 1/\sqrt{1+a^2}$ and moved apart by the distance $2d$ with $d > 2a^2\sqrt{a^2-1}$.

Work is currently under way [**DMM06**] to prove ergodicity for much larger classes of billiards with absolutely focusing arcs.

Afterword

We have presented the theory of two-dimensional chaotic billiards. Our main goal was to cover the fundamentals of this theory, which mostly go back to the original work of Sinai and his school in the 1970s. We also included recent advances in this theory, such as sharp technical estimates on unstable manifolds (large parts of Chapters 4 and 5) and statistical properties (Chapter 7). In fact some *new* results were obtained by one of us (NC) during the preparation of the book (these include Proposition 4.29, estimates in Sections 4.12–4.14 and 5.5, Theorems 5.31 and 7.41, and the ASIP of Section 7.9).

Even though our treatment of Sinai's dispersing billiards was fairly extensive (and at places very deep), our book did not attempt to cover, or even survey, the entire realm of chaotic billiards. We only went some distance analyzing Bunimovich's billiards in Chapter 8, admitting enormous technical difficulties and leaving many issues unresolved. We did include an overview of other planar chaotic billiards with focusing boundary components in Chapter 9, but we left most of the proofs out. The reader can find complete and clearly written proofs in [**Wo86, Mar88, Don91, Bu92**].

Our book did not even touch upon multidimensional chaotic billiards. For the studies of periodic Lorentz gases in space we refer the reader to [**SC87**] (the original proof of ergodicity, which is not entirely complete), [**BCST02**] (the modern, complete proof of ergodicity for scatterers with algebraic boundaries), and [**BCST03**] (a detailed technical analysis of unstable manifolds and singularities, similar to our Chapter 5).

Interesting models of multidimensional chaotic billiards with focusing boundary components (spherical caps) were constructed by Bunimovich and Rehacek [**BR97, BR98a, BR98b**], who also discovered a related phenomenon of astigmatism [**Bu00**].

But the biggest topic we left out was the gas of hard balls. The long and fascinating history of the study of hard ball gases goes back to L. Boltzmann, who in the nineteenth century (if put in modern terms) conjectured their ergodicity to justify the laws of statistical mechanics. In the 1940s N. Krylov pointed out apparent similarities between the dynamics of hard balls and that of geodesic flows on negatively curved manifolds, for which hyperbolicity had already been established. Sinai formalized the problem in the early 1960s, claiming that gases of $N \geq 2$ hard balls on a torus must be hyperbolic and ergodic. Sinai reduced the systems of hard balls to dispersing and semi-dispersing billiards (in the simplest case we did that in Section 4.2).

Sinai himself [**Sin70**] proved ergodicity for $N = 2$ and presented a detailed plan for solving the problem in its generality [**Sin79**]. The main component of his plan, the local ergodic theorem, was obtained by Sinai and Chernov [**SC87**]. Further progress was mainly due to the Hungarian school (A. Krámli, N. Simányi, and D. Szász). We refer the reader to [**KSS90, KSS91, KSS92, BLPS92, Sim99, Sim03, Sim04**] as well as references therein. It appears that the complete solution of this great problem (known today as Boltzmann-Sinai ergodic hypothesis) is within reach. It might be the subject of a future book on chaotic billiards.

APPENDIX A

Measure theory

Here we provide, for the reader's convenience, basic definitions and facts of measure theory that are used in our book.

σ-algebras. Let X be a set. A σ-algebra $\mathfrak{F}$ on X is a nonempty collection of subsets of X with two properties: (i) it is closed under countable unions; i.e. if $A_i \in \mathfrak{F}$ for all $i \geq 1$, then $\cup_{i=1}^{\infty} A_i \in \mathfrak{F}$; and (ii) it is closed under taking complements; i.e. if $A \in \mathfrak{F}$, then $A^c = X \setminus A \in \mathfrak{F}$. A pair $(X, \mathfrak{F})$ is called a measurable space; the sets $A \in \mathfrak{F}$ are said to be measurable.

It is immediate that any σ-algebra is closed under all countable combinations of elementary set-theoretic operations (unions, intersections, differences, symmetric differences, taking complements). Every σ-algebra contains X itself and the empty set $\emptyset$.

Clearly, for any family of σ-algebras $\{\mathfrak{F}_\alpha\}$ their intersection $\mathfrak{F} = \cap_\alpha \mathfrak{F}_\alpha$ is a σ-algebra. For any collection $\mathfrak{C}$ of subsets of X we denote by $\mathbb{F}(\mathfrak{C})$ the minimal σ-algebra containing $\mathfrak{C}$ (it is the intersection of all σ-algebras containing $\mathfrak{C}$). We say that $\mathbb{F}(\mathfrak{C})$ is the σ-algebra generated by $\mathfrak{C}$.

If X is a topological space and $\mathfrak{C}$ denotes the collection of all open subsets of X, then $\mathbb{F}(\mathfrak{C})$ is called the Borel σ-algebra on X. For example, the Borel σ-algebra on $\mathbb{R}$ is the minimal σ-algebra containing all open intervals.

Measures. A measure μ on $(X, \mathfrak{F})$ is a function $\mu : \mathfrak{F} \to \mathbb{R} \cup \{+\infty\}$ with three properties: (i) it is nonnegative, i.e. $\mu(A) \geq 0$ for all $A \in \mathfrak{F}$; (ii) the empty set has zero measure, i.e. $\mu(\emptyset) = 0$; and (iii) μ is σ-additive (or countably additive); i.e. if $\{A_i\}_{i=1}^{\infty} \in \mathfrak{F}$ and $A_i \cap A_j = \emptyset$ for $i \neq j$, then $\mu\big(\cup_{i=1}^{\infty} A_i\big) = \sum_{i=1}^{\infty} \mu(A_i)$.

Since in many cases σ-algebras tend to be rather large, explicit definitions of $\mu(A)$ for all $A \in \mathfrak{F}$ is often an impossible task. Then one can define $\mu(A)$ on a semi-algebra (see below) and use an extension theorem stated in the next subsection.

A semi-algebra is a nonempty collection $\mathfrak{C}$ of subsets of X with two properties: (i) it is closed under intersections; i.e. if $A, B \in \mathfrak{C}$, then $A \cap B \in \mathfrak{C}$; and (ii) if $A \in \mathfrak{C}$, then $A^c = \cup_{i=1}^{n} A_i$, where each $A_i \in \mathfrak{C}$ and $A_1, \ldots, A_n$ are pairwise disjoint subsets of X.

Extension theorem. Suppose $\mathfrak{C}$ is a semi-algebra and a nonnegative function $\mu\colon \mathfrak{C} \to \mathbb{R} \cup \{+\infty\}$ has two properties: (i) $X = \cup_{i=1}^{\infty} B_i$ for some $B_i \in \mathfrak{C}$ such that $\mu(B_i) < \infty$; and (ii) μ is σ-additive; i.e. if $\{A_i\}_{i=1}^{\infty} \in \mathfrak{C}$ with $A_i \cap A_j = \emptyset$ for $i \neq j$, and $\cup_{i=1}^{\infty} A_i \in \mathfrak{C}$, then $\mu\big(\cup_{i=1}^{\infty} A_i\big) = \sum_{i=1}^{\infty} \mu(A_i)$. In this case there is a unique measure on the σ-algebra $\mathbb{F}(\mathfrak{C})$ that agrees with μ on $\mathfrak{C}$.

For example, the collection $\mathfrak{C}$ of all open, closed, and semi-open intervals in $\mathbb{R}$ is a semi-algebra. The length of the interval is a function on $\mathfrak{C}$ satisfying the properties (i) and (ii) above. Thus there is a unique measure $\mathbf{m}$ on the Borel σ-algebra on $\mathbb{R}$ that coincides with the usual length on intervals. It is called the Lebesgue measure; it can also be defined on a larger σ-algebra (see below).

Lebesgue density points. Let $A \subset \mathbb{R}$ be a Borel measurable set and $\mathbf{m}(A) > 0$. Then there is a measurable subset $B \subset A$ with two properties: (i) $\mathbf{m}(A \setminus B) = 0$ and (ii) for every $x \in B$

$$\lim_{\varepsilon \to 0} \frac{\mathbf{m}\big([x-\varepsilon, x+\varepsilon] \cap A\big)}{\mathbf{m}\big([x-\varepsilon, x+\varepsilon]\big)} = 1.$$

Every point $x \in B$ is called a Lebesgue density point of A. The following simple fact is used a few times in our book.

Let $A \subset (a,b)$ be a measurable set. Suppose a sequence $\{\delta_n\}$ of distinct real numbers has two properties: (i) $\delta_n \to 0$ and (ii) the set A is δ_n-translationally invariant; i.e. for any $x \in A \cap (a+\delta_n, b-\delta_n)$ we have $x - \delta_n \in A$ and $x + \delta_n \in A$. Then either $\mu(A) = b - a$ or $\mu(A) = 0$. Indeed, if $0 < \mu(A) < b - a$, then we can pick a Lebesgue density point $x \in A$ and a Lebesgue density point $y \in (a,b) \setminus A$ and translate a small neighborhood of x onto a small neighborhood of y, arriving at a contradiction.

Probability measures on $\mathbb{R}$. We say that μ is a finite measure on $(X, \mathfrak{F})$ if $\mu(X) < \infty$ and a probability measure if $\mu(X) = 1$. The restriction of $\mathbf{m}$ onto the unit interval $[0,1]$ is a probability measure. More generally, any Borel probability measure μ on $\mathbb{R}$ can be defined by

$$\mu((a,b]) = F(b) - F(a),$$

where $F(x)$ is a function that has three properties: (i) it is nondecreasing, (ii) it is right-continuous, and (iii) $\lim_{x\to\infty} F(x) = 1$ and $\lim_{x\to-\infty} F(x) = 0$. We call $F(x)$ the distribution function of μ. Note that $F(x)$ is continuous on $\mathbb{R}$, except possibly countably many points.

Every function $F(x)$ with above properties (i), (ii), and (iii) is a distribution function of a probability measure on $\mathbb{R}$.

We will consider only probability measures.

Lebesgue integral. Let $(X, \mathfrak{F})$ be a measurable space. A function $g\colon X \to \mathbb{R}$ is measurable if for any Borel set $A \subset \mathbb{R}$ we have $g^{-1}(A) \in \mathfrak{F}$. (It is actually enough that preimages of all open intervals $(a,b) \subset \mathbb{R}$ are measurable sets.)

A measurable function g is simple if $g(X)$ is a finite set; in that case let $g(X) = \{c_1, \dots, c_n\}$ and $A_k = g^{-1}(c_k)$ for $1 \le k \le n$. Then we can write $g(x) = \sum_{i=k}^{n} c_k \mathbf{1}_{A_k}(x)$, where

$$\mathbf{1}_A(x) = \begin{cases} 1 & \text{if } x \in A \\ 0 & \text{if } x \notin A \end{cases}$$

denotes the indicator of the set A.

Let μ be a measure on $(X, \mathfrak{F})$ and $g\colon X \to \mathbb{R}$ a measurable function. The (Lebesgue) integral $\int_X g\,d\mu$ is defined as follows. If $g = \mathbf{1}_A$ for some $A \in \mathfrak{F}$, then $\int_X \mathbf{1}_A\,d\mu = \mu(A)$. If $g(x) = \sum_{i=k}^{n} c_k \mathbf{1}_{A_k}(x)$ is a simple function, then $\int_X g\,d\mu = \sum_{k=1}^{n} c_k \mu(A_k)$. If $g \ge 0$, then

$$\int_X g\,d\mu = \sup\bigg\{\int_X h\,d\mu \colon h \text{ is simple and } h \le g\bigg\}.$$

We say that $g \geq 0$ is integrable if the supremum is finite.

For an arbitrary function g, we decompose it as $g = g^+ - g^-$, where $g^+ = \max\{g, 0\}$ and $g^- = \max\{-g, 0\}$. Note that $g^+, g^- \geq 0$ and $|g| = g^+ + g^-$. Now we define $\int_X g\,d\mu = \int_X g^+\,d\mu - \int_X g^-\,d\mu$ (provided both integrals exist). We say that g is integrable if $\int_X g\,d\mu$ exists; the class of integrable (more precisely, μ-integrable) functions on X is denoted by $L^1_\mu(X)$.

More generally, for any $p > 0$ we denote

$$L^p_\mu(X) = \left\{f\colon \int_X |f|^p\,d\mu < \infty\right\}.$$

This is a vector space with norm $\|f\|_p = \left(\int_X |f|^p\,d\mu\right)^{1/p}$ (actually, it becomes a norm if we identify functions that coincide on a set of full measure; see also below). In addition, the space $L^2_\mu(X)$ has an inner product defined by $\langle f, g\rangle = \int_X fg\,d\mu$. Note that if $\mu(X) < \infty$ and $p < q$, then $L^q_\mu(X) \subset L^p_\mu(X)$.

The set of all bounded functions on X, denoted by $L^\infty(X)$, is also a vector space with norm $\|f\|_\infty = \sup_x |f(x)|$.

For any $A \in \mathfrak{F}$ we can define $\int_A f\,d\mu = \int_X \mathbf{1}_A f\,d\mu$.

Riemann-Stieltjes integral. Let μ be a probability measure on $\mathbb{R}$ and $F(x)$ its distribution function; see above. Then for any μ-integrable function $g\colon \mathbb{R} \to \mathbb{R}$

$$\int_{\mathbb{R}} g(x)\,d\mu = \int_{-\infty}^{\infty} g(x)\,dF(x),$$

where the second integral can be defined as follows:

$$\int_{-\infty}^{\infty} g(x)\,dF(x) = \lim_{\substack{b\to\infty \\ a\to-\infty}} \int_a^b g(x)\,dF(x)$$

and

$$\int_a^b g(x)\,dF(x) = \lim_{\|P\|\to 0} \sum_{i=0}^{n} g(x_i^*)\big[F(x_i) - F(x_{i-1})\big] \tag{A.1}$$

where $P = \{a = x_0 < \ldots < x_n = b\}$ is a partition of the interval $[a, b]$, $\|P\| = \max|x_i - x_{i-1}|$, and $x_i^* \in [x_{i-1}, x_i]$ for all $i = 1, \ldots, n$. The integral (A.1) is called the Riemann-Stieltjes integral.

Weak convergence. Let $\{\mu_n\}$ be a sequence of probability measures on $\mathbb{R}$ with distribution functions $\{F_n(x)\}$. We say that $\{\mu_n\}$ weakly converges to a probability measure μ on $\mathbb{R}$ (denoted by $\mu_n \Rightarrow \mu$) if for every bounded continuous function $g\colon \mathbb{R} \to \mathbb{R}$ we have

$$\lim_{n\to\infty} \int_{\mathbb{R}} g\,d\mu_n = \int_{\mathbb{R}} g\,d\mu.$$

Note: $\mu_n \Rightarrow \mu$ if and only if for any interval $[a, b]$ such that $\mu(\{a\}) = \mu(\{b\}) = 0$ we have $\mu_n([a, b]) \to \mu([a, b])$. In terms of distribution functions, this means that $F_n(x) \to F(x)$ for every $x \in \mathbb{R}$ where $F(x)$ is continuous.

Absolute continuity. Let μ and ν be two measures on $(X, \mathfrak{F})$. We say that ν is absolutely continuous with respect to μ, denoted by $\nu \ll \mu$, if $\nu(A) = 0$ for every set with $\mu(A) = 0$. Radon-Nikodym theorem says that $\nu \ll \mu$ if and only if there exists a nonnegative (density) function $f\colon X \to \mathbb{R}$ such that

$$\nu(A) = \int_A f \, d\mu \qquad \forall A \in \mathfrak{F}.$$

The function $f(x)$ is called the Radon-Nikodym derivative, $f = d\nu/d\mu$. The measures μ and ν are said to be equivalent if $\mu \ll \nu$ and $\nu \ll \mu$.

If a probability measure μ on $\mathbb{R}$ with distribution function $F(x)$ is absolutely continuous with respect to the Lebesgue measure $\mathbf{m}$, then its density is $f(x) = F'(x)$ so that

$$\mu(a, b] = F(b) - F(a) = \int_a^b f(x) \, dx.$$

The measure μ is equivalent to $\mathbf{m}$ iff $f(x) > 0$ on a set of full $\mathbf{m}$ measure.

Let $(X, \mathfrak{F}, \mu)$ and $(Y, \mathfrak{G}, \nu)$ be two measure spaces with probability measures. A transformation $T\colon X \to Y$ is said to be measurable if for every $B \in \mathfrak{G}$ we have $T^{-1}B \in \mathfrak{F}$. The image $T\mu$ of the measure μ is defined by $T\mu(B) = \mu(T^{-1}B)$ for $B \in \mathfrak{G}$ or equivalently by

$$\int_Y g \, d(T\mu) = \int_X g \circ T \, d\mu$$

for any bounded measurable function $g\colon Y \to \mathbb{R}$.

The transformation T is said to be absolutely continuous if $\nu \ll T\mu$. Then the density function $f = d\nu/d(T\mu)$ satisfies the 'change of variable formula'

$$\int_Y g \, d\nu = \int_Y gf \, d(T\mu) = \int_X (g \circ T)(f \circ T) \, d\mu$$

for any bounded measurable function $g\colon Y \to \mathbb{R}$. Hence $f \circ T$ is the Jacobian of the map T (with respect to the measures μ and ν).

Two measures μ and ν on $(X, \mathfrak{F})$ are mutually singular (or orthogonal), denoted by $\mu \perp \nu$, if there exists a set $A \in \mathfrak{F}$ such that $\mu(A^c) = 0$ and $\nu(A) = 0$.

The Lebesgue decomposition theorem says that given two measures μ and ν on $(X, \mathfrak{F})$, there is a unique representation $\mu = \mu_{\mathrm{a}} + \mu_{\mathrm{s}}$, where $\mu_{\mathrm{a}} \ll \nu$ and $\mu_{\mathrm{s}} \perp \nu$.

Product measures. Let $(X_1, \mathfrak{F}_1, \mu_1)$ and $(X_2, \mathfrak{F}_2, \mu_2)$ be two measure spaces. The cartesian product $X_1 \times X_2$ is defined by

$$X_1 \times X_2 = \{(x_1, x_2)\colon\ x_1 \in X_1,\ x_2 \in X_2\}.$$

Let $\mathfrak{F}_1 \times \mathfrak{F}_2$ be the σ-algebra on $X_1 \times X_2$ generated by subsets of the form $A_1 \times A_2$, where $A_1 \in \mathfrak{F}_1$ and $A_2 \in \mathfrak{F}_2$. The product measure $\mu_1 \times \mu_2$ is defined to be the unique measure on $\mathfrak{F}_1 \times \mathfrak{F}_2$ satisfying

$$\mu_1 \times \mu_2 \, (A_1 \times A_2) = \mu_1(A_1) \, \mu_2(A_2) \qquad \forall A_1 \in \mathfrak{F}_1,\ A_2 \in \mathfrak{F}_2.$$

The measure space $(X_1 \times X_2, \mathfrak{F}_1 \times \mathfrak{F}_2, \mu_1 \times \mu_2)$ is called the direct product of the spaces $(X_1, \mathfrak{F}_1, \mu_1)$ and $(X_2, \mathfrak{F}_2, \mu_2)$. This construction easily generalizes to finite and countable products of measure spaces.

Complete measures. Let $(X, \mathfrak{F}, \mu)$ be a space with a measure. A measurable set $A \subset X$ is called a null set if $\mu(A) = 0$ and a set of full measure if $\mu(A^c) = 0$. The measure is said to be complete if every subset $B \subset A$ of a null set A is measurable (and then automatically itself a null set).

The Lebesgue measure $\mathbf{m}$ on the Borel σ-algebra of subsets in $\mathbb{R}$ is not complete.

It is a trivial matter to extend a measure μ on $(X, \mathfrak{F})$ to a complete one. Let $\mathfrak{N} = \{N \in \mathfrak{F}\colon \mu(N) = 0\}$. Define a new σ-algebra by

$$\bar{\mathfrak{F}} = \{A \cup E\colon A \in \mathfrak{F},\quad E \subset N \text{ for some } N \in \mathfrak{N}\}$$

and for every $B = A \cup E$ as above set $\bar{\mu}(B) = \mu(A)$. Then $\bar{\mu}$ is a complete measure on $(X, \bar{\mathfrak{F}})$.

The extension of the Lebesgue measure $\mathbf{m}$ on $\mathbb{R}$ to a complete measure is also called the Lebesgue measure, and we still denote it by $\mathbf{m}$. Its restriction onto the unit interval $[0, 1]$ is a complete probability measure.

In the rest of Appendix A, we will consider only complete probability measures.

A fundamental principle consists of disregarding null sets, i.e. working up to sets of zero measure (or modulo sets of zero measure). For example, we identify two functions $f, g\colon X \to \mathbb{R}$ if they differ only on a null set, recall our definition of the norm $\|\cdot\|_p$.

For two measurable sets $A, B \in \mathfrak{F}$ we put

$$A = B \pmod 0 \iff \mu(A \Delta B) = 0;$$

that is, A can be transformed into B by adding and/or removing some null sets. Clearly, $A = B \pmod 0$ is an equivalence relation. Let $\mathfrak{M}$ denote the set of classes of equal (mod 0) measurable sets. We can define on $\mathfrak{M}$ a metric $\rho(A, B) = \mu(A \Delta B)$, which makes it a complete metric space.

Two measure spaces $(X_1, \mathfrak{F}_1, \mu_1)$ and $(X_2, \mathfrak{F}_2, \mu_2)$ are said to be isomorphic (mod 0) if for each $i = 1, 2$ there is a set $B_i \subset X_i$ of full μ_i measure and a bijection $\varphi\colon B_1 \to B_2$ that preserves measurable sets and measures; i.e. for any $A \in \mathfrak{F}_1$ we have $\varphi(A \cap B_1) \in \mathfrak{F}_2$ and $\mu_1(A \cap B_1) = \mu_2\big(\varphi(A \cap B_1)\big)$, and vice versa. The map φ is called an isomorphism.

Generators. Let $(X, \mathfrak{F}, \mu)$ be a space with a complete probability measure. We say that a countable family $\mathfrak{B} = \{E_n\}_{n=1}^\infty$ of measurable subsets is a generator of the space X if it has two properties: (i) the minimal σ-algebra $\mathbb{F}(\mathfrak{B})$ containing $\mathfrak{B}$ includes (mod 0) all measurable sets; i.e. for every $A \subset \mathfrak{F}$ there is $B \subset \mathbb{F}(\mathfrak{B})$ such that $A \subset B$ and $\mu(B \setminus A) = 0$; and (ii) the sets $\{E_n\}$ separate points of X; i.e. for any two distinct points $x, y \in X$ there exists $E_n \subset \mathfrak{B}$ such that either $x \in E_n$ and $y \notin E_n$ or $x \notin E_n$ and $y \in E_n$.

For example, let $X = [0, 1)$ be a semiopen unit interval and μ the Lebesgue measure on it. Then the family of intervals $[a_n, b_n) \subset [0, 1)$ with rational endpoints is a generator of X.

The space X is said to be complete with respect to a generator $\mathfrak{B}$ if every countable intersection $\cap_{n\geq 1} F_n$, where each F_n is one of the two sets E_n and E_n^c, is not empty (note that by property (ii) of a generator this intersection consists of at most one point).

In the above example, $X = [0, 1)$ is not complete with respect to the generator consisting of intervals with rational endpoints. But it is complete with respect to another generator $\{E_n\}_{n=1}^\infty$ defined by

$$E_n = \bigcup_{i=0}^{2^{n-1}-1} \left[\frac{2i}{2^n}, \frac{2i+1}{2^n}\right).$$

The space X is said to be complete (mod 0) with respect to a generator $\mathfrak{B}$ if it can be included into a larger space $X' \supset X$ with a complete probability measure

μ' which agrees with μ on X and such that $\mu'(X' \setminus X) = 0$, and which is complete with respect to a generator $\mathfrak{B}' = \{E'_n\}$ such that $E_n = E'_n \cap X$ for every n. If a space X is complete (mod 0) with respect to one generator, it is complete with respect to all generators.

Lebesgue spaces. A space X with a complete probability measure μ is said to be a Lebesgue space if it is complete (mod 0) with respect to some (and hence all) generators.

The unit interval $X = [0, 1]$ with the Lebesgue measure on it is a Lebesgue space. Generally, a Lebesgue space may contain finitely many or countably many atoms (points of positive measure). In fact, any Lebesgue space is isomorphic (mod 0) to a union of finitely many or countably many atoms of total measure $p \in [0, 1]$ and the interval $[p, 1]$ with the Lebesgue measure on it. (If there are no atoms, then it is isomorphic to the unit interval $[0, 1]$.)

The notion of a Lebesgue space is very useful. On the one hand, it is very broad and includes practically all measure spaces that arise in applications. In particular, any complete separable metric space with a probability measure defined on Borel sets (and properly extended to a complete measure; see above) is a Lebesgue space. A direct product of finitely many or countably many Lebesgue spaces is a Lebesgue space.

On the other hand, Lebesgue spaces have important attributes that more general measure spaces lack. One of the most important features of Lebesgue spaces is the notion of measurable partition; see the next section.

Measurable partitions. This is a rather special topic in measure theory, but it is essential for the proofs of ergodicity and mixing of chaotic billiards. We refer the reader to [**Ro49, Ro67**] for full coverage, as well as [**CFS82**, Appendix 1], [**ME81**, Chapter 1], and [**Bou60**, pp. 57–72] for surveys (the last one focuses on topological aspects of measurable partitions).

Let $(X, \mathfrak{F}, \mu)$ be a Lebesgue space. A partition ξ of X is a collection $\{\xi_\alpha\}$ of disjoint subsets of X such that $\cup_\alpha \xi_\alpha = X$. We denote by $\xi(x)$ the unique element of ξ containing the point $x \in X$. A subset $A \subset X$ is called a ξ-set if it is a union of some elements of ξ. For any partition ξ the collection of measurable ξ-sets makes a σ-algebra.

There are two distinguished partitions that are, in a sense, maximal and minimal. The partition of X into single points is denoted by ε. The (trivial) partition of X whose only element is X is denoted by ν.

A partition ξ is said to be measurable if there exists a countable collection of measurable ξ-sets $\{B_n\}_{n=1}^{\infty}$ such that for any two distinct elements $C_1, C_2 \in \xi$ there is a B_n for which $C_1 \subseteq B_n$ and $C_2 \not\subseteq B_n$, or vice versa. (Note that $C_2 \not\subseteq B_n$ actually implies $C_2 \subseteq B_n^c$, since B_n is a ξ-set.) We call $\{B_n\}$ a generator of ξ.

Every element $C \in \xi$ of a measurable partition ξ can be represented as $C = \cap_{n\geq 1} F_n$, where each F_n is one of the two sets B_n and B_n^c (conversely, every nonempty intersection $\cap_{n\geq 1} F_n$ is exactly one element of C). Hence $C \in \mathfrak{F}$; i.e. every element of a measurable partition is a measurable set. But the converse is far from being true.

Consider, for example, the Lebesgue measure μ on $X = [0, 1)$ and fix an irrational number α. Let ξ be a partition of X such that $x, y \in X$ lie in one element of ξ (i.e. $y \in \xi(x)$) iff $x - y = m + \alpha n$ for some $m, n \in \mathbb{Z}$. Every element $C \in \xi$ is a countable (hence measurable) set. There exist two sequences $\{m_k\}$ and $\{n_k\}$

such that $\delta_k = m_k + \alpha n_k \to 0$ as $k \to \infty$. Now every ξ-set B will be invariant under translations through each δ_k; i.e. $B = \{x + \delta_k \pmod 1,\ x \in B\}$ for every k. Therefore $\mu(B) = 0$ or 1 (cf. our discussion of Lebesgue density points). Thus the partition ξ cannot have a countable generator, so it is not measurable.

Another example of a nonmeasurable partition (arising in the studies of dynamical systems) is the partition of a manifold M into global unstable submanifolds corresponding to a transitive Anosov diffeomorphism acting on M.

A partition ξ is measurable if and only if its elements are level sets of a measurable function $f\colon X \to \mathbb{R}$. The partitions ε and ν are measurable.

Factor (quotient) space. Let ξ be a measurable partition of a Lebesgue space $(X, \mathfrak{F}, \mu)$. The factor space $(X_\xi, \mathfrak{F}_\xi, \mu_\xi)$ is defined as follows. Let X_ξ be the space whose points are elements of ξ and denote by $\varphi\colon X \to X_\xi$ the natural projection $\varphi(x) = \xi(x)$. Let $\mathfrak{F}_\xi$ be the σ-algebra of subsets $A \subset X_\xi$ such that $\varphi^{-1}(A) \in \mathfrak{F}$. Lastly, the factor measure μ_ξ is defined by $\mu_\xi(A) = \mu(\varphi^{-1}A)$ for every $A \in \mathfrak{F}_\xi$. The factor space $(X_\xi, \mathfrak{F}_\xi, \mu_\xi)$ is itself a Lebesgue space.

Conditional measures. Let ξ be a measurable partition of a Lebesgue space $(X, \mathfrak{F}, \mu)$. The most important property of measurable partitions is the existence of conditional measures on its elements. The following fact is sometimes referred to as *disintegration* of the measure μ.

For μ_ξ-almost every element $C \in \xi$ there exists a σ-algebra $\mathfrak{F}_C$ and a probability measure μ_C on $(C, \mathfrak{F}_C)$ with three properties: (i) $(C, \mathfrak{F}_C, \mu_C)$ is a Lebesgue space, (ii) for any $B \in \mathfrak{F}$ we have $B \cap C \in \mathfrak{F}_C$ for μ_ξ-almost every $C \in \xi$, and (iii) for any $B \in \mathfrak{F}$ we have

$$\mu(B) = \int_{X_\xi} \mu_C(B \cap C)\, d\mu_\xi.$$

The measures $\{\mu_C,\ C \in \xi\}$ are called conditional measures induced by μ on $C \in \xi$.

We note that if $\mu(C) > 0$ for some $C \in \xi$, then the conditional measure can be defined in a standard way: $\mu_C(B) = \mu(C \cap B)/\mu(C)$.

The above integral formula is equivalent to a Fubini-type formula

$$\int_X f\, d\mu = \int_{X_\xi} \left(\int_C f\, d\mu_C \right) d\mu_\xi \tag{A.2}$$

for any integrable function f on X. The inner integral $f_\xi = \int_C f\, d\mu_C$ is known as the conditional expectation of the function f with respect to the partition ξ.

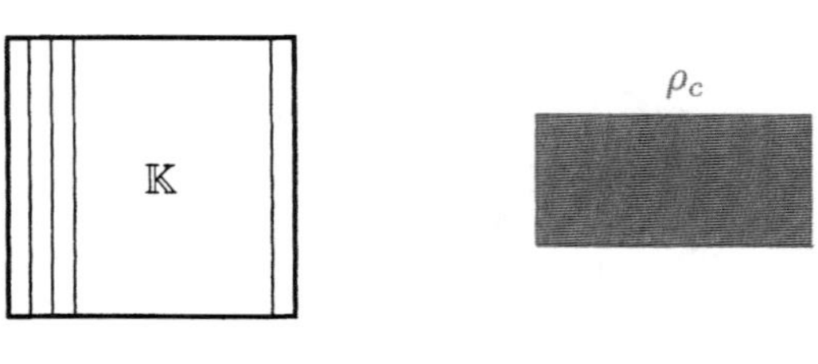

FIGURE A.1. A partition of $\mathbb{K}$ and a conditional density.

Examples. Consider the partition ξ_1 of a unit square $\mathbb{K} = \{(x, y)\colon 0 \le x, y \le 1\}$ into vertical lines $A_c = \{x = c\}$, $c \in [0, 1]$. If μ is a smooth measure on $\mathbb{K}$ with density $\rho(x, y)$ with respect to the Lebesgue measure, then the conditional measure μ_c on A_c has (normalized) density $\rho_c(y) = \rho(c, y)/\int_0^1 \rho(c, y)\, dy$, and the factor measure has density $\rho_\xi(c) = \int_0^1 \rho(c, y)\, dy$ on the unit interval $0 \leq c \leq 1$; now the integral formula (A.2) is simply Fubini's theorem

$$\int_{\mathbb{K}} f(x, y)\rho(x, y)\, dx\, dy = \int_0^1 \left(\int_0^1 f(c, y)\rho_c(y)\, dy \right) \rho_\xi(c)\, dc.$$

In particular, if μ is the Lebesgue measure ($\rho(x, y) = 1$), then μ_c is a uniform measure ($\rho_c(y) = 1$) for every c; see Fig. A.1.

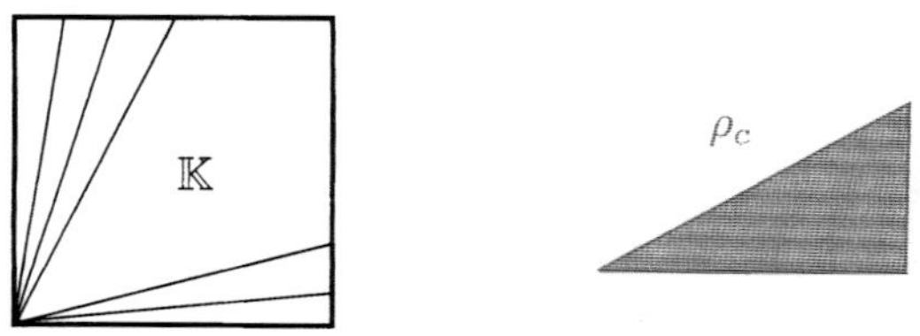

FIGURE A.2. Another partition of $\mathbb{K}$ and its conditional density.

Consider another partition ξ_2 of the unit square $\mathbb{K} = \{(x, y) \colon 0 \leq x, y \leq 1\}$, into lines $y = cx$ coming out of the origin (see Fig. A.2), and let μ be the Lebesgue measure on $\mathbb{K}$. In this case the conditional measure on every line will have a triangular density that comes to zero at the origin; the reader is invited to verify this directly.

Note that the vertical segment $A_0 = \{x = 0\}$ is an element of both partitions, but the corresponding conditional measures differ. This demonstrates that the conditional measure μ_C is not determined by the set $C \subset X$ alone; it depends on the partition ξ that contains C as an element.

Classes of partitions equal (mod 0). We say that two partitions are equal, $\xi = \eta$ (mod 0), if they coincide on a set of full measure; i.e. if there is a set of full measure $D \subset X$ such that for every $A \in \xi$ satisfying $A \cap D \neq \emptyset$. there exists $B \in \eta$ such that $A \cap D = B \cap D$. Clearly this is an equivalence relation. If ξ is a measurable partition and $\eta = \xi$ (mod 0), then η is also a measurable partition. In what follows, by a 'partition' we will mean a class of equal (mod 0) partitions.

There is a natural one-to-one correspondence between measurable partitions and complete sub-σ-algebras that are defined next.

Complete sub-σ-algebras. Let $(X, \mathfrak{F}, \mu)$ be a Lebesgue space. We have denoted by $\mathfrak{M}$ the set of classes of equal (mod 0) measurable subsets of X. The set theoretic operations $\cup$, $\cap$, $\setminus$, Δ, and taking complements naturally carry over from $\mathfrak{F}$ to $\mathfrak{M}$. Then $\mathfrak{M}$ will contain $\emptyset$ (mod 0) and X (mod 0), i.e. the classes of sets equal (mod 0) to $\emptyset$ and X. Also, $\mathfrak{M}$ will be closed under taking complements and countable unions. Thus we can treat it as a σ-algebra.

Let a subset $\mathfrak{N} \subset \mathfrak{M}$ contain the classes $\emptyset$ (mod 0) and X (mod 0) and be closed under taking complements and countable unions. Then measurable subsets of X that belong to classes included in $\mathfrak{N}$ make a σ-algebra; call it $\mathfrak{G}$. Then the restriction of the measure μ to $\mathfrak{G}$ will be a complete probability measure. We call every such $\mathfrak{G}$ a complete sub-σ-algebra of $\mathfrak{F}$.

We have already noted that for every partition ξ the collection of measurable ξ-sets is a σ-algebra; let us call it $\mathfrak{B}_0(\xi)$. Now for every measurable partition ξ the collection of subsets of X defined by

$$\mathfrak{B}(\xi) = \{A \in \mathfrak{F} \colon A = B \pmod 0 \text{ for some } B \in \mathfrak{B}_0(\xi)\}$$

is a complete sub-σ-algebra of $\mathfrak{F}$.

This establishes a one-to-one correspondence between measurable partitions and complete sub-σ-algebras: if $\mathfrak{B}(\xi) = \mathfrak{B}(\eta)$, then $\xi = \eta \pmod 0$, and for every complete sub-σ-algebra $\mathfrak{B} \subset \mathfrak{F}$ there exists a measurable partition ξ such that $\mathfrak{B}(\xi) = \mathfrak{B}$.

The minimal complete sub-σ-algebra consisting of sets equal to $\emptyset \pmod 0$ and $X \pmod 0$ corresponds only to the trivial partition ν. The maximal sub-σ-algebra $\mathfrak{B} = \mathfrak{F}$ corresponds to the partition ε into individual points.

Relations and operations on partitions. Let $(X, \mathfrak{F}, \mu)$ be a Lebesgue space and denote by ξ, η, etc., measurable partitions. We say that ξ is finer than η (or η is coarser than ξ), denoted by $\xi \succcurlyeq \eta \pmod 0$, if there is a set of full measure $D \subset X$ such that

$$\forall A \in \xi \ \exists B \in \eta \colon \ A \cap D \subset B \cap D.$$

We have $\xi \succcurlyeq \eta \pmod 0$ if and only if $\mathfrak{B}(\xi) \supset \mathfrak{B}(\eta)$.

For any family of measurable partitions $\{\xi_\alpha\}$ we define their join $\vee_\alpha \xi_\alpha$ as a measurable partition ξ with two properties: (i) $\xi_\alpha \preccurlyeq \xi$ for every α; and (ii) if ξ' is another measurable partition such that $\xi_\alpha \preccurlyeq \xi'$ for every α, then $\xi \preccurlyeq \xi'$. In other words, ξ is the minimal partition which is finer than every ξ_α.

For any family of complete sub-σ-algebras $\mathfrak{B}_\alpha$ define their join $\vee_\alpha \mathfrak{B}_\alpha$ as the intersection of all complete sub-σ-algebras each of which contains all $\mathfrak{B}_\alpha$'s. Then we have $\mathfrak{B}(\vee_\alpha \xi_\alpha) = \vee_\alpha \mathfrak{B}(\xi_\alpha)$.

For a finite or countable family $\{\xi_n\}$, the join $\xi = \vee_n \xi_n$ is a partition whose elements are (nonempty) intersections $\cap_n C_n$, where $C_n \in \xi_n$ for all n.

For any family of measurable partitions $\{\xi_\alpha\}$ we define their meet $\wedge_\alpha \xi_\alpha$ as a measurable partition ξ with two properties: (i) $\xi_\alpha \succcurlyeq \xi$ for every α; and (ii) if ξ' is another measurable partition such that $\xi_\alpha \succcurlyeq \xi'$ for every α, then $\xi \succcurlyeq \xi'$. In other words, ξ is the maximal partition which is coarser than every ξ_α. We have $\mathfrak{B}(\wedge_\alpha \xi_\alpha) = \cap_\alpha \mathfrak{B}(\xi_\alpha)$.

Given a finite or countable family of measurable partitions $\{\xi_m\}$, one may attempt to construct their meet $\wedge_m \xi_m$ as follows. For any $x, y \in X$ we put $x \sim y$ if there is a finite sequence $A_1, \ldots, A_n \in \cup_m \xi_m$ such that $x \in A_1$, $y \in A_n$, and $A_i \cap A_{i+1} \neq \emptyset$ for all $1 \leq i \leq n-1$ (of course, for each i the sets A_i and A_{i+1} must be elements of different partitions here). Then $\sim$ will be an equivalence relation on X and its classes form a partition of X; we call it $\curlywedge_m \xi_m$.

The partition $\curlywedge_m \xi_m$ corresponds, "logically", to the meet $\wedge_m \xi_m$, and in many cases $\curlywedge_m \xi_m = \wedge_m \xi_m \pmod 0$. But in some cases $\curlywedge_m \xi_m$ may be quite different from $\wedge_m \xi_m$ (in fact, $\curlywedge_m \xi_m$ may not even be measurable), so it is not safe to use it in lieu of $\wedge_m \xi_m$.

For example, let $X = [0,1]$ and μ be the Lebesgue measure on X. Let $\xi_1 = \{[0, 0.5], (0.5, 1]\}$ and ξ_2 be the partition whose elements are the two-point set $\{0, 1\}$ and all one-point subsets of $X \setminus \{0, 1\}$. We invite the reader to verify that $\xi_1 \wedge \xi_2 = \xi_1 \pmod 0$ but $\xi_1 \curlywedge \xi_2 = \nu$, the trivial partition. This problem can be fixed as follows.

Let $D \subset X$ be a full measure set. If we restrict the σ-algebra $\mathfrak{F}$ and the measure μ to D, then D will be a Lebesgue space. If we restrict ξ_m to D (replacing every element $C \in \xi_m$ with $C \cap D$), then we get a measurable partition, $\xi_{m,D}$, of D. We claim that D can be chosen so that $\curlywedge_m \xi_{m,D} \succcurlyeq \wedge_m \xi_{m,D}$; i.e. every element $C \in \wedge_m \xi_{m,D}$ is a $(\curlywedge_m \xi_{m,D})$-set. And of course, if $\curlywedge_m \xi_{m,D}$ is measurable, then $\curlywedge_m \xi_{m,D} = \wedge_m \xi_{m,D}$ (mod 0).

Indeed, let $\{B_n\}_{n=1}^{\infty}$ be a countable generator of the meet $\wedge_m \xi_m$. Then $B_n \in \mathfrak{G}(\xi_m)$ for every n and m. Hence B_n is a ξ_m-set (mod 0); i.e. there exists a ξ_m-set $B'_{n,m} = B_n$ (mod 0). Since $N = \cup_{m,n}(B_n \Delta B'_{n,m})$ is a null set, we can define $D = X \setminus N$. Obviously, $B_n \cap D$ will be a $\xi_{m,D}$-set for every n and m, which proves our claim.

The above fact plays an important role in Section 6.10.

APPENDIX B

Probability theory

Here we provide definitions and facts of probability theory used in the book (mostly in Chapter 7). We use the notions of measure theory introduced in Appendix A.

Random variables. A probability space $(\Omega, \mathfrak{F}, \mathbb{P})$ is a set Ω with a σ-algebra $\mathfrak{F}$ of its subsets and a probability measure $\mathbb{P}$ on $\mathfrak{F}$. Points $\omega \in \Omega$ are called elementary outcomes, and measurable sets $A \in \mathfrak{F}$ are called events.

A random variable is a measurable function $\mathcal{X}\colon \Omega \to \mathbb{R}$. Every random variable $\mathcal{X}$ induces a probability measure $\mu_{\mathcal{X}}$ on $\mathbb{R}$ defined by $\mu_{\mathcal{X}}(A) = \mathbb{P}\{\mathcal{X}^{-1}(A)\}$ for every Borel set $A \subset \mathbb{R}$. The distribution function of that measure (see Appendix A) is called the distribution function $F(x)$ of $\mathcal{X}$ and is denoted by $F_{\mathcal{X}}(x)$. It can be defined by

$$F(x) = \mu_{\mathcal{X}}(-\infty, x] = \mathbb{P}\{\mathcal{X}^{-1}(-\infty, x]\}$$

or slightly less formally by $F(x) = \mathbb{P}(\mathcal{X} \le x)$.

If the measure $\mu_{\mathcal{X}}$ is absolutely continuous with respect to the Lebesgue measure $\mathbf{m}$ on $\mathbb{R}$, then it has a density $f(x) \ge 0$ that satisfies

$$F(b) - F(a) = \int_a^b f(x)\, dx$$

for all $a < b$. In this case the random variable $\mathcal{X}$ is said to be absolutely continuous.

We say that an event $A \in \mathcal{F}$ occurs almost surely if $\mathbb{P}(A) = 1$. For example, two random variables $\mathcal{X}$ and $\mathcal{Y}$ are equal almost surely if $\mathbb{P}(\mathcal{X} = \mathcal{Y}) = 1$.

Mean value, variance, covariance. The mean value (or expectation) of a random variable $\mathcal{X} \in L^1_{\mathbb{P}}(\Omega)$ is

$$\mathbb{E}(\mathcal{X}) = \int_\Omega \mathcal{X}\, d\mathbb{P}.$$

It can also be computed as

$$\mathbb{E}(\mathcal{X}) = \int_{\mathbb{R}} x\, d\mu_{\mathcal{X}} = \int_{-\infty}^{\infty} x\, dF_{\mathcal{X}}(x).$$

The variance of a random variable $\mathcal{X} \in L^2_{\mathbb{P}}(\Omega)$ is defined by

$$\mathrm{Var}(\mathcal{X}) = \mathbb{E}\big[\mathcal{X} - \mathbb{E}(\mathcal{X})\big]^2 = \mathbb{E}\big[\mathcal{X}^2\big] - \big[\mathbb{E}(\mathcal{X})\big]^2,$$

and the standard deviation is $\sigma_{\mathcal{X}} = \sqrt{\mathrm{Var}(\mathcal{X})}$. We have $\mathrm{Var}(\mathcal{X}) \ge 0$, and $\mathrm{Var}(\mathcal{X}) = 0$ if and only if the random variable $\mathcal{X}$ is almost surely constant; i.e. $\mathbb{P}(\mathcal{X} = c) = 1$ for some $c \in \mathbb{R}$.

More generally, for any $m \ge 1$ and $\mathcal{X} \in L^m_{\mathbb{P}}(\Omega)$ the mth moment of $\mathcal{X}$ is defined to be $\mathbb{E}(\mathcal{X}^m)$.

If $\mathcal{X}$ is a random variable with a finite variance $\text{Var}(\mathcal{X}) < \infty$, then its linear transformation

$$\mathcal{Y} = \frac{\mathcal{X} - \mathbb{E}(\mathcal{X})}{\sigma_{\mathcal{X}}}$$

has zero mean, $\mathbb{E}(\mathcal{Y}) = 0$, and unit variance, $\text{Var}(\mathcal{Y}) = 1$. We call $\mathcal{Y}$ the normalization of $\mathcal{X}$.

Chebyshev's inequality says that for any random variable with a finite variance, $\text{Var}(\mathcal{X}) < \infty$, and any $t > 0$

$$\mathbb{P}\{|\mathcal{X} - \mathbb{E}(\mathcal{X})| > t\} \leq t^{-2}\,\text{Var}(\mathcal{X}).$$

Given two random variables $\mathcal{X}$ and $\mathcal{Y}$, their covariance is

$$\begin{aligned}\text{Cov}(\mathcal{X}, \mathcal{Y}) &= \mathbb{E}\big[(\mathcal{X} - \mathbb{E}(\mathcal{X}))(\mathcal{Y} - \mathbb{E}(\mathcal{Y}))\big] \\ &= \mathbb{E}(\mathcal{X}\mathcal{Y}) - \mathbb{E}(\mathcal{X})\,\mathbb{E}(\mathcal{Y}),\end{aligned}$$

and the correlation coefficient is

$$\rho_{\mathcal{X},\mathcal{Y}} = \frac{\text{Cov}(\mathcal{X}, \mathcal{Y})}{\sigma_{\mathcal{X}}\,\sigma_{\mathcal{Y}}}.$$

It follows from the Schwarz inequality that $-1 \leq \rho_{\mathcal{X},\mathcal{Y}} \leq 1$. Note that $\text{Cov}(\mathcal{X}, \mathcal{X}) = \text{Var}(\mathcal{X})$.

Random vectors. A random vector is a finite sequence $\{\mathcal{X}_1, \dots, \mathcal{X}_n\}$ of random variables defined on the same probability space $(\Omega, \mathfrak{F}, \mathbb{P})$; one can consider it as a measurable map $\Omega \to \mathbb{R}^n$. The joint distribution function of such random variables is

$$F(x_1, \dots, x_n) = \mathbb{P}(\mathcal{X}_1 \leq x_1, \dots, \mathcal{X}_n \leq x_n)$$

(more formally, it is the probability of the event $\cap_{i=1}^n \mathcal{X}_i^{-1}(-\infty, x_i])$. A random vector $\{\mathcal{X}_1, \dots, \mathcal{X}_n\}$ is said to be absolutely continuous if there is a (joint) density function $f(x_1, \dots, x_n) \geq 0$ such that

$$F(x_1, \dots, x_n) = \int_{-\infty}^{x_1} \cdots \int_{-\infty}^{x_n} f(s_1, \dots, s_n)\, ds_1 \cdots ds_n.$$

Then for every Borel subset $B \subset \mathbb{R}^n$ we have

$$\mathbb{P}\{(\mathcal{X}_1, \dots, \mathcal{X}_n) \in B\} = \int \cdots \int_B f(x_1, \dots, x_n)\, dx_1 \ \cdots dx_n,$$

where $\mathbf{m}_n$ denotes the Lebesgue measure on $\mathbb{R}^n$.

The mean value of a random vector[1] $\mathbf{X} = (\mathcal{X}_1, \dots, \mathcal{X}_n)^T$ is an n-vector

$$\mathbb{E}(\mathbf{X}) = \big(\mathbb{E}(\mathcal{X}_1), \dots, \mathbb{E}(\mathcal{X}_n)\big)^T.$$

The covariance matrix of a random vector $\mathbf{X} = (\mathcal{X}_1, \dots, \mathcal{X}_n)^T$ is

$$\begin{aligned}\mathbf{V} &= \mathbb{E}\big[(\mathbf{X} - \mathbb{E}(\mathbf{X}))(\mathbf{X} - \mathbb{E}(\mathbf{X}))^T\big] \\ &= \mathbb{E}(\mathbf{X}\mathbf{X}^T) - \mathbb{E}(\mathbf{X})\,\mathbb{E}(\mathbf{X})^T.\end{aligned}$$

Note that its entries are $v_{ij} = \text{Cov}(\mathcal{X}_i, \mathcal{X}_i)$. This matrix is symmetric and positive semidefinite; its diagonal entries are $v_{ii} = \text{Var}(\mathcal{X}_i)$.

[1] We denote vectors by boldface characters and always treat them as column vectors.

Gaussian (normal) random variables. A random variable $\mathcal{X}$ is said to be normal (or Gaussian), denoted by $\mathcal{N}(\mu, \sigma^2)$ if it is absolutely continuous and its density is

$$f(x) = \frac{1}{\sqrt{2\pi}\sigma} e^{-\frac{(x-\mu)^2}{2\sigma^2}}.$$

Here $\mu \in \mathbb{R}$ and $\sigma^2 > 0$ are parameters; in fact $\mu = \mathbb{E}(\mathcal{X})$ and $\sigma^2 = \mathrm{Var}(\mathcal{X})$. A standard normal random variable is $\mathcal{N}(0,1)$, it has zero mean and unit variance. Note that if $\mathcal{X}$ is $\mathcal{N}(\mu, \sigma^2)$, then its normalization $\mathcal{Y} = (cX - \mu)/\sigma$ is $\mathcal{N}(0,1)$.

A random vector $\mathbf{X} = (\mathcal{X}_1, \dots, \mathcal{X}_n)^T$ is said to be normal (or Gaussian) if it is absolutely continuous and its density is

$$f(x_1, \dots, x_n) = \frac{1}{\sqrt{\det(2\pi\mathbf{V})}} e^{-\frac{1}{2}(\mathbf{x}-\boldsymbol{\mu})^T\mathbf{V}^{-1}(\mathbf{x}-\boldsymbol{\mu})}$$

where $\boldsymbol{\mu} \in \mathbb{R}^n$ is a vector and $\mathbf{V}$ is an $n \times n$ symmetric positive definite matrix; we use the notation $\mathbf{x} = (x_1, \dots, x_n)^T$. In fact, $\boldsymbol{\mu} = \mathbb{E}(\mathbf{X})$ and $\mathbf{V}$ is the covariance matrix of $\mathbf{X}$. A standard normal random vector is $\mathcal{N}(\mathbf{0}, I)$; it has zero mean $\boldsymbol{\mu} = \mathbf{0}$ and unit covariance matrix $\mathbf{V} = I$.

Independence. Two events $A, B \subset \Omega$ are said to be independent if $\mathbb{P}(A \cap B) = \mathbb{P}(A)\,\mathbb{P}(B)$. Several events $A_1, \dots, A_n$ are said to be independent if for every sequence $B_1, \dots, B_n$, where each B_i is one of two events A_i and A_i^c, we have

$$\mathbb{P}(B_1 \cap \dots \cap B_n) = \mathbb{P}(B_1) \cdots \mathbb{P}(B_n).$$

Random variables $\mathcal{X}_1, \dots, \mathcal{X}_n$ are said to be independent if for any Borel sets $A_1, \dots, A_n \subset \mathbb{R}$ the events $\mathcal{X}_1^{-1}(A_1), \dots, \mathcal{X}_n^{-1}(A_n)$ are independent.

If random variables $\mathcal{X}_1, \dots, \mathcal{X}_n$ are independent, then their pairwise covariances (and correlation coefficients) vanish, so that the covariance matrix $\mathbf{V}$ is diagonal. For normal vectors, the converse is true: if the covariance matrix of a normal vector $\mathbf{X} = (\mathcal{X}_1, \dots, \mathcal{X}_n)^T$ is diagonal, then its components $\mathcal{X}_1, \dots, \mathcal{X}_n$ are independent. Generally, random variables $\mathcal{X}_1, \dots, \mathcal{X}_n$ are said to be uncorrelated if $\mathrm{Cov}(\mathcal{X}_i, \mathcal{X}_j) = 0$ for $i \neq j$, but they are not necessarily independent.

Random variables $\mathcal{X}_1, \dots, \mathcal{X}_n$ are independent if and only if their joint distribution function is factorized as

$$F_{\mathcal{X}_1,\dots,\mathcal{X}_n}(x_1, \dots, x_n) = F_{\mathcal{X}_1}(x_1) \cdots F_{\mathcal{X}_n}(x_n). \tag{B.1}$$

If random variables $\mathcal{X}_1, \dots, \mathcal{X}_n$ are independent and have finite mean values, then

$$\mathbb{E}(\mathcal{X}_1 \cdots \mathcal{X}_n) = \mathbb{E}(\mathcal{X}_1) \cdots \mathbb{E}(\mathcal{X}_n).$$

In addition, if they have finite variances, then

$$\mathrm{Var}(\mathcal{X}_1 + \dots + \mathcal{X}_n) = \mathrm{Var}(\mathcal{X}_1) + \dots + \mathrm{Var}(\mathcal{X}_n).$$

Let $\{\mathcal{X}_n\}$ be an infinite sequence of random variables. We say that they are independent if independence holds for any finite subcollection $\{\mathcal{X}_{n_1}, \dots, \mathcal{X}_{n_k}\}$. We say that random variables $\{\mathcal{X}_n\}$ have identical distribution (or are identically distributed) if they have a common distribution function $F(x) = F_{\mathcal{X}_n}(x)$. Of course, in this case they have a common mean value and a common variance.

Borel-Cantelli lemma. Let $\{A_n\}$ be a sequence of events. Then

$$A_\infty = \limsup_n A_n := \cap_{m=1}^\infty \cup_{n=m}^\infty A_n$$

is an event consisting of points $\omega \in \Omega$ that belong to infinitely many A_n's. One says that the event A_∞ occurs if and only if infinitely many A_n's occur. The Borel-Cantelli lemma consists of two parts. The first (easy) part says that

$$\sum_{n=1}^{\infty} \mathbb{P}(A_n) < \infty \implies \mathbb{P}(A_\infty) = 0. \tag{B.2}$$

The second (more difficult) part of the Borel-Cantelli lemma says that if A_n's are independent, then the converse of (B.2) is also true.

Convergence. There are three basic ways in which a sequence of random variables $\{\mathcal{X}_n\}$ may converge to another random variable $\mathcal{X}$.

We say that $\{\mathcal{X}_n\}$ converges to $\mathcal{X}$ almost surely (denoted by $\mathcal{X}_n \to \mathcal{X}$ a.s.) if there is an event $D \subset \Omega$ such that $\mathbb{P}(D) = 1$ and $\mathcal{X}_n(\omega) \to \mathcal{X}(\omega)$ for all $\omega \in D$.

We say that $\mathcal{X}_n \to \mathcal{X}$ in probability if for every $\varepsilon > 0$

$$\lim_{n\to\infty} \mathbb{P}\big(|\mathcal{X}_n - \mathcal{X}| > \varepsilon\big) = 0.$$

We say $\{\mathcal{X}_n\}$ converges to $\mathcal{X}$ in distribution (or in law), denoted by $\mathcal{X}_n \Rightarrow \mathcal{X}$, if the measures $\mu_{\mathcal{X}_n}$ on $\mathbb{R}$ weakly converge to the measure $\mu_{\mathcal{X}}$. Equivalently, the distribution functions $F_{\mathcal{X}_n}(x)$ converge to the distribution function $F_{\mathcal{X}}(x)$ at every point $x \in \mathbb{R}$ where $F_{\mathcal{X}}(x)$ is continuous. In this case we say that $\{F_{\mathcal{X}_n}\}$ converges to $F_{\mathcal{X}}$ weakly and write $F_{\mathcal{X}_n} \Rightarrow F_{\mathcal{X}}$.

Almost sure convergence implies convergence in probability. In turn, convergence in probability implies convergence in distribution. If the limit random variable $\mathcal{X}$ is almost surely constant, then convergence in probability is equivalent to convergence in distribution.

Laws of large numbers. Given a sequence of random variables $\{\mathcal{X}_n\}$ consider its partial sums

$$S_n = \mathcal{X}_1 + \cdots + \mathcal{X}_n.$$

We say that $\{\mathcal{X}_n\}$ satisfies the Law of Large Numbers (LLN) if the sequence $\{S_n/n\}$ converges to a constant (i.e., to a constant random variable) in probability (which is, in this case, equivalent to convergence in distribution).

We say that $\{\mathcal{X}_n\}$ satisfies the Strong Law of Large Numbers (SLLN) if the sequence $\{S_n/n\}$ converges to a constant almost surely. Of course, the SLLN implies the LLN.

If $\{\mathcal{X}_n\}$ are independent identically distributed (for brevity, i.i.d.) random variables with a finite mean value, $\mathbb{E}|\mathcal{X}_n| < \infty$, then they satisfy the SLLN. More precisely, $S_n/n \to \mu$ a.s., where $\mu = \mathbb{E}(\mathcal{X}_n)$ is the common mean value of the $\mathcal{X}_n$'s.

Central Limit Theorem. We say that a sequence of random variables $\{\mathcal{X}_n\}$ satisfies the Central Limit Theorem (CLT) if its normalized partial sums $(S_n - \mathbb{E}(S_n))/\sigma_{S_n}$ converge in distribution to the standard normal random variable $\mathcal{N}(0,1)$ (we also say that they converge to the standard normal law). That is, for every $z \in \mathbb{R}$

$$\lim_{n\to\infty} \mathbb{P}\left\{\frac{S_n - \mathbb{E}(S_n)}{\sigma_{S_n}} \le z\right\} = \frac{1}{\sqrt{2\pi}} \int_{-\infty}^{z} e^{-\frac{s^2}{2}}\, ds.$$

If the CLT holds and $\mathrm{Var}(S_n) = \sigma^2 n + \mathcal{O}(n)$ for some $\sigma^2 > 0$ (which is usually the case), then the sequence $(S_n - \mathbb{E}(S_n))/\sqrt{n}$ converges in distribution to a normal law $\mathcal{N}(0, \sigma^2)$.

Let $\{\mathcal{X}_n\}$ be a sequence of i.i.d. random variables that have a finite mean value $\mu = \mathbb{E}(\mathcal{X}_n)$ and a finite variance $\sigma^2 = \mathrm{Var}(\mathcal{X}_n)$. Then it satisfies the CLT. In this case, due to independence, $\mathrm{Var}(S_n) = \sigma^2 n$.

More generally, let $\{\mathcal{X}_n\}$ be a sequence of independent random variables that have finite mean values $\mu_n = \mathbb{E}(\mathcal{X}_n)$ and finite variances $\sigma_n^2 = \mathrm{Var}(\mathcal{X}_n)$. Assume that

$$D_n^2 \colon = \mathrm{Var}(S_n) = \sigma_1^2 + \ldots + \sigma_n^2 \xrightarrow[n\to\infty]{} \infty.$$

Then $\{\mathcal{X}_n\}$ satisfies the CLT under the so called Lindeberg condition: for every $\varepsilon > 0$

$$\lim_{n\to\infty} \frac{1}{D_n^2} \sum_{k=1}^{n} \mathbb{E}\big((\mathcal{X}_k - \mu_n)^2 \cdot \mathbf{1}_{\{|\mathcal{X}_k - \mu_n| > \varepsilon D_n\}}\big) = 0, \tag{B.3}$$

where $\mathbf{1}_A$ denotes the indicator of the event A. The Lindeberg condition is remarkable as it is not only sufficient, but 'almost' necessary. Precisely, if we assume only that $\sigma_n^2/D_n^2 \to 0$ (i.e. the variance σ_n^2 does not grow too abruptly), then the Lindeberg condition (B.3) is actually necessary for the CLT.

The CLT can be extended to random vectors. Let $\{\mathbf{X}_n\}$ be a sequence of i.i.d. random vectors that have a finite mean $\boldsymbol{\mu}$ and a finite covariance matrix $\mathbf{V}$. Consider the partial sums

$$\mathbf{S}_n = \mathbf{X}_1 + \cdots + \mathbf{X}_n.$$

Then the random vectors $(\mathbf{S}_n - n\boldsymbol{\mu})/\sqrt{n}$ converge in distribution to the normal vector (or to the normal law) with zero mean and covariance matrix $\mathbf{V}$, i.e. to $N(\mathbf{0}, \mathbf{V})$.

Characteristic functions. A complex random variable is a measurable function $\mathcal{Z} \colon \Omega \to \mathbb{C}$; one can write $\mathcal{Z} = \mathcal{X} + \mathbf{i}\mathcal{Y}$, where $\mathbf{i} = \sqrt{-1}$, and $\mathcal{X}$ and $\mathcal{Y}$ are the real and imaginary parts of $\mathcal{Z}$. The mean value of $\mathcal{Z}$ is defined by $\mathbb{E}(\mathcal{Z}) = \mathbb{E}(\mathcal{X}) + \mathbf{i}\,\mathbb{E}(\mathcal{Y})$.

A characteristic function of a real random variable $\mathcal{X}$ is a complex valued function of a real argument $t \in \mathbb{R}$ defined by

$$\varphi_{\mathcal{X}}(t) = \mathbb{E}\big[e^{\mathbf{i}t\mathcal{X}}\big] = \mathbb{E}\big[\cos(t\mathcal{X})\big] + \mathbf{i}\,\mathbb{E}\big[\sin(t\mathcal{X})\big].$$

This is actually the Fourier transform of the measure $\mu_{\mathcal{X}}$; it is also the Fourier-Stieltjes transform of the distribution function $F_{\mathcal{X}}(x)$:

$$\begin{aligned} \varphi_{\mathcal{X}}(t) &= \int_{-\infty}^{\infty} e^{\mathbf{i}tx}\, dF_{\mathcal{X}}(x) \\ &= \int_{-\infty}^{\infty} \cos(tx)\, dF_{\mathcal{X}}(x) + \mathbf{i} \int_{-\infty}^{\infty} \sin(tx)\, dF_{\mathcal{X}}(x). \end{aligned} \tag{B.4}$$

It is defined for any $\mathcal{X}$ and all $t \in \mathbb{R}$. In fact, $\varphi_{\mathcal{X}}(0) = 1$ and $|\varphi_{\mathcal{X}}(t)| \le 1$ for all $t \in \mathbb{R}$. The function $\varphi_{\mathcal{X}}(t)$ is uniformly continuous in t and positive semidefinite; i.e.

$$\sum_{i,j=1}^{n} \varphi_{\mathcal{X}}(t_i - t_j)\, z_i \bar{z}_j \ge 0$$

for any $n \ge 2$ complex numbers $z_1, \ldots, z_n$ and real numbers $t_1, \ldots, t_n$.

Moreover, any positive semidefinite function $\varphi(t)$ continuous at $t = 0$ and satisfying $\varphi(0) = 1$ is a characteristic function of a random variable (Bochner's theorem).

For any $n \geq 0$ we have $\varphi^{(n)}(0) = \mathbf{i}^n \mathbb{E}(\mathcal{X}^n)$, provided $\mathcal{X} \in L^n_{\mathbb{P}}(\Omega)$. If random variables $\mathcal{X}_1, \ldots, \mathcal{X}_n$ are independent, then $e^{\mathbf{i}t\mathcal{X}_1}, \ldots, e^{\mathbf{i}t\mathcal{X}_n}$ are also independent; hence

$$\varphi_{\mathcal{X}_1+\cdots+\mathcal{X}_n}(t) = \varphi_{\mathcal{X}_1}(t) \cdots \varphi_{\mathcal{X}_n}(t). \tag{B.5}$$

Conversely, if (B.5) holds for all $t \in \mathbb{R}$, then the random variables $\mathcal{X}_1, \ldots, \mathcal{X}_n$ are independent.

A characteristic function of a real random vector $\mathbf{X} = (\mathcal{X}_1, \ldots, \mathcal{X}_n)^T$ is a complex function of n real arguments $t_1, \ldots, t_n$ defined by

$$\varphi_{\mathbf{X}}(\mathbf{t}) = \mathbb{E}\big[e^{\mathbf{i}\langle \mathbf{t}, \mathbf{X}\rangle}\big] = \mathbb{E}\big[\cos(\langle \mathbf{t}, \mathbf{X}\rangle)\big] + \mathbf{i}\,\mathbb{E}\big[\sin(\langle \mathbf{t}, \mathbf{X}\rangle)\big]$$

where $\mathbf{t} = (t_1, \ldots, t_n)^T$ and $\langle \cdot, \cdot \rangle$ denotes the scalar product of two n-vectors.

The characteristic function of a normal random variable $\mathcal{X} = \mathcal{N}(\mu, \sigma^2)$ with mean μ and variance σ^2 is

$$\varphi_{\mathcal{X}}(t) = e^{\mathbf{i}t\mu - \sigma^2 t^2/2}.$$

The characteristic function of a normal random vector $\mathbf{X} = \mathcal{N}(\boldsymbol{\mu}, \mathbf{V})$ with mean $\boldsymbol{\mu}$ and covariance matrix $\mathbf{V}$ is

$$\varphi_{\mathbf{X}}(\mathbf{t}) = e^{\mathbf{i}\langle \mathbf{t}, \boldsymbol{\mu}\rangle - \mathbf{t}^T \mathbf{V} \mathbf{t}/2}.$$

Continuity theorem. Knowing the distribution function $F_{\mathcal{X}}(x)$ of a random variable $\mathcal{X}$, one can compute its characteristic function by (B.4). There is an inverse formula that gives $F_{\mathcal{X}}(x)$ in terms of $\varphi_{\mathcal{X}}(t)$:

$$F_{\mathcal{X}}(b) - F_{\mathcal{X}}(a) = \lim_{T \to \infty} \frac{1}{2\pi} \int_{-T}^{T} \frac{e^{-\mathbf{i}tb} - e^{-\mathbf{i}ta}}{-\mathbf{i}t} \varphi_{\mathcal{X}}(t)\, dt,$$

which holds for every two points $a < b$ where $F_{\mathcal{X}}$ is continuous. This uniquely determines $F_{\mathcal{X}}$ as its continuity points are dense in $\mathbb{R}$.

Thus, there is a natural one-to-one correspondence between distribution functions $F(x)$ and characteristic functions $\varphi(t)$ of random variables, so that $F_{\mathcal{X}}(x)$ corresponds to $\varphi_{\mathcal{X}}(t)$. It is very important that this correspondence is continuous in the following sense.

Let F and $\{F_n\}_{n=1}^{\infty}$ be distribution functions and φ and $\{\varphi_n\}_{n=1}^{\infty}$ the corresponding characteristic functions. Then $F_n \Rightarrow F$ if and only if the $\varphi_n(t) \to \varphi(t)$ pointwise (Lévy continuity theorem).

The 'if' part of the continuity theorem can be slightly extended. Namely, if $\{\varphi_n(t)\}$ is a sequence of characteristic functions that converges (pointwise) to a function $\varphi(t)$ which is continuous at $t = 0$, then $\varphi(t)$ is a characteristic function of a random variable.

Stationary sequences. Let $\{\mathcal{X}_n\}$, $n \in \mathbb{Z}$, be a sequence of i.i.d. random variables. Then any finite subsequence $\{\mathcal{X}_{n_1}, \ldots, \mathcal{X}_{n_m}\}$ has joint distribution function

$$F_{\mathcal{X}_{n_1}, \ldots, \mathcal{X}_{n_m}}(x_1, \ldots, x_m) = F(x_1) \cdots F(x_m),$$

which is independent of the choice of the indices $n_1, \ldots, n_m$. In particular, it is invariant under translations $(n_1, \ldots, n_m) \mapsto (n_1 + n, \ldots, n_m + n)$ for any $n \in \mathbb{Z}$.

More generally, a sequence of random variables $\{\mathcal{X}_n\}$, $n \in \mathbb{Z}$, is said to be stationary,[2] in the strong sense, if the distribution of its finite subsequences is

[2] We also say that $\{\mathcal{X}_n\}$ makes a stationary random process.

translationally invariant; i.e.

$$F_{\mathcal{X}_{n_1},\dots,\mathcal{X}_{n_m}}(x_1,\dots,x_m) = F_{\mathcal{X}_{n_1+n},\dots,\mathcal{X}_{n_m+n}}(x_1,\dots,x_m)$$

for any $m \geq 1$, any $n, n_1, \dots, n_m \in \mathbb{Z}$ and any $x_1, \dots, x_m \in \mathbb{R}$. Obviously, in this case $\mathcal{X}_n$'s are identically distributed. If $\mathcal{X}_n \in L^1_{\mathbb{P}}(\Omega)$, then they have the same mean value $\mathbb{E}(\mathcal{X}_n) = \mu$. If $\mathcal{X}_n \in L^2_{\mathbb{P}}(\Omega)$, then they have the same mean variance $\mathrm{Var}(\mathcal{X}_n) = \sigma^2$ and the covariance

$$\mathrm{Cov}(\mathcal{X}_n, \mathcal{X}_m) =: \mathbf{C}_{|m-n|} \tag{B.6}$$

depends only on $|m-n|$. A sequence of random variables $\{\mathcal{X}_n\}$, $n \in \mathbb{Z}$, is said to be stationary in the weak sense if the random variables have the same mean value, the same variance, and (B.6) holds (but the finite dimensional distributions are not necessarily translationally invariant).

Stationary sequences of random variables naturally arise in the study of dynamical systems (Chapter 7).

APPENDIX C

Ergodic theory

Here we provide basic definitions and facts of ergodic theory (which studies measure-preserving transformations). We use the notions of measure theory (Appendix A) and probability theory (Appendix B).

Measurable transformations. Let $(X, \mathfrak{F})$ be a measurable space. A transformation $T\colon X \to X$ is said to be measurable if $T^{-1}(B) \in \mathfrak{F}$ for every $B \in \mathfrak{F}$. A transformation $T\colon X \to X$ is called an automorphism if it is a bijection and both T and T^{-1} are measurable. Positive iterations $\{T^n\}$, $n \geq 0$, of a measurable transformation T make a semigroup; all iterations $\{T^n\}$, $n \in \mathbb{Z}$, of an automorphism T make a group. For any point $x \in X$ the sequence $\{T^n x\}$ is called the trajectory (or orbit) of x. Measurable transformations with continuous time (flows) will be described at the end of this appendix.

Let $\mathcal{M}(X)$ denote the set of all probability measures on $(X, \mathfrak{F})$. It is a convex set, as for any $\mu, \nu \in \mathcal{M}(X)$ and $0 < p < 1$ we have $p\mu + (1-p)\nu \in \mathcal{M}(X)$. A measurable transformation $T\colon X \to X$ induces a map $T\colon \mathcal{M}(X) \to \mathcal{M}(X)$ defined by $(T\mu)(B) = \mu(T^{-1}B)$ for every $\mu \in \mathcal{M}(X)$ and $B \in \mathfrak{F}$. (This map is sometimes denoted by T_*, but we prefer T for brevity.)

Invariant measures. Given a measurable transformation $T\colon X \to X$, a measure $\mu \in \mathcal{M}(X)$ is said to be T-invariant if $T\mu = \mu$ (we say that T preserves the measure μ). If T is an automorphism, then $T\mu = \mu$ is equivalent to $T^{-1}\mu = \mu$, so that T and T^{-1} preserve the same measures. Denote by $\mathcal{M}_T(X)$ the set of all T-invariant probability measures on X. It is a convex subset of $\mathcal{M}(X)$.

A measure $\mu \in \mathcal{M}(X)$ is T-invariant if and only if for any measurable function $f\colon X \to \mathbb{R}$ we have

$$\int_X f \circ T \, d\mu = \int_X f \, d\mu$$

(in the sense that if one integral exists, then so does the other and they are equal). More generally, for any $\mu \in \mathcal{M}(X)$ its image $\mu_1 = T\mu$ is characterized by

$$\int_X f \circ T \, d\mu = \int_X f \, d\mu_1.$$

A measurable transformation $T\colon X \to X$ induces a linear map U_T on the space of measurable functions $f\colon X \to \mathbb{R}$ defined by

$$(U_T f)(x) = (f \circ T)(x) = f(T(x)).$$

For any T-invariant measure μ and $p > 0$ the map $U_T\colon L^p_\mu(X) \to L^p_\mu(X)$ preserves norm $\|\cdot\|_p$; it also preserves the inner product in $L^2_\mu(X)$. If T is an automorphism, then U_T is a bijection; hence it is a unitary operator on $L^2_\mu(X)$.

In what follows we mostly deal with a measurable transformation (usually, an automorphism) $T\colon X \to X$ preserving a measure $\mu \in \mathcal{M}_T(X)$. We refer to a quadruple $(X, \mathfrak{F}, T, \mu)$ as a measure-preserving transformation.

Poincaré recurrence theorem. Let $T\colon X \to X$ preserve a measure $\mu \in \mathcal{M}_T(X)$ and $\mu(A) > 0$ for some measurable set $A \subset X$. Then for μ-almost every point $x \in A$ we have

$$T^{n_i}(x) \in A \qquad \text{for some sequence } n_1 < n_2 < \cdots .$$

Under the conditions of the above theorem, the map

$$T_A(x) = T^{n_A(x)}(x), \qquad n_A(x) = \min\{n \geq 1\colon T^n(x) \in A\}$$

is defined a.e. on A; it is called the Poincaré return map. It preserves the conditional measure μ_A on A defined by $\mu_A(B) = \mu(A \cap B)/\mu(A)$.

Invariant sets and functions. We say that a measurable set $B \subset X$ is T-invariant if $T^{-1}(B) = B$. Note that this happens if $T(B) \subset B$ and $T(B^c) \subset B^c$.

Suppose T preserves a measure μ. Then a measurable set B is said to be T-invariant (mod 0) if $B = T^{-1}B$ (mod 0). If B is T-invariant (mod 0), then there exist a T-invariant set $\tilde{B}$ such that $B = \tilde{B}$ (mod 0).

A function $f\colon X \to \mathbb{R}$ is T-invariant if $U_T f = f$, i.e. $f = f \circ T$. In this case f is constant on every trajectory of the map T. If T preserves a measure μ, then we say that a function $f\colon X \to \mathbb{R}$ is T-invariant (mod 0) if $f(x) = f(Tx)$ for μ-a.e. point $x \in X$. In that case there exists a T-invariant function $\tilde{f}$ such that $f = \tilde{f}$ (mod 0).

Ergodic measures. A T-invariant measure $\mu \in \mathcal{M}_T(X)$ is said to be ergodic if for any T-invariant set $B \subset X$ we have $\mu(B) = 0$ or $\mu(B) = 1$. Equivalently, for any T-invariant (mod 0) set $B \subset X$ we have $\mu(B) = 0$ or $\mu(B) = 1$.

A T-invariant measure μ is ergodic if and only if any T-invariant function $f\colon X \to \mathbb{R}$ is a.e. constant; i.e. $\mu(x\colon f(x) = c) = 1$ for some $c \in \mathbb{R}$. Equivalently, μ is ergodic iff any T-invariant (mod 0) function $f\colon X \to \mathbb{R}$ is a.e. constant; i.e. $\mu(x\colon f(x) = c) = 1$ for some $c \in \mathbb{R}$.

We also say that T is ergodic if it is clear from the context which invariant measure μ is associated with T.

A T-invariant measure $\mu \in \mathcal{M}_T(X)$ is ergodic iff it is an extremal point in the convex set $\mathcal{M}_T(X)$. Any two distinct ergodic measures μ_1, μ_2 are mutually singular (orthogonal).

If a measurable transformation $T\colon X \to X$ has a unique invariant measure μ, it will be automatically ergodic. Then T is said to be uniquely ergodic.

Ergodic decomposition. Assume that $(X, \mathfrak{F}, \mu)$ is a Lebesgue space, so that the notion of measurable partitions will apply. If $T\colon X \to X$ preserves the measure μ, then the collection of T-invariant (mod 0) subsets $A \subset X$ forms a complete sub-σ-algebra $\mathfrak{G} \subset \mathfrak{F}$. It corresponds to a unique (mod 0) measurable partition ξ of X. Denote by μ_C, $C \in \xi$, the conditional measures on the elements of C.

In this case almost every element $C \in \xi$ is T-invariant (mod 0), i.e. $\mu_C(C \cap T^{-1}C) = 1$. Hence the restriction of T onto C is defined μ_C-a.e. on C; we call it T_C. The map $T_C\colon C \to C$ preserves the measure μ_C and is ergodic for a.e. $C \in \xi$. The elements $C \in \xi$ with conditional measures μ_C are called ergodic components of the map T.

Isomorphisms. Two measure-preserving transformations $(X_1, \mathfrak{F}_1, T_1, \mu_1)$ and $(X_2, \mathfrak{F}_2, T_2, \mu_2)$ are said to be isomorphic if for each $i = 1, 2$ there is a T_i-invariant set $B_i \subset X_i$ of full μ_i measure and a bijection $\varphi\colon B_1 \to B_2$ such that (i) φ preserves measurable sets and measures, and (ii) φ preserves dynamics; i.e. $\varphi \circ T_1 = T_2 \circ \varphi$ on B_1. The map φ is called an isomorphism.

Many properties of measure-preserving transformations are invariant under isomorphisms. For example, μ_1 is ergodic iff μ_2 is also.

Ergodic sums and averages. Let $(X, \mathfrak{F}, T, \mu)$ be a measure-preserving transformation and $f\colon X \to X$ a measurable function. Then for every $x \in X$ the sequence $\{f(T^n x)\}$ of values of f on the trajectory of x plays an important role. We think of f as an observable quantity and of $f(T^n x)$ as its value at time n; hence $\{f(T^n x)\}$ can regarded as a time series. Its partial sums

$$S_n(x) = f(x) + f(Tx) + \cdots + f(T^{n-1}x)$$

are sometimes called ergodic sums, and the limit

$$f_+(x) = \lim_{n\to+\infty} \frac{1}{n} S_n(x) \tag{C.1}$$

(if it exists) is called the forward time average (or ergodic average) of the function f along the orbit of x. If T is an automorphism, one can define the backward time average by

$$f_-(x) = \lim_{n\to+\infty} \frac{1}{n} S_{-n}(x) \tag{C.2}$$

where $S_{-n}(x) = f(x) + f(T^{-1}x) + \cdots + f(T^{-n+1}x)$.

Birkhoff Ergodic Theorem. Let $(X, \mathfrak{F}, T, \mu)$ be a measure-preserving transformation and $f \in L^1_\mu(X)$. Then
(a) for almost every point $x \in X$ the limit $f_+(x)$ defined by (C.1) exists;
(b) the function $f_+(x)$ is T-invariant; more precisely, if $f_+(x)$ exists, then $f_+(T^n x)$ exists for all n and $f_+(T^n x) = f_+(x)$;
(c) f_+ is integrable and $\int_X f_+\, d\mu = \int_X f\, d\mu$;
(d) if μ is ergodic, then $f_+(x)$ is a.e. constant and its value is $\int_X f\, d\mu$.

If $T : X \to X$ is an automorphism, then for almost every point $x \in X$ the limit $f_-(x)$ defined by (C.2) exists as well. Moreover, the limits (C.1) and (C.2) coincide a.e.: $f_+ = f_-$ (mod 0).

The integral $\int_X f\, d\mu$ is regarded as the space average of the function f. Part (d) of the ergodic theorem asserts that if μ is ergodic, then the time averages are equal to the space averages.

Another version of the ergodic theorem (called the L^p von Neumann ergodic theorem) asserts that for every $p \geq 1$ and $f \in L^p_\mu(X)$ we have $\|S_n/n - f_+\|_p \to 0$ as $n \to \infty$.

Frequencies of returns. Let $A \subset X$ and $x \in X$. The limit

$$r_A(x) = \lim_{n\to+\infty} \frac{\#\{0 \leq i \leq n-1 :\ T^i(x) \in A\}}{n}$$

(if it exists) is called the (asymptotic) frequency of visits (or returns) of the point x to the set A.

It follows from the ergodic theorem that $r_A(x)$ exists for a.e. $x \in X$. Moreover, $r_A(x) > 0$ for a.e. $x \in A$. In addition, if μ is ergodic, then $r_A(x) = \mu(A)$ for a.e.

$x \in X$. Hence the orbit of a.e. point $x \in X$ spends time in the set A proportional to its measure $\mu(A)$. In this sense, the ergodic measure μ describes the asymptotic distribution of almost every orbit $\{T^n(x)\}$, $n \geq 0$, in the space X.

Mixing. We say that a map $T : X \to X$ preserving a measure μ is mixing (or strongly mixing) if for all pairs of measurable subsets $A, B \subset X$

$$\lim_{n\to\infty} \mu(T^{-n}A \cap B) = \mu(A)\,\mu(B).$$

As $\mu(A) = \mu(T^{-n}A)$, this can be written as

$$\lim_{n\to\infty} \bigl|\mu(T^{-n}A \cap B) - \mu(T^{-n}A)\,\mu(B)\bigr| = 0;$$

i.e. the events $T^{-n}A$ and B become asymptotically independent, as $n \to \infty$. Note that $x \in T^{-n}A$ is equivalent to $T^n(x) \in A$; i.e. we are speaking about the events $x \in B$ (characterizing x time 0) and $T^n(x) \in A$ (characterizing the image of x at time n). Thus mixing is commonly interpreted as asymptotic independence of the distant future from the present.

Mixing is equivalent to

$$\lim_{n\to\infty} \langle f \cdot (g \circ T^n)\rangle = \langle f\rangle\langle g\rangle \qquad \forall f, g \in L^2_\mu(X)$$

where $\langle f\rangle = \int_X f\,d\mu$, for brevity. Given f and g, the quantity

$$\mathbf{C}_{f,g}(n) = \langle f \cdot (g \circ T^n)\rangle - \langle f\rangle\langle g\rangle$$

is called the correlation (between f and g) at time n (note that it is actually the covariance of the random variables f and $g \circ T^n$; see Appendix B). Thus mixing is equivalent to the convergence of correlations to zero ($\mathbf{C}_{f,g}(n) \to 0$); this property is called the decay of correlations.

Weak mixing and multiple mixing. We say that T is weakly mixing (with respect to the invariant measure μ) if for all pairs of measurable subsets $A, B \subset X$

$$\lim_{n\to\infty} \frac{1}{n} \sum_{i=0}^{n-1} |\mu(T^{-i}A \cap B) - \mu(A)\,\mu(B)| = 0.$$

Equivalently, for any $f, g \in L^2_\mu(X)$

$$\lim_{n\to\infty} \frac{1}{n} \sum_{i=0}^{n-1} |\mathbf{C}_{f,g}(n)| = 0.$$

Let $m \geq 2$. We say that T is m-mixing, or mixing of multiplicity m, if for any measurable subsets $A_1, A_2, \dots, A_m \subset X$ and $0 \leq n_1 < n_2 < \cdots < n_m$ we have

$$\mu(T^{-n_1}A_1 \cap T^{-n_2}A_2 \cap \cdots \cap T^{-n_m}A_m) \to \mu(A_1)\,\mu(A_2)\cdots\mu(A_m)$$

whenever

$$\min\{n_2 - n_1, n_3 - n_2, \dots, n_m - n_{m-1}\} \to \infty \tag{C.3}$$

(m-mixing means asymptotic independence of events occurring at m different moments of time if the time intervals between them grow).

Equivalently, for any $f_1, \dots, f_m \in L^m_\mu(X)$ with $\langle f_i\rangle = 0$ we have

$$\langle (f_1 \circ T^{n_1}) \cdot (f_2 \circ T^{n_2}) \cdots (f_m \circ T^{n_m})\rangle \to 0 \tag{C.4}$$

whenever (C.3) holds. The quantity (C.4) is called the multiple correlation.

Mixing is equivalent to 2-mixing. If T is m-mixing, then it is k-mixing for every $2 \le k \le m$.

All mixing properties (weak mixing, strong mixing, multiple mixing) are invariant under isomorphisms. Multiple mixing with $m \ge 3$ implies strong mixing (but not vice versa). Strong mixing implies weak mixing (but not vice versa). Weak mixing implies ergodicity (but not vice versa).

Circle rotation. Perhaps the most popular example of a measure-preserving transformation is a circle rotation.

Let $X = \mathbb{R}/\mathbb{Z}$ be the unit 1-torus, or a circle of length one, with a cyclic (angular) coordinate $x \in [0,1]$ (here the points 0 and 1 are identified). The rotation through an angle α is defined by

$$T(x) = x + \alpha \qquad (\text{mod } 1).$$

It preserves the Lebesgue measure $\mathbf{m}$ on X.

If $\alpha = p/q$ is rational (assuming that p and q are relatively prime), then every point $x \in X$ is periodic with the same period q.

If α is irrational, then the trajectory of every point $x \in X$ is dense and uniformly distributed in X; i.e. for any interval $A \subset X$ we have $r_A(x) = \mathbf{m}(A)$ (Weyl's theorem). In this case the Lebesgue measure is ergodic but not mixing (not even weakly mixing). Moreover, $\mathbf{m}$ is the only invariant measure for the map T; hence T is uniquely ergodic.

Linear translations of tori. These generalize circle rotations. Let $d \ge 2$ and $X = \mathbb{R}^d/\mathbb{Z}^d$ be the unit d-torus with cyclic (angular) coordinates $\mathbf{x} = (x_1, \dots, x_d) \in [0,1]^d$. Fix a vector $\mathbf{a} = (a_1, \dots, a_d) \in \mathbb{R}^d$. The translation of X through the vector $\mathbf{a}$ is defined by

$$T_{\mathbf{a}}(\mathbf{x}) = \mathbf{x} + \mathbf{a} \qquad (\text{mod } 1).$$

It preserves the d-dimensional Lebesgue measure on X.

The translation $T_{\mathbf{a}}$ is ergodic if and only if the components $(a_1, \dots, a_d)$ of the vector $\mathbf{a}$ are rationally independent of 1; i.e.

$$m_0 + m_1 a_1 + m_2 a_2 + \cdots + m_d a_d \ne 0$$

for any integers $m_0, m_1, \dots, m_d \in \mathbb{Z}$, unless $m_0 = m_1 = \cdots = m_d = 0$. The map $T_{\mathbf{a}}$ is never weakly mixing.

Symbolic space. Let $S = \{1, \dots, r\}$ be a finite set. We call S an alphabet and its elements letters. Let $\Sigma_+ = \Sigma_{+,r} = S^{\mathbb{Z}_+}$ denote the space of infinite sequences of letters, so a point $\underline{\omega} \in \Sigma_+$ is a sequence $\underline{\omega} = \{\omega_n\}_{n=0}^{\infty}$ with $\omega_n \in S$ for each $n \ge 0$.

Also, let $\Sigma = \Sigma_r = S^{\mathbb{Z}}$ be the space of double infinite sequences of letters; i.e. Σ consists of sequences $\underline{\omega} = \{\omega_n\}_{n=-\infty}^{\infty}$ with $\omega_n \in S$ for each $n \in \mathbb{Z}$. We call Σ_+ and Σ symbolic spaces. (For brevity, we suppress the index r.)

We equip the finite set S with the discrete topology (in which all the subsets of S are open) and the spaces Σ and Σ_+ with the corresponding product topology. The corresponding Borel σ-algebras are denoted by $\mathfrak{F}$ and $\mathfrak{F}_+$.

The (left) shift homeomorphism $\sigma\colon \Sigma \to \Sigma$ is defined by $\underline{\omega}' = \sigma(\underline{\omega})$ with $\omega_i' = \omega_{i+1}$ for all $i \in \mathbb{Z}$. Similarly, the (left) shift $\sigma_+\colon \Sigma_+ \to \Sigma_+$ is defined by $\underline{\omega}' = \sigma_+(\underline{\omega})$ with $\omega_i' = \omega_{i+1}$ for all $i \ge 0$; it a continuous r-to-1 map on Σ_+.

Let μ_0 be a probability measure on the finite set S (not entirely concentrated at one point). Denote by μ the corresponding product measure $\mu_0^{\mathbb{Z}}$ on Σ and by μ_+ the corresponding product measure $\mu_0^{\mathbb{Z}_+}$ on Σ_+. The space $(\Sigma, \mathfrak{F}, \mu)$ corresponds to a sequence of independent identically distributed random variables each of which takes finitely many values, a classical object of study in probability theory.

The shift σ preserves the measure μ on Σ. The shift σ_+ preserves the measure μ_+ on Σ_+. Both shifts are ergodic, mixing, and m-mixing for all $m \geq 2$.

The system $(\Sigma, \mathfrak{F}, \sigma, \mu)$ is said to be a Bernoulli shift. Note that it is completely characterized by the finite probability distribution μ_0 on the alphabet S.

Symbolic representation. Let $(X, \mathfrak{F}, T, \mu)$ be a measure-preserving transformation. Let $X = A_1 \cup \cdots \cup A_r$ be a finite partition of X into disjoint measurable subsets. Given a map $T\colon X \to X$, for every point $x \in X$ we define its itinerary by

$$\underline{\omega}(x) = \{\omega_n\}_{n=0}^{\infty} \in \Sigma_+ : \qquad T^n(x) \in A_{\omega_n} \quad \forall n \geq 0.$$

We say that $X = \cup_i A_i$ is a generating partition if distinct points have distinct itineraries (equivalently, for any $x \neq y$ there is an $n \geq 0$ such that $T^n(x) \in A_i$ and $T^n(y) \in A_j$ with some $i \neq j$). Then the map $\varphi\colon X \to \Sigma_+$ defined by $\varphi(x) = \underline{\omega}(x)$ is one-to-one; it induces a measure $\mu_X = \varphi(\mu)$ on Σ_+ which is σ_+-invariant. The map φ is an isomorphism between the given system $(X, \mathfrak{F}, T, \mu)$ and $(\Sigma_+, \mathfrak{F}_+, \sigma_+, \mu_X)$; the latter is called the symbolic representation of the former.

If T is an automorphism, then its symbolic representation should be constructed differently. First, the itinerary of $x \in X$ is defined by

$$\underline{\omega}(x) = \{\omega_n\}_{n=-\infty}^{\infty} \in \Sigma : \qquad T^n(x) \in A_{\omega_n} \quad \forall n \in \mathbb{Z}.$$

Again, a generating partition means that distinct points have distinct itineraries (equivalently, for any $x \neq y$ there is an $n \in \mathbb{Z}$ such that $T^n(x) \in A_i$ and $T^n(y) \in A_j$ with some $i \neq j$). Then the map $\varphi\colon X \to \Sigma$ defined by $\varphi(x) = \underline{\omega}(x)$ is one-to-one. It induces a measure $\mu_X = \varphi(\mu)$ on Σ which is σ-invariant, and we obtain an isomorphism between the given system $(X, \mathfrak{F}, T, \mu)$ and its symbolic representation $(\Sigma, \mathfrak{F}, \sigma, \mu_X)$.

Bernoulli property. An automorphism $T\colon X \to X$ preserving a measure μ is said to be Bernoulli (or have Bernoulli property, or B-property) if it is isomorphic to a Bernoulli shift.

Equivalently, there is a finite generating partition $\xi = \{A_1, \ldots, A_r\}$ of X such that the corresponding symbolic representation of T is a Bernoulli shift (i.e., the induced measure μ_X on Σ (see above) is a product measure). In this case the partitions $T^n\xi = \{T^n(A_1), \ldots, T^n(A_r)\}$ are independent; i.e.

$$\mu\big(T^n(A_i) \cap T^m(A_j)\big) = \mu\big(T^n(A_i)\big)\,\mu\big(T^m(A_j)\big)$$

for all $m \neq n$ and $1 \leq i, j \leq r$.

Kolmogorov automorphisms (K-mixing). An automorphism $T\colon X \to X$ preserving a measure μ is said to be Kolmogorov (or be a K-automorphism, or have K-property) if there exists a measurable partition ξ of X with three properties: (i) it is monotonically increasing, i.e. $\xi \preccurlyeq T\xi$; (ii) its forward limit is the partition into individual points, i.e. $\vee_{n=0}^{\infty} T^n\xi = \varepsilon$; and (iii) its backward limit is the trivial partition, i.e. $\wedge_{n=0}^{\infty} T^{-n}\xi = \nu$.

Equivalently, $T\colon X \to X$ is a K-automorphism if for any measurable subsets $B, A_1, \ldots, A_r \subset X$ we have

$$\lim_{n\to\infty} \sup_{A \in \mathfrak{G}_n^{\infty}(A_1,\ldots,A_r)} |\mu(A \cap B) - \mu(A)\,\mu(B)| = 0,$$

where $\mathfrak{G}_n^\infty(A_1,\ldots,A_r)$ denotes the sub-σ-algebra generated by the sets $T^m A_i$ for all $m \geq n$ and $i = 1,\ldots,r$.

This is an asymptotic independence between the present event B and all distant future events in $\mathfrak{G}_n^\infty(A_1,\ldots,A_r)$. This property is stronger ('more uniform') than multiple mixing but weaker than Bernoulli. K-property is often called K-mixing. Bernoulli property implies K-property, but not vice versa. K-property implies multiple mixing, but not vice versa. For smooth or piecewise smooth maps (and flows; see below) with smooth, or at least SRB, invariant measures (that includes all billiards) the K-property usually implies Bernoulli; see [**CH96, OW98**].

The K-property is invariant under isomorphisms. The K-property can be characterized in terms of the Pinsker partition; see below.

Entropy. Let $(X,\mathfrak{F},T,\mu)$ be a measure-preserving transformation. The entropy of a finite partition $\xi = \{A_1,\ldots,A_r\}$ of X is defined by

$$H(\xi) = -\sum_{i=1}^{r} \mu(A_i)\,\ln\mu(A_i)$$

(with the convention $0\ln 0 = 0$). Note that $H(\xi) = -\int_X \ln\mu(\xi(x))\,d\mu$. We have $0 \leq H(\xi) \leq \ln r$, with the minimum (0) attained on the trivial partition $\xi = \nu$ and the maximum ($\ln r$) attained on equipartitions, which are characterized by $\mu(A_1) = \cdots = \mu(A_r) = 1/r$.

Since the measure μ is invariant, the partition $T^{-n}\xi = \{T^{-n}A_1,\ldots,T^{-n}A_r\}$ has the same entropy, $H(T^{-n}\xi) = H(\xi)$, for every $n \geq 1$. If T is an automorphism, this is true for all $n \in \mathbb{Z}$.

The entropy of T with respect to a finite partition ξ is

$$h(T,\xi) = \lim_{n\to\infty} \tfrac{1}{n} H\bigl(\vee_{i=0}^{n-1} T^{-i}\xi\bigr).$$

This limit always exists and is nonnegative; in fact the sequence on the right-hand side decreases monotonically. Lastly, the entropy of T is

$$h(T) = \sup_\xi h(T,\xi),$$

where the supremum is taken over all finite partitions ξ of X. Clearly, $0 \leq h(T) \leq \infty$. The entropy is invariant under isomorphisms.

If ξ is a generating partition, i.e. $\vee_{i=0}^{\infty} T^{-i}\xi = \varepsilon$, then $h(T,\xi) = h(T)$. If T is an automorphism, then the existence of a one-sided generator (a partition ξ satisfying $\vee_{i=0}^{\infty} T^{-i}\xi = \varepsilon$) implies $h(T) = 0$. In that case $h(T) = h(T,\xi)$ for any two-sided generator ξ, i.e. a partition ξ satisfying $\vee_{i=-\infty}^{\infty} T^{-i}\xi = \varepsilon$.

We have $h(T^n) = n\,h(T)$ for every $n \geq 1$. If T is an automorphism, then $h(T^n) = |n| h(T)$ for every $n \in \mathbb{Z}$, in particular $h(T^{-1}) = h(T)$.

Pinsker partition. Let $(X,\mathfrak{F},T,\mu)$ be a measure-preserving transformation of a Lebesgue space. Then

$$\pi(T) = \vee\{\xi\colon \xi \ \text{ finite}, \quad h(T,\xi) = 0\}$$

is called the Pinsker partition. Its sub-σ-algebra $\mathfrak{P}(T) = \mathfrak{G}(\pi(T))$ is called the Pinsker σ-algebra.

If T is an automorphism, then the Pinsker σ-algebra is T-invariant, that is $T^{-1}\mathfrak{P}(T) = \mathfrak{P}(T)$. It is the maximal sub-$\sigma$-algebra of $\mathfrak{F}$ such that T restricted to

$(X, \mathfrak{P}(T))$ has zero entropy. One often says, somewhat informally, that $\pi(T)$ is the 'maximal partition with zero entropy'.

Suppose T is an automorphism and a measurable partition ξ of X has two properties: (i) it is monotonically increasing, i.e. $\xi \preccurlyeq T\xi$; (ii) its forward limit is the partition into individual points, i.e. $\vee_{n=0}^{\infty} T^n\xi = \varepsilon$. Then $\wedge_{n=0}^{\infty} T^{-n}\xi \succcurlyeq \pi(T)$.

An automorphism T is Kolmogorov if and only if its Pinsker partition is trivial, $\pi(T) = \nu$. Thus, T is a K-automorphism if and only if its entropy is positive, $h(T, \xi) > 0$, with respect to every nontrivial finite partition ξ.

Flows. Let $(X, \mathfrak{F})$ be a measurable space. A dynamical system with continuous time (a flow) is a one-parameter family $\{S^t\}$, $t \in \mathbb{R}$, of measurable transformations $S^t \colon X \to X$ that satisfies two properties: (i) $S^{t+s} = S^t \circ S^s$ (the group property), in particular S^0 is the identity; (ii) the map $X \times \mathbb{R} \to X$ defined by $(x, t) \mapsto S^t x$ is measurable (the σ-algebra on $X \times \mathbb{R}$ is the product of $\mathfrak{F}$ and the Borel σ-algebra on $\mathbb{R}$).

Condition (ii) imposes some regularity on the family $\{S^t\}$ with respect to t. It is equivalent to the following: for every measurable function $F \colon X \to \mathbb{R}$ the function $G(x, t) = F(S^t x)$ is a measurable map $X \times \mathbb{R} \to \mathbb{R}$.

For every point $x \in X$ the set $\{S^t x\}$, $t \in \mathbb{R}$, is called the orbit (or trajectory) of x. In many applications, including billiards, X is a topological space and $\{S^t x\}$ is a continuous curve for every $x \in X$.

The flow $\{S^t\}$ preserves a measure $\mu \in \mathcal{M}(X)$ if $\mu(S^t A) = \mu(A)$ for all measurable subsets $A \subset X$ and all $t \in \mathbb{R}$. That is, μ is a common invariant measure for all the automorphisms S^t included in the flow.

Invariant measures for flows. The properties of automorphisms extend to flows with some (usually trivial) modifications. A measurable set $B \subset X$ is invariant under a flow $\{S^t\}$ if $B = S^t B$ for every $t \in \mathbb{R}$. If the flow $\{S^t\}$ preserves a measure μ, then a measurable set B is said to be invariant (mod 0) under the flow if $B = S^t B$ (mod 0) for every $t \in \mathbb{R}$. If B is invariant (mod 0), then there exists an invariant set $\tilde{B}$ such that $B = \tilde{B}$ (mod 0).

A function $f \colon X \to \mathbb{R}$ is invariant under $\{S^t\}$ if $f = f \circ S^t$ for all $t \in \mathbb{R}$. In this case f is constant on every orbit of the flow $\{S^t\}$. If $\{S^t\}$ preserves a measure μ, then we say that a function $f \colon X \to \mathbb{R}$ is invariant (mod 0) under the flow if for every $t \in \mathbb{R}$ we have $f(x) = f(S^t x)$ for μ-a.e. point $x \in X$. In that case there exists an invariant function $\tilde{f}$ such that $f = \tilde{f}$ (mod 0).

A flow $\{S^t\}$ is ergodic with respect to an invariant measure μ if any $\{S^t\}$-invariant (mod 0) set $A \subset X$ has measure 0 or 1. Equivalently, a flow $\{S^t\}$ is ergodic if any invariant (mod 0) function f is a.e. constant; i.e. $\mu(x \colon f(x) = c) = 1$ for some $c \in \mathbb{R}$.

If at least one automorphism S^t in the flow is ergodic, then the flow is ergodic. Conversely, if the flow $\{S^t\}$ is ergodic, then the automorphism S^t is ergodic for all but countably many $t \in \mathbb{R}$.

Birkhoff Ergodic Theorem for flows. Given a measurable function $f \colon X \to \mathbb{R}$ its (forward and backward) time averages are defined by

$$f_{\pm}(x) = \lim_{T \to \pm\infty} \frac{1}{T} \int_0^T f(S^t(x))\, dt.$$

Suppose the flow $\{S^t\}$ preserves a measure μ and $f \in L^1_\mu(X)$. Then
(a) for almost every point $x \in X$ the above limits exist and $f_+(x) = f_-(x)$;
(b) the function $f_\pm(x)$ is T-invariant; more precisely, if $f_\pm(x)$ exists, then $f_\pm(S^t x)$ exists for all $t \in \mathbb{R}$ and $f_\pm(S^t x) = f_\pm(x)$;
(c) $f_\pm$ is integrable and $\int_X f_\pm \, d\mu = \int_X f \, d\mu$;
(d) if $\{S^t\}$ is ergodic, then $f_\pm(x)$ is a.e. constant and its value is $\int_X f \, d\mu$.

Mixing flows. A flow $S^t : X \to X$ is mixing with respect to an invariant measure μ if for any measurable sets $A, B \subset X$ we have

$$\lim_{t \to \pm\infty} \mu\big(A \cap S^t(B)\big) = \mu(A)\, \mu(B).$$

If a flow $\{S^t\}$ is mixing, then every map S^t, $t \neq 0$, is also mixing.

A flow $\{S^t\}$ is Kolmogorov (or K-flow, or has K-mixing property) if there exists a measurable partition ξ of X with three properties: (i) it is monotonically increasing, i.e. $\xi \preccurlyeq S^t \xi$ for every $t > 0$; (ii) its forward limit is the partition into individual points, i.e. $\vee_{t=0}^{\infty} S^t \xi = \varepsilon$; and (iii) its backward limit is the trivial partition, i.e. $\wedge_{t=0}^{\infty} S^{-t} \xi = \nu$.

If the map S^t is a K-automorphism for at least one $t \in \mathbb{R}$, then every S^t, $t \neq 0$, is a K-automorphism. The flow $\{S^t\}$ is a K-flow iff any (and hence every) map S^t, $t \neq 0$, is a K-automorphism.

If at least one automorphism S^t in the flow is Bernoulli, then every other S^t, $t \neq 0$, is also Bernoulli. In this case we say that the flow $\{S^t\}$ is Bernoulli.

Bernoulli property implies K-mixing property, but not vice versa. K-mixing property implies mixing, but not vice versa. Mixing implies ergodicity, but not vice versa.

Linear flows on tori. Let $d \geq 2$ and $X = \mathbb{R}^d/\mathbb{Z}^d$ be the unit d-torus with cyclic (angular) coordinates $\mathbf{x} = (x_1, \ldots, x_d) \in [0,1]^d$. Fix a vector $\mathbf{a} = (a_1, \ldots, a_d) \in \mathbb{R}^d$ and define a flow $\{S^t_{\mathbf{a}}\}$ on X by

$$S^t_{\mathbf{a}}(\mathbf{x}) = \mathbf{x} + t\mathbf{a} \pmod{1}.$$

Alternatively, $\{S^t_{\mathbf{a}}\}$ can be defined by the differential equation $\dot{\mathbf{x}}(t) = \mathbf{a}$. This flow preserves the d-dimensional Lebesgue measure on X.

The flow $\{S^t_{\mathbf{a}}\}$ is ergodic if and only if the components $a_1, \ldots, a_d$ of the vector $\mathbf{a}$ are rationally independent; i.e.

$$m_1 a_1 + m_2 a_2 + \cdots + m_d a_d \neq 0$$

for any integers $m_1, \ldots, m_d \in \mathbb{Z}$, unless $m_1 = \cdots = m_d = 0$. The flow $\{S^t_{\mathbf{a}}\}$ is never mixing. Also, the flow $\{S^t_{\mathbf{a}}\}$ is ergodic if and only if for every $x \in X$ the orbit $\{S^t x\}$ is dense in X.

If $d = 2$, then there is an alternative: either a_1/a_2 is irrational, in which case the flow $\{S^t_{\mathbf{a}}\}$ is ergodic and its every orbit is dense; or a_1/a_2 is rational, in which case $S^t_{\mathbf{a}}$ is the identity for some $t \neq 0$ (hence every orbit is periodic).

Entropy and Pinsker partitions for flows. The entropy of the map S^t of any flow $\{S^t\}$ is a linear function of time: $h(S^t) = |t| h(S^1)$. Thus the entropy of the flow is defined by $h(\{S^t\}) = h(S^1)$.

The Pinsker partition $\pi(S^t)$ is the same for every map S^t, $t \neq 0$; it is called the Pinsker partition of the flow, $\pi(\{S^t\})$. The Pinsker σ-algebra $\mathfrak{P}(\{S^t\}) = \mathfrak{S}\big(\pi(\{S^t\})\big)$ is S^t-invariant for every $t \in \mathbb{R}$.

Suppose a measurable partition ξ of X has two properties: (i) it is monotonically increasing, i.e. $\xi \preccurlyeq S^t\xi$ for $t > 0$; (ii) its forward limit is the partition into individual points, i.e. $\vee_{t=0}^{\infty} S^t\xi = \varepsilon$. Then $\wedge_{t=0}^{\infty} S^{-t}\xi \succcurlyeq \pi(\{S^t\})$. A flow $\{S^t\}$ is Kolmogorov if and only if its Pinsker partition is trivial, i.e. $\pi(\{S^t\}) = \nu$.

Bibliography

[Ab59] L. M. Abramov, *On the entropy of a flow* (in Russian), Dokl. Akad. Nauk SSSR **128** (1959), 873–875.

[Al69] V. M. Alekseev, *Quasi random dynamical systems*, Mat. USSR Sbornik **7** (1969), 1–43.

[An67] D. V. Anosov, *Geodesic flows on closed Riemannian manifolds with negative curvature*, Proc. Steklov Inst. Math. **90** (1967).

[Ar89] V. I. Arnold, *Mathematical methods of classical mechanics*, Springer, New York, 1989.

[AS67] D. V. Anosov & Ya. G. Sinai, *Some smooth ergodic systems*, Russ. Math. Surv. **22** (1967), 103–167.

[AW67] R. Adler & B. Weiss, *Entropy, a complete metric invariant for automorphisms of the torus*, Proc. Nat. Acad. Sci. USA **57** (1967), 1573–1576.

[Ba00] V. Baladi, *Positive transfer operators and decay of correlations*, World Scientific, River Edge, NY, 2000.

[BCST02] P. Balint, N. Chernov, D. Szasz, and I. P. Toth, *Multi-dimensional semi-dispersing billiards: singularities and the fundamental theorem*, Ann. H. Poincaré **3** (2002), 451–482.

[BCST03] P. Balint, N. Chernov, D. Szasz, and I. P. Toth, *Geometry of multidimensional dispersing billiards*, Astérisque **286** (2003), 119–150.

[BD87] R. Burton & M. Denker, *On the central limit theorem for dynamical systems*, Trans. AMS **302** (1987), 715–726.

[Be01] A. Berger, *Chaos and chance*, Walter de Gruyter, Berlin, 2001.

[BG06] P. Balint & S. Gouezel, *Limit theorems in the stadium billiard*, preprint.

[Bi27] G. D. Birkhoff, *Dynamical systems*, A.M.S. Colloquium Publications, New York (1927).

[Bl92] P. M. Bleher, *Statistical properties of two-dimensional periodic Lorentz gas with infinite horizon*, J. Stat. Phys. **66** (1992), 315–373.

[BLPS92] L. A. Bunimovich, C. Liverani, A. Pellegrinotti & Y. Suhov, *Ergodic systems of n balls in a billiard table*, Comm. Math. Phys. **146** (1992), 357–396.

[BL02] X. Bressaud & C. Liverani, *Anosov diffeomorphism and coupling*, Ergod. Th. Dynam. Syst. **22** (2002), 129–152.

[Bou60] N. Bourbaki, *Éléments de mathématique.* XXV. Part 1. Livre VI: Intégration. Chap. 6: Intégration vectorielle. Actualites scientifiques et industrielles, **1281**, Paris: Hermann (1960).

[Bow70] R. Bowen, *Markov partitions for Axiom A diffeomorphisms*, Amer. J. Math. **92** (1970), 725–747.

[Bow75] R. Bowen, *Equilibrium states and the ergodic theory of Anosov diffeomorphisms*, Lect. Notes Math. **470**, Springer, Berlin, 1975.

[BP01] L. Barreira & Ya. Pesin, *Lyapunov exponents and smooth ergodic theory*, Univ. Lect. Series, **23**, AMS, Providence, RI, 2001.

[BrS02] M. Brin & G. Stuck, *Introduction to Dynamical Systems*, Cambridge U. Press, 2002.

[BS73] L. A. Bunimovich & Ya. G. Sinai, *On a fundamental theorem in the theory of dispersing billiards*, Math. USSR Sbornik **19** (1973), 407–423.

[BS80] L. A. Bunimovich & Ya. G. Sinai, *Markov partitions for dispersed billiards*, Comm. Math. Phys. **78** (1980), 247–280.

[BS81] L. A. Bunimovich & Ya. G. Sinai, *Statistical properties of Lorentz gas with periodic configuration of scatterers*, Comm. Math. Phys. **78** (1981), 479–497.

[BS96] L. A. Bunimovich & H. Spohn, *Viscosity for a periodic two disk fluid: an existence proof*, Comm. Math. Phys. **176** (1996), 661–680.

[BR97] L. A. Bunimovich & J. Rehácek, *Nowhere dispersing 3D billiards with non-vanishing Lyapunov exponents*, Comm. Math. Phys. **189** (1997), 729–757.

[BR98a] L. A. Bunimovich & J. Rehácek, *On the ergodicity of many-dimensional focusing billiards. Classical and quantum chaos*, Ann. H. Poincaré **68** (1998), 421–448.

[BR98b] L. A. Bunimovich & J. Rehácek, *How high-dimensional stadia look like*, Comm. Math. Phys. **197** (1998), 277–301.

[BSC90] L. A. Bunimovich, Ya. G. Sinai & N. I. Chernov, *Markov partitions for two-dimensional billiards*, Russ. Math. Surv. **45** (1990), 105–152.

[BSC91] L. A. Bunimovich, Ya. G. Sinai & N. I. Chernov, *Statistical properties of two-dimensional hyperbolic billiards*, Russ. Math. Surv. **46** (1991), 47–106.

[Bu74a] L. A. Bunimovich, *On billiards close to dispersing*, Math. USSR. Sb. **23** (1974), 45–67.

[Bu74b] L. A. Bunimovich, *The ergodic properties of certain billiards*, Funk. Anal. Prilozh. **8** (1974), 73–74.

[Bu79] L. A. Bunimovich, *On ergodic properties of nowhere dispersing billiards*, Comm. Math. Phys. **65** (1979), 295–312.

[Bu90] L. A. Bunimovich, *A theorem on ergodicity of two-dimensional hyperbolic billiards*, Comm. Math. Phys. **130** (1990), 599–621.

[Bu92] L. A. Bunimovich, *On absolutely focusing mirrors*, In *Ergodic Theory and related topics, III (Güstrow, 1990)*. Edited by U. Krengel et al., Lecture Notes in Math. **1514**, Springer, Berlin (1992), 62–82.

[Bu00] L. A. Bunimovich, *Hyperbolicity and astigmatism*, J. Stat. Phys. **101** (2000), 373–384.

[BY00] M. Benedicks & L.-S. Young, *Markov extensions and decay of correlations for certain Hénon maps*, Astérisque **261** (2000), 13–56.

[CCG] G. Casati, G. Comparin & I. Guarneri, *Decay of correlations in certain hyperbolic systems*, Phys. Rev. A **26** (1982), 717–719.

[CD06] N. Chernov & D. Dolgopyat, *Brownian Brownian motion – I*, to appear in Memoir. AMS.

[CFS82] I. P. Cornfeld, S. V. Fomin & Ya. G. Sinai, *Ergodic theory*, Fundamental Principles of Mathematical Sciences, **245**, Springer-Verlag, New York, 1982.

[Ch93] N. I. Chernov, *On local ergodicity in hyperbolic systems with singularities*, Funct. Anal. Appl. **27** (1993), 51–54.

[Ch95] N. Chernov, *Limit theorems and Markov approximations for chaotic dynamical systems*, Prob. Th. Rel. Fields **101** (1995), 321–362.

[CH96] N. Chernov & C. Haskell, *Nonuniformly hyperbolic K-systems are Bernoulli*, Ergod. Th. Dynam. Syst. **16** (1996), 19–44.

[Ch97] N. Chernov, *Entropy, Lyapunov exponents, and mean free path for billiards*, J. Stat. Phys. **88** (1997), 1–29.

[Ch99] N. Chernov, *Decay of correlations and dispersing billiards*, J. Stat. Phys. **94** (1999), 513–556.

[Ch01] N. Chernov, *Sinai billiards under small external forces*, Ann. H. Poincaré **2** (2001), 197–236.

[Ch02] N. Chernov, *Invariant measures for hyperbolic dynamical systems*, In: Handbook of Dynamical Systems, **1A**, 321–407, North-Holland, Amsterdam, 2002.

[Ch06a] N. Chernov, *Regularity of local manifolds in dispersing billiards*, Math. Phys. Electr. J. **12** (2006), No. 1, 54 pp.

[Ch06b] N. Chernov, *Advanced statistical properties of dispersing billiards*, to appear in J. Stat. Phys. (2006).

[CM92] N. I. Chernov & R. Markarian, *Entropy of non-uniformly hyperbolic plane billiards*, Bol. Soc. Bras. Mat. **23** (1992) 121–135.

[CM03] N. Chernov & R. Markarian, *Introduction to the ergodic theory of chaotic billiards*, 2nd Ed., IMPA, Rio de Janeiro, Brasil, 2003.

[CM06] N. Chernov & R. Markarian, *Dispersing billiards with cusps: slow decay of correlations*, submitted.

[CY00] N. I. Chernov & L.-S. Young, *Decay of correlations of Lorentz gases and hard balls*. In *Hard Ball Systems and Lorentz Gas*, D. Szász, editor, Springer, Berlin (2000), 89–120.

[CZ05] N. Chernov & H.-K. Zhang, *Billiards with polynomial mixing rates*, Nonlinearity **18** (2005), 1527–1553.

[Den89] M. Denker, *The central limit theorem for dynamical systems*, Dyn. Syst. Ergod. Th. Banach Center Publ., **23**, PWN–Polish Sci. Publ., Warsaw, 1989.

[Dev89] R. Devaney, *An introduction to chaotic dynamical systems*, 2nd Ed., Addison-Wesley, New York, 1989.

[DM01] G. Del Magno, *Ergodicity of a class of truncated elliptical billiards*, Nonlinearity **14** (2001), 1761–1786.

[DMM03] G. Del Magno & R. Markarian, *Bernoulli elliptical stadia*, Comm. Math. Phys. **233** (2003), 211–230.

[DMM06] G. Del Magno & R. Markarian, *On the Bernoulli property of planar hyperbolic billiards*, preprint (2006).

[Dob68a] R. L. Dobrushin, *The description of a random field by means of conditional probabilities and conditions of its regularity*, Th. Prob. Appl. **13** (1968), 197–224.

[Dob68b] R. L. Dobrushin, *Gibbsian random fields for lattice systems with pairwise interaction*, Funct. Anal. Appl. **2** (1968), 292–301.

[Dob68c] R. L. Dobrushin, *The problem of uniqueness of a Gibbsian random field and the problem of phase transitions*, Funct. Anal. Appl. **2** (1968), 302–312.

[Don91] V. Donnay, *Using integrability to produce chaos: billiards with positive entropy*, Comm. Math. Phys. **141** (1991), 225–257.

[DPS86] *Dependence in probability and statistics. A survey of recent results*, Eds. E. Eberlein and M. Taqqu, Progress in Probability and Statistics, **11**, Birkhäuser, Boston, MA, 1986.

[DS00] *Dynamical Systems, Ergodic Theory and Applications*, Encycl. Math. Sciences, **100**, Ed., Ya. G. Sinai, Springer, Berlin, 2000.

[FM88] B. Friedman & R. F. Martin, *Behavior of the velocity autocorrelation function for the periodic Lorentz gas*, Physica D **30** (1988), 219–227.

[Ga74] G. Gallavotti, *Lectures on the billiard.* In *Dynamical systems, theory and applications (Rencontres, Battelle Res. Inst., Seattle, Wash., 1974)*, Lect. Notes Phys. **38**, Springer, Berlin (1975), 236–295.

[GG94] P. Garrido & G. Gallavotti, *Billiards correlation function*, J. Stat. Phys. **76** (1994), 549–585.

[GKT95] G. Galperin, T. Krüger, & S. Troubetzkoy, *Local instability of orbits in polygonal and polyhedral billiards*, Comm. Math. Phys. **169** (1995), 463–473.

[GO74] G. Gallavotti & D. S. Ornstein, *Billiards and Bernoulli schemes*, Comm. Math. Phys. **38** (1974), 83-101.

[Gur65] B. M. Gurevich, *Construction of increasing partitions for special flows*, Th. Prob. Appl. **10** (1965), 627–645.

[Gut86] E. Gutkin, *Billiards in polygons*, Physica D **19** (1986), 311–333.

[Gut96] E. Gutkin, *Billiards in polygons: survey of recent results*, J. Stat. Phys. **83** (1996), 7–26.

[GZ90] G. A. Galperin & A. N. Zemlyakov, *Mathematical billiards*, Kvant **77**, Nauka, Moscow, 1990.

[Had01] J. Hadamard, *Sur l'itération et les solutions asymptotiques des équations différentielles*, Bull. Soc. Math. France **29** (1901), 224–228.

[Hal77] B. Halpern, *Strange billiard tables*, Trans. AMS **232** (1977), 297–305.

[HB00] *Hard Ball Systems and Lorentz Gas*, D. Szász, editor, Springer, Berlin, 2000.

[Ho39] E. Hopf, *Statistik der geodetischen Linien in Mannigfaltigkeiten negativer Krümmung*, Ber. Verh. Sächs. Akad. Wiss. Leipzig **91** (1939), 261–304.

[Ho40] E. Hopf, *Statistik der Lösungen geodätischer Probleme vom unstabilen Typus, II*, Math. Annalen **117** (1940), 590–608.

[HP70] M. Hirsch & C. Pugh, *Stable manifolds and hyperbolic sets.* In: Global Analysis, Proc. Symp. in Pure Math., **14**, AMS, Providence, RI, 1970, pp. 133–163.

[Hu01] H. Hu, *Statistical properties of some almost hyperbolic systems* In: Smooth Ergodic Theory and Its Applications (Seattle, WA, 1999), 367–384, Proc. Sympos. Pure Math., **69**, Amer. Math. Soc., Providence, RI, 2001.

[IL71] I. A. Ibragimov & Y. V. Linnik, *Independent and stationary sequences of random variables*, Wolters-Noordhoff, Gröningen, 1971.

[KB94] A. Katok & K. Burns, *Infinitesimal Lyapunov functions, invariant cone families and stochastic properties of smooth dynamical systems*, Ergod. Th. Dynam. Syst. **14** (1994), 757–785.

[KH95] A. Katok & B. Hasselblatt, *Introduction to the modern theory of dynamical systems*, Cambridge U. Press (1995).

[KM81] I. Kubo & H. Murata, *Perturbed billiard systems II, Bernoulli properties*, Nagoya Math. J. **81** (1981), 1–25.

[KMS86] S. Kerckhoff, H. Masur & J. Smillie, *Ergodicity of billiard flows and quadratic differentials*, Ann. Math. **124** (1986), 293–311.

[KS86] A. Katok & J.-M. Strelcyn, with the collaboration of F. Ledrappier & F. Przytycki, *Invariant manifolds, entropy and billiards; smooth maps with singularities*, Lect. Notes Math., **1222**, Springer, New York (1986).

[KSS90] A. Krámli, N. Simányi & D. Szász, *A "transversal" fundamental theorem for semi-dispersing billiards*, Comm. Math. Phys. **129** (1990), 535–560

[KSS91] A. Krámli, N. Simányi & D. Szász, *The K–property of three billiard balls*, Ann. Math. **133** (1991), 37–72

[KSS92] A. Krámli, N. Simányi & D. Szász, *The K-property of four billiard balls*, Comm. Math. Phys. **144** (1992), 107–142.

[KT91] V. V. Kozlov & D. V. Treshchëv, *Billiards. A Genetic Introduction to the Dynamics of Systems with Impacts* Translations of Mathematical Monographs, **89**, AMS, Providence, RI (1991).

[La73] V. F. Lazutkin, *On the existence of caustics for the billiard ball problem in a convex domain*, Math. USSR Izv. **7** (1973), 185–215.

[Led84] F. Ledrappier, *Propriétés ergodiques des mesures de Sinai*, IHES Public. Math., **59** (1984), 163–188.

[Leo61] V. P. Leonov, *On the dispersion of time-dependent means of a stationary stochastic process*, Th. Probab. Appl., **6** (1961), 87–93.

[Ll92] R. de la Llave, *Smooth conjugacy and S-R-B measures for uniformly and non-uniformly hyperbolic systems*, Comm. Math. Phys. **150** (1992), 289–320.

[LMM86] R. de la Llave, J. Marco & R. Moriyón, *Canonical perturbation theory of Anosov systems and regularity results for the Livsic cohomology equation*, Ann. Math. **123** (1986), 537–611.

[Lo05] H. A. Lorentz, *The motion of electrons in metallic bodies*, Proc. Amst. Acad. **7** (1905), 438–453.

[LR69] O. E. Lanford & D. Ruelle, *Observables at infinity and states with short range correlations in statistical mechanics*, Comm. Math. Phys. **13** (1969), 194–215.

[LS82] F. Ledrappier & J.-M. Strelcyn, *A proof of the estimation from below in Pesin's entropy formula*, Erg. Th. Dynam. Syst. **2** (1982), 203–219.

[LW95] C. Liverani & M. Wojtkowski, *Ergodicity in Hamiltonian systems*, Dynamics reported, Dynam. Report. Expositions Dynam. Systems (N.S.) **4**, Springer, Berlin (1995), 130–202.

[Mac83] J. Machta, *Power law decay of correlations in a billiard problem*, J. Stat. Phys. **32** (1983), 555–564.

[Man83] R. Mañe, *Introducao à Teoria Ergódica*. IMPA, Rio de Janeiro (1983). English Ed., *Ergodic Theory and Differentiable Dynamics*, Springer, Berlin, Heidelberg, New York (1987).

[Mar88] R. Markarian, *Billiards with Pesin region of measure one*, Comm. Math. Phys. **118** (1988), 87–97.

[Mar93a] R. Markarian, *The fundamental theorem of Sinai-Chernov for dynamical systems with singularities*. In *Dynamical Systems. Santiago de Chile 1990*, R. Bamón, R. Labarca, J. Lewowicz, and J. Palis, editors, Longman, Harlow (1993), 131–158.

[Mar93b] R. Markarian, *New ergodic billiards: exact results*, Nonlinearity **6** (1993), 819–841.

[Mar94] R. Markarian, *Non-uniformly hyperbolic billiards*, Ann. Fac. Sci. Toulouse Math. **3** (1994), 223–257.

[Mar04] R. Markarian, *Billiards with polynomial decay of correlations*, Erg. Th. Dynam. Syst. **24** (2004), 177–197.

[Mat75] G. Matheron, *Random sets and integral geometry*, J. Wiley & Sons, New York, 1975.

[ME81] N. Martin & J. England, *Mathematical theory of entropy*, Addison-Wesley, Reading, Mass., 1981.

[MOP96] R. Markarian, S. Oliffson & S. Pinto, *Chaotic properties of the elliptical stadium*, Comm. Math. Phys. **174** (1996), 661–679.

[Os68] V. I. Oseledec, *A multiplicative ergodic theorem*, Trans. Moskow Math. Soc. **19** (1968), 197–231.

[OW98] D. S. Ornstein & B. Weiss, *On the Bernoulli nature of systems with some hyperbolic structure*, Erg. Th. Dynam. Syst. **18** (1998), 441–456.

[Pes76] Ya. B. Pesin, *Families of invariant manifolds corresponding to nonzero characteristic exponents*, Math. USSR Izvest. **10** (1976), 1261–1305.

[Pes77a] Ya. B. Pesin, *Characteristic Lyapunov exponents and smooth ergodic theory*, Russ. Math. Surv. **32:4** (1977), 55–114.

[Pes77b] Ya. B. Pesin, *Description of π-partition of a diffeomorphism with invariant measure*, Math. Notes **22** (1977), 506–515.

[Pes92] Ya. B. Pesin, *Dynamical systems with generalized hyperbolic attractors: hyperbolic, ergodic and topological properties*, Ergod. Th. Dynam. Sys. **12** (1992), 123–151.

[PeS82] Ya. B. Pesin & Ya. G. Sinai, *Gibbs measures for partially hyperbolic attractors*, Erg. Th. Dynam. Sys. **2** (1982), 417–438.

[Pet83] K. Petersen, *Ergodic Theory*, Cambridge U. Press, 1983.

[PhS75] W. Philipp & W. Stout, *Almost sure invariance principles for partial sums of weakly dependent random variables*, Memoir. AMS **161** (1975).

[PP90] W. Parry & M. Pollicott, *Zeta functions and the periodic orbit structure of hyperbolic dynamics*, Astérisque **187–188** (1990), 268 pp.

[Re95] J. Rehácek, *On the ergodicity of dispersing billiards*, Rand. Comput. Dynam. **3** (1995), 35–55.

[Ro49] V. Rohlin, *On the fundamental ideas of measure theory*, AMS Transl. Ser. 1, **10**, 1–52, AMS, Providence, RI, 1952.

[Ro67] V. Rohlin, *Lectures on the entropy theory of transformations with invariant measure*, Uspehi Mat. Nauk **22** (1967), 3–56.

[Ru76] D. Ruelle, *A measure associated with Axiom-A attractors*, Amer. J. Math. **98** (1976), 619–654.

[Ru78] D. Ruelle, *Thermodynamic Formalism*, Addison-Wesley, Reading, Mass., 1978.

[Ru79] D. Ruelle, *Ergodic theory of differentiable dynamical systems*, Publ. Math. IHES **50** (1979), 27–58.

[Sa76] L. A. Santaló, *Integral Geometry and Geometric Probability*, Addison Wesley, Reading, Mass., 1976.

[SC87] Ya. G. Sinai & N. I. Chernov, *Ergodic properties of some systems of 2-dimensional discs and 3-dimensional spheres*, Russ. Math. Surv. **42** (1987), 181–207.

[Sim99] N. Simányi, *Ergodicity of hard spheres in a box*, Ergod. Th. Dynam. Syst. **19** (1999), 741–766.

[Sim03] N. Simányi, *Proof of the Boltzmann-Sinai ergodic hypothesis for typical hard disk systems*, Invent. Math. **154** (2003), 123–178.

[Sim04] N. Simányi, *Proof of the ergodic hypothesis for typical hard ball systems*, Ann. H. Poincaré **5** (2004), 203–233.

[Sin66] Ya. G. Sinai, *Classical dynamic systems with countably-multiple Lebesgue spectrum. II*, Izv. Akad. Nauk SSSR Ser. Mat. **30** (1966), 15–68; also AMS Trans. **68** (1968), 34–88.

[Sin68a] Ya. G. Sinai, *Markov partitions and C-diffeomorphisms*, Funct. Anal. Appl. **2** (1968), 61–82.

[Sin68b] Ya. G. Sinai, *Construction of Markov partitions*, Funct. Anal. Appl. **2** (1968), 245–253.

[Sin70] Ya. G. Sinai, *Dynamical systems with elastic reflections. Ergodic properties of dispersing billiards*, Russ. Math. Surv. **25** (1970), 137–189.

[Sin72] Ya. G. Sinai, *Gibbs measures in ergodic theory*, Russ. Math. Surv. **27** (1972), 21–69.

[Sin76] Ya. G. Sinai, *Introduction to Ergodic Theory*. Princeton U. Press, Princeton (1976).

[Sin79] Ya. G. Sinai, *Development of Krylov's ideas*, Afterword to N. S. Krylov, *Works on the Foundations of Statistical Physics*. Princeton U. Press, Princeton, NJ, (1979), 239–281.

[St86] J.-M. Strelcyn, *Plane billiards as smooth dynamical systems with singularities*, In: Lect. Notes Math., **1222**, Springer, New York (1986), pp. 199–278.

[SV04] D. Szász & T. Varjú, *Markov towers and stochastic properties of billiards*. In: Modern Dynamical Systems and Applications, 433–445, Cambridge U. Press, Cambridge, 2004.

[Sz92] D. Szász, *On the K-property of some planar hyperbolic billiards*, Comm. Math. Phys. **145** (1992), 595–604.

[Ta95] S. Tabachnikov, *Billiards*, Panor. Synth. No. 1, SMF, Paris (1995).

[Va92] S. Vaienti, *Ergodic properties of the discontinuous sawtooth map*, J. Stat. Phys. **67** (1992), 251–269.

[Wa82] P. Walters, *An Introduction to Ergodic Theory*, Springer, New York (1982).

[Wo85] M. Wojtkowski, *Invariant families of cones and Lyapunov exponents*, Ergod. Th. Dynam. Syst. **5** (1985), 145–161.

[Wo86] M. Wojtkowski, *Principles for the design of billiards with nonvanishing Lyapunov exponents*, Comm. Math. Phys. **105** (1986), 391–414.

[Wo94] M. Wojtkowski, *Two applications of Jacobi fields to the billiard ball problem*, J. Differ. Geom. **40** (1994), 155–164.

[Yo98] L.-S. Young, *Statistical properties of dynamical systems with some hyperbolicity*, Ann. Math. **147** (1998), 585–650.

[Yo99] L.-S. Young, *Recurrence times and rates of mixing*, Israel J. Math. **110** (1999), 153–188.

[ZE97] G. M. Zaslavsy & M. Edelman, *Maxwell's demon as a dynamical system*, Phys. Rev. E **56** (1997), 5310–5320.

Index